ADVANCES IN COMPUTATIONAL MULTIBODY SYSTEMS

Computational Methods in Applied Sciences

Volume 2

Series Editor

E. Oñate

Advances in Computational Multibody Systems

Edited by

JORGE A.C. AMBRÓSIO
IDMEC, Instituto Superior Técnico,
Lisbon, Portugal

Springer

A C.I.P. Catalogue record for this book is available from the Library of Congress.

ISBN-10 1-4020-3392-3 (HB)
ISBN-10 1-4020-3393-1 (e-book)
ISBN-13 978-1-4020-3392-6 (HB)
ISBN-13 978-1-4020-3393-3 (e-book)

Published by Springer,
P.O. Box 17, 3300 AA Dordrecht, The Netherlands.

www.springeronline.com

Printed on acid-free paper

Table of Contents

Preface

The area of Multibody Dynamics is a part of the Computational Mechanics scientific field associated to solid mechanics. It can be argued that among all the areas in solid mechanics the methodologies and applications associated to multibody dynamics are those that provide a better framework to aggregate different disciplines. This idea is clearly reflected in the multidisciplinary applications in biomechanics that use multibody dynamics to described the motion of the biological entities, in finite elements where multibody dynamics provide powerful tools to describe large motion and kinematic restrictions between system components, in system control where the methodologies used in multibody dynamics are the prime form of describing the systems under analysis, or even in many applications that involve fluid-structures interaction or aeroelasticity.

The ECCOMAS thematic conference Multibody Dynamics 2003 that took place in Lisbon, Portugal was organized in turn of special sessions dedicated to multibody dynamics in Biomechanics, Vehicle Dynamics, Contact and Impact, Optimization and Design Sensitivity, Flexible Multibody Systems, Education of Computational Kinematics, Dynamics and Multibody Systems, Multidisciplinary Applications and Real-Time Applications. These sessions were organized by recognized experts in each of these areas and gathered together 127 participants from 22 countries, including Japan, Korea, India, USA, Mexico, Canada and many of the European nations, who delivered 90 communications during the 4 days of the conference.

This book contains the contributions of the special session organizers, or of participants selected by the organizers, that reflect the State-of-Art in the application of Multibody Dynamics to different areas of engineering. The chapters of this book are enlarged and revised versions of the communications, delivered at the conference, which were enhanced in terms of self-containment and tutorial quality by the authors. The result is a comprehensive text that constitutes a valuable reference for researchers and design engineers which helps to appraise the potential for the application of multibody dynamics methodologies to a wide range of areas of scientific and engineering relevance.

Lisbon, Portugal
Jorge Alberto Cadete Ambrósio
Chairman

A Fast and Simple Semi-Recursive Formulation for Multi-Rigid-Body Systems

J. García de Jalón[1], E. Álvarez[1], F.A. de Ribera[1], I. Rodríguez[2] and F.J. Funes[3]

[1] Escuela Técnica Sup. de Ingenieros Industriales, Univ. Poli. de Madrid, Spain
[2] STT Engineering and Systems, S.L, San Sebastián, Spain
[3] Telefónica de España, Madrid, Spain

This work describes a topological semi-recursive formulation for multibody dynamics that is very simple and efficient. This formulation is called "semi-recursive" because it uses recursivity, but at the end it needs to solve a system of n linear equations, with n the numbers of degrees of freedom. With relative coordinates the formulation shall include the closure-of-the-loop constraint equations. These constraint equations are more complicated to take into account with fully recursive O(n) formulations, which are the fastest for long, open-chain systems. For this reason, several semi-recursive formulations simpler and easier to implement have been developed in the last few years. In this paper, some semi-recursive formulations are reviewed and a new variant, that is simpler and more general, is described in detail. A simple way to introduce the topology of the spanning tree is presented. Special attention is paid to closed-loop multibody systems with rods, and the benefits of opening the loops by removing these rods • while keeping its inertia forces exactly• are explained. Some examples and numerical results illustrate the aforesaid theoretical developments.

1 Introduction

Complex multibody systems arise in many areas of engineering: automobiles, machinery, robotics, aerospace, biomechanics, etc. Although the motion differential equations that govern their dynamic behavior have been known since the times of Newton, Euler and Lagrange, their practical application started 40 years ago when space and robotics problems demanded more precise mathematical models, and digital computers provided the means to numerically integrate these differential equations in an acceptable elapsed time.

The first practical systems studied (spacecraft and robots) were open-chain systems, so relative coordinates were more appropriate than Cartesian coordinates. In addition to this, relative coordinates had fewer storage requirements. In the seventies and eighties the main emphasis switched to automobile applications, which are inherently closed-chain systems. So, software packages that used highly constrained Cartesian coordinates, such as ADAMS and DADS, were developed.

J.A.C. Ambrósio (ed.), Advances in Computational Multibody Systems, 1–23.

These programs are based on *global formulations*, in the sense that they consider all mechanisms –open-chain and closed-chain; loosely or severely constrained– in exactly the same way. As a consequence, the efficiency was low. In contrast with the global formulations are the *topological formulations*, which try to take advantage of the system topology to improve the efficiency of the dynamic simulations. The topological formulations tend to use relative coordinates or special sparse matrix techniques.

In this paper, after a review of some of the improvements published in the last few years for global formulations, a simple semi-recursive formulation based on a double velocity transformation will be described. It is known that the presence of rods (slender elements with two spherical joints) present some difficulties when relative coordinates are used. So, a very interesting option is to open the closed chains by eliminating these rods (and perhaps also some joints in the system). For kinematics, a rod element can be replaced by a single constraint equation, in fact a constant distance condition, but for dynamics its inertia forces should be kept in an exact way. Later on, the procedure followed to take into account the rod inertia will be explained in detail.

In the descriptor form, using Cartesian dependent coordinates, the motion differential equations take the form,

$$\mathbf{M}(\mathbf{q})\ddot{\mathbf{q}} - \mathbf{\Phi}_{\mathbf{q}}^{T}\boldsymbol{\lambda} = \mathbf{Q}(\mathbf{q},\dot{\mathbf{q}}) \tag{1}$$

where $\mathbf{q}$ is the vector of Cartesian coordinates that defines the system position, $\dot{\mathbf{q}}$ and $\ddot{\mathbf{q}}$ are its first and second order time derivatives, $\mathbf{M}$ is the inertia or mass matrix, $\mathbf{Q}$ is a vector that includes the external and velocity dependent inertia forces, $\mathbf{\Phi}_{\mathbf{q}}$ is the Jacobian matrix of the kinematic constraint equations and $\boldsymbol{\lambda}$ the vector of Lagrange multipliers. The position, velocity and acceleration vectors in Equation (1) must satisfy the corresponding constraint equations,

$$\mathbf{\Phi}(\mathbf{q}) = \mathbf{0} \tag{2}$$

$$\dot{\mathbf{\Phi}}(\mathbf{q}) = \mathbf{\Phi}_{\mathbf{q}}\dot{\mathbf{q}} + \mathbf{\Phi}_{t} = \mathbf{0} \tag{3}$$

$$\ddot{\mathbf{\Phi}}(\mathbf{q}) = \mathbf{\Phi}_{\mathbf{q}}\ddot{\mathbf{q}} + \dot{\mathbf{\Phi}}_{\mathbf{q}}\dot{\mathbf{q}} + \dot{\mathbf{\Phi}}_{t} = \mathbf{0} \tag{4}$$

Equations (1) and (4) constitute a system of index 3 DAEs. If only Equations (1) and (4) are considered, the following index 1 DAE system –equivalent to an ODE system– is obtained:

$$\begin{bmatrix} \mathbf{M}(\mathbf{q}) & \mathbf{\Phi}_{\mathbf{q}}^{T} \\ \mathbf{\Phi}_{\mathbf{q}} & \mathbf{0} \end{bmatrix} \begin{Bmatrix} \ddot{\mathbf{q}} \\ \boldsymbol{\lambda} \end{Bmatrix} = \begin{Bmatrix} \mathbf{Q}(\mathbf{q},\dot{\mathbf{q}}) \\ -\dot{\mathbf{\Phi}}_{\mathbf{q}}\dot{\mathbf{q}} - \dot{\mathbf{\Phi}}_{t} \end{Bmatrix} \tag{5}$$

The matrix in this system of linear equations is known as the *augmented matrix* [26] or a matrix with *optimization structure* [26,33]. The system of differential equations (5) presents a constraint stabilization problem. As only the acceleration constraint equations have been imposed, the positions and velocities provided by the integrator suffer from the "drift" phenomenon. Two popular solutions to this problem are the Baumgarte stabilization method [5,14] and the mass-orthogonal projections of position and velocity vectors [25].

Another way to solve the constraint stabilization problem is to use *velocity transformations*, which map the dependent Cartesian velocities $\dot{\mathbf{q}}$ on a minimal set $\dot{\mathbf{z}}$ of truly independent velocities. Let matrix **R** be the orthogonal complement of the Jacobian matrix $\mathbf{\Phi_q}$, that is an $n\times f$ matrix whose columns are a basis of the nullspace of $\mathbf{\Phi_q}$. The dependent velocities $\dot{\mathbf{q}}$ can be expressed as a linear combination of the columns of matrix **R**. The coefficients of this linear combination are the independent velocities $\dot{\mathbf{z}}$,

$$\dot{\mathbf{q}} = \mathbf{R}_1\dot{z}_1 + \mathbf{R}_2\dot{z}_2 + \ldots + \mathbf{R}_f\dot{z}_f = \mathbf{R}\dot{\mathbf{z}} \tag{6}$$

Matrix **R** can be computed very easily by a coordinate partition of vector $\dot{\mathbf{q}}$ on *dependent* and *independent* velocities. The dependent velocities are those velocities related with the columns of the pivots in the Gauss factorization of matrix $\mathbf{\Phi_q}$. The independent velocities $\dot{\mathbf{z}}$ can be expressed as the projections of the dependent ones on the rows of a full rank ($f\times n$) constant matrix **B** in the form,

$$\dot{\mathbf{z}} = \mathbf{B}\dot{\mathbf{q}} \tag{7}$$

The rows of matrix **B** shall be linearly independent of the rows of the Jacobian matrix $\mathbf{\Phi_q}$. Equations (3) and (7) can be expressed together in the form,

$$\begin{bmatrix} \mathbf{\Phi_q} \\ \mathbf{B} \end{bmatrix} \dot{\mathbf{q}} = \begin{Bmatrix} \mathbf{b} \\ \dot{\mathbf{z}} \end{Bmatrix}, \qquad \mathbf{b} \equiv -\mathbf{\Phi}_t \tag{8}$$

Because of the conditions established for matrix **B**, the matrix in this system of linear equations can be inverted. Consider this inverse matrix in partitioned form,

$$\begin{bmatrix} \mathbf{\Phi_q} \\ \mathbf{B} \end{bmatrix} \begin{bmatrix} \mathbf{\Phi_q} \\ \mathbf{B} \end{bmatrix}^{-1} = \begin{bmatrix} \mathbf{\Phi_q} \\ \mathbf{B} \end{bmatrix} \begin{bmatrix} \mathbf{S} & \mathbf{R} \end{bmatrix} = \begin{bmatrix} \mathbf{I}_m & \mathbf{0}_{m\times f} \\ \mathbf{0}_{f\times m} & \mathbf{I}_f \end{bmatrix} \tag{9}$$

This expression is used to define matrices **S** and **R**, which are part of the referred inverse matrix. By introducing the result of Equation (9) in Equation (8), the following result is obtained for the velocity transformation,

$$\dot{\mathbf{q}} = \begin{bmatrix} \mathbf{\Phi_q} \\ \mathbf{B} \end{bmatrix}^{-1} \begin{Bmatrix} \mathbf{b} \\ \dot{\mathbf{z}} \end{Bmatrix} = \begin{bmatrix} \mathbf{S} & \mathbf{R} \end{bmatrix} \begin{Bmatrix} \mathbf{b} \\ \dot{\mathbf{z}} \end{Bmatrix} = \mathbf{S}\mathbf{b} + \mathbf{R}\dot{\mathbf{z}} \qquad \left(\mathbf{b} \equiv -\mathbf{\Phi}_t\right) \tag{10}$$

Matrix **S** in Equations (9) and (10) is never computed explicitly: only the product (**Sb**) need to be computed. According to Equation (10), this product is given by the dependent velocities $\dot{\mathbf{q}}$ computed with null independent velocities ($\dot{\mathbf{z}} = \mathbf{0}$).

The constant matrix **B** can be computed in several ways. It can be computed by the QR factorization (or the SVD) as the orthogonal complement of the Jacobian matrix in a previous position. The matrix **B** so computed is valid as far as the inverse matrix in Equation (9) exists and is well conditioned. However, there is a simpler and cheaper procedure based on the Gaussian elimination with full pivoting, that is stable enough for most applications. This method coincides with the coordinate partitioning method introduced by Wehage and Haug [34], and with the velocity transformation method used by Serna et al. [30]. In these methods matrix **B** is chosen as a Boolean matrix that defines the independent velocities $\dot{\mathbf{z}}$ as a subset of velocities $\dot{\mathbf{q}}$. In partitioned form,

$$\begin{bmatrix} \boldsymbol{\Phi}_\mathbf{q} \\ \mathbf{B} \end{bmatrix} \dot{\mathbf{q}} = \begin{bmatrix} \boldsymbol{\Phi}_\mathbf{q}^d & \boldsymbol{\Phi}_\mathbf{q}^i \\ \mathbf{0}_{f\times m} & \mathbf{I}_f \end{bmatrix} \begin{Bmatrix} \dot{\mathbf{q}}^d \\ \dot{\mathbf{q}}^i \end{Bmatrix} = \begin{Bmatrix} \mathbf{0} \\ \dot{\mathbf{z}} \end{Bmatrix} \tag{11}$$

Matrix $\boldsymbol{\Phi}_\mathbf{q}{}^d$ is non-singular because its columns contain the pivots of $\boldsymbol{\Phi}_\mathbf{q}$. So, the full matrix in Equation (11) is invertible. Considering this inverse in partitioned form,

$$\begin{bmatrix} \boldsymbol{\Phi}_\mathbf{q} \\ \mathbf{B} \end{bmatrix} \begin{bmatrix} \boldsymbol{\Phi}_\mathbf{q} \\ \mathbf{B} \end{bmatrix}^{-1} = \begin{bmatrix} \boldsymbol{\Phi}_\mathbf{q} \\ \mathbf{B} \end{bmatrix} \begin{bmatrix} \mathbf{S} & \mathbf{R} \end{bmatrix} = \begin{bmatrix} \boldsymbol{\Phi}_\mathbf{q}^d & \boldsymbol{\Phi}_\mathbf{q}^i \\ \mathbf{0}_{f\times m} & \mathbf{I}_f \end{bmatrix} \begin{bmatrix} \mathbf{S}^d & \mathbf{R}^d \\ \mathbf{S}^i & \mathbf{R}^i \end{bmatrix} = \begin{bmatrix} \mathbf{I}_m & \mathbf{0}_{m\times f} \\ \mathbf{0}_{f\times m} & \mathbf{I}_f \end{bmatrix} \tag{12}$$

By identification of terms on the left and on the right-hand sides, it is concluded that matrices **S** and **R** and its partitions are given by,

$$\mathbf{S} = \begin{bmatrix} \mathbf{S}^d \\ \mathbf{S}^i \end{bmatrix} = \begin{bmatrix} \left(\boldsymbol{\Phi}_\mathbf{q}^d\right)^{-1} \\ \mathbf{0}_{f\times m} \end{bmatrix}, \quad \mathbf{R} = \begin{bmatrix} \mathbf{R}^d \\ \mathbf{R}^i \end{bmatrix} = \begin{bmatrix} -\left(\boldsymbol{\Phi}_\mathbf{q}^d\right)^{-1} \boldsymbol{\Phi}_\mathbf{q}^i \\ \mathbf{I}_f \end{bmatrix} \tag{13}$$

For accelerations it is possible to express together the acceleration Equations (4) and the time derivative of Equation (7). The following is obtained,

$$\begin{bmatrix} \boldsymbol{\Phi}_\mathbf{q} \\ \mathbf{B} \end{bmatrix} \ddot{\mathbf{q}} = \begin{Bmatrix} \mathbf{c} \\ \ddot{\mathbf{z}} \end{Bmatrix}, \quad \ddot{\mathbf{q}} = \begin{bmatrix} \boldsymbol{\Phi}_\mathbf{q} \\ \mathbf{B} \end{bmatrix}^{-1} \begin{Bmatrix} \mathbf{c} \\ \ddot{\mathbf{z}} \end{Bmatrix} = \begin{bmatrix} \mathbf{S} & \mathbf{R} \end{bmatrix} \begin{Bmatrix} \mathbf{c} \\ \ddot{\mathbf{z}} \end{Bmatrix} = \mathbf{Sc} + \mathbf{R}\ddot{\mathbf{z}}$$
$$\left(\mathbf{c} \equiv -\dot{\boldsymbol{\Phi}}_\mathbf{q}\dot{\mathbf{q}} - \dot{\boldsymbol{\Phi}}_t\right) \tag{14}$$

where the product (**Sc**) can be computed as the part of the dependent accelerations that depends on the velocities $\dot{\mathbf{q}}$ and on the rheonomic constraints, i. e., computed with null independent accelerations ($\ddot{\mathbf{z}} = \mathbf{0}$).

Substituting this result in the equations of motion (1), pre-multiplying by matrix $\mathbf{R}^T$ and taking into account that $\boldsymbol{\Phi}_\mathbf{q}\mathbf{R}=\mathbf{0}$, the following expression is obtained for the dynamic equations in independent coordinates,

$$\mathbf{R}^T\mathbf{M}\mathbf{R}\ddot{\mathbf{z}} = \mathbf{R}^T\left(\mathbf{Q}(\mathbf{q},\dot{\mathbf{q}}) - \mathbf{MSc}\right) \tag{15}$$

This is an ODE system that does not suffer from the difficulties of the DAEs in Equation (5). Observe that the Lagrange multipliers no longer appear. Equations (1) to (15) are classical in MBS formulations and they are the basis of the improved formulations that will be described next.

This improved formulations, in one way or another, try to take advantage of the particular characteristics of the descriptor form, Equation (5) or the state space (projected) form, Equation (15). In the last few years some authors have described ways to improve the efficiency of global formulations in descriptor form. For instance, von Schwerin [33] proposed a formulation that uses recursivity and a $O(n)$ Adams-Bashforth-Moulton integrator. Cuadrado et al. [15] have obtained high efficiency by using Cartesian Natural coordinates [19,20] combined with an index 3 penalty method. Although both formulations are global, the ideas behind their improvements can be used to enhance the topological formulations where they are applied to close-loop multibody systems.

2 Topological semi-recursive formulations

Open-chain systems, when formulated using relative coordinates, avoid constraint equations. For these systems there are several fully recursive formulations able to solve the direct and inverse dynamic problems with $O(n)$ arithmetic operations [2,4,6,18]. Fully recursive formulations are the most efficient ones for chains with large number of bodies. A careful analysis carried out by Stelzle, Kecskeméthy and Hiller [32] demonstrated that these methods are the most efficient for a number of bodies $n>7$.

If the system has closed loops the relative coordinates are not independent. The usual way to deal with these systems is to open the loops by cutting some joints and to carry out the dynamic analysis by applying the constraint forces and imposing the corresponding constraint equations. In this case the $O(n)$ dynamic formulation becomes much more involved [1,3,8,29], and many authors prefer to keep it simpler with a partial application of recursivity, leading to *semi-recursive* formulations. At the end, all semi-recursive formulations arrive at systems of linear or nonlinear equations, depending on the kind of numerical integrators they use, explicit or implicit. Solving these equations is an $O(n^3)$ arithmetic operations process, assuming that no provision is taken to take advantage of the zeroes or sparsity pattern of the system matrix.

Most semi-recursive formulations use Cartesian coordinates in the first stage of the analysis. The closed loops are opened by removing some joints (or bodies). The dynamic equations of the spanning tree and the closure of the loop constraint equations are formulated in Cartesian coordinates, and then they switch to relative coordinates using a *velocity transformation.* A difference between the formulations is the *reference point* they use to define the Cartesian velocities and accelerations of each body. An obvious choice can be the center of gravity, that simplifies the setting of the Newton-Euler dynamic equations; another possibility is the origin of the reference frame attached to the moving body. A third possibility is to use the point pertaining to the moving body that instantaneously coincides with the origin of the global or inertial reference frame; as the position of this point is the same for all bodies, some recursive expressions become very simple.

Among the semi-recursive formulations reported in the bibliography it is necessary to mention the method of Bae et al [11,13] This formulation is very different from the others, because it is closely tied to the stiff numerical integrator described by Yen [35]. Bae uses as reference point the origin of the moving reference frame. To perform a time step, Bae sets a system of nonlinear equations whose unknowns are the new relative positions, velocities, accelerations and Lagrange multipliers. These equations come from the dynamic equations, the kinematic constraints for positions, velocities and accelerations, and from the backward differentiation formula used by the numerical integrator. These nonlinear equations are solved by a Newton-Raphson method whose tangent or Jacobian matrix is computed exactly using forward and backward recursive formulae for kinematic and force terms, respectively. This method is relatively simple and seems to be very robust and efficient, particularly when the system of linear equations has a small or moderate size. Because of the BDF formula used by the integrator, it is able to deal with both stiff and non-stiff problems.

The first semi-recursive formulation based on a velocity transformation between Cartesian and relative velocities was due to Jerkovsky [24]. These ideas were subsequently extended by other authors, such as Kim and Vanderploeg [23], Nikravesh and Gim [27], García de Jalón et al. [21], and Bae and Won [9]. More recently, other recursive and semi-recursive formulations have been developed by Negrut, Serban and Potra [26], Saha and Schiehlen [29], Saha [28] and Kim [22]. In the following sections a new variant that is very simple and general will be described in detail.

2.1 Semi-recursive formulation for open-chain systems

As many authors do, this method starts with the dynamic equations set in Cartesian coordinates and then applies two velocity transformations that lead to the motion differential equations using a set of *independent* relative coordinates.

This method is able to deal with any kind of multibody systems. It adapts well both to explicit and implicit integrators. If the system has closed loops, it is first converted to an open-chain system by the cut-joint method or –in some cases– by removing some bodies with a particular geometry and mass distribution (rods). The motion equations are first formulated in Cartesian coordinates. Also the closure of the loop constraints are formulated in Cartesian coordinates; the corresponding constraint forces are introduced through Lagrange multipliers.

Then, a first velocity transformation switches from the Cartesian velocities to the relative velocities corresponding to the open-chain system. After this transformation only the closure of the loop constraints remain; they are transformed to the relative velocity space. In order to get a system of ordinary differential equations, a second velocity transformation that keeps only a subset of independent relative velocities is applied.

In the formulation that is described here, the geometry of each moving body is defined in a reference frame attached to the moving body by using *natural coordinates*, i. e., by defining a set of points and unit vectors that describe the geometry of the body and its joints, as can be seen in Figure 1. In this way, the geometry is simpler and clearer than using multiple "markers" or additional reference frames attached to the moving bodies [13,22]. When needed, this geometric information is easily transformed to the global reference frame using the body *position variables*, that are the position vector of the origin of the moving reference frame $\mathbf{r}_i$ and the transformation matrix $\mathbf{A}_i$.

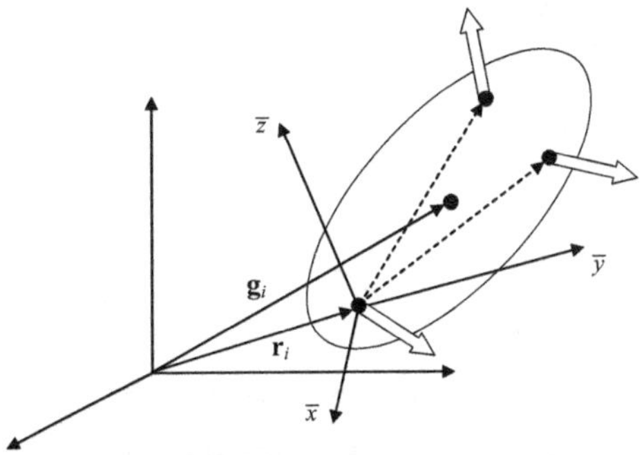

Fig. 1. Body geometry defined with points and unit vectors

For reasons that will become apparent later, the Cartesian velocities $\mathbf{Z}_i$ include the velocity $\dot{\mathbf{s}}_i$ of the point attached to body i that instantaneously coincides with the origin of the inertial reference frame. These Cartesian velocities and accelerations are also used by Negrut et al. [26] and Kim [22]. The Cartesian velocities and accelerations are defined by the vectors,

$$\mathbf{Z}_i \equiv \begin{Bmatrix} \dot{\mathbf{s}}_i \\ \boldsymbol{\omega}_i \end{Bmatrix}, \quad \dot{\mathbf{Z}}_i \equiv \begin{Bmatrix} \ddot{\mathbf{s}}_i \\ \dot{\boldsymbol{\omega}}_i \end{Bmatrix} \tag{16}$$

Vectors $\mathbf{Z}$ and $\dot{\mathbf{Z}}$ are respectively the vectors that contains the Cartesian velocities and accelerations of all the bodies:

$$\mathbf{Z}^T = \left\{ \mathbf{Z}_1^T \;\; \mathbf{Z}_2^T \;\; \cdots \;\; \mathbf{Z}_n^T \right\}, \quad \dot{\mathbf{Z}}^T = \left\{ \dot{\mathbf{Z}}_1^T \;\; \dot{\mathbf{Z}}_2^T \;\; \cdots \;\; \dot{\mathbf{Z}}_n^T \right\} \tag{17}$$

Using points and unit vectors, joints between contiguous bodies are modeled very easily. For instance, in a revolute joint between bodies i-1 and i (see Figure 2), an output point and a unit vector of element i-1 coincide respectively with the input point and unit vector of element i. For a prismatic joint both elements share a unit vector, and the input point of element i is located on the line defined by the output point and unit vector of element i-1 (see Figure 3); in this case both elements share the same transformation matrix.

For dynamics, it is useful to have expressions for the Cartesian velocities $\mathbf{Y}$ and accelerations $\dot{\mathbf{Y}}$ based on the center of gravity, that are defined in the form:

$$\mathbf{Y}_i = \begin{Bmatrix} \dot{\mathbf{g}}_i \\ \boldsymbol{\omega}_i \end{Bmatrix} = \begin{bmatrix} \mathbf{I}_3 & -\tilde{\mathbf{g}}_i \\ \mathbf{0} & \mathbf{I}_3 \end{bmatrix} \begin{Bmatrix} \dot{\mathbf{s}}_i \\ \boldsymbol{\omega}_i \end{Bmatrix} = \mathbf{D}_i \mathbf{Z}_i \tag{18}$$

$$\dot{\mathbf{Y}}_i = \begin{Bmatrix} \ddot{\mathbf{g}}_i \\ \dot{\boldsymbol{\omega}}_i \end{Bmatrix} = \begin{bmatrix} \mathbf{I}_3 & -\tilde{\mathbf{g}}_i \\ \mathbf{0} & \mathbf{I}_3 \end{bmatrix} \begin{Bmatrix} \ddot{\mathbf{s}}_i \\ \dot{\boldsymbol{\omega}}_i \end{Bmatrix} + \begin{Bmatrix} \tilde{\boldsymbol{\omega}}_i \tilde{\boldsymbol{\omega}}_i \mathbf{g}_i \\ \mathbf{0} \end{Bmatrix} = \mathbf{D}_i \dot{\mathbf{Z}}_i + \mathbf{e}_i \tag{19}$$

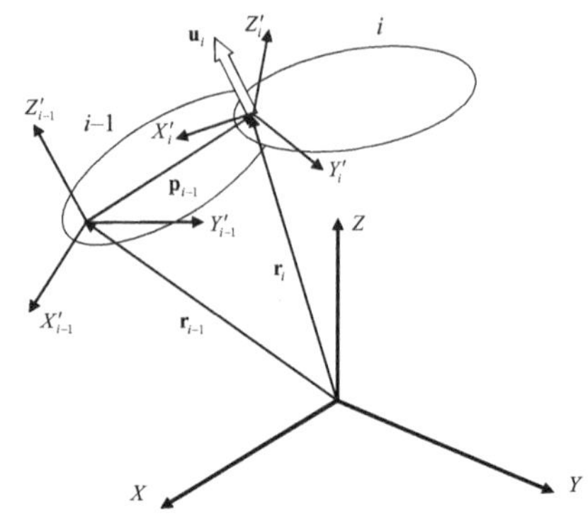

Fig. 2. Revolute joint

Equations (18) and (19) constitute the definition of matrix $\mathbf{D}_i$ and vector $\mathbf{e}_i$. In these expressions, $\tilde{\mathbf{g}}_i$ and $\tilde{\boldsymbol{\omega}}_i$ are the skew-symmetric matrices associated with vectors $\mathbf{g}_i$ and $\boldsymbol{\omega}_i$.

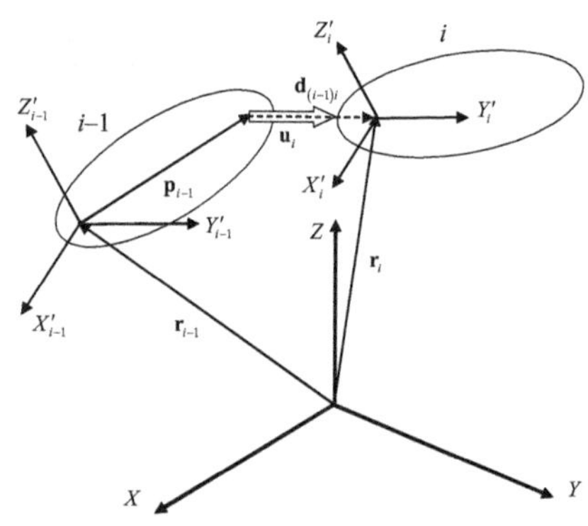

Fig. 3. Prismatic joint

For open-chain systems (and for closed-loop systems, after removing some joints or bodies) the Cartesian positions, velocities and accelerations can be computed recursively upwards from the relative coordinates, velocities and accelerations. The recursive calculations for positions are straightforward and are not included here because they are not important for the first velocity transformation.

The recursive expressions for velocities and accelerations are,

$$\mathbf{Z}_i = \mathbf{Z}_{i-1} + \mathbf{b}_i \dot{z}_i \tag{20}$$

$$\dot{\mathbf{Z}}_i = \dot{\mathbf{Z}}_{i-1} + \mathbf{b}_i \ddot{z}_i + \mathbf{d}_i \tag{21}$$

where z_i are the relative coordinates, and vectors $\mathbf{b}_i$ and $\mathbf{d}_i$ have simple expressions that depend on the kind of joint i. Note that if different reference points are used for bodies i and i-1, the Equations (20) and (21) should include a transformation matrix $\mathbf{B}_i$ [28,29],

$$\mathbf{Y}_i = \mathbf{B}_i \mathbf{Y}_{i-1} + \mathbf{b}_i \dot{z}_i \tag{22}$$

$$\dot{\mathbf{Y}}_i = \mathbf{B}_i \dot{\mathbf{Y}}_{i-1} + \mathbf{b}_i \ddot{z}_i + \mathbf{d}_i \tag{23}$$

The simpler form of Equations (20) and (21), as compared with Equations (22) and (23), has important advantages in some *accumulation expressions* that will appear later. For open-chain systems the velocity transformation defined by Equation (6) can be set directly. It is possible to consider only revolute and prismatic joints, because other joints can be decomposed on a combination of revolute and prismatic joints with massless intermediate bodies. In this case, the column j of matrix $\mathbf{R}$ can be computed directly, because its elements are the Cartesian velocities of the bodies that are upwards in the tree, originated by a unit relative velocity in the joint j and null relative velocities in other joints. As all bodies share the same reference point, all have the same velocity $\mathbf{b}_i$ given by Equation (20).

This velocity transformation and the way the system topology is taken into account is better explained with an example. Figure 4 shows an open-chain, tree configured multibody system. As suggested by Negrut et al. [26] the bodies have been numbered from the leaves to the root, in such a way that each body has a number lower than its parent.

This numbering avoids the later filling-in in the Gauss elimination process. Each joint has the same number as the body that is upwards when the tree is traversed from the root to the leaves. For this example, the velocity transformation matrix corresponding to Equation (6) has the following form:

$$\mathbf{R} = \begin{bmatrix} \mathbf{b}_1 & \mathbf{b}_2 & 0 & 0 & \mathbf{b}_5 \\ 0 & \mathbf{b}_2 & 0 & 0 & \mathbf{b}_5 \\ 0 & 0 & \mathbf{b}_3 & \mathbf{b}_4 & \mathbf{b}_5 \\ 0 & 0 & 0 & \mathbf{b}_4 & \mathbf{b}_5 \\ 0 & 0 & 0 & 0 & \mathbf{b}_5 \end{bmatrix} = \begin{bmatrix} \mathbf{I} & \mathbf{I} & 0 & 0 & \mathbf{I} \\ 0 & \mathbf{I} & 0 & 0 & \mathbf{I} \\ 0 & 0 & \mathbf{I} & \mathbf{I} & \mathbf{I} \\ 0 & 0 & 0 & \mathbf{I} & \mathbf{I} \\ 0 & 0 & 0 & 0 & \mathbf{I} \end{bmatrix} \begin{bmatrix} \mathbf{b}_1 & 0 & 0 & 0 & 0 \\ 0 & \mathbf{b}_2 & 0 & 0 & 0 \\ 0 & 0 & \mathbf{b}_3 & 0 & 0 \\ 0 & 0 & 0 & \mathbf{b}_4 & 0 \\ 0 & 0 & 0 & 0 & \mathbf{b}_5 \end{bmatrix} \equiv \mathbf{T}\mathbf{R}_d \tag{24}$$

where $\mathbf{I}$ is the identity matrix of size 6×6, $\mathbf{T}$ is the *path matrix* that defines the connectivity of the mechanism and $\mathbf{R}_d$ is a diagonal matrix whose elements are the vectors $\mathbf{b}_i$ defined in Equation (20). Remember that vector $\mathbf{b}_i$ represents the velocity of the point that coincides with the inertial frame origin, induced by a unit relative velocity in joint i.

The introduction of the path matrix $\mathbf{T}$ is a key point of this formulation. This allows the topology of the spanning tree to be taken into account in a straightforward way. Other authors, as Negrut el al. [26], need to introduce complicated expressions to explain the recursive processes on different branches that start from a common junction body. Observe that the k row of the path matrix $\mathbf{T}$ gives the joints or relative coordinates that are below body k, while the k column gives the bodies that are upwards of joint k.

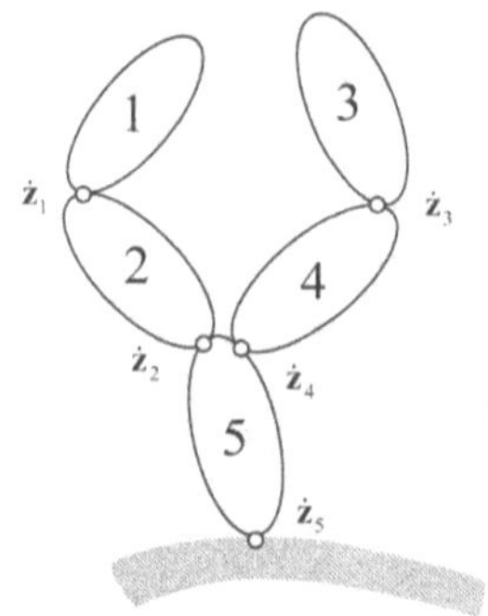

Fig. 4. An example of an open-chain system

Taking into account Equations (18) and (19), the virtual power of the inertia and external forces acting on the whole system can be expressed as,

$$\begin{aligned}\sum_{i=1}^{n} \mathbf{Y}_i^{*T}\left(\mathbf{M}_i\dot{\mathbf{Y}}_i-\mathbf{Q}_i\right)&=\sum_{i=1}^{n} \mathbf{Z}_i^{*T}\mathbf{D}_i^T\left(\mathbf{M}_i\mathbf{D}_i\dot{\mathbf{Z}}_i+\mathbf{M}_i\mathbf{e}_i-\mathbf{Q}_i\right)=\\ &=\sum_{i=1}^{n} \mathbf{Z}_i^{*T}\left(\bar{\mathbf{M}}_i\dot{\mathbf{Z}}_i-\bar{\mathbf{Q}}_i\right)=0\end{aligned} \tag{25}$$

where the virtual velocities have been denoted with an asterisk (*). The matrices appearing in Equation (25) are,

$$\begin{gathered}\mathbf{M}_i=\begin{bmatrix} m_i\mathbf{I}_3 & \mathbf{0}\\ \mathbf{0} & \mathbf{J}_i\end{bmatrix},\quad \bar{\mathbf{M}}_i=\mathbf{D}_i^T\mathbf{M}_i\mathbf{D}_i=\begin{bmatrix} m_i\mathbf{I}_3 & -m_i\tilde{\mathbf{g}}_i\\ m_i\tilde{\mathbf{g}}_i & \mathbf{J}_i-m_i\tilde{\mathbf{g}}_i\tilde{\mathbf{g}}_i\end{bmatrix}\\ \bar{\mathbf{Q}}_i=\mathbf{D}_i^T\left(\mathbf{M}_i\mathbf{e}_i-\mathbf{Q}_i\right)\end{gathered} \tag{26}$$

where m_i is the mass of body i and $\mathbf{J}_i$ is its inertia tensor; matrix $\mathbf{D}_i$ and vector $\mathbf{e}_i$ are defined in Equation (19). In Equations (25) and (26) the inertia matrices and vector forces denoted with an upper bar refer to the origin of the global reference frame. The advantage of referring everything to the origin of the global frame is that vectors and matrices from different bodies can be added directly, without any further transformation.

By defining the system inertia matrix $\bar{\mathbf{M}}$, the force vector $\bar{\mathbf{Q}}$ and the acceleration vector $\dot{\mathbf{Z}}$ in the form,

$$\begin{aligned}\bar{\mathbf{M}}&\equiv \operatorname{diag}\left(\bar{\mathbf{M}}_1,\bar{\mathbf{M}}_2,\ldots,\bar{\mathbf{M}}_n\right)\\ \bar{\mathbf{Q}}^T&=\left[\bar{\mathbf{Q}}_1^T,\bar{\mathbf{Q}}_2^T,\ldots,\bar{\mathbf{Q}}_n^T\right]\\ \dot{\mathbf{Z}}^T&=\left[\dot{\mathbf{Z}}_1^T,\dot{\mathbf{Z}}_2^T,\ldots,\dot{\mathbf{Z}}_n^T\right]\end{aligned} \tag{27}$$

the dynamic Equations (25) can be written as:

$$\mathbf{Z}^{*T}\left(\bar{\mathbf{M}}\dot{\mathbf{Z}}-\bar{\mathbf{Q}}\right)=0 \tag{28}$$

Using Equation (24), the velocity transformation and its time derivative, can be written for the whole system in the form:

$$\mathbf{Z} = \mathbf{R}\dot{\mathbf{z}} = \mathbf{T}\mathbf{R}_d\dot{\mathbf{z}} \tag{29}$$

$$\dot{\mathbf{Z}} = \mathbf{T}\mathbf{R}_d\ddot{\mathbf{z}} + \mathbf{T}\dot{\mathbf{R}}_d\dot{\mathbf{z}} \tag{30}$$

These expressions are very similar to the expressions obtained by Saha [28] in his DeNOC (Decoupled Natural Orthogonal Complement) method. The Saha $\mathbf{N}_l$ matrix is triangular when serial manipulators are considered, and is a function of the $\mathbf{B}_i$ matrices in Equation (22) because of the reference point he uses. Matrix $\mathbf{R}_d$ is analogous to the Saha $\mathbf{N}_d$ matrix.

Substituting Equations (29) and (30) in Equation (28) and taking into account that the relative virtual velocities are independent, a new set of equations is obtained,

$$\mathbf{R}_d^T\left(\mathbf{T}^T\bar{\mathbf{M}}\mathbf{T}\right)\mathbf{R}_d\ddot{\mathbf{z}} = \mathbf{R}_d^T\mathbf{T}^T\left(\bar{\mathbf{Q}} - \bar{\mathbf{M}}\mathbf{T}\dot{\mathbf{R}}_d\dot{\mathbf{z}}\right) \tag{31}$$

That are analogous to Equations (15). It is interesting to visualize the pattern of the inertia matrix in Equation (31), for the open-chain example in Figure 4,

$$\mathbf{R}_d^T\mathbf{T}^T\bar{\mathbf{M}}\mathbf{T}\mathbf{R}_d = \mathbf{R}_d^T\mathbf{M}^{\Sigma}\mathbf{R}_d =$$
$$= \begin{bmatrix} \mathbf{b}_1^T\mathbf{M}_1^{\Sigma}\mathbf{b}_1 & \mathbf{b}_1^T\mathbf{M}_1^{\Sigma}\mathbf{b}_2 & \mathbf{0} & \mathbf{0} & \mathbf{b}_1^T\mathbf{M}_1^{\Sigma}\mathbf{b}_5 \\ \mathbf{b}_2^T\mathbf{M}_1^{\Sigma}\mathbf{b}_1 & \mathbf{b}_2^T\mathbf{M}_2^{\Sigma}\mathbf{b}_2 & \mathbf{0} & \mathbf{0} & \mathbf{b}_2^T\mathbf{M}_2^{\Sigma}\mathbf{b}_5 \\ \mathbf{0} & \mathbf{0} & \mathbf{b}_3^T\mathbf{M}_3^{\Sigma}\mathbf{b}_3 & \mathbf{b}_3^T\mathbf{M}_3^{\Sigma}\mathbf{b}_4 & \mathbf{b}_3^T\mathbf{M}_3^{\Sigma}\mathbf{b}_5 \\ \mathbf{0} & \mathbf{0} & \mathbf{b}_4^T\mathbf{M}_3^{\Sigma}\mathbf{b}_3 & \mathbf{b}_4^T\mathbf{M}_4^{\Sigma}\mathbf{b}_4 & \mathbf{b}_4^T\mathbf{M}_4^{\Sigma}\mathbf{b}_5 \\ \mathbf{b}_5^T\mathbf{M}_1^{\Sigma}\mathbf{b}_1 & \mathbf{b}_5^T\mathbf{M}_2^{\Sigma}\mathbf{b}_2 & \mathbf{b}_5^T\mathbf{M}_3^{\Sigma}\mathbf{b}_3 & \mathbf{b}_5^T\mathbf{M}_4^{\Sigma}\mathbf{b}_4 & \mathbf{b}_5^T\mathbf{M}_5^{\Sigma}\mathbf{b}_5 \end{bmatrix} \tag{32}$$

where,

$$\begin{gathered} \mathbf{M}_1^{\Sigma} = \bar{\mathbf{M}}_1, \qquad \mathbf{M}_2^{\Sigma} = \bar{\mathbf{M}}_2 + \mathbf{M}_1^{\Sigma}, \qquad \mathbf{M}_3^{\Sigma} = \bar{\mathbf{M}}_3 \\ \mathbf{M}_4^{\Sigma} = \bar{\mathbf{M}}_4 + \mathbf{M}_3^{\Sigma}, \qquad \mathbf{M}_5^{\Sigma} = \bar{\mathbf{M}}_5 + \mathbf{M}_2^{\Sigma} + \mathbf{M}_4^{\Sigma} \end{gathered} \tag{33}$$

Matrices $\mathbf{M}_i^{\Sigma}$ are the *composite inertia matrices* described by many authors. They represent the *accumulation* of the inertia matrices of all the elements that are upwards of joint *i*. This is a result of the introduction of the path matrix $\mathbf{T}$ in the formulation.

In an analogous way, the *accumulated* external forces $\mathbf{Q}^{\Sigma}$ and velocity dependent accumulated inertia forces $\mathbf{P}^{\Sigma}$ can be computed from the right-hand side of Equation (31),

$$\begin{gathered} \mathbf{Q}^{\Sigma} \equiv \left\{\mathbf{Q}_1^{\Sigma T}, \mathbf{Q}_2^{\Sigma T}, \mathbf{Q}_3^{\Sigma T}, \mathbf{Q}_4^{\Sigma T}, \mathbf{Q}_5^{\Sigma T}\right\}^T = \mathbf{T}^T\bar{\mathbf{Q}} \\ \mathbf{P}^{\Sigma} \equiv \left\{\mathbf{P}_1^{\Sigma T}, \mathbf{P}_2^{\Sigma T}, \mathbf{P}_3^{\Sigma T}, \mathbf{P}_4^{\Sigma T}, \mathbf{P}_5^{\Sigma T}\right\}^T = -\mathbf{T}^T\bar{\mathbf{M}}\mathbf{T}\dot{\mathbf{R}}_d\dot{\mathbf{z}} \end{gathered} \tag{34}$$

where,

$$\begin{gathered}
\mathbf{Q}_1^\Sigma = \bar{\mathbf{Q}}_1; \quad \mathbf{Q}_2^\Sigma = \bar{\mathbf{Q}}_2 + \mathbf{Q}_1^\Sigma; \quad \mathbf{Q}_3^\Sigma = \bar{\mathbf{Q}}_3 \\
\mathbf{Q}_4^\Sigma = \bar{\mathbf{Q}}_4 + \mathbf{Q}_3^\Sigma; \quad \mathbf{Q}_5^\Sigma = \bar{\mathbf{Q}}_5 + \mathbf{Q}_2^\Sigma + \mathbf{Q}_4^\Sigma \\
\mathbf{P}_1^\Sigma = -\bar{\mathbf{M}}_1\left(\mathbf{T}\dot{\mathbf{R}}_d\dot{\mathbf{z}}\right)_1; \quad \mathbf{P}_2^\Sigma = -\bar{\mathbf{M}}_2\left(\mathbf{T}\dot{\mathbf{R}}_d\dot{\mathbf{z}}\right)_2 + \mathbf{P}_1^\Sigma; \quad \mathbf{P}_3^\Sigma = -\bar{\mathbf{M}}_3\left(\mathbf{T}\dot{\mathbf{R}}_d\dot{\mathbf{z}}\right)_3 \\
\mathbf{P}_4^\Sigma = -\bar{\mathbf{M}}_4\left(\mathbf{T}\dot{\mathbf{R}}_d\dot{\mathbf{z}}\right)_4 + \mathbf{P}_3^\Sigma, \quad \mathbf{P}_5^\Sigma = -\bar{\mathbf{M}}_5\left(\mathbf{T}\dot{\mathbf{R}}_d\dot{\mathbf{z}}\right)_5 + \mathbf{P}_2^\Sigma + \mathbf{P}_4^\Sigma
\end{gathered} \tag{35}$$

The meaning of the *accumulation* of external forces is clear. With respect to the velocity dependent inertia forces, Equation (35) shall be related with Equation (30). The term ($\mathbf{T}\dot{\mathbf{R}}_d\dot{\mathbf{z}}$) represents the velocity dependent Cartesian accelerations, i.e., the Cartesian accelerations computed with the true velocities $\dot{\mathbf{z}}$ and null relative accelerations ($\ddot{\mathbf{z}} = \mathbf{0}$). This vector is the forward *accumulation* of vectors $\mathbf{d}_j$ in Equation (21), from the fixed base body to the joint considered.

The matrix in Equation (32) shows the advantages of numbering the bodies and joints from the leaves to the root: the Gaussian elimination or the LU factorization keeps the pattern of zeros in the matrix, i. e., it maintains the skyline or the sparsity of this matrix, avoiding some arithmetic operations. For instance, for a tree model of the human body, with 42 bodies and 41 joints, up to a 13% reduction of the total CPU time has been obtained, using Matlab. The advantages of this numeration method were pointed out by Negrut et al. [26]. The Equations (31) constitute a system of ordinary differential equations whose coefficient matrix and right-hand side vector can be computed recursively in a very efficient way.

2.2 Semi-recursive formulation for closed-loop systems

The dynamics of closed-loop multibody systems can be formulated by adding the constraint equations to the dynamic equations corresponding to the open-chain system obtained by opening the closed loops. Here the methods quoted previously to improve the global methods that use the descriptor form can be applied with analogous advantages, although the problem size is much lower with relative coordinates. There are three main possibilities:

1. To introduce the constraints by the Lagrange multiplier method and to use the most appropriate solution method for the augmented matrix: the RSM or the NSM.
2. To follow the methodology of Cuadrado et al. [17] and to introduce the constraints by the penalty method with augmented Lagrangian. The integration is carried out with the trapezoidal rule and the velocities and accelerations are corrected with mass orthogonal projections. This procedure has been explored quite recently by Cuadrado and Dopico [16].
3. It is also possible to select an independent subset of relative coordinates, in such a way that a set of ordinary differential equations will be obtained again. This is carried out by a new velocity transformation similar to the one introduced to arrive at Equation (15). In this case the transformation matrix **R** will

be obtained numerically. This is the method that will be explained in the rest of this paper. It is faster than method 1), it is free from constraint stabilization problems, and it is not tied to a particular integrator as method 2).

The closing-loop constraint equations are first formulated in Cartesian coordinates and then transformed to relative coordinates. In this paper two ways to set the closed-loop constraint equations will be considered. The first one is the *cutting joint* method, that is very common in the literature. The second method to open the loops consists of the elimination of one or more *rods* (slender bodies with two spherical joints and a negligible moment of inertia around the direction of the axis). This second procedure has not been found in the bibliography and it is very interesting in applications, as car suspension systems, where rods are very common. *Rods* are rigid bodies difficult to analyze: sometimes they are considered as distance constraints, neglecting their inertia forces [26]; many commercial computer programs do not allow the introduction of bodies with two spherical joints and oblige the replacement of a spherical joint by a universal joint. To open the loops by removing a rod introduces one constraint equation, but keeping the rod and opening the loop by cutting a spherical joint introduces three constraint equations and keeps the two relative coordinates of the universal joint. On the other hand, a rod can be eliminated easily, but to consider exactly its inertia properties is more involved.

Figure 5 shows a revolute joint that is defined with natural coordinates as the sharing of a point and a unit vector. To take into account the constraints of such a joint, two points and two unit vectors, belonging to different bodies, must be compelled to coincide.

To set the closed-loop condition for the system shown in Figure 6 the following equations must be established:

$$\mathbf{r}_j - \mathbf{r}_k = \mathbf{0} \quad \left(3 \text{ independent equations}\right) \tag{36}$$

$$\mathbf{u}_j - \mathbf{u}_k = \mathbf{0} \quad \left(\text{only 2 independent equations}\right) \tag{37}$$

For the rod element in Figure 6 only a constant distance condition is necessary:

$$\left(\mathbf{r}_j - \mathbf{r}_k\right)^T \left(\mathbf{r}_j - \mathbf{r}_k\right) - l_{jk}^2 = 0 \tag{38}$$

The constraint equations (36-38) shall be expressed in terms of the relative coordinates $\mathbf{z}$. This is not difficult, because points $\mathbf{r}_i$ and $\mathbf{r}_k$, and unit vectors $\mathbf{u}_i$ and $\mathbf{u}_k$ can be expressed as functions of the relative coordinates of the joints in their respective branches of the open-chain system.

It is also necessary to compute the Jacobian matrix of constraints (36-38) with respect to relative coordinates $\mathbf{z}$. As the aforesaid constraints are expressed as a function of Cartesian coordinates, the chain derivative rule shall be used. For instance, for the constant distance constraint, given by Equation (38),

$$\mathbf{\Phi}_{\mathbf{z}} = \mathbf{\Phi}_{\mathbf{r}_j} \frac{\partial \mathbf{r}_j}{\partial \mathbf{z}} + \mathbf{\Phi}_{\mathbf{r}_k} \frac{\partial \mathbf{r}_k}{\partial \mathbf{z}} = \mathbf{\Phi}_{\mathbf{r}_j} \frac{\partial \dot{\mathbf{r}}_j}{\partial \dot{\mathbf{z}}} + \mathbf{\Phi}_{\mathbf{r}_k} \frac{\partial \dot{\mathbf{r}}_k}{\partial \dot{\mathbf{z}}} \tag{39}$$

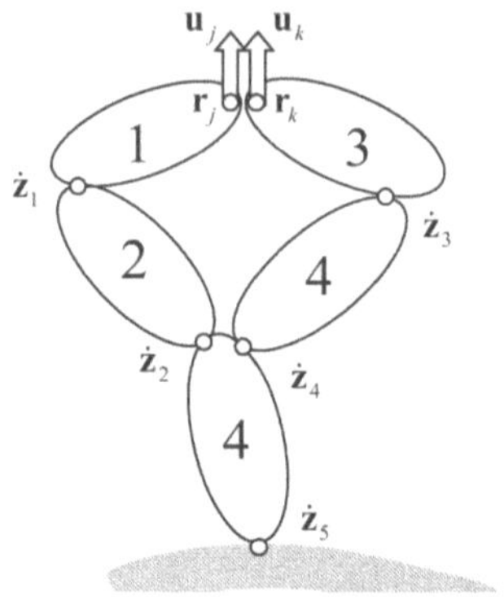

Fig. 5. Chain closed with a revolute joint

The derivatives with respect to the coordinates $\mathbf{r}_i$ and $\mathbf{r}_k$ in Equation (38) are easy to find:

$$\mathbf{\Phi}_{\mathbf{r}_j} = 2\left(\mathbf{r}_j^T - \mathbf{r}_k^T\right), \qquad \mathbf{\Phi}_{\mathbf{r}_k} = -2\left(\mathbf{r}_j^T - \mathbf{r}_k^T\right) \tag{40}$$

The derivatives of the position vectors $\mathbf{r}_i$ and $\mathbf{r}_k$ with respect to the relative coordinates $\mathbf{z}$ can be computed from the velocities of these points induced by unit relative velocities in the joints between the fixed body and bodies j and k, respectively. For instance, if the joint i is a revolute joint determined by a point $\mathbf{r}_i$ and a unit vector $\mathbf{r}_i$, located between the base body and point $\mathbf{r}_i$, the velocity of point j originated by a unit velocity in joint i can be set as,

$$\frac{\partial \dot{\mathbf{r}}_j}{\partial \dot{z}_i} = \mathbf{u}_i \times \left(\mathbf{r}_j - \mathbf{r}_i\right) = \tilde{\mathbf{u}}_i \left(\mathbf{r}_j - \mathbf{r}_i\right) \tag{41}$$

So, it can be assumed that the closure of the loop constraint equations $\mathbf{\Phi}(\mathbf{z})=\mathbf{0}$ and their Jacobian matrix $\mathbf{\Phi}_\mathbf{z}$ are known or easy to compute. Using the coordinate partitioning method based on Gaussian elimination with full pivoting as in Equation (11), it is possible to arrive at the following partitioned velocity equation,

$$\begin{bmatrix} \mathbf{\Phi}_\mathbf{z}^d & \mathbf{\Phi}_\mathbf{z}^i \end{bmatrix} \begin{Bmatrix} \dot{\mathbf{z}}^d \\ \dot{\mathbf{z}}^i \end{Bmatrix} = \mathbf{0}, \qquad \dot{\mathbf{z}}^d = -\left(\mathbf{\Phi}_\mathbf{z}^d\right)^{-1} \mathbf{\Phi}_\mathbf{z}^i \dot{\mathbf{z}}^i \tag{42}$$

where it is assumed that matrix $\mathbf{\Phi}^d_\mathbf{z}$ is invertible. Equation (42) allows an easy calculation of the transformation matrix $\mathbf{R}_\mathbf{z}$ that relates dependent and independent relative velocities,

$$\dot{\mathbf{z}} = \mathbf{R}_\mathbf{z} \dot{\mathbf{z}}^i, \qquad \begin{Bmatrix} \dot{\mathbf{z}}^d \\ \dot{\mathbf{z}}^i \end{Bmatrix} = \begin{bmatrix} -\left(\mathbf{\Phi}_\mathbf{z}^d\right)^{-1} \mathbf{\Phi}_\mathbf{z}^i \\ \mathbf{I} \end{bmatrix} \dot{\mathbf{z}}^i, \qquad \mathbf{R}_\mathbf{z} = \begin{bmatrix} -\left(\mathbf{\Phi}_\mathbf{z}^d\right)^{-1} \mathbf{\Phi}_\mathbf{z}^i \\ \mathbf{I} \end{bmatrix} \tag{43}$$

The differentiation of this equation with respect to time leads to

$$\ddot{\mathbf{z}} = \mathbf{R}_\mathbf{z} \ddot{\mathbf{z}}^i + \dot{\mathbf{R}}_\mathbf{z} \dot{\mathbf{z}}^i \tag{44}$$

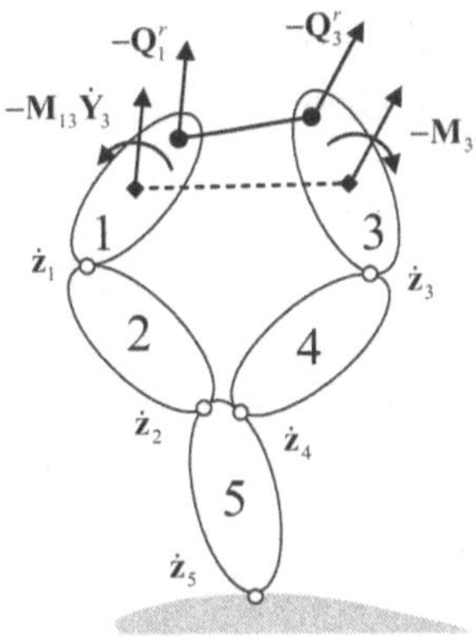

Fig. 6. Chain closed with a rod element

The velocity transformation defined by Equations (43) and (44) can be introduced in the motion differential equations (31). By pre-multiplication by matrix $\mathbf{R}^T_{\mathbf{z}}$ the following is obtained,

$$\mathbf{R}_{\mathbf{z}}^T\mathbf{R}_d^T\mathbf{M}^{\Sigma}\mathbf{R}_d\mathbf{R}_{\mathbf{z}}\ddot{\mathbf{z}}^i = \mathbf{R}_{\mathbf{z}}^T\mathbf{R}_d^T\mathbf{Q}^{\Sigma} - \mathbf{R}_{\mathbf{z}}^T\mathbf{R}_d^T\mathbf{M}^{\Sigma}\left(\dot{\mathbf{R}}_d\dot{\mathbf{z}} + \mathbf{R}_d\dot{\mathbf{R}}_{\mathbf{z}}\dot{\mathbf{z}}^i\right) \tag{45}$$

All the terms in this equation are known, except the parenthesis that contains the derivatives of the transformation matrices. It is simpler to compute the two terms jointly. Considering Equation (43), the parenthesis in Equation (45) is,

$$\begin{aligned}\dot{\mathbf{R}}_d\dot{\mathbf{z}} + \mathbf{R}_d\dot{\mathbf{R}}_{\mathbf{z}}\dot{\mathbf{z}}^i &= \dot{\mathbf{R}}_d\mathbf{R}_{\mathbf{z}}\dot{\mathbf{z}}^i + \mathbf{R}_d\dot{\mathbf{R}}_{\mathbf{z}}\dot{\mathbf{z}}^i = \\ &= \left(\dot{\mathbf{R}}_d\mathbf{R}_{\mathbf{z}} + \mathbf{R}_d\dot{\mathbf{R}}_{\mathbf{z}}\right)\dot{\mathbf{z}}^i = \frac{d\left(\mathbf{R}_d\mathbf{R}_{\mathbf{z}}\right)}{dt}\dot{\mathbf{z}}^i\end{aligned} \tag{46}$$

This derivative can be computed from the product of velocity transformations that relates Cartesian and independent relative velocities,

$$\mathbf{Z} = \mathbf{R}\dot{\mathbf{z}} = \mathbf{T}\mathbf{R}_d\dot{\mathbf{z}} = \mathbf{T}\mathbf{R}_d\mathbf{R}_{\mathbf{z}}\dot{\mathbf{z}}^i \tag{47}$$

Taking the time derivative of this equation,

$$\dot{\mathbf{Z}} = \mathbf{T}\mathbf{R}_d\mathbf{R}_{\mathbf{z}}\ddot{\mathbf{z}}^i + \mathbf{T}\frac{d\left(\mathbf{R}_d\mathbf{R}_{\mathbf{z}}\right)}{dt}\dot{\mathbf{z}}^i \tag{48}$$

In this equation, it can be seen that the product of the path matrix $\mathbf{T}$ times the derivative looked for, can be computed as the Cartesian accelerations $\dot{\mathbf{Z}}$ that would be produced by the true velocities $\dot{\mathbf{z}}$ and null relative independent accelerations ($\ddot{\mathbf{z}}^i = \mathbf{0}$). The dynamic equations (45) can be written in the form,

$$\mathbf{R}_{\mathbf{z}}^T\mathbf{R}_d^T\mathbf{M}^{\Sigma}\mathbf{R}_d\mathbf{R}_{\mathbf{z}}\ddot{\mathbf{z}}^i = \mathbf{R}_{\mathbf{z}}^T\mathbf{R}_d^T\left(\mathbf{Q}^{\Sigma} - \mathbf{T}^T\bar{\mathbf{M}}\frac{d\left(\mathbf{T}\mathbf{R}_d\mathbf{R}_{\mathbf{z}}\right)}{dt}\dot{\mathbf{z}}^i\right) \tag{49}$$

To compute the derivative in this expression as explained before, the dependent relative accelerations corresponding with null $\ddot{\mathbf{z}}^i$ must be computed. Taking time derivatives of Equation (42):

$$\begin{bmatrix}\mathbf{\Phi}_\mathbf{z}^d & \mathbf{\Phi}_\mathbf{z}^i\end{bmatrix}\begin{Bmatrix}\ddot{\mathbf{z}}^d \\ \ddot{\mathbf{z}}^i\end{Bmatrix}+\begin{bmatrix}\dot{\mathbf{\Phi}}_\mathbf{z}^d & \dot{\mathbf{\Phi}}_\mathbf{z}^i\end{bmatrix}\begin{Bmatrix}\dot{\mathbf{z}}^d \\ \dot{\mathbf{z}}^i\end{Bmatrix}=\mathbf{0},\quad \ddot{\mathbf{z}}^d=-\left(\mathbf{\Phi}_\mathbf{z}^d\right)^{-1}\begin{bmatrix}\dot{\mathbf{\Phi}}_\mathbf{z}^d & \dot{\mathbf{\Phi}}_\mathbf{z}^i\end{bmatrix}\begin{Bmatrix}\dot{\mathbf{z}}^d \\ \dot{\mathbf{z}}^i\end{Bmatrix} \tag{50}$$

where the product ($\dot{\mathbf{\Phi}}_\mathbf{z}\dot{\mathbf{z}}$) can be computed as,

$$\dot{\mathbf{\Phi}}_\mathbf{z}\dot{\mathbf{z}}=\dot{\mathbf{\Phi}}_{\mathbf{r}_j}\frac{\partial \mathbf{r}_j}{\partial \mathbf{z}}\dot{\mathbf{z}}+\mathbf{\Phi}_{\mathbf{r}_j}\frac{\partial \dot{\mathbf{r}}_j}{\partial \mathbf{z}}\dot{\mathbf{z}}+\dot{\mathbf{\Phi}}_{\mathbf{r}_k}\frac{\partial \mathbf{r}_k}{\partial \mathbf{z}}\dot{\mathbf{z}}+\mathbf{\Phi}_{\mathbf{r}_k}\frac{\partial \dot{\mathbf{r}}_k}{\partial \mathbf{z}}\dot{\mathbf{z}} \tag{51}$$

$$\mathbf{\Phi}_{\mathbf{r}_j}=2\left(\mathbf{r}_j^T-\mathbf{r}_k^T\right)\qquad \dot{\mathbf{\Phi}}_{\mathbf{r}_j}=2\left(\dot{\mathbf{r}}_j^T-\dot{\mathbf{r}}_k^T\right) \tag{52}$$

So, a way to compute all the terms in the ODEs set described by Equation (49) has been completed. Two velocity transformations have been introduced. The first one, from Cartesian to open-chain relative velocities, can be applied directly and leads to an accumulation of forces and inertias. The second one is applied in a fully numerical way to a system of –usually– a much smaller size.

It remains to explain how the inertia of the removed rods is taken into account in the motion differential Equations (49). For this, Figure 2 shall be considered again. Assuming a uniform mass distribution, the rod's inertia matrix is,

$$\mathbf{M}^r=\frac{m}{6}\begin{bmatrix}2\mathbf{I}_3 & \mathbf{I}_3 \\ \mathbf{I}_3 & 2\mathbf{I}_3\end{bmatrix}=\begin{bmatrix}\alpha\mathbf{I}_3 & \beta\mathbf{I}_3 \\ \beta\mathbf{I}_3 & \alpha\mathbf{I}_3\end{bmatrix}\qquad \alpha\equiv\frac{m}{3},\ \beta\equiv\frac{m}{6} \tag{53}$$

The dynamic equilibrium of the rod allows the computation of the forces transmitted by the rod to the neighbor elements as a function of its inertia and external forces,

$$\mathbf{M}^r\begin{Bmatrix}\ddot{\mathbf{r}}_j \\ \ddot{\mathbf{r}}_k\end{Bmatrix}=\begin{Bmatrix}\mathbf{Q}_j^r \\ \mathbf{Q}_k^r\end{Bmatrix}+\begin{Bmatrix}\mathbf{f}_j^r \\ \mathbf{f}_k^r\end{Bmatrix} \tag{54}$$

where $\mathbf{Q}_i^r$ and $\mathbf{Q}_k^r$ are the forces applied by the rod to the neighbor bodies on points j and k, and $\mathbf{f}_i^r$ and $\mathbf{f}_k^r$ are the external forces applied to the rod transmitted to points j and k. The forces acting on elements j and k shall be translated to the reference point –the origin of the global frame– by adding the corresponding torque. So, the forces transmitted to bodies j and k, at their reference points, are:

$$\begin{Bmatrix}\bar{\mathbf{Q}}_j \\ \bar{\mathbf{Q}}_k\end{Bmatrix}=-\begin{bmatrix}\mathbf{I}_3 & \mathbf{0} \\ \tilde{\mathbf{r}}_j & \mathbf{0} \\ \mathbf{0} & \mathbf{I}_3 \\ \mathbf{0} & \tilde{\mathbf{r}}_k\end{bmatrix}\left(\begin{bmatrix}\alpha\mathbf{I}_3 & \beta\mathbf{I}_3 \\ \beta\mathbf{I}_3 & \alpha\mathbf{I}_3\end{bmatrix}\begin{Bmatrix}\ddot{\mathbf{r}}_j \\ \ddot{\mathbf{r}}_k\end{Bmatrix}-\begin{Bmatrix}\mathbf{f}_j^r \\ \mathbf{f}_k^r\end{Bmatrix}\right)=-\begin{bmatrix}\alpha\mathbf{I}_3 & \beta\mathbf{I}_3 \\ \alpha\tilde{\mathbf{r}}_j & \beta\tilde{\mathbf{r}}_j \\ \beta\mathbf{I}_3 & \alpha\mathbf{I}_3 \\ \beta\tilde{\mathbf{r}}_k & \alpha\tilde{\mathbf{r}}_k\end{bmatrix}\begin{Bmatrix}\ddot{\mathbf{r}}_j \\ \ddot{\mathbf{r}}_k\end{Bmatrix}+\begin{Bmatrix}\mathbf{f}_j^r \\ \tilde{\mathbf{r}}_j\mathbf{f}_j^r \\ \mathbf{f}_k^r \\ \tilde{\mathbf{r}}_k\mathbf{f}_k^r\end{Bmatrix} \tag{55}$$

The acceleration of points *j* and *k* can be expressed in terms of the respective bodies' accelerations $\dot{\mathbf{Z}}_j$ and $\dot{\mathbf{Z}}_k$,

$$\ddot{\mathbf{r}}_j = \ddot{\mathbf{s}}_j + \dot{\boldsymbol{\omega}}_j \times \mathbf{r}_j + \boldsymbol{\omega}_j \times (\boldsymbol{\omega}_j \times \mathbf{r}_j) = \begin{bmatrix} \mathbf{I}_3 & -\tilde{\mathbf{r}}_j \end{bmatrix} \dot{\mathbf{Z}}_j + \tilde{\boldsymbol{\omega}}_j \tilde{\boldsymbol{\omega}}_j \mathbf{r}_j \tag{56}$$

$$\ddot{\mathbf{r}}_k = \ddot{\mathbf{s}}_k + \dot{\boldsymbol{\omega}}_k \times \mathbf{r}_k + \boldsymbol{\omega}_k \times (\boldsymbol{\omega}_k \times \mathbf{r}_k) = \begin{bmatrix} \mathbf{I}_3 & -\tilde{\mathbf{r}}_k \end{bmatrix} \dot{\mathbf{Z}}_k + \tilde{\boldsymbol{\omega}}_k \tilde{\boldsymbol{\omega}}_k \mathbf{r}_k \tag{57}$$

Substituting these expressions in Equation (55) the following equation is obtained,

$$\begin{Bmatrix} \bar{\mathbf{Q}}_j \\ \bar{\mathbf{Q}}_k \end{Bmatrix} = -\begin{bmatrix} \bar{\mathbf{M}}^r_{jj} & \bar{\mathbf{M}}^r_{jk} \\ \bar{\mathbf{M}}^r_{kj} & \bar{\mathbf{M}}^r_{kk} \end{bmatrix} \begin{Bmatrix} \dot{\mathbf{Z}}_j \\ \dot{\mathbf{Z}}_k \end{Bmatrix} - \begin{Bmatrix} \begin{bmatrix} \mathbf{I}_3 \\ \tilde{\mathbf{r}}_j \end{bmatrix} (\alpha \tilde{\boldsymbol{\omega}}_j \tilde{\boldsymbol{\omega}}_j \mathbf{r}_j + \beta \tilde{\boldsymbol{\omega}}_k \tilde{\boldsymbol{\omega}}_k \mathbf{r}_k) \\ \begin{bmatrix} \mathbf{I}_3 \\ \tilde{\mathbf{r}}_k \end{bmatrix} (\beta \tilde{\boldsymbol{\omega}}_j \tilde{\boldsymbol{\omega}}_j \mathbf{r}_j + \alpha \tilde{\boldsymbol{\omega}}_k \tilde{\boldsymbol{\omega}}_k \mathbf{r}_k) \end{Bmatrix} + \begin{Bmatrix} \mathbf{f}^r_j \\ \tilde{\mathbf{r}}_j \mathbf{f}^r_j \\ \mathbf{f}^r_k \\ \tilde{\mathbf{r}}_k \mathbf{f}^r_k \end{Bmatrix} \tag{58}$$

where,

$$\bar{\mathbf{M}}^r_{jj} = \alpha \begin{bmatrix} \mathbf{I}_3 & \tilde{\mathbf{r}}^T_j \\ \tilde{\mathbf{r}}_j & \tilde{\mathbf{r}}_j \tilde{\mathbf{r}}^T_j \end{bmatrix}, \quad \bar{\mathbf{M}}^r_{kk} = \alpha \begin{bmatrix} \mathbf{I}_3 & \tilde{\mathbf{r}}^T_k \\ \tilde{\mathbf{r}}_k & \tilde{\mathbf{r}}_k \tilde{\mathbf{r}}^T_k \end{bmatrix}$$
$$\bar{\mathbf{M}}^r_{jk} = \bar{\mathbf{M}}^{rT}_{kj} = \beta \begin{bmatrix} \mathbf{I}_3 & \tilde{\mathbf{r}}^T_k \\ \tilde{\mathbf{r}}_j & \tilde{\mathbf{r}}_j \tilde{\mathbf{r}}^T_k \end{bmatrix} \tag{59}$$

The three terms on the right-hand side of Equation (58) represent, respectively, the acceleration dependent inertia forces, the velocity dependent inertia forces and the contribution of external forces (such as the weight) acting on the rod. These forces shall be added at the element level in the motion differential Equations (49). The external forces and the velocity dependent inertia forces are directly added to the corresponding forces of bodies *j* and *k*.

The acceleration dependent inertia forces are a little more involved because of the coupling terms $\bar{\mathbf{M}}^r_{jk}$ and $\bar{\mathbf{M}}^r_{kj}$ in Equation (58). The diagonal blocks $\bar{\mathbf{M}}^r_{jj}$ and $\bar{\mathbf{M}}^r_{kk}$ are directly added to the inertia matrices of bodies *j* and *k*, according to the expressions,

$$\bar{\mathbf{M}}'_j = \bar{\mathbf{M}}_j + \bar{\mathbf{M}}^r_{jj} \qquad \bar{\mathbf{M}}'_k = \bar{\mathbf{M}}_k + \bar{\mathbf{M}}^r_{kk} \tag{60}$$

In Figure 6 the coupling inertia forces are represented in the respective centers of gravity. The force $-\mathbf{M}^r_{13}\dot{\mathbf{Y}}_3$ acts on body *j*=1, but it depends on the acceleration of body *k*=3; it will be propagated backwards on the branch of body 1, but it depends on the relative accelerations in the branch of body 3. To see this with more detail, the modified inertia matrix for the whole system in Figure 6 is written as,

$$\bar{\mathbf{M}} = \begin{bmatrix} \bar{\mathbf{M}}_1' & 0 & \bar{\mathbf{M}}_{13}^r & 0 & 0 \\ 0 & \bar{\mathbf{M}}_2 & 0 & 0 & 0 \\ \bar{\mathbf{M}}_{31}^r & 0 & \bar{\mathbf{M}}_3' & 0 & 0 \\ 0 & 0 & 0 & \bar{\mathbf{M}}_4 & 0 \\ 0 & 0 & 0 & 0 & \bar{\mathbf{M}}_5 \end{bmatrix} \tag{61}$$

When this inertia matrix is transformed with the path matrix $\mathbf{T}$ the following result is obtained (in italics the matrix elements modified by the coupling terms),

$$\mathbf{T}^T\mathbf{M}\mathbf{T} = \begin{bmatrix} \mathbf{I} & \mathbf{I} & 0 & 0 & \mathbf{I} \\ 0 & \mathbf{I} & 0 & 0 & \mathbf{I} \\ 0 & 0 & \mathbf{I} & \mathbf{I} & \mathbf{I} \\ 0 & 0 & 0 & \mathbf{I} & \mathbf{I} \\ 0 & 0 & 0 & 0 & \mathbf{I} \end{bmatrix}^T \begin{bmatrix} \bar{\mathbf{M}}_1' & 0 & \bar{M}_{13}^r & 0 & 0 \\ 0 & \bar{\mathbf{M}}_2 & 0 & 0 & 0 \\ \bar{M}_{31}^r & 0 & \bar{\mathbf{M}}_3' & 0 & 0 \\ 0 & 0 & 0 & \bar{\mathbf{M}}_4 & 0 \\ 0 & 0 & 0 & 0 & \bar{\mathbf{M}}_5 \end{bmatrix} \begin{bmatrix} \mathbf{I} & \mathbf{I} & 0 & 0 & \mathbf{I} \\ 0 & \mathbf{I} & 0 & 0 & \mathbf{I} \\ 0 & 0 & \mathbf{I} & \mathbf{I} & \mathbf{I} \\ 0 & 0 & 0 & \mathbf{I} & \mathbf{I} \\ 0 & 0 & 0 & 0 & \mathbf{I} \end{bmatrix} =$$
$$= \begin{bmatrix} \mathbf{M}_1'^{\Sigma} & \mathbf{M}_1'^{\Sigma} & \bar{M}_{13}^r & \bar{M}_{13}^r & \bar{\mathbf{M}}_1' + \bar{M}_{13}^r \\ \mathbf{M}_1'^{\Sigma} & \mathbf{M}_2'^{\Sigma} & \bar{M}_{13}^r & \bar{M}_{13}^r & \mathbf{M}_2'^{\Sigma} + \bar{M}_{13}^r \\ \bar{M}_{31}^r & \bar{M}_{31}^r & \mathbf{M}_3'^{\Sigma} & \mathbf{M}_3'^{\Sigma} & \mathbf{M}_3'^{\Sigma} + \bar{M}_{31}^r \\ \bar{M}_{31}^r & \bar{M}_{31}^r & \mathbf{M}_3'^{\Sigma} & \mathbf{M}_4'^{\Sigma} & \mathbf{M}_4'^{\Sigma} + \bar{M}_{31}^r \\ \bar{\mathbf{M}}_1' + \bar{M}_{31}^r & \mathbf{M}_2'^{\Sigma} + \bar{M}_{31}^r & \mathbf{M}_3'^{\Sigma} + \bar{M}_{13}^r & \mathbf{M}_4'^{\Sigma} + \bar{M}_{13}^r & \mathbf{M}_5'^{\Sigma} + \bar{M}_{13}^r + \bar{M}_{31}^r \end{bmatrix} \tag{62}$$

After the first velocity transformation, the following open-chain inertia matrix is obtained,

$$\mathbf{R}_d^T\mathbf{T}^T\bar{\mathbf{M}}\mathbf{T}\mathbf{R}_d =$$
$$= \begin{bmatrix} \mathbf{b}_1^T\mathbf{M}_1^{\Sigma}\mathbf{b}_1 & \mathbf{b}_1^T\mathbf{M}_1^{\Sigma}\mathbf{b}_2 & \mathbf{b}_1^T\bar{M}_{13}^r\mathbf{b}_3 & \mathbf{b}_1^T\bar{M}_{13}^r\mathbf{b}_4 & \mathbf{b}_1^T\left(\mathbf{M}_1^{\Sigma}+\bar{M}_{13}^r\right)\mathbf{b}_5 \\ \mathbf{b}_2^T\mathbf{M}_1^{\Sigma}\mathbf{b}_1 & \mathbf{b}_2^T\mathbf{M}_2^{\Sigma}\mathbf{b}_2 & \mathbf{b}_2^T\bar{M}_{13}^r\mathbf{b}_3 & \mathbf{b}_2^T\bar{M}_{13}^r\mathbf{b}_4 & \mathbf{b}_2^T\left(\mathbf{M}_2'^{\Sigma}+\bar{M}_{13}^r\right)\mathbf{b}_5 \\ \mathbf{b}_3^T\bar{M}_{31}^r\mathbf{b}_1 & \mathbf{b}_3^T\bar{M}_{31}^r\mathbf{b}_2 & \mathbf{b}_3^T\mathbf{M}_3^{\Sigma}\mathbf{b}_3 & \mathbf{b}_3^T\mathbf{M}_3^{\Sigma}\mathbf{b}_4 & \mathbf{b}_3^T\left(\mathbf{M}_3^{\Sigma}+\bar{M}_{31}^r\right)\mathbf{b}_5 \\ \mathbf{b}_4^T\bar{M}_{31}^r\mathbf{b}_1 & \mathbf{b}_4^T\bar{M}_{31}^r\mathbf{b}_2 & \mathbf{b}_4^T\mathbf{M}_3^{\Sigma}\mathbf{b}_3 & \mathbf{b}_4^T\mathbf{M}_4^{\Sigma}\mathbf{b}_4 & \mathbf{b}_4^T\left(\mathbf{M}_4^{\Sigma}+\bar{M}_{31}^r\right)\mathbf{b}_5 \\ \mathbf{b}_5^T\left(\mathbf{M}_1^{\Sigma}+\bar{M}_{31}^r\right)\mathbf{b}_1 & \mathbf{b}_5^T\left(\mathbf{M}_2^{\Sigma}+\bar{M}_{31}^r\right)\mathbf{b}_2 & \mathbf{b}_5^T\left(\mathbf{M}_3^{\Sigma}+\bar{M}_{13}^r\right)\mathbf{b}_3 & \mathbf{b}_5^T\left(\mathbf{M}_4^{\Sigma}+\bar{M}_{13}^r\right)\mathbf{b}_4 & \mathbf{b}_5^T\left(\mathbf{M}_5'^{\Sigma}+\bar{M}_{31}^r+\bar{M}_{13}^r\right)\mathbf{b}_5 \end{bmatrix} \tag{63}$$

The inertia matrices in Equations (62) and (63) clearly show the accumulation pattern for matrices $\bar{\mathbf{M}}_{13}^r$ and $\bar{\mathbf{M}}_{31}^r$. Matrix $\bar{\mathbf{M}}_{13}^r$ appears in rows {1, 2, 5} and in columns {3, 4, 5}. Remembering that in the dynamic equations (31) the rows of the inertia matrix are related to forces and the columns to relative accelerations, it can be seen that this coupling matrix appears in the rows corresponding to the joints that are located backwards in the branch of body 1 (the elements and joints that receive the accumulated inertia forces), and in the columns of the joints that are located backwards of body 3, i. e., the joints whose relative acceleration determine the Cartesian acceleration of body 3. This topological information, the bodies or joints that are behind bodies 1 and 3 in the open-chain system, can be found from rows 1 and 3 of the path matrix $\mathbf{T}$ introduced in Equation (24).

3 Numerical examples

In this section two numerical examples solved with the semi-recursive formulation described in this paper will be presented. This formulation has been implemented in Matlab and Fortran. The Matlab implementation is available in the address *http://mec21.etsii.upm.es/mbs*.

In both implementations only revolute and prismatic joints are allowed. Other joints, as the cylindrical, spherical, universal, etc., are modeled as a series of revolute and prismatic joints with intermediate fictitious massless bodies. For multibody systems that do not have joints with the ground or fixed body, a floating joint with six degrees of freedom is defined between the ground and the body that is chosen as the *base body*. This floating joint is decomposed in three revolute and three prismatic joints.

3.1 Five-point rear suspension

The five-point rear suspension of a car (see Figure 7) has been considered as a test by many authors, von Schwerin [33] among them. It has nine rigid bodies: the fixed body, the frame, the carrier, the wheel and five rods.

According to von Schwerin, without including the frame, this two degrees of freedom system was modeled with 19 relative coordinates by Hiller and Frik, and with 14 coordinates by Grupp and Simeon. von Schwerin used 45 natural coordinates and 43 constraint equations. In this paper the closed loops have been opened by eliminating the five rods. The resulting open-chain system –that includes the frame– has eight degrees of freedom and can be modeled with only *eight* relative coordinates. The closed-chain system has five constraint equations that correspond to the five constant distance conditions of the rods.

Table 1 shows the results of CPU time obtained with the Compaq Fortran 90 and Matlab implementations. The programs have been run on a AMD K7 Athlon 1200 computer with 256 Mbytes. The maneuver simulated is an echelon and its duration is 2 seconds. The numerical integrators have been the Shampine and Gordon DE routine (written in Fortran) and the *ode113* Matlab integrator. Both are based on the Adams-Bashforth-Moulton method. It must be remarked that Matlab is a very good environment to develop the program, but not to execute it. Once the Matlab version runs correctly, the Fortran version can be developed and debugged very easily. The results of exporting C/C++ code from Matlab using the Matlab Compiler have been in this case very poor, from the point of view of the numerical efficiency.

Table 1. CPU times for the five-point rear suspension in an AMD K7 1200 Mhz computer

	Fortran 90	Matlab 6.5
Total simulation time (s)	0.12	23.874
Number of function calls	1165	624
Time per function call (ms)	0.1013	38.42

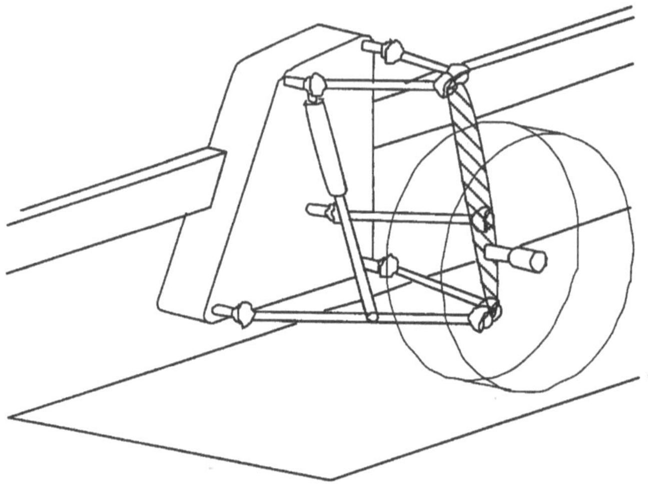

Fig. 7. Five point rear suspension

If instead of opening the closed loops by removing the five rods, the conventional method of cutting five spherical joints were used, this model would have 18 relative coordinates and 15 constraint equations because the five spherical joints removed. With this conventional modeling the Matlab code needs 26.5% more CPU time than by eliminating the rods.

3.2 Complete car model

Figure 8 shows the model of a car with McPherson suspension in the front and a five point suspension in the rear. This system has 26 rigid bodies (12 rods), 36 joints (6 revolute, 2 prismatic and 28 spherical) and 14 degrees of freedom. This system can be converted into an open-chain system by removing the 12 rods and two spherical joints in the front suspension. The open-chain system has three floating joints, one for the chassis and two for the two carriers in the rear suspensions.

Table 2. shows the CPU time and number of function evaluations achieved with the Fortran and the Matlab implementations. The computer used has been the same as in the previous example. The results obtained show that real-time simulations are possible with rather old PCs.

Table 2. CPU times and No. of function evaluations for the complete car model

	Fortran 90 (RK4)	Fortran 90 (DE)	Matlab 6.5 (ode113)
No of steps	100		100
No of function evaluations	400	600	400
Ms/function evaluation	0.625	0.6414	169,37
Total CPU time	0.25	0.39	67,75

4. Conclusions

Global formulations or formulations based on Cartesian coordinates allow simple and general ways to deal with complex multibody systems. However, its efficiency can be low. On the other hand, topological formulations based on velocity transformations can share the simplicity and generality of global formulations with the higher efficiency that results from the use of relative coordinates and some recursive processes. These formulations are called *semi-recursive* because they don't eliminate the need to solve systems of linear equations.

If the system is open-chain the velocity transformation from Cartesian to relative coordinates is straightforward. The introduction of the *path matrix* allows an easy way to take into account the topology for tree configured systems. If the systems have closed loops they can be opened by cutting some joints and/or removing some rigid bodies (rods). The closure of the loop constraints are formulated in Cartesian coordinates and then transformed to relative coordinates to be added to the open-chain dynamic equations. At this point, the improved methods for the descriptor form in Cartesian coordinates published in the last few years can be used to enforce the closure of the loop constraints. The result is a family of methods which are fast, simple and robust. Some benefits in the simplicity can be obtained by defining the geometry of the moving bodies using natural coordinates. The appropriate selection of the reference point for the Cartesian velocities and accelerations also contribute to the simplification of the recursive expressions for kinematic variables and for the inertia matrices. With these simplified expressions, the path matrix can be easily defined to introduce the system topology in the analysis. Finally, if rods are present in the model, its elimination to open the closed loops reduces the number of relative coordinates and constraint equations. A method to take into account its inertia properties exactly has been described in detail.

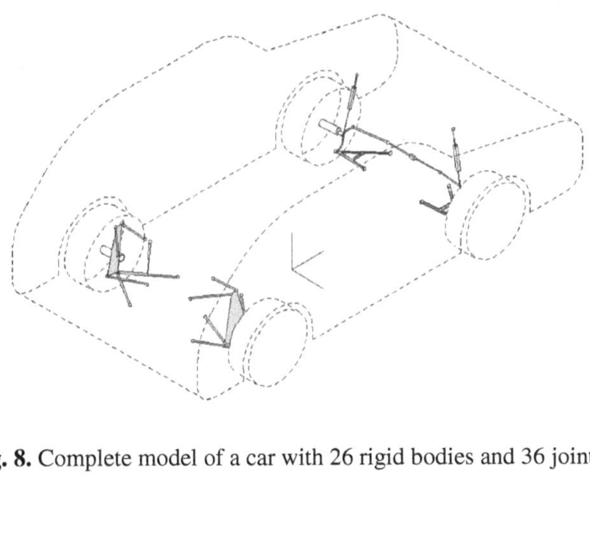

Fig. 8. Complete model of a car with 26 rigid bodies and 36 joints

References

1. Angeles J (2002) Fundamentals of robotic mechanical systems 2nd edn, Springer, New York
2. Anderson KS (1992) An order-N formulation for motion simulation of general constrained multirigid-body systems. Comp Struct 43:565-572
3. Anderson KS, Critchley JH (2003) Improved 'order-N' performance algorithm for the simulation of constrained multi-rigid-body dynamic systems. Mult Syst Dynamics, 9:185–212
4. Armstrong WW (1979) Recursive solution to the equations of motion of an N-link manipulator. In: Proc 5th World Congress on the Theory of Machines and Mechanisms Montreal, Canada 2:1343-1346
5. Ascher U, Chin H, Petzold L, Reich S (1995) Stabilization of constrained mechanical systems with DAEs and invariant manifolds. Mech of Struct Mach 23:135-157
6. Bae DS, Haug EJ (1987) A recursive formulation for constrained mechanical system dynamics. Part I: Open-loop systems. Mech of Struct Mach 15:359-382
7. Bae DS, Haug EJ (1987-88) A recursive formulation for constrained mechanical system dynamics. Part II: Closed-loop systems. Mech of Struct Mach 15:481-506
8. Bae DS, Haug EJ (1987-88) A recursive formulation for constrained mechanical system dynamics. Part II: Closed-loop systems. Mech of Struct Mach 15:481-506
9. Bae DS, Won YS, (1990) A hamiltonian equation of motion for real-time vehicle simulation. In: Ravani B (ed) Advances in design and automation 1990. ASME Press 2:151-157
10. Bae DS, Han JM, Choi JH (2000) An implementation method for constrained flexible multibody dynamics using a virtual body and joint. Mult Sys Dynamics, 4:297-315
11. Bae DS, Lee JK, Cho HJ, Yae H, (2000) An explicit integration method for real-time simulation of multibody vehicle models. Comp Meth Appl Mech Eng 187:337-350
12. Bae DS, Cho H, Lee S, Moon W, (2001) Recursive formulas for design sensitivity analysis of mechanical systems. Comp Meth Appl Mech Eng 190:3865-3879
13. Bae DS, Han JM, Choi JH, Yang SM (2001) A generalized recursive formulation for constrained multibody dynamics. Int J Num Meth Eng 50:1841-1859
14. Baumgarte J (1972) Stabilization of constraints and integrals of motion in dynamical systems. Comp Meth Appl Mech Eng 1:1-16
15. Cardenal J, Cuadrado J, Morer P, Bayo E (1999) A multi-index variable time step method for the dynamic simulation of multibody systems. Int J Num Meth Eng 44:1579-1598
16. Cuadrado J, Dopico D (2003) A hybrid global-topological real-time formulation for multibody systems. Fourth Symposium on Multibody Dynamics and Vibration, at the ASME Nineteenth Biennial Conference on Mechanical Vibration and Noise, Chicago, Illinois, USA, September 2-6
17. Cuadrado J, Cardenal J, Morer P, Bayo E (2000) Intelligent simulation of multibody dynamics: space-state and descriptor methods in sequential and parallel computing environments. Mult Sys Dynamics 4:55-73
18. Featherstone R (1987) Robot dynamics algorithms. Kluwer Academic, Boston

19. García de Jalón J, Bayo E (1994) Kinematic and dynamic simulation of multi-body systems – the real-time challenge. Springer, New York
20. García de Jalón J, Unda J, Avello A (1986) Natural coordinates for the computer analysis of multibody systems. Comp Meth Appl Mech Eng 56:309-327
21. García de Jalón J, Avello A, Jiménez JM, Martín F, Cuadrado J, (1991) Real time simulation of complex 3-D multibody systems with realistic graphics. In: Haug EJ, Deyo RC (eds) Real-time integration methods for mechanical system simulation. NATO ASI Series 69:265-292. Springer-Verlag, Berlin, New York
22. Kim SS (2002) A subsystem synthesis method for efficient vehicle multibody dynamics. Mult Sys Dynamics 7:189-207
23. Kim SS, Vanderploeg MJ (1986) A general and efficient method for dynamic analysis of mechanical systems using velocity transformations. J Mech Trans Autom Design 108:176-182
24. Jerkovsky W (1978) The structure of multibody dynamic equations. J Guid Control 1:173-182
25. Lubich C (1991) Extrapolation integrators for constrained multibody systems. Imp Comp Sci Eng 3:213-234
26. Negrut D, Serban R, Potra FA (1997) A topology based approach for exploiting sparsity in multibody dynamics. Joint formulation. Mech of Struct Mach 25:379-396
27. Nikravesh PE, Gim G (1989) Systematic construction of the equations of motion for multibody systems containing closed kinematic loops. In: Ravani B (ed) Advances in Design Automation 1989. ASME Press 3:27-33
28. Saha SK (2002) Modeling constrained systems with the natural orthogonal complement: recursive algorithms. Notes Workshop T3, IEEE Conf. on Robotics and Automation. Washington DC. http://www.angelfire.com/sc/saha/Ira02tut.pdf
29. Saha SK, Schiehlen WO (2001) Recursive kinematics and dynamics for parallel structured closed-loop multibody systems. Mech of Struct Mach 29:143-175
30. Serna MA, Avilés R, García de Jalón J (1982) Dynamic analysis of plane mechanisms with lower pairs in basic coordinates. Mech Mach Theory 17:397-403
31. Serban R, Negrut D, Potra FA, Haug EJ (1997) A topology based approach for exploiting sparsity in multibody dynamics in Cartesian formulation. Mech of Struct Mach 25:379-396
32. Stelzle W, Kecskeméthy A, Hiller M (1995) A comparative study of recursive methods. Arch Appl Mechanics 66:9-19
33. von Schwerin R (1999) Multibody system simulation. Numerical methods, algorithms and software. Springer-Verlag, Berlin
34. Wehage R, Haug EJ (1982) Generalized coordinate partitioning for dimension reduction in analysis of constrained mechanical systems. J Mech Design 104:247-255
35. Yen J (1993) Constrained equations of motion in multibody dynamics as ODEs on manifolds. SIAM J Num Analysis 30:553-568

Hardware-in-the-Loop for Quality Assurance of an Active Automotive Safety System

Oliver Öttgen and Manfred Hiller

University Duisburg-Essen, Institute of Mechatronics and System Dynamics
Lotharstr. 1, 47057 Duisburg, Germany
{oettgen,hiller}@mechatronik.uni-duisburg.de

1 Introduction

During the last few years vehicle safety has been improved significantly by introducing active and passive automotive safety systems, which are typical examples of mechatronic systems. Passive safety systems are developed for crashworthiness, whereby active safety systems are developed to reduce the number of severe accidents. Whereas complexity, interdependency and demands concerning safety systems continuously increase, a trend to shorter development cycles with consistent quality occurs. Subject of the considered development process is the vehicle dynamics control system Electronic Stability Program (ESP, Robert Bosch GmbH), which increases the vehicle handling during critical situations by influencing the brake system and drive train of a vehicle. The ESP is relevant to security of the occupants because of its active interventions on the vehicle dynamics. Thus, the development of the monitoring software as a part of the ESP self-diagnosis is an important element and has to be verified systematically by tests for quality assurance in all development phases.

The combination of vehicle dynamics simulations and road tests is considered as the optimal solution for the monitoring software verification. Software tests on the Electronic Control Unit (ECU) under reproducible boundary conditions are possible using a vehicle model in a Hardware-in-the-Loop (HIL) simulation environment and especially in early development phases the availability of vehicle prototypes is limited. But even the best simulation environment cannot exactly represent reality and results are only as good as the vehicle model, whereby road tests are still indispensable A generic real-time vehicle model with one parameter set for a class of vehicles is used to test the ESP monitoring software by HIL simulations, whereby the generic model can be improved by both varying the model structure and parameter adaptation.

In section 2, the development process of mechatronic systems based on the V-shaped model and spiral model is explained. The considered active automotive safety system ESP and its mode of operation is demonstrated in section 3. Fur-

J.A.C. Ambrósio (ed.), Advances in Computational Multibody Systems, 25–44.

thermore, an overview in the field of the monitoring software as a part of the ESP self-diagnosis is given and the monitoring software verification is mentioned. In section 4, a three-dimensional, complex and validated vehicle model is introduced, which is used as a reference. The HIL simulation environment and its generic real-time vehicle model are shortly described in section 5. In section 6, the real-time vehicle model is analyzed using the reference model. Alternatives to improve the model and enhance its field of application are discussed.

2 Development Process of Mechatronic Systems

In this section, the development process of mechatronic systems is presented. Firstly, the development process is described on the basis of an adapted V-shaped model. During the process the V-shaped model is cycled completely or partially in dependency of the requirements. This cyclical procedure is similar to the spiral model (section 2.2).

2.1. Development Process Based on the V-shaped Model

The V-shaped model has its seeds in the software development and can be adapted to the requirements of a mechatronic development process [1], [2]. Fig. 1 generally describes the development procedure of mechatronic systems from the requirements of a product up to the system in use or serial production. The development project starts with a development request on the left part of the V-shaped model. The conceptual formulation has to be specified and the components of the mechatronic project have to be defined within the system specification. Thus, the requirements on the hardware and software are derived from the specified request and provide the basis to evaluate the final product. In the field of system design the overall system is divided into modular components, whose sub-functions and interfaces are specified. The formulation of the structure of the modular components takes place separately by the module design. The specified functions are realized according to the formulated structures by the domain specific design, which completes the conversion of the system specification.

The system integration part of the V-shaped model is used to verify and validate the previous process steps for quality assurance. First of all, the module design is verified by a module test with Software-in-the-Loop simulations. Thereby, the compliance of the effective and desired properties or characteristics of the modules is tested. If all modules satisfy their specifications, the components are combined to the overall system. By the integration test the system design with regard to the interfaces of the modules and their desired cooperation is verified.

Extensive tests of the overall system ensure the system specification and fulfillment of the development request within the system validation. Both the integration test and system validation are executed by HIL simulations and road tests. The result is the system in use.

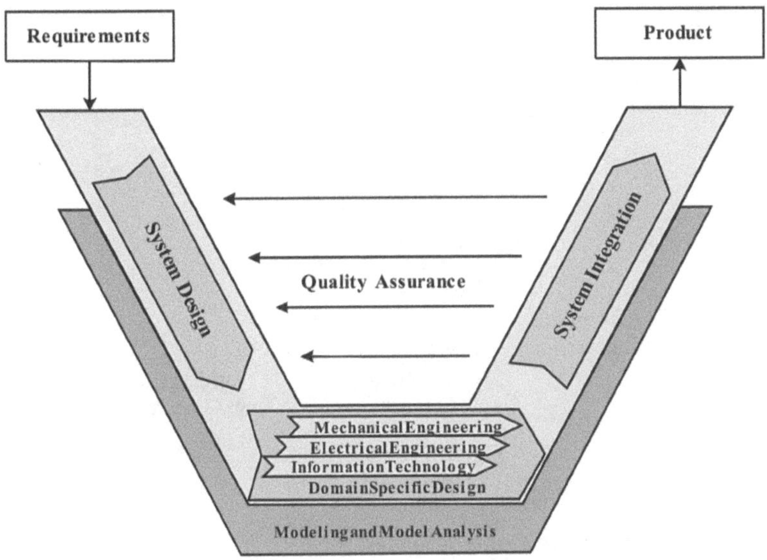

Fig. 1. The adapted V-shaped model [1], [2]

2.2 Cyclical Development Process

During a mechatronic development process the V-shaped model is completely or partially cycled because of frequently changing requests and requirements, respectively. This cyclical procedure is very similar to the spiral model, whereby the achieved progress and the cumulative costs with every cycle can be described by the angular and radial spiral dimension (Fig. 2). The spiral model is ascribed to Boehm [3] and is initialized with the specification of aims, solution alternatives and constraints. The solution alternatives are analyzed and evaluated in the next step within the spiral cycle. Furthermore, the risks have to be determined and then reduced or eliminated. Now the next level product is realized and validated in terms of a functional model, a prototype or the final product. The spiral cycle is ending with the product or because of poor evaluation at this point. Otherwise, after making a review, the development team plans the next phase and specifies its commitments. Subject of the development process is the active automotive safety system ESP.

3 Verification of the ESP Self-diagnosis

In this section, the functionality of the considered active automotive safety system ESP is explained as a typical example of a mechatronic system. After giving a short overview in the field of the monitoring software as a part of the ESP self-diagnosis (section 3.2), the monitoring software verification is mentioned in section 3.3.

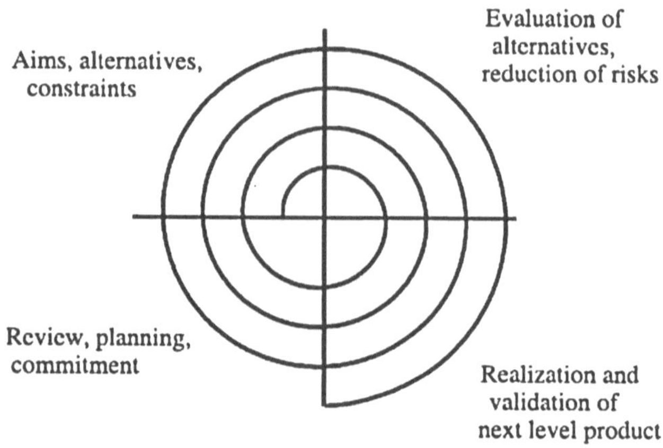

Fig. 2. The spiral model

3.1 The Vehicle Dynamics Control System ESP

During the last years the safety of vehicles has increased significantly by introducing passive and active safety systems. Passive safety systems are developed for crashworthiness, whereby active safety systems are developed to reduce the number of severe accidents. The Vehicle Dynamics Control (VDC) system ESP increases the vehicle handling during critical situations by influencing the brake system and drive train [4, 5]. Consequently, it influences longitudinal, lateral and yaw dynamics of a vehicle.

The Antilock Brake System (ABS) and Traction Control System (TCS) are subsystems of the ESP. The ABS actively reduces the brake pressure at each wheel, when the pressure induced by the driver causes a locked wheel while vehicle decelerations. The control variable is the longitudinal wheel slip, which is defined as the relative wheel velocity difference to the vehicle velocity. Thus, the vehicle is steerable while decelerating, because changing the steering angle has no effect on the forces at a locked wheel. Similar to the ABS the TCS avoids wheel spinning while vehicle accelerations.

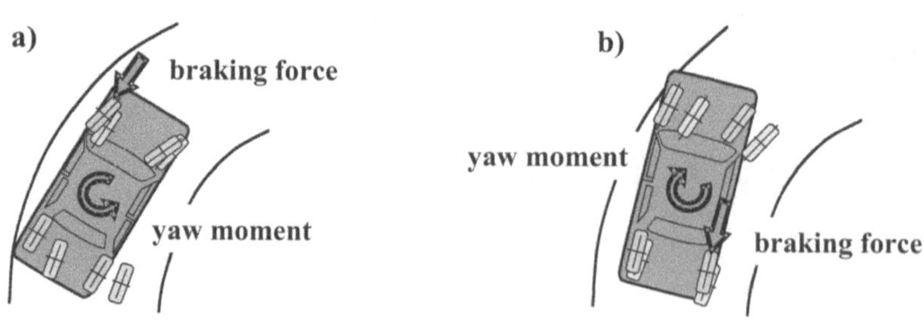

Fig. 3. Mode of ESP operation with oversteering a) and understeering b) vehicle behavior

The spinning wheels are slown down by controlling the longitudinal slip through reducing the engine torque or applying a brake pressure to the specific wheels. The superimposed VDC system controls the vehicle yaw moment at the vertical axis by controlling the longitudinal slip at each wheel, whereby the main task is to limit the vehicle side-slip angle. The side-slip angle describes the rotation of the vehicle motion direction relative to the longitudinal vehicle axis. In Fig. 3 the ESP operation with oversteering and understeering vehicle behavior is shown (Robert Bosch GmbH). In Fig. 3 a) oversteering vehicle behavior occurs, because the vehicle leaves the nominal trajectory and skids at the rear axle. By braking at the outer front wheel, a yaw moment is applied to stabilize the vehicle motion and keep it manageable for the driver. Fig. 3 b) shows an understeering vehicle behavior with drifting at the front axle. In that case a brake pressure at the inner rear wheel is applied to stabilize the vehicle behavior.

For yaw moment control with considering a limitation of the vehicle side-slip angle the intended and current vehicle behavior have to be compared. The driver inputs steering wheel angle, accelerator pedal position and brake pressure are used to obtain the intended behavior. The knowledge of the current vehicle behavior is based on wheel speed, lateral acceleration and yaw rate sensors. A challenge within the development process of mechatronic systems with the complexity of the VDC system ESP is to ensure the safety requirements on the system. Therefore, the development of the monitoring software as a part of the ESP self-diagnosis is an important and weightily part of the ESP development.

3.2 ESP Self-diagnosis

The ESP is relevant to security of the occupants because of its active interventions on the vehicle dynamics. Therefore, the development of the monitoring software as a part of the ESP self-diagnosis is an important element within the development process. The failure detection logic within the monitoring software is divided into hardware close, signal based and model based monitoring. The first part monitors the electrical functions of the valve and pump motor, the power supply for the ECU, electronic components inside the ECU, the sensor wires and connectors, the Controller Area Network (CAN) bus, etc. [6].

In addition, the signal based monitoring checks the physical plausibility of one signal. For example, a fault is detected, if a sensor signal is out of range or has an implausible signal gradient. Within the model based monitoring, signals are estimated with the sensor signals using several models. For example, together with the measured yaw rate, four yaw rates are derived with the steering wheel angle, lateral acceleration and wheel speeds [6]. Using a selection algorithm a reference yaw rate is determined, which is insensitive to a sensor failure. With inverted models the determination of a reference steering wheel angle and lateral acceleration is possible as well.

The system behavior after a fault detection by any monitoring is similar, whereby the failure handling manages the switch off strategy, failure storage and the information to the driver or other ECUs, respectively. Within the switch off strategy the system is switched into a specific backup level by a predefined strategy for each fault. For example, a steering wheel angle sensor failure has no effect on the ABS and TCS functionality of the ESP, meaning that only the superimposed VDC system has to be shut down. In the Electrically Erasable Programmable Read-Only Memory (EEPROM) a unique failure code for each fault is stored, which provides details for diagnosis functionality. Furthermore, the driver gets information on the ESP backup level by signal lamps in the vehicle cockpit and the ESP status is transmitted to other ECUs, since a network of several relevant ECUs exists. To ensure the functionality the monitoring software has to be verified.

3.3 Verification of the Monitoring Software

As described before the ESP monitoring software is fundamental, because of its active interventions on the vehicle dynamics. To ensure the monitoring software quality, it has to be tested and verified systematically (Fig. 1).

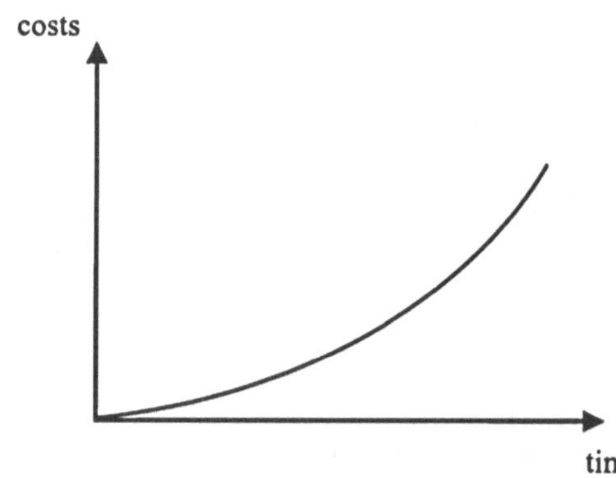

Fig. 4. Correction costs dependency on fault detection time

Faults within the development of huge software projects are not avoidable, but it is essential to detect the faults as soon as possible. The later the detection the more expensive is the fault correction, which is principally described by a nonlinear dependency in Fig. 4. Thus, monitoring software tests are important during all phases of development. The combination of vehicle dynamics simulations and road tests is considered as the optimal solution for the monitoring software verification. Integration tests on the ECU or a prototype are possible using a vehicle model in a HIL simulation environment (section 2.1).

The advantage of real-time simulations is the reproducibility by defining all boundary conditions and events exactly. Moreover, the realization of life-threatening or vehicle destructing driving maneuvers is possible without any consequences and the simulation environment supplies state variables, which are not available in the real vehicle. The real-time simulation provides time and cost savings since many simulations are shortly practicable, the test environment is rapidly changeable and a test automation is possible. Especially in early development phases the availability of vehicle prototypes is limited. But even the best simulation environment cannot exactly represent the reality and the results are only as good as the vehicle model. Therefore, road tests are still indispensable to analyze effects, which cannot be simulated by a HIL simulation.

4 Reference Vehicle Model

The structure of the reference vehicle model is shown in the following sections. Firstly, the modular simulation environment is described. The abstraction process from a real vehicle to the topological multibody systems structure is subject of section 4.2. Section 4.3 exemplary shows the topological structure of the four link wheel suspension at the front axle of the reference model.

4.1 Simulation Environment

The modular vehicle dynamics simulation environment FASIM_C++ [7] is used for the analysis of the generic real-time model in section 6. The simulation environment is developed in cooperation with the Robert Bosch GmbH and the University Duisburg-Essen in a timeframe of about 15 years. The object-oriented programming language is C++ and the components are modeled by rigid bodies, because the lower frequent vehicle motions are primarily of interest.

The three-dimensional reference vehicle model of an Audi A8 [8] (Audi internal name D2) is built up by mechanical (chassis, suspension, drive train, etc.) and non-mechanical (hydraulics, driver, environment, etc.) modules (Fig. 5). The model is validated through numerous road tests and the structure enables facile changes of particular components.

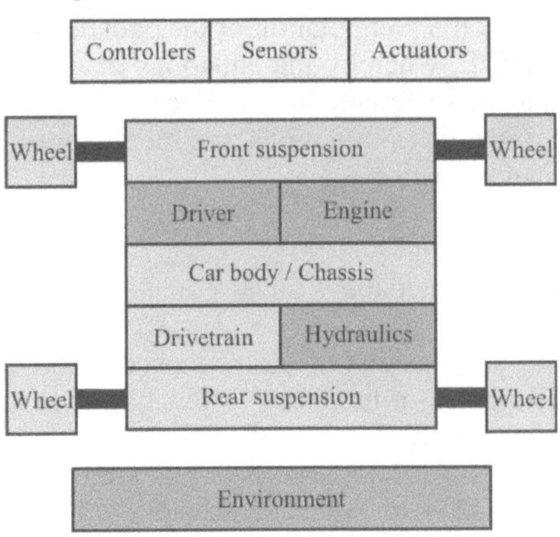

Fig. 5. Modular model structure

4.2 Abstraction Process

The reference vehicle model is based on a multibody systems approach including closed kinematical loops for wheel suspensions and drive train. Fig. 6 shows the abstraction and modeling process of the detailed reference model.

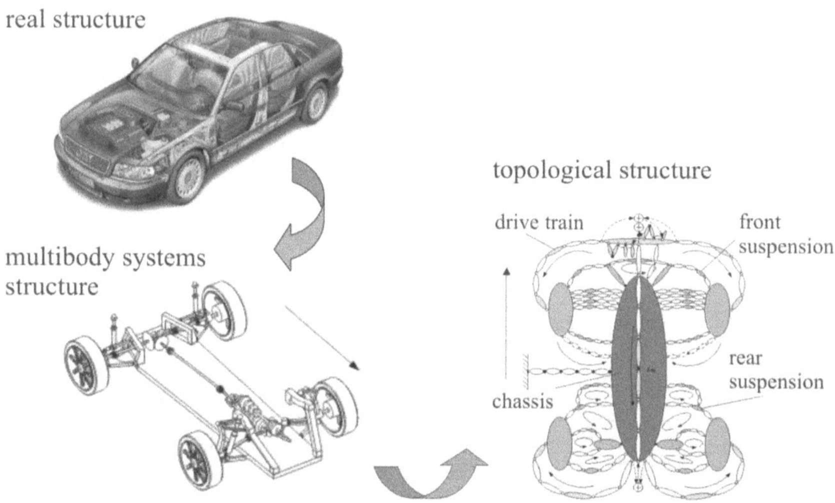

Fig. 6. Vehicle abstraction

The abstraction process starts with the real structure of a vehicle. As mentioned in section 4.1, the vehicle is modeled by rigid bodies and is described by a multibody systems structure. Based on the multibody systems modeling approach the topological structure of the vehicle matches to the representation in Fig. 6. The kinematics calculation and the computation of the equations of motion are realized with the software MOBILE, whereby the equations of motion are generated with the method of kinematical differentials [9]. Because the wheel suspensions are important for vehicle dynamics simulations, the modeling of the D2 front suspension is exemplary shown in section 4.3.

4.3 Independent Wheel Suspension

The independent wheel suspensions of the reference model are modeled exactly. The front axle is equivalent to a model of a four link axle and the rear axle to a model of a trapezium arm axle. Besides the wishbones, steering link and wheel carrier, kinematical models for the elastic bearings are considered. Following, the modeling process of the D2 front wheel suspension is exemplary explained. The four link wheel suspension is shown in Fig. 7.

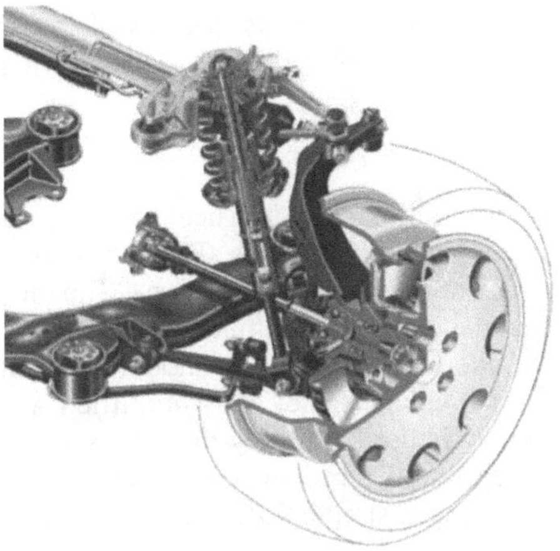

Fig. 7. Four link wheel suspension [8]

Firstly, the wheel suspension is schematically demonstrated by a multibody systems structure in Fig. 8 a), whereby the elements are represented by rigid bodies. Because the model consists of $n_B = 7$ bodies and $n_J = 11$ joints, 4 independent kinematical loops exist without considering the antiroll bar:

$$n_L = n_J - n_B = 4 \tag{1}$$

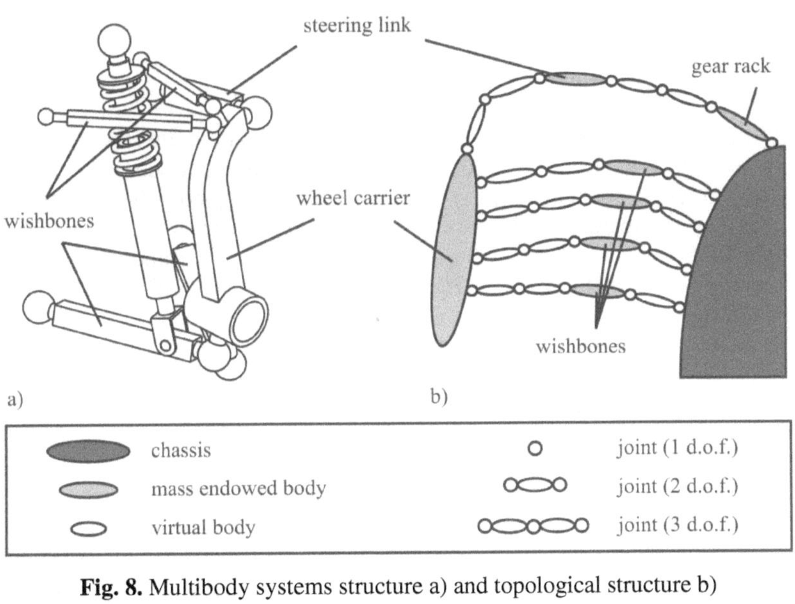

Fig. 8. Multibody systems structure a) and topological structure b)

With the overall number of $f_J = 27$ joints degrees of freedom (DOF), the four link wheel suspension without the isolated DOF of the steering link $f_I = 1$ has $f = 2$ DOF:

$$f = f_J - 6n_L - f_I = 2. \qquad (2)$$

The topological structure of the four link wheel suspension in Fig. 8 b) clarifies number and location of the kinematical loops. The spring-and-damper unit has no influence on the kinematics of the wheel suspensions. The modeling of the D2 trapezium arm suspension at the rear axle is similar.

Due to its complexity, the reference vehicle model is unsuitable for a real-time usage. Therefore, a generic vehicle model with simplified wheel suspensions in a HIL simulation environment is used to verify the ESP self-diagnosis in section 3. But the complex model represents a proper replacement for an analysis of the generic real-time model, because it supplies state variables, which are not available in the real vehicle.

5 HIL Simulation Environment

The HIL simulation environment for the ESP monitoring software verification is presented in the following section. Afterwards the used generic real-time vehicle model is described and its field of application within the software verification mentioned.

5.1 HIL Test Environment

For testing the ESP monitoring software the HIL simulation environment LabCar NG [10] is used, which is shown in Fig. 9. The environment consists of software and hardware components, whereby the software primary consists of the components Developer, Operator and the vehicle model. The LabCar Developer is based on ASCET-SD and used for development of the software components. The LabCar Operator manages the LabCar Hardware, the vehicle model downloads to the real-time target and the access to all calibration and measurement variables of the vehicle model, which is described in section 5.2 later on.

In addition to the operation PC as user interface to the test system and a measurement, respectively, diagnosis PC the environment hardware consists of the components signal box, load box, breakout box and emulator. An emulator represents the ECU and is extended by the possibility to measure internal signals of the ECU. The vehicle model including hydraulics, sensor and actuator models is downloaded to the signal box with a VMEbus/VME64 standard support and simulated on a Motorola PowerPC. Instead of a vehicle model the use of a stimulus set is possible, whereby open-loop driving maneuvers can be reproduced through time-dependent signal demands. The discrete sensor signal outputs of the vehicle model to the ECU are converted and prepared to analog electronic signals by D/A converters of the load box. The signals from the ECU are transferred to discrete signals by A/D converters and returned to the vehicle model. The interface between signal and load box is a VME bus system.

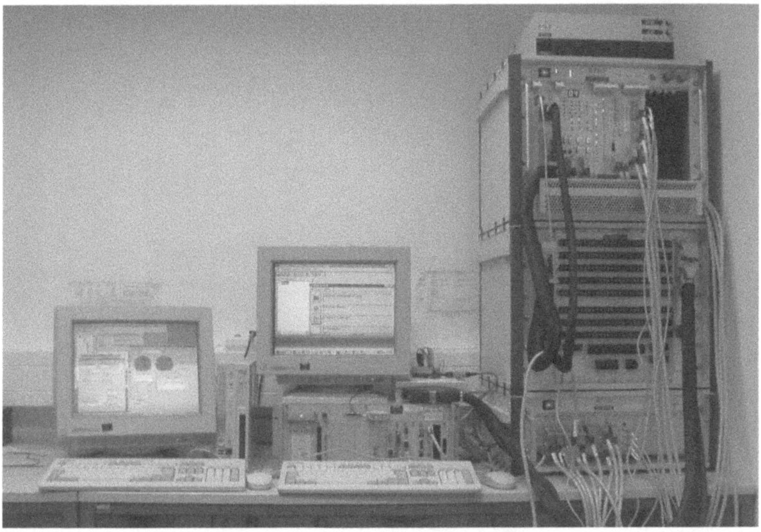

Fig. 9. Test environment LabCar NG

For electronic signal manipulations like interruptions and short-circuits the breakout box can be used. The signal plausibility for example is tested with a software manipulation of the sensor models.

5.2 Real-time Vehicle Model

Because the detailed reference model in section 4 is not real-time capable, a more abstract vehicle model is required to verify the ESP self-diagnosis. A generic real-time vehicle model within the HIL simulation environment LabCar NG is used for the ESP monitoring software verification. The three-dimensional vehicle model has mechanical and non-mechanical components, whereby the focus is on the suspension model for simulating longitudinal, lateral and vertical dynamics [11]. The real-time model is a multibody system with an open kinematical structure and modeled by mainly non-interacting transmission blocks in the development environment ASCET-SD. Besides springs and dampers, anti-roll bars are considered as force elements between the chassis and wheel carriers.

Furthermore, the model contains the Highway Safety Research Institute (HSRI) tire model [12] and includes 18 mechanical DOF, 6 DOF for the vehicle movement in space, 4 DOF for the wheel rotations, 4 DOF for the vertical and 4 DOF for the longitudinal wheel carrier movements (Fig. 10). The independent wheel suspensions of the generic model are modeled as semi-trailing arms to approximate the exact wheel suspensions (Fig. 11). The experiences in the field of the

ESP monitoring software verification point out, that the greater part of the testable monitoring software can be verified by the generic model.

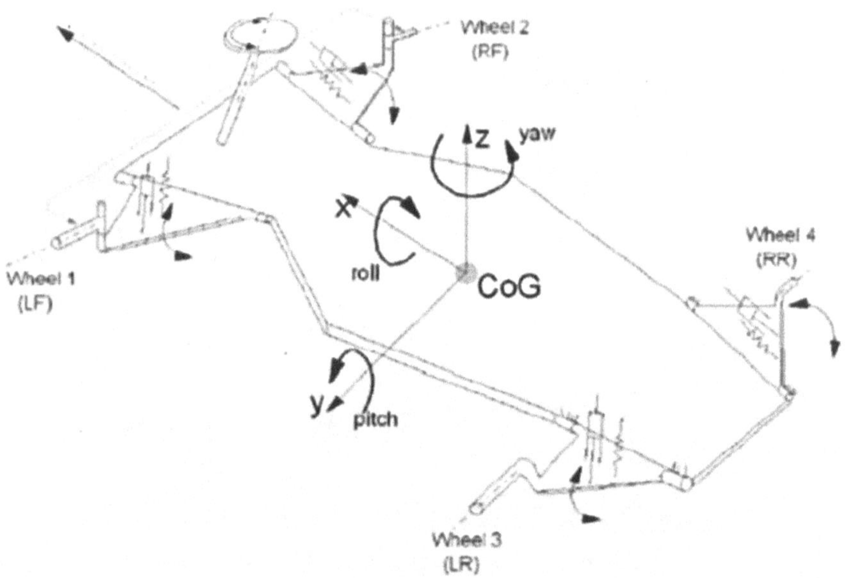

Fig. 10. Three-dimensional suspension model [11]

The generic vehicle model with only one parameter set for a class of vehicles is cost-effective and adequate for testing the most of the hardware close and signal based monitoring (section 3.2). A verification of the accurate switch off strategy for correlative faults is executable as well, whereby tests in the vicinity of the release thresholds and the model based monitoring require a higher vehicle model quality.

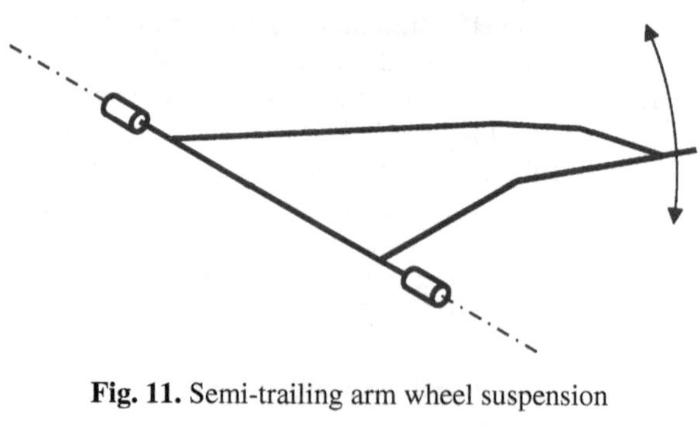

Fig. 11. Semi-trailing arm wheel suspension

6 Improvement of the Real-time Vehicle Model

In this section, the scope is on the analysis of the generic real-time vehicle model. In Fig. 12 a scale for the vehicle model quality is shown. An extended single track model [13] represents the lower bound of the scale and the detailed reference model the upper bound. The structure of the real-time model in section 5.2 is placed between the single track and reference model, whereby the parameters of the real-time model influences the quality.

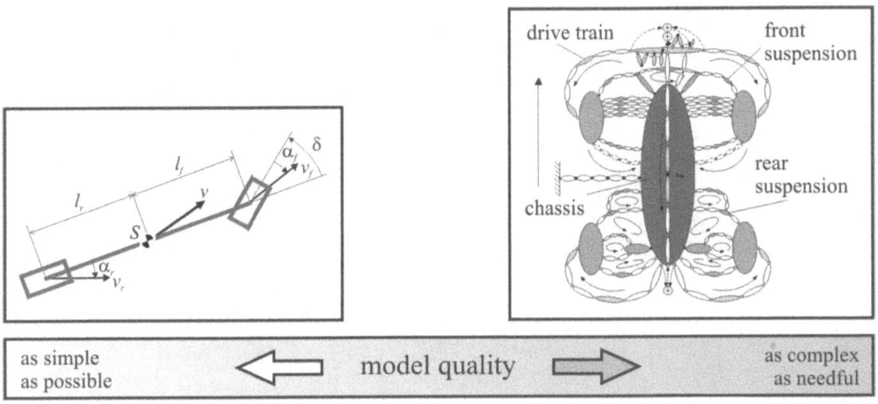

Fig. 12. Classification of the vehicle model

Because of real-time capability the model has to be as simple as possible, but for vehicle dynamics analyses it has to be as complex as needful. Following, in section 6.1, the generic real-time model with only one parameter set for a class of vehicles is compared to the reference model. Afterwards improvement alternatives to enhance the field of application are discussed (section 6.2).

6.1 Analysis of the Real-time Vehicle Model

In this section, lateral and yaw dynamics of the generic real-time vehicle model with one parameter set for a class of vehicles is analyzed with the complex and validated reference model. Especially the modules tire, wheel suspension and the force elements spring, damper and anti-roll bar influence the vehicle dynamics.

First of all, the tire characteristics of the reference vehicle model and the generic model are identified, because the tire-road contact is of particular importance for vehicle dynamics simulations. Therefore, steering angle steps with both vehicle models are simulated up to a lateral acceleration range of $a_y \approx 7$ m/s^2. The total lateral tire force dependency on the slip angle of one axle is identified using a single-track model [14]. The slip angle describes the rotation of the wheel velocity direction relative to the longitudinal wheel axis. The nonlinear dependency is exemplary shown at the front axle in Fig. 13. Whereas the linearized single-track model is based on a linear tire model approach, the tire characteristic of the generic model is qualitatively conform to the reference model. The trajectories of the generic vehicle model (dashed) with one parameter set for a class of vehicles in comparison to the reference model (solid) is shown in Fig. 14.

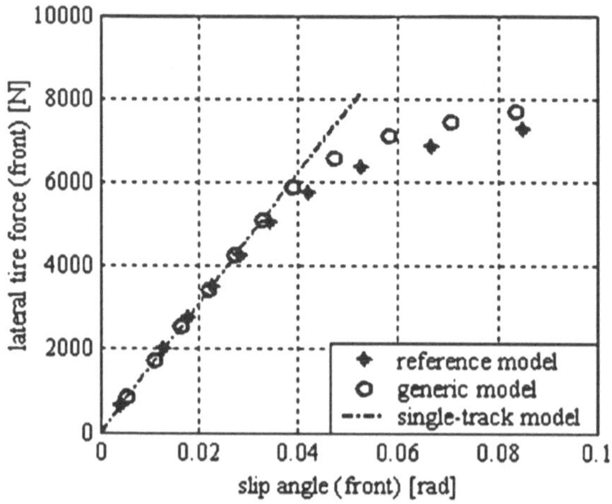

Fig. 13. Identified lateral tire force dependency on slip angle (front axle)

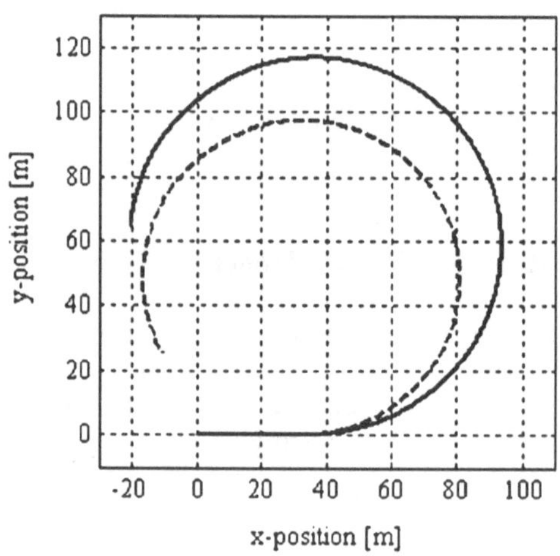

Fig. 14. Trajectories of reference model (solid) and generic model (dashed)

The exemplary driving maneuver is a steering angle step with $\delta_{end} = 60°$, a steering angle velocity of $d\delta/dt = 200°/s$ and a friction coefficient of $\mu = 1$. The initial longitudinal vehicle velocity is $v_{x,0} = 15$ m/s, whereby the throttle pedal is constantly hold during the maneuver. The driven radius of the reference model is $r_{ref} = 58.5$ m, whereby the generic model follows a circle with a radius of $r_{gen} = 48.7$ m. The relative radius deviation of $\Delta r = 16.7$ % points out, that the generic vehicle with only one parameter set for a class of vehicles is unsuitable for vehicle dynamics analyses of the considered vehicle. The relative deviation is computed as follows:

$$\Delta r = \frac{\left| r_{\text{ref}} - r_{\text{gen}} \right|}{r_{\text{ref}}}. \qquad (3)$$

The lateral acceleration and yaw velocity characteristic in Fig. 15 documents this conclusion. The maximal values of the lateral acceleration and yaw velocity indicate a relative deviation of $\Delta a_{y,\,max} = 9.1$ % and $\Delta(\mathrm{d}\psi/\mathrm{d}t)_{max} = 22.1$ %, which are build by:

$$\Delta a_{y,\max} = \frac{\left| a_{y,\text{ref},\max} - a_{y,\text{gen},\max} \right|}{a_{y,\text{ref},\max}}, \qquad (4)$$

$$\Delta(\mathrm{d}\psi/\mathrm{d}t)_{\max} = \frac{\left| (\mathrm{d}\psi_{\text{ref}}/\mathrm{d}t)_{\max} - (\mathrm{d}\psi_{\text{gen}}/\mathrm{d}t)_{\max} \right|}{(\mathrm{d}\psi_{\text{ref}}/\mathrm{d}t)_{\max}}. \qquad (5)$$

Following, improvement alternatives for the generic vehicle model to enhance the field of application are discussed.

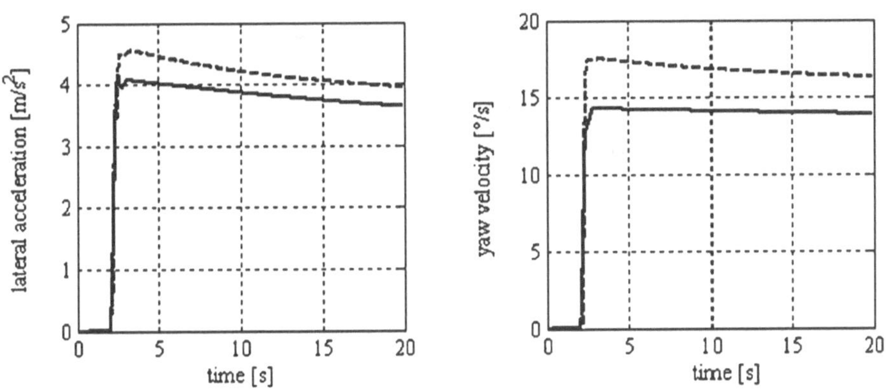

Fig. 15. Lateral acceleration and yaw velocity of reference model (solid) and generic model (dashed)

6.2 Alternatives to Improve the Real-time Vehicle Model

An improvement of the real-time model to enhance the field of monitoring software verification is possible by both varying the model structure and parameter adaptation. But the vehicle model quality always is limited by the processor performance, because the simulation has to run in real-time (Fig. 16). Another model quality limit is set by the economic efficiency of the vehicle model validation. The model quality dependency on the development costs is nonlinear, so that the costs rise over-proportionally with increasing quality.

Besides the tire model the module wheel suspension has special influence on the vehicle dynamics. But the integration of a complex wheel suspension model with closed kinematical loops would cause problems with the real-time capability of the model as denoted in Fig. 16. Furthermore, the development environment ASCET-SD embarks on a strategy with mainly non-interacting transmission blocks, which is unsuitable for modeling of kinematical loops. Consequently, an improvement of the generic model by parameter adaptation is aimed.

In Fig. 17 the trajectories of the reference model (solid) and an adapted generic vehicle model (dotted) are exemplary shown at a steering angle step with $\delta_{end} = 60°$ on a friction coefficient of $\mu = 1$. Firstly, the adaptation is reduced to the lateral tire stiffness of the used HSRI tire model to compensate all modeling simplifications. Fig. 18 presents the lateral accelerations and yaw velocities of both vehicle models, whereby the deviations are decreased to $\Delta r = 4.6\ \%$, $\Delta a_{y,\,max} = 4.8\ \%$ and $\Delta(d\psi/dt)_{max} = 6.8\ \%$ in comparison to the generic model with the standard parameter set (section 5.2).

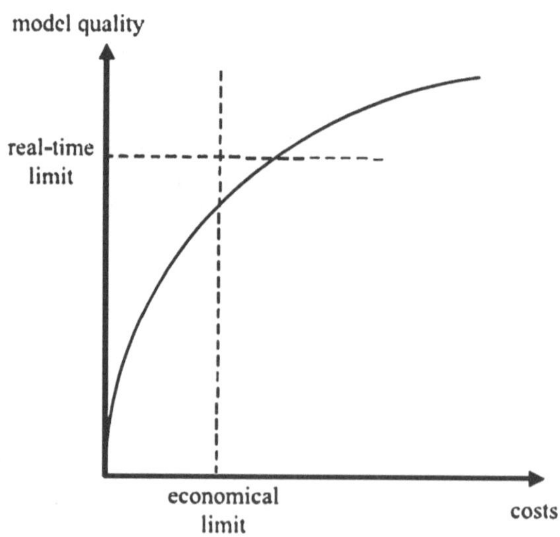

Fig. 16. Model quality dependency on costs

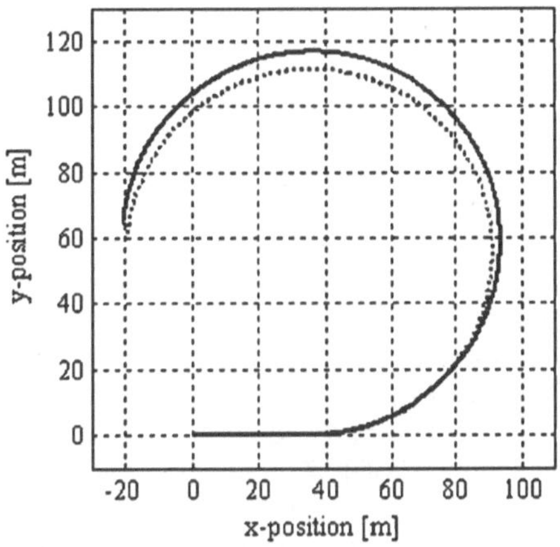

Fig. 17. Trajectories of reference model (solid) and adapted model (dotted)

The simulation results in Fig. 17 and Fig. 18 point out the potential to improve lateral and yaw dynamics of the generic model by parameter adaptation. Besides the tire model adaptation the wheel suspension parameters angle and length are suitable to adapt the semi-trailing arms of the generic model to the exact wheel suspension of a vehicle. Especially in higher lateral acceleration ranges the influence of the force elements spring, damper and anti-roll bar on the vehicle dynamics has to be considered. Therefore, an optimal parameter set for the modules tire, wheel suspension and force elements has to be found to simulate lateral and yaw dynamics accurately. But the economic limit as exemplary shown in Fig. 16 has to be considered for an efficient improvement of the generic vehicle model in relation to the profit for the ESP monitoring software verification.

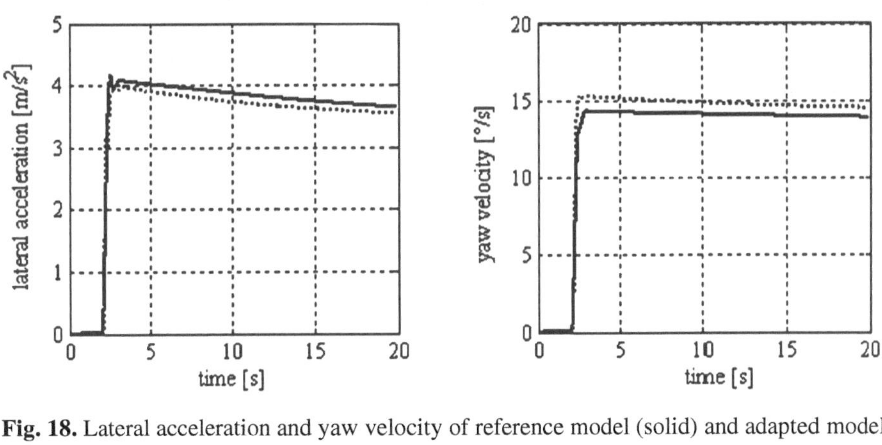

Fig. 18. Lateral acceleration and yaw velocity of reference model (solid) and adapted model (dotted)

7 Conclusions

Vehicle safety has been improved significantly by introducing active and passive automotive safety systems, which are typical examples of mechatronic systems. Whereas the complexity, interdependency and demands concerning safety systems continuously increase, a trend to shorter development cycles with consistent quality occurs. Simulation, therefore, has meanwhile become an important role in this development process, contributing also to a better understanding of the physical properties of the system. Complex mechatronic systems like the vehicle dynamics control system ESP are developed using an adapted V-shaped model, which is completely or partially cycled similar to the spiral model. The ESP is relevant to security of the occupants because of its active interventions on the vehicle dynamics. Therefore, the development of the monitoring software as a part of the ESP self-diagnosis is an important element within the ESP development process and has to be verified systematically by tests in all development phases.

The combination of vehicle dynamics simulations and road tests is the optimal solution for the monitoring software verification. The HIL application offers time and cost savings, reproducibility, performance of life-threatening driving maneuvers, the possibility for a test automation and the supply of state variables, which are not available in the real vehicle. But even the best simulation environment cannot exactly represent reality and results are only as good as the vehicle model, whereby road tests are still indispensable. The greater part of the ESP monitoring software can be verified by using a generic real-time model with one parameter set for a class of vehicles within the HIL simulation. Tests in the vicinity of the thresholds and the model based monitoring require a higher vehicle model quality. But generally in early phases of development a vehicle prototype is not available and the generic vehicle model covers a considerable part of the monitoring software verification.

An improvement of the real-time model to enhance the field of monitoring software verification is possible by both varying the model structure and parameter adaptation. But for example, varying the structure by integrating a complex wheel suspension model with closed kinematical loops would cause problems with the real-time capability of the vehicle model. Therefore instead, an improvement by parameter adaptation is aimed. Considering lateral and yaw dynamics, the simulation results with adapted tire model parameters, to compensate all modeling simplifications, point out the model improvement potential.

Besides the tire model adaptation the wheel suspension parameters – angle and length – are suitable to adapt the semi-trailing arms of the generic model to the exact wheel suspension of a vehicle. Especially, in higher lateral acceleration ranges the influence of the force elements spring, damper and anti-roll bar on the vehicle dynamics has to be regarded. Thus, an optimal parameter set for the modules tire, wheel suspension and force elements has to be found to simulate lateral and yaw dynamics accurately, whereby the economic limit of parameter validation in relation to the improvement profit has to be considered. Furthermore, a complex and validated reference model is not always available for an improvement of the models of actual vehicles. Therefore, a parameter adaptation procedure based on measurable vehicle state variables has to be worked out.

Acknowledgment

The authors give special thanks to the Robert Bosch GmbH, who enables us the development of the vehicle dynamics simulation environment FASIM_C++ for already more than 15 years. Furthermore, we are grateful for the possibility to work with the HIL simulation environment LabCar NG within the ESP monitoring software development.

References

[1] VDI 2206 (2003) Entwicklungsmethodik für mechatronische Systeme, Berlin: Beuth Verlag, Germany

[2] Gausemeier J, Moehringer S (2002) VDI2206 – A New Guideline for the Design of Mechatronic Systems. In: IFAC Conference on Mechatronic Systems. Berkeley, USA

[3] Boehm BW (1988) A Spiral Model of Software Development and Enhancement. Computer 21(5,):61-72

[4] van Zanten AT, Erhardt R, Pfaff G. (1995) VDC The Vehicle Dynamics Control System of Bosch, SAE paper no. 950759

[5] van Zanten AT (2002) Evolution of Electronic Control Systems for Improving the Vehicle Dynamic Behavior. In: International Symposium on Advanced Vehicle Control, Hiroshima, Japan

[6] van Zanten AT, Erhardt R, Landesfeind K, Pfaff G (1998) VDC Systems Development and Perspective, SAE paper no. 980235

[7] Hiller M, Schuster C, Adamski D (1996) Fasim_C++ - A Versatile Developing Environment for Vehicle Dynamics Simulation. In Ambrosio J, Pereira M, Pina da Silva F (eds)Proceedings of the NATO-ASI in Crashworthiness of Transportation Systems: Structural Impact and Occupant Protection, Troia, Portugal, pp. 219-235

[8] Müller E, Heißing B (1994) Der neue Audi A8 Teil2, Automobiltechnische Zeitschrift 96. Stuttgart: Franckh-Kosmos Verlag, Germany

[9] Hiller M, Kecskeméthy A (1989) Equations of Motion of Complex Multibody Systems Using Kinematical Differentials. In: Proc. of 9th Symposium on Engineering Applications of Mechanics, Vol. 13, London, Canada, pp.113-121

[10] ETAS GmbH (1999) LabCar White Paper, Hardware-in-the-Loop Test Systems for Electronic Control Units. Stuttgart, Germany

[11] ETAS GmbH (2002) Vehicle Dynamics Simulation Model V1.2, User's Guide, Stuttgart, Germany

[12] Dugoff H, Fancher PS, Segel L (1970) An analysis of tire traction properties and their influence on vehicle dynamic performance. SAE paper no. 700377

[13] Öttgen O, Bertram T (2002) Influencing Vehicle Handling through Active Roll Moment Distribution. In: International Symposium on Advanced Vehicle Control, Hiroshima, Japan
[14] Gillespie TD (1992) Fundamentals of Vehicle Dynamics. Society of Automotive Engineers

Multibody Aspects of Motorcycle Modelling with Special Reference to Autosim

Robin S. Sharp, Simos Evangelou and David J. N. Limebeer

Department of Electrical and Electronic Engineering, Imperial College London, South Kensington Campus, London SW7 2AZ, UK;
robin.sharp@imperial.ac.uk

Modelling of the ride and handling dynamics of motorcycles using the symbolic mechanical multibody system package Autosim has been carried out since 1995. Motorcycles are principally of tree structure but their geometry is complex in relation to the tyre to road contact and tyre force and moment descriptions and to the chain drive system. They may contain closed kinematic loops, according to common suspension and steering design variations. Various aspects of the modelling problem are discussed and some implications, from a multibody standpoint, of choosing different options are exposed. Simulation results illustrate the "anti-squat" behaviour of a chain drive transmission and the "anti-dive" behaviour of a Telelever front suspension system.

1 Introduction

Mathematical modelling and simulation of the responses of motorcycles to control inputs from throttle, brake and steering are contemporary standard activities [16]. The purposes include increasing understanding of behaviour and the relationships between behaviour, design and operating conditions [8,16,18] and the exploration of new design features [5,15,19]. Models need to be understandable, fully defined and documented, and to have some attractive balance between realism and computational speed. For some applications, e. g. hardware-in-the-loop, speed is essential [1,2,11,12,22].

Motorcycle dynamics can be considered at each of two distinct levels. The lower level involves confining the motion to small perturbations from straight running, in which case in-plane, ride dynamics are decoupled at first order from out-of-plane handling dynamics and relatively simple models will suffice for many purposes. Such models can be built by hand [14]. Indeed, chronologically, there was no choice. The higher level involves general motions, with potentially strong interactions between in-plane and out-of-plane motions [8,16,18,19], for which automated multibody modelling is currently the only sensible approach.

The capabilities and features of Autosim [7] are very well matched to the motorcycle problem, as long as the relatively low frequency behaviour is of interest

J.A.C. Ambrósio (ed.), Advances in Computational Multibody Systems, 45–68.

(<12Hz say) and a multi-rigid body representation of the system is considered adequate. Autosim is basically a multibody description language written in LISP, with the means to automatically convert such a description into an executable simulation model, among other things. The modelling problem becomes one in harmonizing the finite command set of the multibody language chosen with the concepts and parameters of the system of interest.

In this work the architectural aspects of modern motorcycles are discussed. Some relevant information about using Autosim is provided. The mechanical features of motorcycles and the description capabilities of the language are brought together to define a number of alternative models. Some model features and behavioural results are shown next and conclusions are finally drawn.

2 Contemporary Motorcycle Architecture

At the most basic level, motorcycles all have the same historic layout, involving a relatively rigid centre section, in which the masses are concentrated, drive to the rear wheel via a chain, belt or shaft and steering of the front frame including the front wheel and round section pneumatic tyres. The rider sits on top of the centre section and steers by direct connection through the handlebars with the front frame. Many current production motorcycles have the long-established telescopic front fork and rear swing arm suspension arrangements, Figure 1. It has been clearly established that the main frame normally needs to be considered torsionally flexible in the steering head region to properly represent the behaviour of real machines [6,16-20].

Important variants of these standard arrangements include the so-called "Telelever" front suspension, Figure 2, and "Monoshock" rear spring / damper unit linkages, Figures 3, 4 and 5. The basic system shown in Figure 1 has "tree" structure, while the other designs involve closed kinematic loops, which are interesting from a multibody viewpoint [13]. A shaft drive system is straightforward to model, but a chain drive is more difficult, literal modelling being problematic due to the number of links in a typical chain and to the fact that each link changes its function as it progresses through its duty cycle, Figure 6.

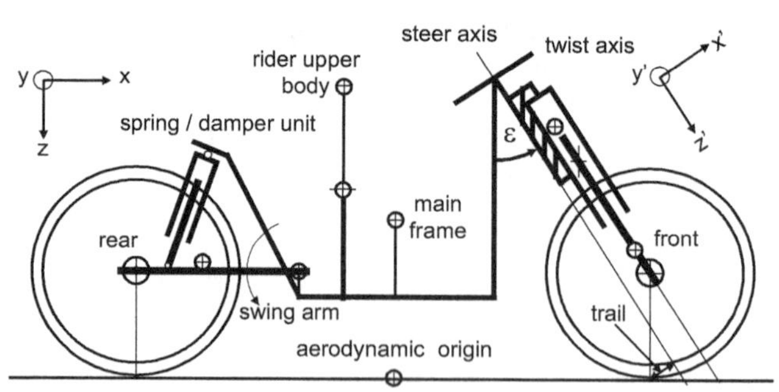

Fig. 1. Diagrammatic motorcycle with telescopic front forks and swing arm rear suspension.

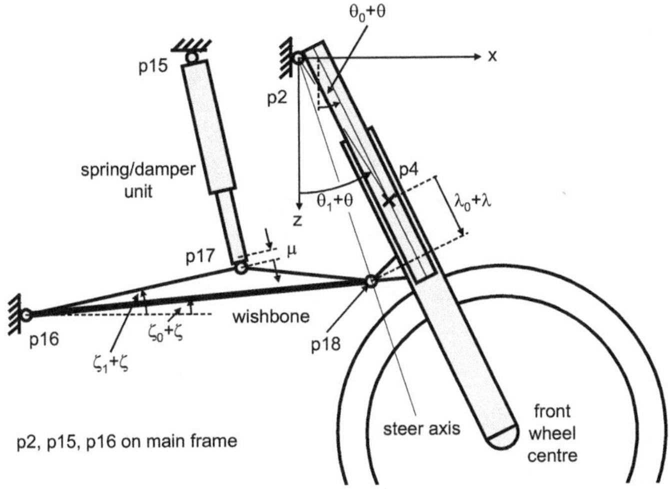

Fig. 2. Telelever front suspension system with frame torsional compliance.

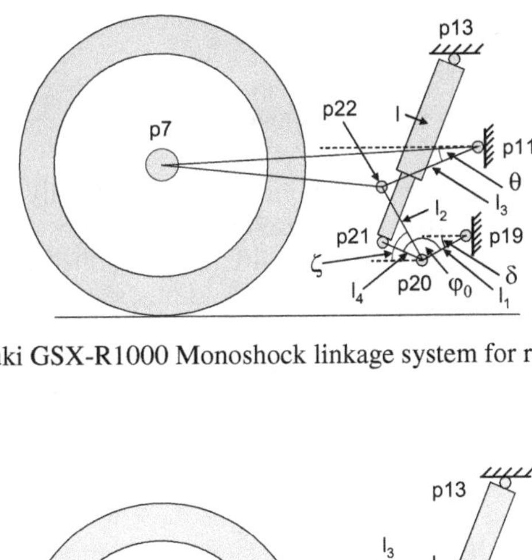

Fig. 3. Suzuki GSX-R1000 Monoshock linkage system for rear suspension.

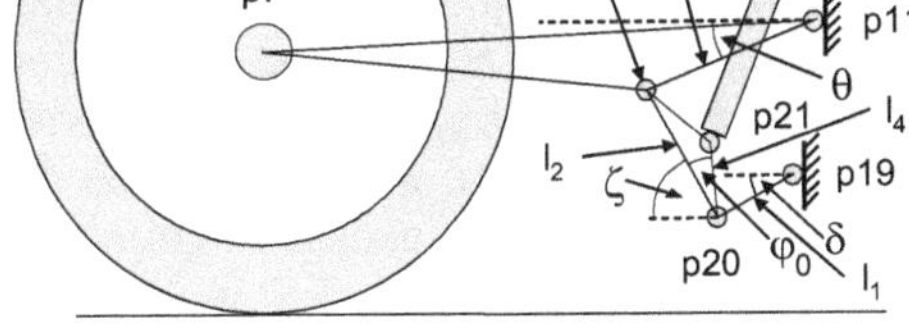

Fig. 4. Honda Pro-link Monoshock rear suspension system.

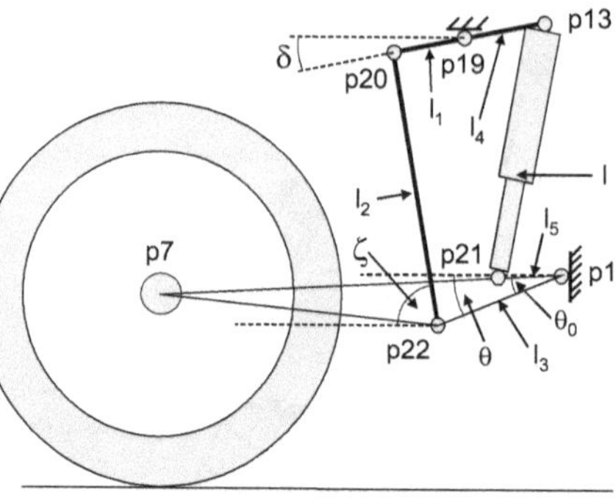

Fig. 5. Full floater Monoshock rear suspension system.

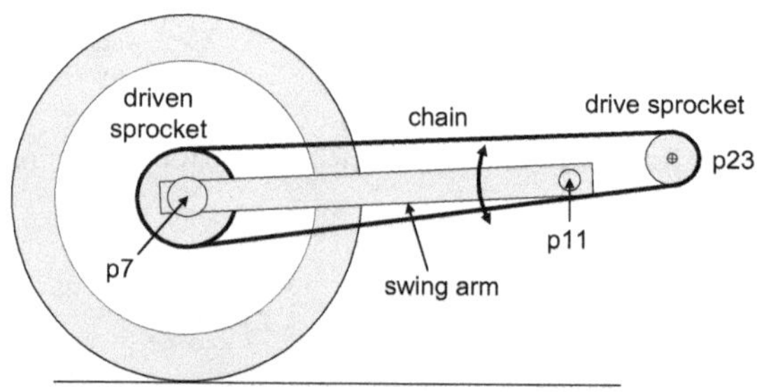

Fig. 6. Chain drive associated with swing arm rear suspension.

External forces and moments are applied aerodynamically and through frictional coupling between the tyre and the ground in the contact patches. The aerodynamic influences included are drag and lift forces applied at the aerodynamic origin, Figure 1, as well as a pitching moment as follows:

$$F_x = -0.5C_D \rho A v^2 \tag{1}$$

$$F_z = -0.5.C_L \rho A v^2 \tag{2}$$

$$M_y = 0.5.C_P \rho A L v^2 \tag{3}$$

with ρ the air density, A the frontal area, L the wheelbase and ν the longitudinal wind speed relative to the motorcycle. The tyre forces and moments are best described by the "Magic Formulae" as designed and developed specifically for motorcycles [3,4,9,19,21]. These formulae need to be supplied with the tyre load (normal to the ground), the longitudinal slip, the lateral slip and the wheel camber angle relative to the ground and they return the longitudinal force and the lateral force acting at the centre of the contact area and an aligning moment. If the contact geometry is described in detail, Figures 7 and 8, with the contact centre moving round the tyre circumference and laterally round the tyre sidewall, this force and moment system treatment is complete. A detailed discussion with parameter sets is given in [19].

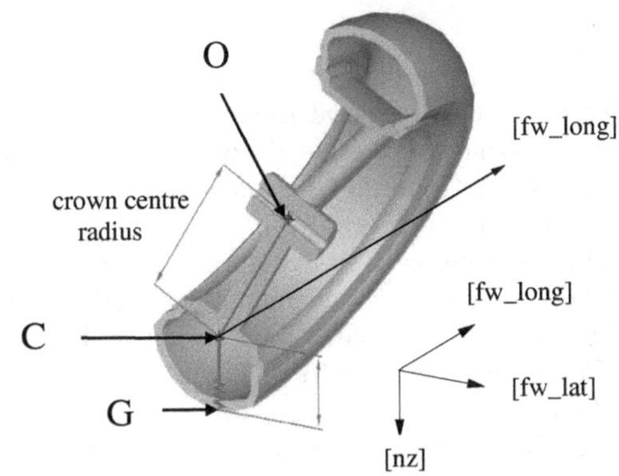

Fig. 7. Three dimensional tyre to ground contact geometry, showing the possible movement of the contact centre circumferentially and laterally.

The main joints are for steering and the two suspensions. The steering head joint, in particular, is designed and made to be virtually frictionless, taper roller bearings being common. Proper maintenance demands care in adjusting the head bearings, so that the friction is very small but the clearance is nil. The suspension bearings also are designed to operate with very little friction, so that the joints can be considered ideal. This is important from a multibody modelling point of view [13].

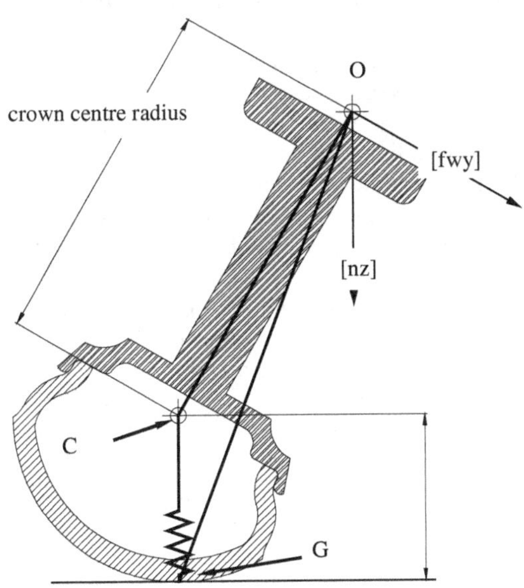

Fig. 8. Two dimensional tyre to ground contact geometry, showing the lateral movement of the contact centre and the vertical flexibility of the tyre structure.

3 Autosim Features

The starting point for any Autosim-based model building is the inertial reference frame “n”. The first body, whatever it is, is allowed to move with up to six degrees of freedom with respect to n. Children of this, or any other body present, can be added, with a description of the freedoms permitted to this new body, relative to its parent. In the nominal configuration of the system, the point common to both parent and child, names for the mass and inertia elements etc. need to be defined. The equations of motion are derived via Kane’s equations, which are based on the principle of virtual power [13]. Points can be defined in bodies. These are most often fixed points in the bodies to which they belong but they may be specified as “moving”, with their locations specified by coordinates in a defined reference system.

Forces with given magnitudes and directions can be applied to points and reacted on other points. Alternatively, a strut following a force law can be defined as acting between two points, in which case, the direction of the force is in the direction of the line joining the points. Moments with magnitudes and directions can be defined as acting between bodies.

Position constraints, velocity constraints or both together can be added. The analyst has the option to specify which freedom should be eliminated from the problem as a result of each constraint added. Constraints may require points on two bodies to have no relative movement, for example, or they may consist of symbolic expressions which must have zero value. To deal with a closed kinematic loop, this is first opened by an imagined cut in such a way that the longest chain of bodies contributing to the loop is as short as possible. The loop closure constraints are then added. On loading the model, Autosim forms the displacement loop equations [10], expressing the condition that traversing the loop in circular fashion yields a zero vector resultant. Each of the three resulting scalar equations is differentiated with respect to time to give the velocity level constraint equations, which are always linear in velocities. Thus dependent velocities can be chosen automatically and written as linear functions of the remaining independent velocities. The dependent velocities are temporarily lost from the problem, while the independent ones remain as generalised speeds to be evaluated through time by numerical integration.

At each integration step, the dependent speeds are known from the independent ones and approximations to the dependent coordinates can be found by integration of the dependent speeds. Displacement drift will normally occur in this process, implying opening of the closed kinematic loop. To avoid this drift, the displacement level loop equations are utilised in a Newton-Raphson iterative scheme to update the values of the dependent coordinates, until these equations are satisfied. The Newton-Raphson procedure requires repeated solution of the linear simultaneous equation update problem by matrix inversion. Usually, since the approximate solution from integration is close to the true solution of the displacement loop equations, very few updates are needed. However, some economy may be possible by relying only on the leading diagonal terms in the coefficient matrix for the update calculations[1]. User experience suggests this to be an excellent way of dealing with closed kinematic loops.

New state variables can be added to a model and they can be made integral functions of existing variables. Differentials with respect to time and partial differentials of variables can be defined. Max, min, sign and ifthen functions can be used, allowing discontinuous actions, like tyres leaving the ground and limit stops being contacted, to be modelled easily. If necessary, the analyst can define how discontinuous functions are to be differentiated.

Equations of motion come in the form:

$$\begin{aligned} \left[S(\underline{q},t)\right]\dot{\underline{q}} &= \underline{v}(\underline{q},\underline{u},t) \quad \text{kinematics} \\ \left[M(\underline{q},t)\right]\dot{\underline{u}} &= \underline{f}(\underline{q},\underline{u},t) \quad \text{dynamics} \end{aligned} \tag{4}$$

in which

$$\begin{bmatrix} S & 0 \\ 0 & M \end{bmatrix}$$

is the speed coefficient matrix, $\underline{q}$ is the generalised displacement vector, $\underline{u}$ is the generalised speed vector and $\underline{f}$ the force vector. Straightforward integration of the equations in this form implies the need to invert the speed coefficient matrix at each time step. There is an option to LU decompose this matrix symbolically, following which, at each time step, 2n purely sequential arithmetic operations allow the computation of all the rates of change of generalised displacements and speeds. Whether or not there is advantage in taking this option is case dependent. The probability of its being advantageous increases in alignment with the number of zeros in the matrix. On loading the model, its form and the numbers of multiply/divide, add/subtract and function evaluations per time step are declared. Consequently, symbolic LU decomposition can be compared with the alternative numerical process on a case by case basis. The operation count can be reduced, leading to faster simulation, by declaring variables to be small. In particular, small angle theory can be employed where it is appropriate.

4 Modelling the Motorcycle in Autosim

It is convenient to define all the key points in the model, Figure 1, in the inertial reference frame, n. As the bodies are added, the points which belong to them can be located with respect to n. Bodies and freedoms are added as shown in Figure 9. Children are specified relative to parents.

---

[1] Autosim contains a switch *fast-computed-coordinates*. When this is "on", only the leading diagonal terms are used in the updating calculations. A larger number of iterative steps of simpler form is implied. This is faster in some cases.

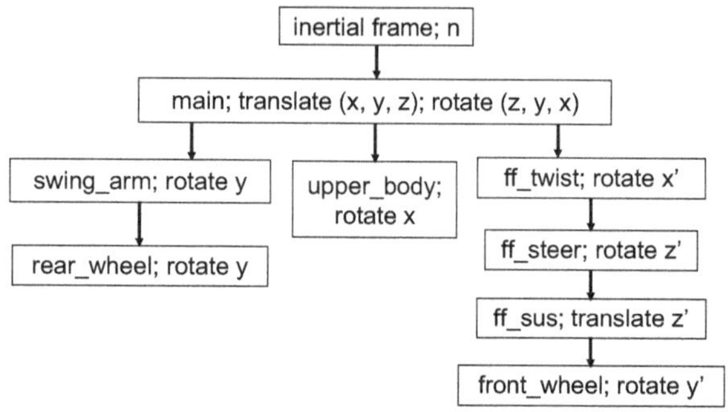

Fig. 9. Parent/child body structure for model of motorcycle with telescopic front forks and swing arm rear suspension.

Tyre to ground contact points are specified as follows: In any general configuration, the wheel longitudinal direction is perpendicular to both the wheel spindle vector and the vertical, Figure 10. The special radial direction to the tyre crown centre circle in the lateral plane of the contact point is perpendicular to both the longitudinal and the vertical directions. The vector $\underline{OC}$, from the wheel centre to the tyre crown centre, see also Figures 7 and 8, is of fixed length given by the tyre crown radius and in the radial direction defined. Where a vertical through C meets the ground plane is the contact centre G. The contact centre is defined as a moving point in the wheel, located by its coordinates in the wheel axis system, derived from the analysis above.

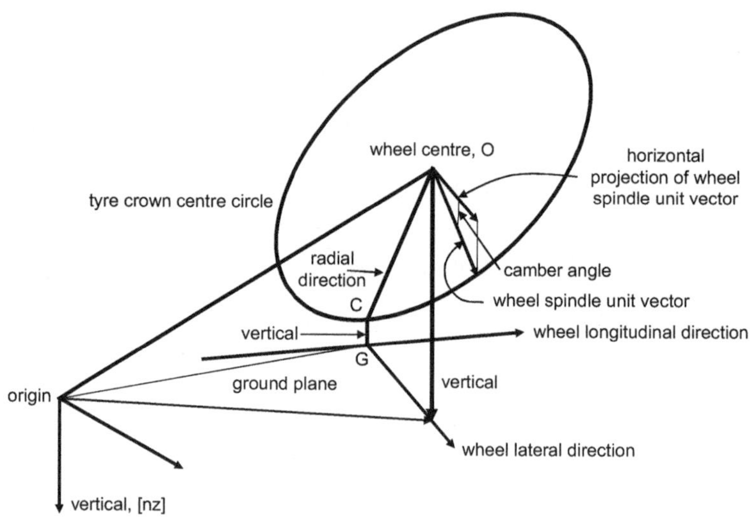

Fig. 10. Geometry of the front wheel, enabling to define the tyre/ground contact conditions.

The height of the wheel centre from the ground is the component in the vertical direction of the vector joining the global origin (at the rear wheel contact centre in the motorcycle's nominal state) to the wheel centre. The vertical component of OC is the tyre crown radius multiplied by the cosine of the camber angle, so that the distance from C to G can be found. The change in this distance from the nominal state, through the assumption of a linear elastic tyre structure, gives the change in the tyre load from the nominal load and therefore the load itself.

The wheel camber angle is clear in Figures 8 and 10. The longitudinal slip is given by the ratio of the rearward longitudinal velocity of the tyre tread base material to the absolute value of the rolling velocity. This is the longitudinal velocity of the tread base with its spin component taken away. The lateral slip is the ratio of the negative lateral component of the velocity of the tread base material to the absolute value of the rolling velocity [18,19].

In the tyre model, there is a time lag, σ/v, between changes in lateral slip and the corresponding changes in the lateral distortion of the tyre carcass and in the forces which depend on it [4,16]. The tyre relaxation length, σ, is proportional to the tyre cornering stiffness and also dependent on speed. The load, the camber angle, the longitudinal slip and the lagged lateral slip are passed to the tyre algorithm, which returns with the forces and moments.

4.1 Telelever Front Suspension

When the motorcycle has a Telelever front arrangement, the bodies down the right side of Figure 9, from ff_twst through ff_steer, ff_sus and front_wheel, can be replaced as shown in Figure 11. The newly required points p15 to p18, Figure 2, are defined first in n and later copied to each relevant body. The upper forks are added with three rotational freedoms to point p2 on the main frame, its local axes being aligned with x', y and z'. The lower forks are added to the upper forks with a single translational freedom. The wishbone is added with rotational freedoms in yaw and pitch to the main frame with joint at p16. Position and speed constraints are added to disallow relative motion between the wishbone end and the lower forks at p18 in each of three orthogonal directions. The spring/damper unit is described as a strut acting between p15 on the main frame and p17 on the wishbone. A spring/damper moment is applied to the upper fork rolling motion (about x') to represent the frame twist compliance.

Table 1. Telelever example geometric data. The x and z coordinates in n of the relevant points, Figure 2, are shown. The steering head angle used is 0.2971 rad.

P2	P4	P15	P16	P17	P18
1.233	1.32	1.103	0.853	1.133	1.313
-0.7488	-0.576	-0.729	-0.509	-0.569	-0.559

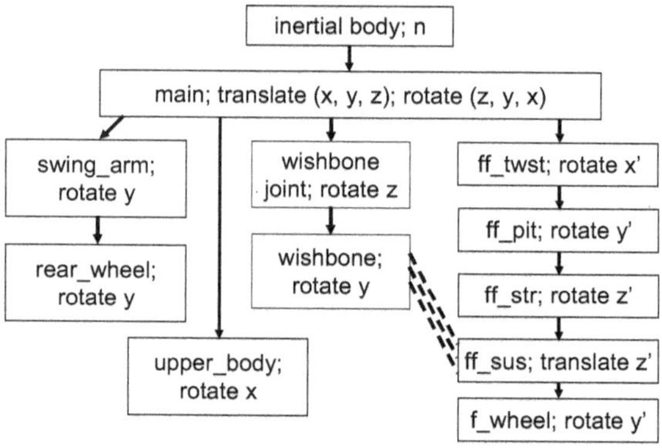

Fig. 11. Parent/child body structure for model of motorcycle with Telelever front forks and swing arm rear suspension. The dashed lines represent constraints at the ball joint between the wishbone and the lower forks.

As a second alternative, the model of the Telelever system, Figure 2, can be solved kinematically by simulation for a prescribed motion of the lower forks. The numerical results for the upper fork pitching displacement and for the spring/damper unit extension can be fitted by simple polynomials and these can be used in the motorcycle model as constraint functions. Numerical results for a layout defined in Table 1 are shown in Figure 12.

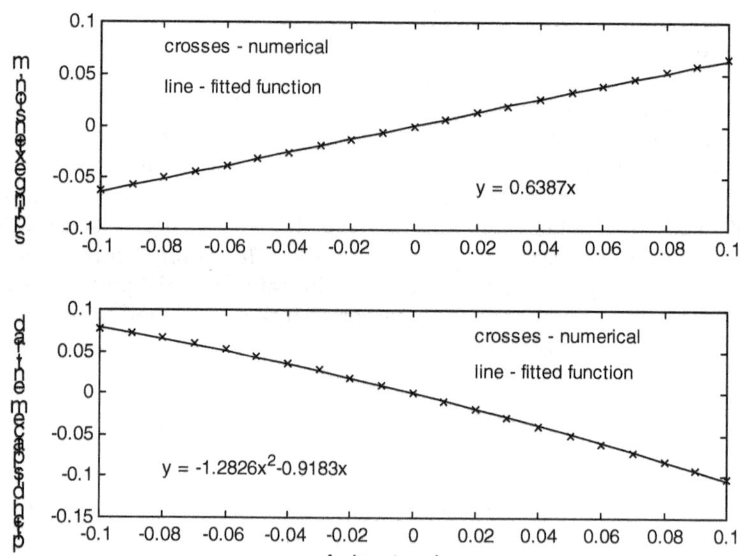

Fig. 12. Example Telelever kinematic simulation results with polynomial fits. The simple polynomials are virtually perfect representations of the numerical results.

A third alternative is to analyse the Telelever geometrically and to use the symbolic relationships between θ and λ and between μ and λ obtained in constraint functions. In either of these two alternative treatments, the wishbone and the constraints at p18 are omitted, while the pitching displacement of the upper forks is given as either an approximate or an exact function of the fork extension. The strut force is easily converted via the virtual power principle assuming ideal joints [19] to an equivalent force acting between the upper and lower forks, as:

$$F_f = f_s\{f_f(e_f), \frac{df_f(e_f)}{de_f}\dot{e}_f\}\frac{df_f(e_f)}{de_f} \tag{5}$$

In which the spring/damper unit force is represented by:

$$F_s = f_s(e_s, \dot{e}_s) \tag{6}$$

with the nominal spring deflection, e_s, being related to the fork extension, e_f, by:

$$e_s = f_f(e_f) \tag{7}$$

The loop closure equations, for the circuit P2, P4, P18, P16, P2, Figure 2, are:

$$\begin{aligned} &l_2\sin(\theta_1+\theta)+(\lambda_0+\lambda)\sin(\theta_0+\theta)-l_4\cos(\theta_0+\theta)- \\ &l_5\cos(\zeta_0+\zeta)-x_{16}+x_2=0 \end{aligned} \tag{8}$$

$$\begin{aligned} &l_2\cos(\theta_1+\theta)+(\lambda_0+\lambda)\cos(\theta_0+\theta)+l_4\sin(\theta_0+\theta)+ \\ &l_5\sin(\zeta_0+\zeta)-z_{16}+z_2=0 \end{aligned} \tag{9}$$

in which l_2 is P2-P4, l_4 is the projection of P4-P16 perpendicular to the fork line and l_5 is the wishbone length P16-P18. The symbols θ_0, θ_1, λ_0 and ζ_0, relate to the nominal configuration and their numerical values are known from the positions of the key points.

Equations 8 and 9 can be combined to give l^2_5 as $l^2_5\cos^2(\zeta_0+\zeta)+l^2_5\sin^2(\zeta_0+\zeta)$, which expression can be expanded and simplified to give the form:

$$c_1+c_2\cos\theta+c_3\sin\theta=0 \tag{10}$$

where

$$\begin{aligned} c_1 &= 2\lambda_0\lambda-l_5^2+x_{16}^2+z_{16}^2-2x_2x_{16}-2z_2z_{16}+x_2^2+z_2^2+l_2^2 \\ &+\lambda_0^2+\lambda^2+l_4^2+2l_2(\lambda_0+\lambda)\cos(\theta_0-\theta_1) \\ &+2l_2l_4\sin(\theta_0-\theta_1) \end{aligned} \tag{11}$$

$$\begin{aligned} c_2 &= 2l_2\sin\theta_1(x_2-x_{16})+2(\lambda_0+\lambda)\sin\theta_0(x_2-x_{16})+ \\ &2l_4\cos\theta_0(x_{16}-x_2)+2l_2\cos\theta_1(z_2-z_{16})+ \\ &2(\lambda_0+\lambda)\cos\theta_0(z_2-z_{16})-l_5^2+2l_4\sin\theta_0(z_2-z_{16}) \end{aligned} \tag{12}$$

$$\begin{aligned} c_3 &= 2l_2\cos\theta_1(x_2-x_{16})+2(\lambda_0+\lambda)\cos\theta_0(x_2-x_{16})+ \\ &2l_4\sin\theta_0(x_2-x_{16})+2l_2\sin\theta_1(z_{16}-z_2)+ \\ &2(\lambda_0+\lambda)\sin\theta_0(z_{16}-z_2)+2l_4\cos\theta_0(z_2-z_{16}) \end{aligned} \tag{13}$$

From Equation 10,

$$\theta = -\arcsin\frac{c_1}{\sqrt{c_2^2 + c_3^2}} - \arctan\frac{c_2}{c_3} \tag{14}$$

with θ known, ζ can be found from Equation 9. The coordinates of P17 are:

$$x_{17} = x_{16} + l_6\cos(\zeta_1 + \zeta), \quad z_{17} = z_{16} - l_6\sin(\zeta_1 + \zeta)$$

in which l_6 is the distance P16-P17 and ζ_1 relates to the nominal state, Figure 2, so that the change in the spring unit length can be found as a function of λ, the fork extension, through:

$$\mu = \sqrt{(x_{15} - x_{17})^2 + (z_{17} - z_{15})^2} - l_0 \tag{15}$$

where l_0 is the nominal length of the spring/damper unit.

Evaluation of these expressions, Equations 14 and 15, numerically yields the same results as obtained from the previous kinematic analysis, Figure 12, confirming their accuracy.

Equation 14 for θ is used directly in Autosim in an "add-position-constraint" command. A corresponding "add-speed-constraint" is needed, requiring the determination of $\dot{\theta}$, which is found as:

$$\dot{\theta} = \left(\frac{\partial\theta}{\partial c_1}\frac{\partial c_1}{\partial\lambda} + \frac{\partial\theta}{\partial c_2}\frac{\partial c_2}{\partial\lambda} + \frac{\partial\theta}{\partial c_3}\frac{\partial c_3}{\partial\lambda}\right)\dot{\lambda} \tag{16}$$

In similar fashion, description of the spring/damper unit force effective at the forks requires Equation 15 to give the spring extension in relation to the fork movement and also the derivative, needed for the damper velocity and the motion ratio:

$$\frac{d\mu}{d\lambda} = \left(\frac{\partial\mu}{\partial x_{17}}\frac{\partial x_{17}}{\partial\zeta} + \frac{\partial\mu}{\partial z_{17}}\frac{\partial z_{17}}{\partial\zeta}\right)\frac{\partial\zeta}{\partial\lambda} \tag{17}$$

The effective force at the forks is then given, as Equation 5, by:

$$f_s(\mu,\dot{\mu})\frac{d\mu}{d\lambda} = f_s\left(f_e(\lambda), \frac{df_e(\lambda)}{d\lambda}\dot{\lambda}\right)\frac{d\mu}{d\lambda} \tag{18}$$

4.2 Monoshock Rear Suspension

Monoshock linkages take several different forms, as illustrated by Figures 3-5. The linkage couples the single spring/damper unit to the swing arm through a varying leverage ratio, performing no function in relation to the suspension geometry. The links can be modelled literally or preferably analytically [19]. In the latter case, some economy of computation at simulation time will result. Three of the most common designs are examined below.

4.2.1 Monoshock rear suspension as on the GSX-R1000

This Suzuki system is shown in Figure 3 and a geometric analysis of it has been published [19]. The spring/damper unit length was derived as a function of the swing arm displacement from nominal. That function can be used as in Equations 5-7 above, via Autosim's partial derivative capability, to represent the linkage system and the spring/damper unit force law by an equivalent moment about the swing arm joint. In the notation of Equations 5-7, this moment is:

$$M = f_s\{f_f(\theta), \frac{df_f(\theta)}{d\theta}\dot{\theta}\}\frac{df_f(\theta)}{d\theta} \tag{19}$$

4.2.2 Monoshock rear suspension used by Honda

The so-called Pro-link system, as shown in Figure 4, is a small variation of the GSX-R1000 design. The spring/damper unit acts on the connecting link instead of the rocking lever. A geometric analysis is very similar to that done before. Referring to Figure 4, the displacement loop equations for the circuits p11, p22, p20, p19, p11 and p11, p22, p20, p21, p11 yield:

$$x_{11} - l_3\cos\theta + l_2\cos\zeta + l_1\cos\delta = x_{19} \tag{20}$$

$$z_{11} + l_3\sin\theta + l_2\sin\zeta - l_1\sin\delta = z_{19} \tag{21}$$

$$x_{11} - l_3\cos\theta + l_2\cos\zeta + l_4\cos(\pi - \zeta - \phi_0) = x_{21} \tag{22}$$

$$z_{11} + l_3\sin\theta + l_2\sin\zeta - l_4\sin(\pi - \zeta - \phi_0) = z_{21} \tag{23}$$

Forming l^2_1 as $l^2_1\cos^2\delta + l^2_1\sin^2\delta$, from Equations 20 and 21, putting $c_1 = -x_{11} + x_{19} + l_3\cos\theta$ and $c_2 = -z_{11} + z_{19} - l_3\sin\theta$ and letting $c_1 = A\cos\phi$ and $c_2 = -A\sin\phi$:

$$\zeta = \cos^{-1}\left(\frac{-l_1^2 + l_2^2 + c_1^2 + c_2^2}{2l_2\sqrt{c_1^2 + c_2^2}}\right) + \tan^{-1}\left(\frac{c_2}{c_1}\right) \tag{24}$$

The spring/damper unit length is:

$$l = \sqrt{(x_{13} - x_{21})^2 + (z_{13} - z_{21})^2} \tag{25}$$

with x_{21} and z_{21} given by Equations 22 and 23 and ζ given by Equation 21 as a function of θ. Thus, the suspension unit extension is a known function of the swing arm displacement and the moment M in Equation 19 can be defined.

4.2.3 Fully floating Monoshock design

Referring to Figure 5 and using a similar approach, the loop equations are:

$$x_{19} - l_1\cos\delta + l_2\cos\zeta + l_3\cos\theta = x_{11} \tag{26}$$

$$z_{19} + l_1\sin\delta + l_2\sin\zeta - l_3\sin\theta = z_{11} \tag{27}$$

Forming l^2_2 as $l^2_2 \cos^2 \zeta + + l^2_1 \sin^2 \zeta$, introducing $c_1 = -x_{11} + x_{19} + l_3 \cos\theta$ and $c_2 = -z_{11} + z_{19} - l_3 \sin\theta$, and letting letting $c_1 = A \cos\phi$ and $c_2 = -A \sin\phi$, it can be shown that:

$$\delta = -\cos^{-1}\left(\frac{l_1^2 - l_2^2 + c_1^2 + c_2^2}{2l_1\sqrt{c_1^2 + c_2^2}}\right) - \tan^{-1}\left(\frac{c_2}{c_1}\right) \tag{28}$$

Also:

$$x_{13} - l_4 \cos\delta = x_{19} \tag{29}$$

$$z_{13} + l_4 \sin\delta = z_{19} \tag{30}$$

$$x_{11} - l_5 \cos(\theta - \theta_0) = x_{21} \tag{31}$$

$$z_{11} + l_5 \sin(\theta - \theta_0) = z_{21} \tag{32}$$

$$l = \sqrt{(x_{13} - x_{21})^2 + (z_{13} - z_{21})^2} \tag{33}$$

leading to the suspension unit extension being definable as a function of the swing arm displacement, with x_{13} and z_{13} from Equations 29 and 30 and x_{21} and z_{21} from Equations 31 and 32.

4.3 Chain Drive

The chain functions as a tension link between the points p_{gu} on the gearbox sprocket and p_{wu} on the rear wheel sprocket when the engine is driving the wheel and between p_{wl} and p_{gl} when the wheel is driving the engine, Figure 13. Some slack or clearance is allowed between driving and overrunning, and the chain is treated as elastic and damped. The force transmitted between the relevant points thus depends on some logic, to decide between drive and overrun, and the relative angular displacements and velocities of the gearbox and rear wheel sprockets. The points themselves need to be defined as moving points in the appropriate sprockets. The point definitions are derived via a geometric analysis which follows.

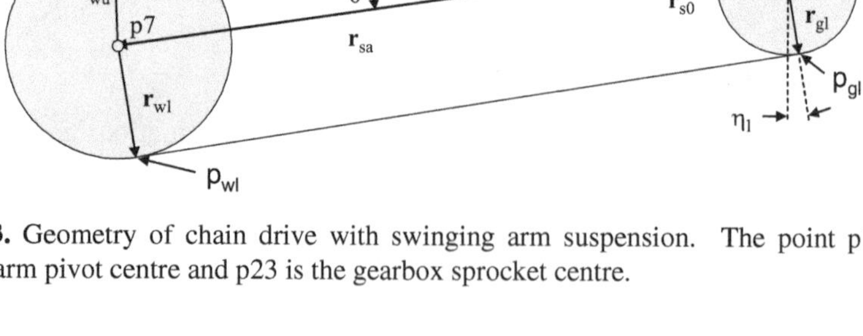

Fig. 13. Geometry of chain drive with swinging arm suspension. The point p11 is the swing arm pivot centre and p23 is the gearbox sprocket centre.

In Figure 13, these relationships are apparent:

$$\mathbf{r}_{sa} = -r_{sa}\cos\theta \;\; \mathbf{i} + r_{sa}\sin\theta \;\; \mathbf{k} \tag{34}$$

$$\mathbf{r}_{s0} = (x_{23} - x_{11}) \;\; \mathbf{i} + (z_{23} - z_{11}) \;\; \mathbf{k} \tag{35}$$

$$\mathbf{r}_{wu} = -r_w \sin\eta_u \;\; \mathbf{i} - r_w \cos\eta_u \;\; \mathbf{k} \tag{36}$$

$$\mathbf{r}_{gu} = -r_g \sin\eta_u \;\; \mathbf{i} - r_g \cos\eta_u \;\; \mathbf{k} \tag{37}$$

$$\mathbf{r}_{wl} = r_w \sin\eta_l \;\; \mathbf{i} + r_w \cos\eta_l \;\; \mathbf{k} \tag{38}$$

$$\mathbf{r}_{gl} = r_g \sin\eta_l \;\; \mathbf{i} + r_g \cos\eta_l \;\; \mathbf{k} \tag{39}$$

The chain lines are perpendicular to the radius vectors, so that:

$$(-\mathbf{r}_{wu} - \mathbf{r}_{sa} + \mathbf{r}_{s0} + \mathbf{r}_{gu}).\mathbf{r}_{gu} = 0 \tag{40}$$

$$(-\mathbf{r}_{wl} - \mathbf{r}_{sa} + \mathbf{r}_{s0} + \mathbf{r}_{gl}).\mathbf{r}_{gl} = 0 \tag{41}$$

Through some algebra, these equations yield:

$$\begin{aligned}\eta_u = {} & \sin^{-1}\left(\frac{r_g - r_w}{\sqrt{(r_{sa}\cos\theta + x_{23} - x_{11})^2 + (r_{sa}\sin\theta - z_{23} + z_{11})^2}}\right) \\ & + \tan^{-1}\left(\frac{r_{sa}\sin\theta - z_{23} + z_{11}}{r_{sa}\cos\theta + x_{23} - x_{11}}\right)\end{aligned} \tag{42}$$

$$\begin{aligned}\eta_l = {} & \sin^{-1}\left(\frac{-r_g + r_w}{\sqrt{(r_{sa}\cos\theta + x_{23} - x_{11})^2 + (r_{sa}\sin\theta - z_{23} + z_{11})^2}}\right) \\ & + \tan^{-1}\left(\frac{r_{sa}\sin\theta - z_{23} + z_{11}}{r_{sa}\cos\theta + x_{23} - x_{11}}\right)\end{aligned} \tag{43}$$

expressing the angles η_u and η_l as functions of the swing arm angle to the horizontal, θ, obtained by summing the nominal value with the displacement from nominal. For any swing arm position, the coordinates of the points p_{gu}, p_{wu}, p_{gl} and p_{wl} can therefore be specified.

The upper chain run is in tension if $(-r_g\varphi_g + r_w\,\varphi_w + \delta\,l_u)$ is positive, the first two terms accounting for the spinning of the sprockets and the third for the increase in separation of p_{gu} and p_{wu} due to suspension motion. Similarly, the lower run is in tension if $(r_g\varphi_g - r_w\,\varphi_w + \delta\,l_u - s_l)$ is positive, in which s_l represents the chain slack. If the upper chain run is in tension, it applies a force between the end points of the form (*-kx-cv*) where *k* is the spring rate and *c* the damping coefficient, *x* is the extension and *v* the extension speed. In such a case, the lower run force is zero and conversely.

5 Observations on the Resulting Models

5.1 Tuning for Fast Simulation

Obtaining a fast simulation model depends partly on controlling the number of arithmetic operations per time step necessary to evaluate the rates of change of generalised speeds. The fewer these are, the better it is. Preparing the motorcycle model in various guises yielded the results of Table 2, with the variants defined in Table 3.

Table 2. Operation counts for various analysis options

Model variant	Multiply/divides	Add/subtracts	Function evaluations
A	6515	3261	99
B	6909	3631	99
C	6481	3241	93
D	6559	3306	99
E	6464	3267	97
F	6467	3246	97
G	6484	3219	96
H	6732	3508	113
I	11170	5451	107
J	9413	4779	104
K	9435	4631	99
L	5432	2969	96
M	5247	2937	92
N	10221	4797	101
O	9341	4570	97
P	5227	2907	94
Q	10902	5039	109
R	9378	4620	105
S	5264	2915	102
T	5169	2890	98

In the nominal case, the Monoshock rear suspension is modelled analytically and the LU decomposition is done numerically at simulation time.

From these results, it can be concluded that: (i) symbolic LU decomposition is unhelpful in the case of the motorcycle. Although only one comparison is included, this has been a general feature over many trials: (ii) literal modelling of the Monoshock mechanism is the most extravagant, while using off-line generated fitted functions is the most economical. Using analytically obtained functions lies between the extremes: (iii) the same is true for the modelling of the Telelever front suspension mechanism: (iv) the differences in operation counts between literal, function fit and analytical representations are only modest in the cases of both Monoshock and Telelever mechanisms: (v) restricting the main frame pitch angle to small reduces the model size substantially (vi) restricting the rider upper body lean and the swing arm angles to small brings very modest benefit to the operation counts: (vii) modelling the Telelever suspension in full detail is costly: (viii) using small angle theory in dealing with the Telelever constraints, by considering wishbone yaw and pitch, frame twist and steering head pitch displacements to be small, reduces the model size substantially. It also reduces the model building time, in the context of Autosim's loading operation, considerably.

Table 3. Key to Table 2

System	Description
A	Nominal configuration with telescopic front forks, Monoshock rear , shaft drive
B	A with symbolic LU decomposition prior to simulating
C	A with the Monoshock system modelled by fitted functions
D	A with the Monoshock system modelled literally
E	A with the swing arm angular displacement declared small
F	A with the rider upper body angular displacement declared small
G	A with the main frame pitch angular displacement declared small
H	A with chain drive
I	A with Telelever front suspension modelled literally
J	I with the main frame pitch angular displacement declared small
K	I with wishbone yaw and pitch, steering head pitch and frame twist angles small
L	K with the main frame pitch angular displacement also declared small
M	L with rider upper body lean and swing arm angles declared small
N	A with Telelever front suspension modelled by fitted functions
O	N with wishbone yaw and pitch, steering head pitch and frame twist angles small
P	O with the main frame pitch angular displacement also declared small
Q	A with Telelever front suspension modelled analytically
R	Q with wishbone yaw and pitch, steering head pitch and frame twist angles small
S	R with the main frame pitch angular displacement also declared small
T	S with the rider upper body and swing arm angles also declared small

For straight running at constant speed, all the models give effectively the same behaviour. For more complex but ordinary manoeuvring, it can be expected that varying any particular model will not influence the behaviour to any significant degree, provided that it is used within its compass. Clearly, a proper connection between model building and model application must be maintained.

5.2 The Influences of Chain Drive on the Accelerating Motorcycle

The nominal geometry of the chain drive is shown in Figure 14. p23 is at (0.629, 0.3708) and the sprocket radii are 0.041 and 0.104m. To demonstrate the good operation of the chain drive model, simulation results of the motorcycle in constant acceleration of $5m/s^2$ from 5m/s initial speed up to about 65m/s are included, without consideration of whether or not this is feasible with the standard power plant. The opportunity is also taken to compare the behaviour for shaft drive and for chain drive and to see the influences of raising or lowering the driving sprocket by 0.03m from its nominal position, Figures 15-17.

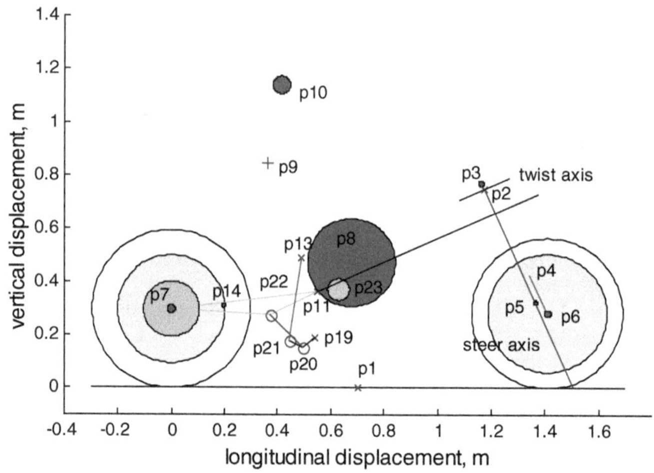

Fig. 14. Scaled motorcycle showing masses and chain sprockets in nominal state. Masses shown in proportion. Points fixed to main frame shown by crosses.

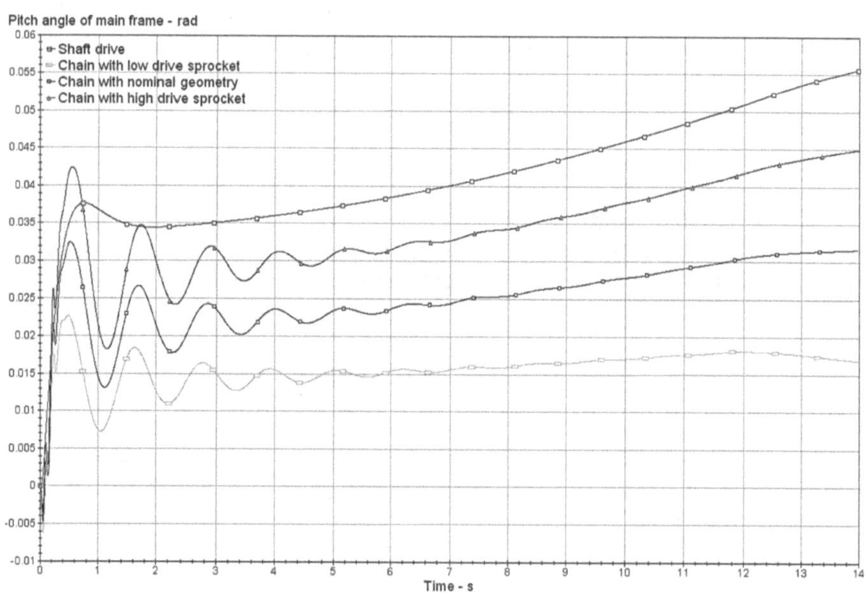

Fig. 15. Pitch up angle of main frame for shaft drive and chain drive systems.

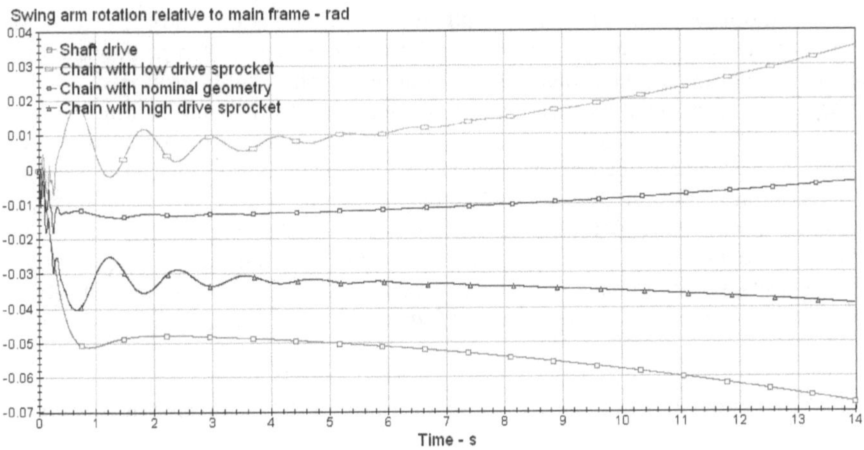

Fig. 16. Swing arm angular displacement (positive in rebound) for shaft drive and chain drive systems.

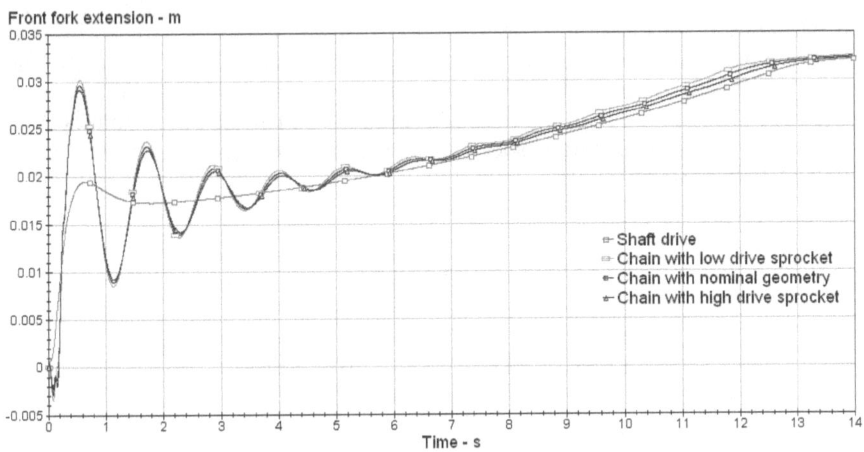

Fig. 17. Front fork extension for shaft drive and chain drive systems.

It can be observed in these results that the chain transmission system introduces a longitudinal oscillatory mode, by virtue of the chain compliance, which is not present in the case of shaft drive. The chain compliance also includes a contribution from the shock cushioning system included in the connection of the sprocket to the rear wheel. The damping of this mode can be controlled by any damping included in parallel with the chain "spring" and by a derivative term included in the rider's speed control strategy. In the virtual world, this damping can be increased or decreased at will by parameter adjustments and in the simulations shown, no chain damping has been included. The derivative term in the speed controller has been made effective only at very low speeds. In reality, the extent of the excitation of the mode will depend on the skill and attention of the rider. In demanding a constant $5m/s^2$ acceleration here, there is no attempt at including rider finesse in the computations. The oscillations are inevitably provoked.

The "squat" response of the shaft driven machine to the application of drive torque is strong and the squat suppression properties of the chain drive system are apparent. The standard geometry leads to very little swing arm motion relative to the main frame. With the raised drive sprocket, the swing arm moves in bump (as with shaft drive) while, with the lowered sprocket, the swing arm moves in rebound and the anti-squat influence is most marked. The main frame pitches nose-up in all cases, since the front forks extend. This non-oscillatory component of the extension is substantially the same for all four cases.

5.3 The Influence of Telelever Front Suspension on the Braking Motorcycle

The nominal geometry of the notional Telelever equipped motorcycle is shown in Figure 18. As in Section 5.2, the operation of the Telelever system model is demonstrated by some simulation results. These are also used to illustrate the particular properties of the Telelever suspension as a function of the height of the rearward wishbone pivot location. The pivot is first lowered and then raised by 4cm from the nominal for the "low pivot" and "high pivot" cases respectively. The front braking torque employed is near the limit of what is possible without causing a "stoppie" (the rear tyre lifting from the ground) and it slows the motorcycle nearly to a stop from an initial speed of 75m/s in 10s. Simulation results are shown in Figures 19 – 22.

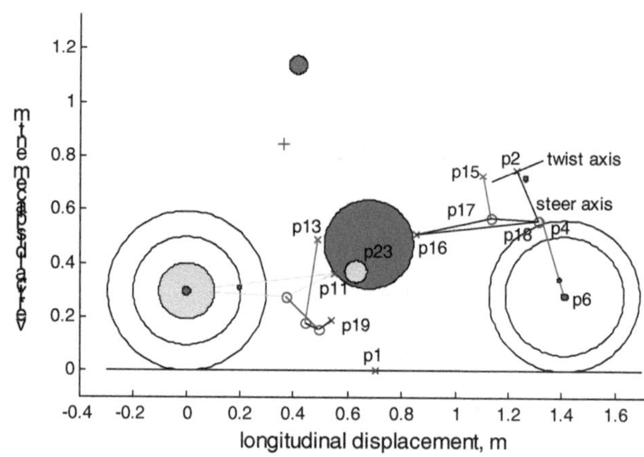

Fig. 18. Illustration of standard Telelever geometry. Masses shown in proportion. Points fixed to main frame shown by crosses.

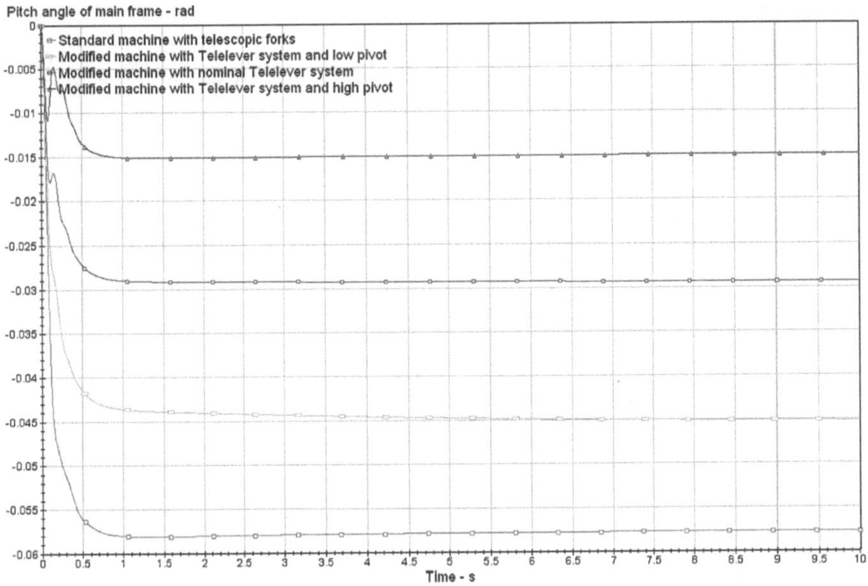

Figure 19. Main frame displacement in pitch in front wheel braking for standard and Telelever equipped motorcycles.

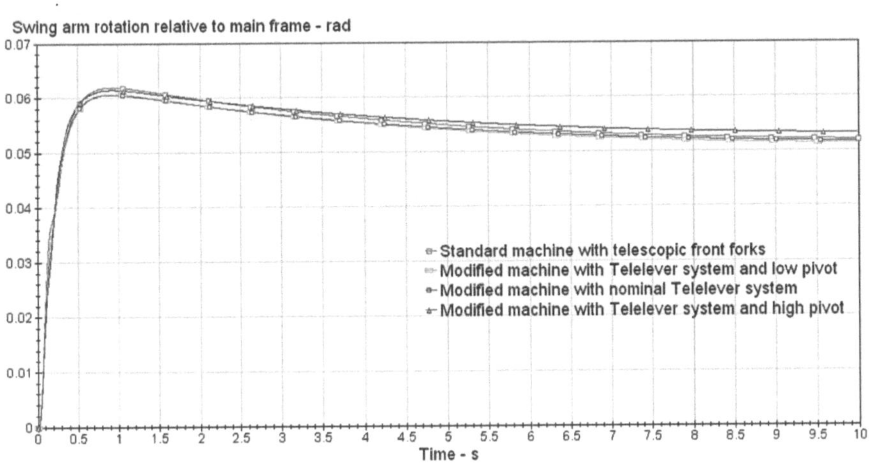

Fig. 20. Swing arm angular displacement (positive in rebound) in front wheel braking for standard and Telelever equipped motorcycles.

The motorcycle pitches nose-down in each case, the angle reached varying by a factor of about 4 between the standard machine and the Telelever equipped machine with the high wishbone pivot location. In this last case, the wishbone and front fork movements are quite small. Then, most of the pitch attitude change is accounted for by swing arm movement in rebound. These swing arm movements are nearly the same for all the cases. The anti-dive property of the Telelever system is increased by raising the wishbone rearward joint and is accompanied by a more rapid change in loading of the front tyre at the commencement of the braking manoeuvre. Although not shown, the variation in the front tyre loading follows the same pattern, although it does not, of course, start from zero. Again, a skilled and attentive rider can be expected to apply a smoother torque than that simulated here, in order to control the transients well, but the control problem appears to be made more difficult by the employment of anti-dive kinematics.

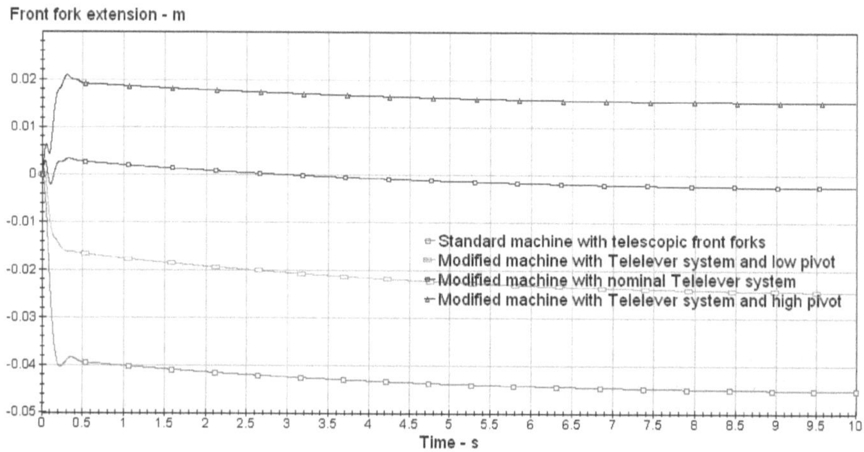

Fig. 21. Front fork extension in front wheel braking for standard and Telelever equipped motorcycles.

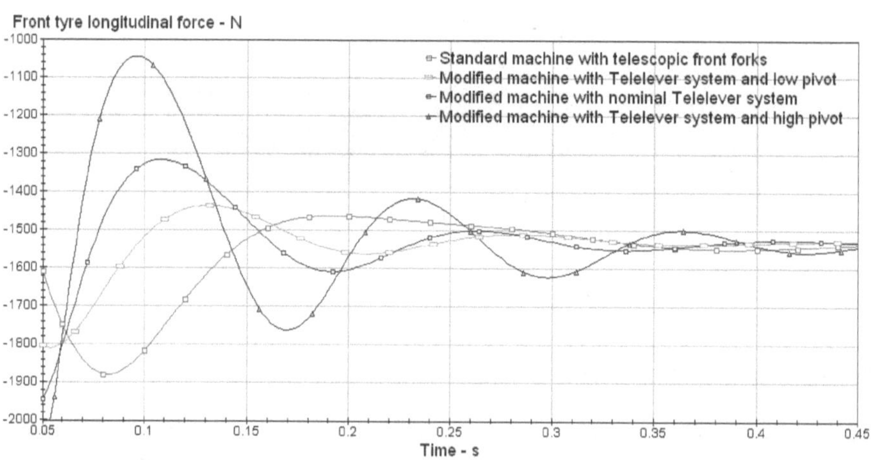

Fig. 22. Transient build up of braking force at the front tyre over the initial stages of the simulation runs.

6 Conclusions

Many general issues in the modelling of motorcycle dynamics have been reviewed and the problem has been related to the use of Autosim as an aid. New analyses of two variants of the Monoshock rear suspension system and of a Telelever front suspension have been included.

It has been explained how the availability of analytic or numerical kinematic solutions to a mechanism makes possible the modelling of the system in alternative ways, which are essentially the same in terms of behaviour. The options have been explored in detail for both Monoshock and Telelever suspensions and operation counts relating to each time step of the integration of the equations of motion have been used to show the implications of choosing different alternatives. Although the literal modelling is the easiest and the analytical modelling often much more difficult and more resource consuming in connection with model building, the resulting models are surprisingly similar in terms of these numbers of operations. Using small angle theory in some instances has little influence on the operation count, while in other cases, it makes a large difference.

A chain drive model has been developed and explained. Compliance in the chain introduces a new longitudinal mode of motion which is likely to be quite oscillatory. However, relevant parameter values are not known with any precision at this stage.

Simulation results have illustrated the operation of the chain drive and Telelever models in particular, with the former showing "anti-squat" properties under hard acceleration and the latter demonstrating "anti-dive" behaviour, both depending on the geometric details of the design.

References

1. Cossalter V, Lot R (2002) A motorcycle multi-body model for real time simulations based on the natural coordinates approach. Vehicle System Dynamics 37(6):423-447
2. Cuadrado J, Cardenal J, Bayo E (1997) Modeling and solution methods for efficient real-time simulation of multibody dynamics. Multibody System Dynamics 1(3):259-280
3. de Vries EJH, Pacejka HB (1997) Motorcycle tyre measurements and models. In Palkovics L (ed) Proc. 15th IAVSD Symposium on the Dynamics of Vehicles on Roads and on Tracks, Suppl. Vehicle System Dynamics 28:280-298
4. de Vries EJH, Pacejka HB (1998).The effect of tyre modeling on the stability analysis of a motorcycle. In Proc. AVEC '98, Nagoya, SAE of Japan, pp 355-360
5. Evangelou S, Limebeer DJN, Sharp RS, Smith MC (2004) Influence of a steering inerter on motorcycle stability and response. CDC conference paper, in press

6. Giles CG, Sharp RS (1983) Static and dynamic stiffness and deflection mode measurements on a motorcycle, with particular reference to steering behaviour. In Proc. I. Mech. E./MIRA Conference on Road Vehicle Handling, London, Mech. Eng. Publ., pp 185-192
7. Kortüm W (1993) Review of multibody computer codes for vehicle system dynamics. In Kortüm W and Sharp RS (eds) Veh. Syst. Dyn. Supplement 22 pp 3-31
8. Limebeer DJN, Sharp RS, Evangelou S (2002) Motorcycle steering oscillations due to road profiling. Proc. ASME, J. App. Mech. 69(6):724-739.
9. Pacejka HB (2002) Tyre and Vehicle Dynamics. Butterworth Heinemann, Oxford, England
10. Paul B (1979) Kinematics and dynamics of planar machinery. Prentice Hall, Englewood Cliffs, New Jersey
11. Sayers MW, Mousseau CW (1990) Real-time vehicle dynamic simulation obtained with a symbolic multibody program. In Proc., Symposium on Transportation Systems: Automotive component modelling and control, ASME, Dallas TX, pp 51-58
12. Sayers MW (1999) Vehicle models for RTS applications. Vehicle System Dynamics 32(4-5):421-438
13. Schiehlen W (1997) Multibody System Dynamics: Roots and perspectives. Multibody System Dynamics, 1(2):149-188
14. Sharp RS (1971) The stability and control of motorcycles. J. Mech. Eng. Sci. 13(5):316-329
15. Sharp RS (2000) Variable geometry active rear suspension for motorcycles. In Proc. AVEC 2000, 5th International Symposium on Advanced Vehicle Control, Ann Arbor, Michigan, August 22-24 pp 389-395
16. Sharp RS (2001) Stability, control and steering responses of motorcycles. Veh. Syst. Dyn. 35(4-5):291-318
17. Sharp RS, Alstead CJ (1980) The influence of structural flexibilities on the straight running stability of motorcycles. Veh. Syst. Dyn. 9(6), 327-357
18. Sharp RS, Limebeer DJN (2001) A motorcycle model for stability and control analysis. Multibody System Dynamics 6(2):123-142
19. Sharp RS, Evangelou S, Limebeer DJN (2004) Advances in the modelling of motorcycle dynamics. Multibody System Dynamics 12(3):251-283
20. Spierings PTJ (1981) The effects of lateral front fork flexibility on the vibrational modes of straight-running single track vehicles. Veh. Syst. Dyn. 10(1):37-38
21. Tezuka Y, Ishii H, Kiyota S (2001) Application of the magic formula tire model to motorcycle maneuverability analysis. JSAE Review 22:305-310
22. Watanabe Y, Sayers MW (2002) Extending vehicle dynamics software for analysis, design, control and real-time testing. In Proc. AVEC '02, Sept. 9-13, Hiroshima, Japan (CD-ROM).

Gluing for Dynamic Simulation of Distributed Mechanical Systems

Gregory M. Hulbert, Zheng-Dong Ma and Jinzhong Wang

Department of Mechanical Engineering, The University of Michigan, 2250 G.G. Brown, 2350 Hayward, Ann Arbor, MI 48109-2125
{hulbert,mazd}@umich.edu

A new concept of gluing, also known as co-simulation, for dynamic simulation of distributed mechanical systems is presented to couple subsystem models in a distributed computing environment. Using the new gluing algorithm, subsystem models can be analyzed at their distributed locations, using their own independent solvers, and on their own platforms. The gluing algorithm relies only on information available at the subsystem interfaces. This not only enables efficient integration of subsystem models, but also engenders model security by limiting model access only to the exposed interface information. These features make the new gluing approach suitable for practical use in a distributed simulation environment. The new concept system has been implemented with a standardized model description using XML and a logical distribution architecture including model database layout, wrapping methods for simulation codes, and a Web-based user interface. Examples using the developed simulation system demonstrate the potential for the application of the new methodology to general mechanical system simulation problems, including the design and analysis of globally distributed automotive systems.

1 Introduction

The automotive industry relies extensively on a multi-layered supply chain model. As such, the components of a vehicle are designed and tested in different component supply companies; thus, the associated analysis and design models are distributed in different locations. This work addresses the need for a simulation environment that can incorporate distributed heterogeneous mechanical systems models and couple them together to perform dynamics simulations to assist the virtual prototyping process. As shown in Figure 1, design and manufacturing of an automotive vehicle may be distributed amongst an OEM and its tiered suppliers. The simulation models of the different components often are developed using different computing platforms and different software tools. Traditionally, in order to study the behavior of the whole vehicle, all the subsystem models need to be brought together to form a monolithic all-in-one model. This results in a lengthy

J.A.C. Ambrósio (ed.), Advances in Computational Multibody Systems, 69–94.

and complicated process to collect and integrate subsystem models with little flexibility to incorporate design changes of the subsystems made by the suppliers in the later stages of vehicle design.

In order to simulate more efficiently the vehicle system and to deal with the distributed nature of the problem, two significant challenges must be addressed. First, the distributed subsystems models might be developed using different software packages, running on different computers, and/or residing at different geographical locations. Second, the model developers, to protect proprietary information, often are not willing to share their models directly. This suggests that only minimal information should be exchanged during the coupled simulation, and the local model developer/supplier should be able to control the accessibility of the model developed. These requirements call for a new integration algorithm that can efficiently and sufficiently integrate subsystems models in a distributed environment and that does not require internal details of the component models.

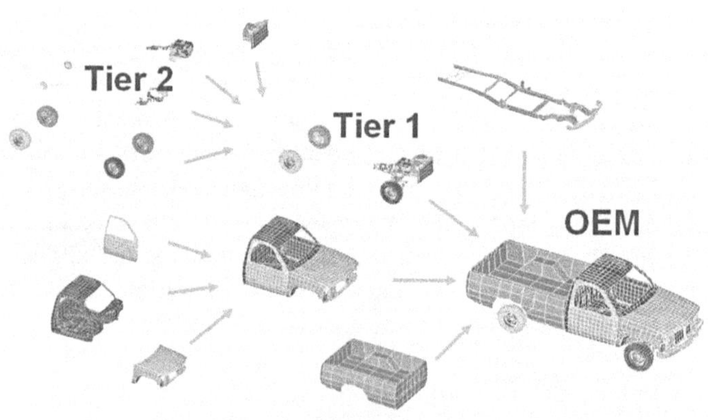

Fig. 1. A distributed mechanical system

Researchers have studied the decomposition of large mechanical systems with a primary focus on the decomposition strategy. Arising from the field of parallel computing applied to finite element analysis, many have explored extending the concept of parallel algorithms to distributed simulation. A primitive version of a distributed finite element simulation is presented in [1], in which stiffness matrices and load vectors are generated concurrently on clients and sent to a central server to be assembled. Other researchers adopted a different approach, using parallel solution of the system of equations. The parallelization of both direct solution methods [2] and iterative methods [3] has been studied extensively. A substructuring method, FETI, is presented in [4], introducing extra traction variables and exhibiting more flexibility for model reduction and coupling, compared to existing competing schemes. A similar algorithm, IGI, is presented in [5], which provides a practical way to update the interface force and in which the subsystem models can be solved independently. As discussed in [6], these efforts all focus on active partitioning of an existing large-scale system rather than coupling an already distributed system, employing, e.g., substructuring [7] or domain decomposition methods [8], which usually requires the internal details of the subsystem models.

In the multibody dynamics arena, researchers have also studied how to partition and parallelize systems [6, 9-20]. One strategy adopted by researchers is similar to substructuring in FEM, i.e., a small global problem is formed by incorporating condensed subsystem models. This smaller problem then is solved to provide necessary information to subsystem models. The subsystem models subsequently can be solved based on this information. In [9], a subsystem synthesis method was proposed for dynamic analysis of vehicle multibody systems, in which each subsystem is independently analyzed with a virtual reference body and the overall vehicle system analysis is formed by synthesizing the effective inertia matrix and force vector from the virtual reference body of each subsystem. A divide-and-conquer algorithm is presented in [10,11] for rigid body dynamics, which reduces the system to an "articulated-body" by recursively applying a formula that constructs the articulated-body equations of motion of the system from those of its constituent parts. Both the inputs and outputs of the formula are equations of motion. Another approach is given in [12,13], in which the equations of the subsystem models are evaluated in parallel, and the results are loaded into a single system wide equation to explicitly calculate the constraint forces. The strategy adopted in [14] follows a similar path.

Treating the subsystem models as control blocks and taking advantage of many sophisticated control-based simulation software packages is another common modeling approach. In [15], a modular formulation for multibody systems is proposed, based on the block representation of a multibody system with corresponding input and output quantities. This "block diagram" representation of the system can then be embedded into appropriate simulation packages, e.g., SIMULINK. In [16], "Co-simulation," is presented, which employs a new discrete time sliding mode controller (DTSM) to satisfy the algebraic constraints among the subsystem models and to solve the causal conflicts associated with the algebraic constraints.

The methods reviewed above either involve the active decomposition of the full system and require more information than just that associated with the subsystem interfaces or mandate specific requirements or structures on the formulation of the subsystems. In the context of coupling already distributed subsystems, the gluing perspective is preferred. A study is presented in [6], in which the terminology "gluing algorithm" is first suggested to describe a class of algorithms that can be used to glue distributed component models for use in dynamics simulations [17-19].

We have been developing a concept platform for simulating general distributed mechanical systems. Here, the mechanical system may have a large number of components represented by either finite element models and/or multibody dynamics models. The goal is to fill the gap between state-of-the-art simulation techniques and the practical collaborative product development systems described previously. Our efforts include three aspects as follows, with a focus on developing the gluing algorithm.

1. a standardized description of models implemented in XML;
2. a distribution architecture that can be realized using available computational technologies;
3. a gluing algorithm that can couple component models without requiring sending the complete models or modifications to the model internals.

2 Basic Concept of Gluing

2.1 Subsystem Models

Standard simulation practice involves the use of a dataset, which includes, e.g., structural geometry, material data, loading conditions. This dataset describes the simulation model being used to represent a specific mechanical component along with the simulation scenario in which to exercise the model. To obtain simulation results, the dataset must be input to a specific simulation code. Using the nomenclature of data trees, this combination of dataset and simulation code is defined as a *leaf model.* A leaf model represents the minimum information required to directly execute a component simulation. Thus, in the usual design and simulation environment, the analyst works with leaf models. Instead of using the term parent model, we use integrated model to refer to models that are assembled from leaf models and/or lower-level integrated models, i.e., children models. As such, integrated models include the information required to couple/assemble its children models. Within the proposed framework, an integrated model also contains a strategy for gluing together its children models.

An example of an integrated model is a truck chassis frame shown in Figure 2, which is formed by gluing six leaf models: right rail, left rail and four cross-members. In the glue integration, the leaf models are "glued" together to form an integrated model — the complete chassis frame. This integrated model can be then integrated with other subsystem models to form a higher level integrated model. This process can be repeated, obtaining a full vehicle model as shown in Figure 2.

2.2 Gluing Principles

The proposed gluing algorithm only relies on the information at the interfaces of the models that are to be coupled. Here, interface refers to the connections or common surfaces of two models. An interface can be represented by a set of interface nodes in a finite element model, or by a set of connecting joints in a multi-body dynamics model. The typical information available at the interface can be classified, in the present context, as kinematic information and force information. The kinematic information may contain displacements, velocities, and/or accelerations of the interface. Force information refers to action-reaction forces at the interface.

Mechanics principles require that at any interface the force quantities, namely action-reaction forces, satisfy the equilibrium equations, and the kinematic quantities satisfy the compatibility conditions, where it is assumed that the equilibrium[1] and compatibility conditions in the internal domain of each subsystem are satisfied *a priori*. The proposed gluing algorithm employs an iterative process, starting with an initial guess of some of the interface quantities. These interface quantities are then updated using a prescribed iteration process to satisfy the equilibrium and/or compatibility conditions at the interface.

In general, if a proper set of interface force variables is defined such that that the equilibrium conditions are satisfied, then only the compatibility conditions need to be considered during the iteration process. In this case, the interface force variables can be considered as functions of the interface kinematic quantities, and these interface force variables can be updated using the kinematic information and compatibility conditions. Similarly, if a proper set of the interface kinematic variables is defined such that the compatibility conditions are satisfied, then only the equilibrium conditions need to be considered during the iteration process. In this latter case, the interface kinematic variables are functions of the force quantities at the interface, and they can be updated by satisfying the equilibrium conditions. Different gluing algorithms ensue, depending on which group of interface quantities is considered as the defined input.

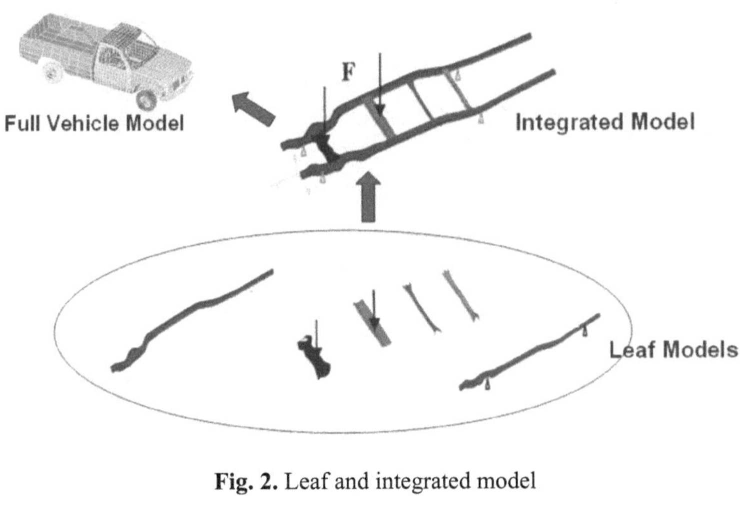

Fig. 2. Leaf and integrated model

[1] Here the equilibrium will also consider the inertial forces and dynamic loads in the case of a dynamic problem.

2.3 Gluing Strategies

Figure 3 illustrates three typical coupling strategies, which are candidates for a gluing algorithm. Here, **X** represents the vector of kinematic quantities, and **T** the vector of force quantities at the interfaces of the two models to be glued together. Figure 3a illustrates the T-T coupling strategy, for a two-subsystem case. In this strategy, kinematic quantities at the interfaces of both subsystem models, i.e., $\mathbf{X}_n^I$ and $\mathbf{X}_n^{II}$, are used as inputs to the coordinator. The interface force vectors of the two models, $\mathbf{T}_n^I$ and $\mathbf{T}_n^{II}$, are coordinator outputs, which will be applied to the subsystem models for next time step calculations. Figure 3b illustrates the X-X coupling strategy. In this method, interface force vectors $\mathbf{T}_n^I$ and $\mathbf{T}_n^{II}$ of the two models are used as inputs to the coordinator. The kinematic quantity vectors $\mathbf{X}_n^I$ and $\mathbf{X}_n^{II}$ are the coordinator outputs. The MEPI algorithm developed in [6, 18] has the form of an X-X method.

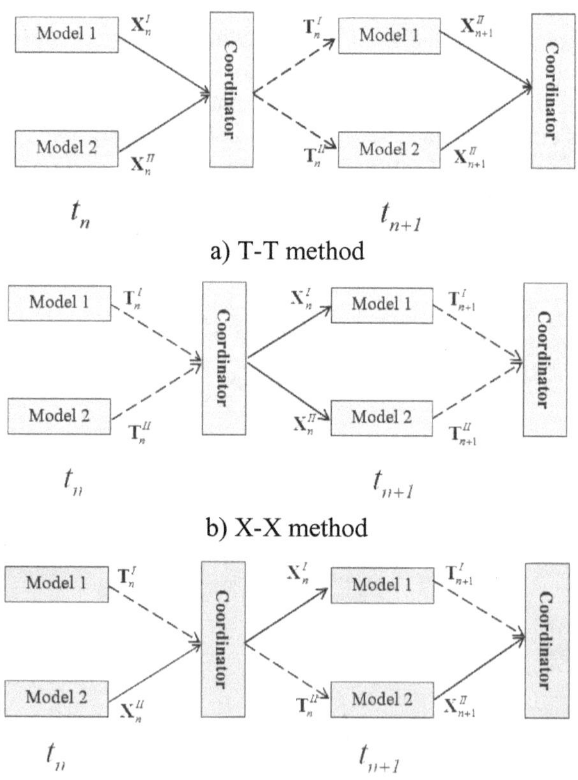

Fig. 3. Coupling strategies (**T**: force quantity vector, **X**: kinematic quantity vector, t_n: time at the nth step.)

Figure 3c illustrates a mixed coupling strategy, the X-T method, in which the interface force vector $\mathbf{T}_n^I$ of Model 1 and kinematic quantity vector $\mathbf{X}_n^{II}$ of Model 2 are used as inputs to the coordinator, while the kinematic quantity vector $\mathbf{X}_n^I$ of Model 1 and the interface force vector $\mathbf{T}_n^{II}$ of Model 2 become outputs from the coordinator. This strategy was adopted in [19] to simulate the behavior of an army tank using a distributed computing facility. This strategy is adopted by SIMULINK, and describes the so-called "across and through" variables method adopted in 20-Sim, which is employed in the control block strategy of [15].

Clearly the defined structure of the coordinator plays an important role in the problem and must be different for different coupling strategies. In the following, we will focus only on the T-T method. The major advantage of using the T-T strategy in a general problem of mechanical system simulation is that the forces can be easily applied to the subsystems when solving the subsystems equations, compared with the need to prescribe the kinematic conditions at the interfaces. The T-T strategy is more suitable to the standard setup of simulation codes that are employed in the subsystems analyses, and thus it improves the efficiency of the integration process and enhances the independency of the subsystems models.

2.4 Time Stepping Methods

Various time stepping methods to advance the subsystem solutions can be considered depending on whether or not there is a leading subsystem and how the time steps are arranged for the information exchange between the different models. Figure 4 illustrates three typical iteration methods, namely, parallel, leading and walking. For leading and walking, the compatibility of the whole system is not explicitly required as the simulation moves forward. Thus, they actually represent sequential methods. In the parallel method, components exchange information within the time step, and the coupled subsystems are compatible at the end of each time step. In this paper, we consider only the parallel method.

3 The T-T Method

3.1 Fundamental Concept

Assume that **F** is a properly defined interface force vector; that is, **F** contains the necessary and sufficient set of variables that can represent the force space at the interfaces considered and **F** is self-balanced, i.e., the equilibrium conditions at the interfaces are automatically satisfied if **F** is employed. Let **e** be an error measure vector that represents the violation of the compatibility conditions at the interfaces, where **e**=0 indicates that the compatibility conditions are fully satisfied. In the general case, **e** can be considered as a function of **F**, namely

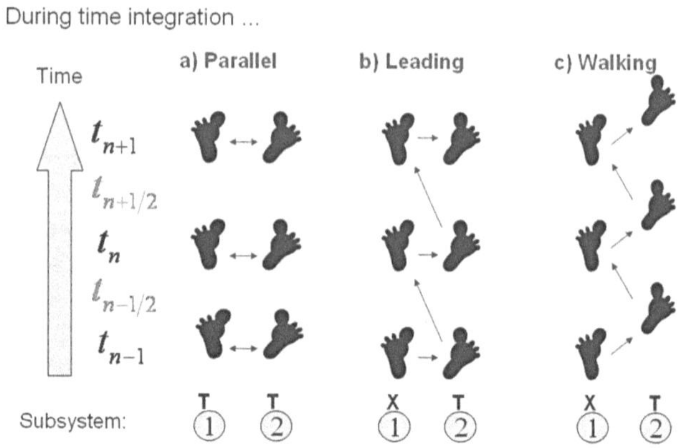

Fig. 4. Time stepping methods

$$\mathbf{e} = \mathbf{e}(\mathbf{F}) \tag{1}$$

Since $\mathbf{F}$ is defined in such way that the equilibrium conditions can be automatically satisfied, the objective of the gluing algorithm is to bring $\mathbf{e}$ to zero, that is, find $\mathbf{F}$ such that

$$\mathbf{e} = \mathbf{0} \tag{2}$$

Equation (2) defines a set of (linear or nonlinear) equations, which can be solved by a properly chosen algorithm of (linear or nonlinear) equation solvers.

Assuming an initial guess $\mathbf{F} = \mathbf{F}^{(i)}, (i = 0)$, we have $\mathbf{e}^{(i)} = \mathbf{e}\left(\mathbf{F}^{(i)}\right)$, then in the general case, a gluing algorithm (T-T method) is proposed as

$$\mathbf{F}^{(i+1)} = \mathbf{F}^{(i)} + \mathbf{\Lambda}\left(-\mathbf{e}^{(i)}\right) \tag{3}$$

where $\mathbf{\Lambda}$ is called the *gluing matrix* or *lambda matrix*, which will be constant if Equation (2) is linear, or a function of $\mathbf{F}$ if Equation (2) is nonlinear. The gluing matrix can be obtained, for example, by using a standard Newton-Raphson method, which engenders

$$\mathbf{\Lambda} = \left(\frac{\partial \mathbf{e}}{\partial \mathbf{F}}\right)^{-1}_{\mathbf{F}=\mathbf{F}^{(i)}} \tag{4}$$

Equation (3) simply implies that the interface forces can be updated (to satisfy the compatibility conditions) using only the kinematic information at the interface, provided that the gluing matrix is obtained. Therefore, the key issue becomes how to obtain the lambda matrix in a systematic and efficient way based on the interface information.

3.2 Gluing Matrix Calculation

In general, the equations of motion of each subsystem model can be written as

$$\mathbf{L}_i^i \begin{Bmatrix} \mathbf{q}_o^i \\ \mathbf{q}_c^i \end{Bmatrix} = \begin{Bmatrix} \mathbf{f}_o^i \\ \mathbf{f}_c^i \end{Bmatrix}, (i = 1, 2, \cdots, n) \tag{5}$$

where i is the subsystem number, and $\mathbf{L}^i$ represents a generalized operator, which results in a set of linear or non-linear, static or dynamic, constrained or unconstrained equations, depending on the physics of the subsystem considered. For example, Equation (5) can be a set of equations associated with a finite element model or a multibody dynamics model. In the latter case, $\mathbf{L}^i$ includes internal (joint) constraints. Internal kinematic information of the *i-th* subsystem is denoted by, $\mathbf{q}^i_o$; $\mathbf{f}^i_o$ denotes internal forces associated with $\mathbf{q}^i_o$; $\mathbf{q}^i_c$ denotes interface kinematic information; $\mathbf{f}^i_o$ denotes action-reaction forces at the interface; n is the total number of the subsystems to be coupled. The problem here is to couple the subsystems equations in Equation (5) so that the equilibrium and compatibility conditions at the subsystems interfaces are satisfied while individual equations in Equation (5) are solved independently.

In the general case, the subsystem interface force vector $\mathbf{f}^i_c$ in Equation (5) can be represented by a subset of the variables in $\mathbf{F}$; therefore $\mathbf{f}^i_c$ can be written as

$$\mathbf{f}_c^i = \mathbf{f}_c^i(\mathbf{F}), \quad (i = 1, 2, \cdots, n) \tag{6}$$

We define a matrix $\mathbf{C}_i$, for the *i-th* subsystem as

$$\mathbf{C}_i = \frac{\partial \mathbf{f}_c^i}{\partial \mathbf{F}}, \quad (i = 1, 2, \cdots, n) \tag{7}$$

Typically, $\mathbf{C}_i$ will be a simple, constant matrix.

Assume that $\mathbf{U}$ is an assembly of the interface kinematic variables (typically displacements) of all subsystems, namely,

$$\mathbf{U} = \left\{\mathbf{q}_c^i\right\}^T = \begin{Bmatrix} \mathbf{q}_c^1 \\ \\ \mathbf{q}_c^n \end{Bmatrix} \tag{8}$$

In the general case, the error measure $\mathbf{e}$ can be written as

$$\mathbf{e} = \mathbf{e}(\mathbf{U}) \tag{9}$$

We now define a matrix, $\mathbf{B}_i$, for the *i-th* subsystem as

$$\mathbf{B}_i = \frac{\partial \mathbf{e}}{\partial \mathbf{q}_c^i}, (i = 1, 2, \cdots, n) \tag{10}$$

Using Equations (7) and (10), Equation (4) can be rewritten as

$$\mathbf{\Lambda}^{-1} = \sum_{i=1}^{n} \mathbf{B}_i \mathbf{G}^i \mathbf{C}_i \tag{11}$$

and

$$\mathbf{G}^i = \frac{\partial \mathbf{q}_c^i}{\partial \mathbf{f}_c^i} \tag{12}$$

$\mathbf{G}^i$ is called interface flexibility matrix of subsystem i, which is related only to the subsystem i and can be calculated by solving Equation (5) independently for each subsystem, where $i=1,2,...,n$. It is important to note that $\mathbf{\Lambda}$ is essentially the inverse matrix of an assembly of the subsystem interface flexibility matrices defined in Equation (12). Therefore, $\mathbf{C}^i$ and $\mathbf{B}^i$ are called the *assembly matrices* of subsystem i, and we will discuss them further below. Note that $\mathbf{G}^i$ in Equation (12) can be approximated as

$$\mathbf{G}^i \cong \left[\frac{\Delta q_k}{\Delta f_j} \right], \text{ (for small } \Delta f_j) \tag{13}$$

where q_k is the *k-th* component of $\mathbf{q}^i_c$ and f_i is the *j-th* component of $\mathbf{f}^i_c$, Δq_k is the change of $q_k$with respect to an incremental interface force Δf_i. For a linear system, Equation (13) is exact. In other words, for a linear system, $\mathbf{G}^i$ is independent of the external force, and $\Delta q_k/\Delta f_i$ is independent of the amplitude of the Δf_i used, therefore we can assume

$$\Delta f_j = 1 \tag{14}$$

and,

$$\Delta q_k = q_k^1 - q_k^0 \tag{15}$$

where q^0_k is calculated by solving Equation (5) without applying any interface force, and q^1_k is calculated by applying a unit interface force at the *j-th* interface degree of freedom. Note that for each Δf_j we obtain a vector $\Delta \mathbf{q}_j = \{\Delta q_k\}^T$.

In summary, the procedure to calculate $\mathbf{G}^i$ is as follows:

1. Calculate initial $\mathbf{q}^0$ without applying any interface force.
2. Apply a unit force to a degree of freedom, j, in the interface.
3. Obtain $\Delta \mathbf{q}_j$ by solving the subsystem's Equation (5) (using its own independent solver).
4. Repeat steps 2 and 3 for all interface degrees of freedom to obtain $\mathbf{G}^i$=[$\Delta \mathbf{q}_1$, $\Delta \mathbf{q}_2$,..., $\Delta \mathbf{q}_N$], where N is the total number of interface degrees of freedom.

Algorithm 1: Calculating interface matrix

Usually, the number of interface degrees of freedom is much smaller than the number of total degrees of freedom of the subsystem model. Therefore, $\mathbf{\Lambda}$ can be easily calculated when $\mathbf{G}^i$, $(i=1,2,...,n)$ are obtained. The proposed approach treats each subsystem as a black box without accessing its internal information. Subsystem interface matrices, $\mathbf{G}^i$, can be then calculated by calling the independent solvers associated with the subsystem models, and the subsystems can be glued together using only the interface information. Note that for a linear system, there is no need for iteration when Equation (3) is used. Thus, the gluing process converges in one iteration. Also, for linear problems, if the time step size is constant, the gluing matrix can be pre- calculated and stored because no update of the gluing matrix is needed. However, for non-linear problems or problems with variable time step size, update of the gluing matrix is usually necessary. Instead of calculating the gluing matrix in each time step, more efficient quasi-Newton methods, such as the DFP or BFGS method [21], can be employed to update the gluing matrix. For example, if the DFP method is employed, the gluing matrix can be updated using

$$\mathbf{\Lambda}_{i+1} = \mathbf{\Lambda}_i + \frac{\Delta\mathbf{F}_i \otimes \Delta\mathbf{F}_i}{\Delta\mathbf{F}_i \cdot \Delta\mathbf{e}_i} - \frac{(\mathbf{\Lambda}_i \cdot \Delta\mathbf{e}_i) \otimes (\mathbf{\Lambda}_i \cdot \Delta\mathbf{e}_i)}{\Delta\mathbf{e}_i \cdot \mathbf{\Lambda}_i \cdot \Delta\mathbf{e}_i} \tag{16}$$

where i denotes the time step. Furthermore, note that multi-step methods can also be used to improve the performance of the numerical integration.

4 Assembly Process

4.1 Type of Connections

In order to apply the gluing algorithm to a general mechanical system simulation, various types of interfaces must be considered. These connections can be classified, from the simulation viewpoint, into four categories as shown in Table 1.

Only rigid connections are considered in this paper. The gluing algorithm, however, can be extended to treat the other connections. Another important matter at the interface is whether the connected subsystems have matching nodes or not. This usually depends on the meshing or discretization processes adopted for the different subsystems. A general formulation will be provided in the following for both matching and non-matching interfaces.

Table 1: Types of Connections

Type	Examples
Rigid Connections	Fixed and Joints (including Spherical, Universal, Revolute, Cylindrical, Translational joints)
Compliant Connections	Bushings, Mounts, Springs/Dampers
Contact and Impact	Point to Point, Point to Surface, Surface to Surface
Actuators	Applied Force/Torque, Motion Driver

4.2 Assembly Matrices

To show more specific structure of the assembly matrices, $\mathbf{C}_i$ and $\mathbf{B}_i$, defined in Equations (7) and (10), we discuss some typical cases. First, if components in the independent force vector $\mathbf{F}$ can be directly selected from $\mathbf{f}^i_c,(i=1,2,...,n)$, then

$$\mathbf{F}=\left\{\left(\mathbf{f}_A^1\right)^T,\left(\mathbf{f}_A^2\right)^T,\cdots,\left(\mathbf{f}_A^n\right)^T\right\}^T \tag{17}$$

where $\mathbf{f}^i_A$ is the subset of independent force variables selected from $\mathbf{f}^i_c$, and thus

$$\mathbf{f}_c^i=\begin{Bmatrix}\mathbf{f}_A^i\\ \mathbf{f}_R^i\end{Bmatrix} \tag{18}$$

where $\mathbf{f}^i_R$ contains the remaining force components in $\mathbf{f}^i_c$. For convenience, we call $\mathbf{f}^i_A$ "active" or "action" forces and $\mathbf{f}^i_R$ "passive" or "reaction" forces. Passive forces are in general determined by the active forces of the adjacent subsystems. For example, for a system with the simple connections shown in Figure 5, we have

$$\mathbf{f}_R^i=-\mathbf{f}_A^{i+1},(i=1,2,\cdots,n-1) \tag{19}$$

Here, in Equation (19), it is assumed that the interface between the subsystems i and $i+1$ has matching discretization, namely, subsystems i and $i+1$ have the same nodes at the interface in the case a finite element, boundary element, or finite difference method is used.

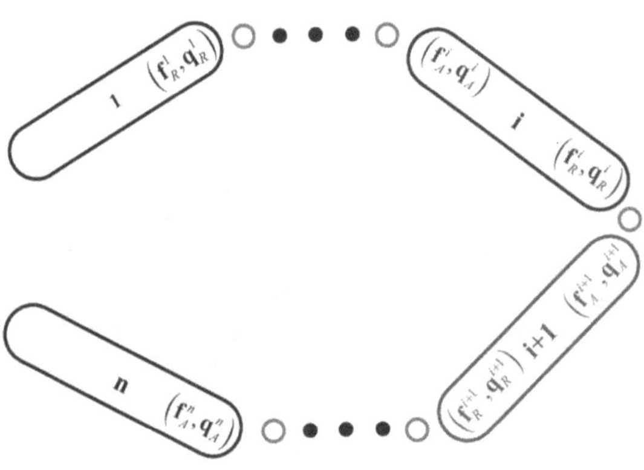

Fig. 5. Gluing simple-connected components

Using Equations (17)-(19), $\mathbf{C}_i$ in Equation (7) becomes a matrix with only terms of 1, -1 or 0. To be more specific, for the subsystem i, we have

$$\mathbf{C}_i=\begin{bmatrix}\mathbf{0} & \cdots & \overset{i}{\mathbf{I}} & \overset{i+1}{\mathbf{0}} & \cdots & \mathbf{0}\\ \mathbf{0} & \cdots & \mathbf{0} & -\mathbf{I} & \cdots & \mathbf{0}\end{bmatrix} \tag{20}$$

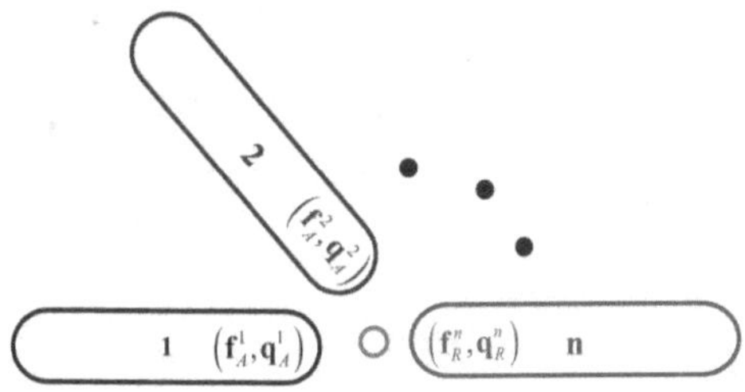

Fig. 6. Gluing multiple-connected components

For the same problem, assume the error measure is

$$\mathbf{e} = \begin{Bmatrix} \vdots \\ \mathbf{q}_A^i - \mathbf{q}_R^{i-1} \\ \mathbf{q}_A^{i+1} - \mathbf{q}_R^i \\ \vdots \end{Bmatrix} \Big\} i - th\ subsystem \tag{21}$$

Then from Equation (10) we have

$$\mathbf{B}_i = \mathbf{C}_i^T = \begin{bmatrix} \mathbf{0} & \mathbf{0} \\ \vdots & \vdots \\ \mathbf{I} & \mathbf{0} \\ \mathbf{0} & -\mathbf{I} \\ \vdots & \vdots \\ \mathbf{0} & \mathbf{0} \end{bmatrix} \Big\} i - th\ subsystem \tag{22}$$

Another typical case is when the system has multiple components being connected at the same interface. For example, for the system shown in Figure 6, an error measure can be taken as

$$\mathbf{e} = \begin{Bmatrix} \mathbf{q}_A^1 - \mathbf{q}_R^n \\ \mathbf{q}_A^2 - \mathbf{q}_R^n \\ \vdots \\ \mathbf{q}_A^{n-1} - \mathbf{q}_R^n \end{Bmatrix} \tag{23}$$

and we can select **F** as

$$\mathbf{F} = \left\{ (\mathbf{f}_A^1)^T, (\mathbf{f}_A^2)^T, \cdots, (\mathbf{f}_A^{n-1})^T \right\}^T \tag{24}$$

Then from Equation (7) we have

$$\mathbf{C}_i = \begin{cases} \begin{bmatrix} \cdots & \underset{i-th}{\mathbf{I}} & \cdots \end{bmatrix} & (for\ i = 1, 2, \cdots, n-1) \\ \begin{bmatrix} -\mathbf{I} & -\mathbf{I} & \cdots & -\mathbf{I} \end{bmatrix} & (for\ i = n) \end{cases} \tag{25}$$

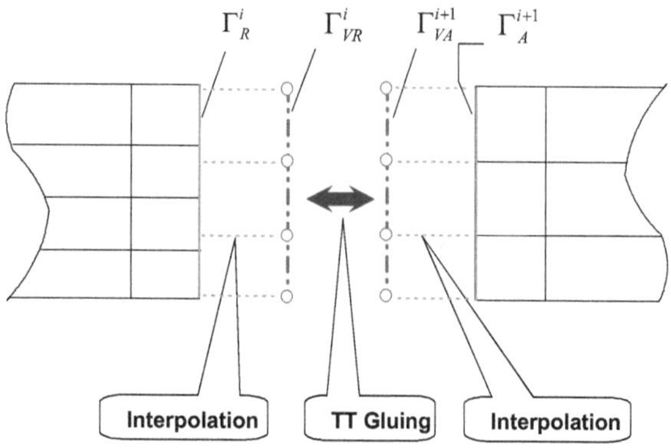

Fig. 7. Virtual interface for non-matching interfaces

From Equation (10), we have

$$\mathbf{B}_i = \mathbf{C}_i^T, (i = 1, 2, \cdots, n) \tag{26}$$

Note that subsystem assembly matrices $\mathbf{C}_i$ and $\mathbf{B}_i$ defined in Equations (7) and (10) are general; they can have more complex form than that shown in Equations (20) and (22) and Equations (25) and (26), depending on how the independent force vector **F** and the compatibility conditions are defined. This generality provides a means to deal with various kinds of connections for the subsystems, for example, connections without a matching interface on the adjacent subsystems, and various kinds of mechanical joints. In the case of a non-matching interface, a virtual interface between the non-matching interfaces can be defined; the gluing algorithm is then applied on the virtual interface. Figure 7 shows a schematic for the concept of the virtual interface. As shown in Figure 7, Γ^i_{VR} denotes a virtual interface connected to the interface Γ^i_R of subsystem i, and Γ^{i+1}_{VA} denotes that connected to the interface Γ^{i+1}_A of subsystem $i+1$. Between Γ^i_{VR} and Γ^i_R, we have

$$\mathbf{f}^i_R = \mathbf{S}^i_R \mathbf{p}^i_{VR} \text{ and } \mathbf{v}^i_{VR} = \left(\mathbf{S}^i_R\right)^T \mathbf{q}^i_R \tag{27}$$

where $\mathbf{p}^i_{VR}$ and $\mathbf{v}^i_{VR}$ denote force and kinematic quantity vectors at the virtual interface Γ^i_{VR}, and $\mathbf{S}^i_R$ is the interpolation matrix between the Γ^i_{VR} and Γ^i_R. Similarly, between Γ^{i+1}_{VA} and Γ^i_A, we have

$$\mathbf{f}^{i+1}_A = \mathbf{S}^{i+1}_A \mathbf{p}^{i+1}_{VA} \text{ and } \mathbf{v}^{i+1}_{VA} = \left(\mathbf{S}^{i+1}_A\right)^T \mathbf{q}^{i+1}_A \tag{28}$$

where, $\mathbf{p}^{i+1}_{VA}$ and $\mathbf{v}^{i+1}_{VA}$ denote force and kinematic quantity vectors at the virtual interface Γ^{i+1}_{VA}, and $\mathbf{S}^{i+1}_A$ is the interpolation matrix between the Γ^{i+1}_A and Γ^{i+1}_A. Since Γ^i_{VR} and Γ^{i+1}_A are defined as matching interfaces, the previous formulations apply to the force and kinematic quantity vectors at Γ^{i}_{VR} and Γ^{i+1}_{A}. Using Equations (27) and (28), Equations (20) and (22) now can be generalized as

$$\mathbf{C}_i = \mathbf{B}_i^T = \begin{bmatrix} \mathbf{0} & \cdots & \overbrace{\mathbf{S}_A^i}^{i} & \overbrace{\mathbf{0}}^{i+1} & \cdots & \mathbf{0} \\ \mathbf{0} & \cdots & \mathbf{0} & -\mathbf{S}_R^i & \cdots & \mathbf{0} \end{bmatrix} \tag{29}$$

In the same way, using the virtual interface, Equations (25) and (26) can be generalized as

$$\mathbf{C}_i = \mathbf{B}_i^T = \begin{cases} \begin{bmatrix} \cdots & \underbrace{\mathbf{S}_A^i}_{i-th} & \cdots \end{bmatrix} & (for\ i = 1, 2, \cdots, n-1) \\ \begin{bmatrix} -\mathbf{S}_R^1 & -\mathbf{S}_R^2 & \cdots & -\mathbf{S}_R^n \end{bmatrix} & (for\ i = n) \end{cases} \tag{30}$$

5 Implementation

To build an executable platform based on the gluing algorithm developed, we need to deal with two other critical issues, namely, model description and logical distributed architecture design.

5.1 Model Description Using XML

Figure 8 shows the outline of a standardized model description developed using XML for describing subsystem models in the glue integration. The root element has three child elements, which correspond to the three information classes for the glue integration in the general platform, namely, *General Information*, *Assembly Information*, and *Simulation Information*. Here, General Information provides: 1) model identification, including unique ID, identifications in the network, etc., for data management; 2) search information; 3) contact information of the developer and maintainer. Example items in the *General Information* include: model ID, model name, category, model image, developer name and location, dates, contact information, application range of the model, simulation time, fidelity.

Assembly Information enables: 1) instantiating integrated models and defining its subsystems; 2) matching interfaces of the subsystem models and defining the connection context; 3) providing information for gluing algorithms to couple the simulation models. Basically, two categories of information are included in the *Assembly Information*: interface definition information and information on how the model is assembled. Example items include: interface name, belonging, interface type and context (nodes, joints, contacts, etc.), geometry and other detailed modeling information for the interface (if it is a leaf model), links to the subsystem models (if it is an integrated model), and matching information of the interfaces of component models.

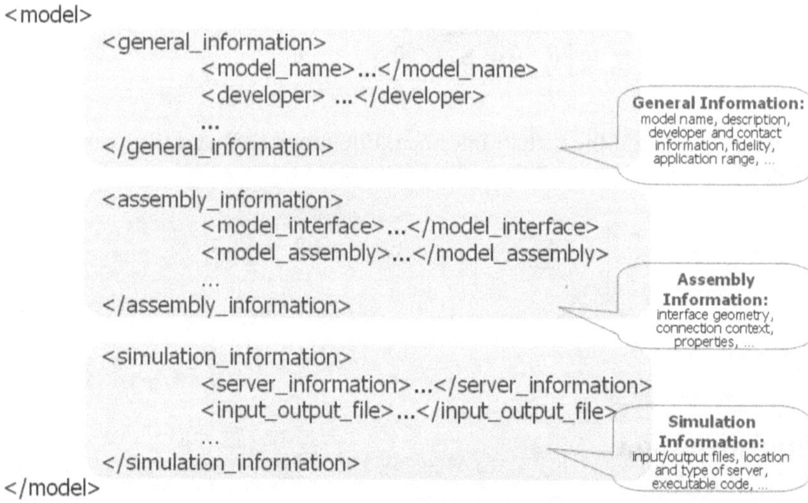

Fig. 8. Structure of the XML description

Simulation Information is for the purpose of defining code execution information for the simulation. Example items in the *Simulation Information* include: server type, location, input/output file names and locations, simulation code and the execution pass to the simulation code, simulation parameters (e.g., time step, error tolerances, and simulation time).

5.2 Logical Distributed Architecture

Figure 9 shows the proposed logical distributed architecture. Each ellipse represents a model simulation server, which is either a wrapped simulation code that can be accessed through the network or an implementation of the gluing algorithm. A user accesses the system through a web browser and sends the XML description of a model to the server to conduct a simulation based upon the description. The web server creates an integrated model object and sends the XML file to the model object. The integrated model object parses the XML file and creates subsystem model objects according to the description in the `<model_assembly>` element. Then, the integrated model object sends the XML descriptions of subsystem models to the corresponding objects. If a component model is also an integrated model, the above procedure is followed for the integrated model. In the model database, each leaf model can have several different designs. Therefore, the proposed simulation platform can be used to quickly assess different design options and their influence on the integrated system.

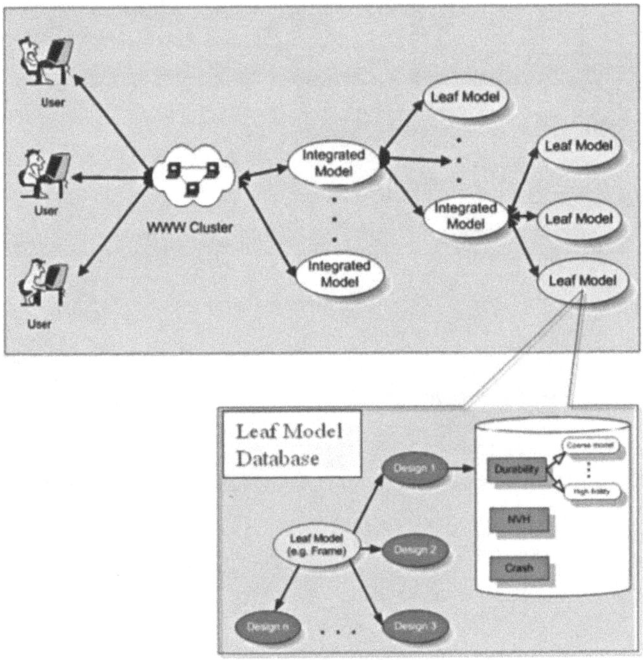

Fig. 9. Logical distribution architecture and model database

The leaf model object exposes the standard interface methods to the outside world; on the other hand it also serves as the wrapper of a simulation code. In our demonstration system, a legacy finite element analysis code was wrapped to serve as the leaf model simulation server. The legacy code was written in FORTRAN and was converted into a DLL (dynamic link library). As shown in Figure 10, the leaf model object is a .Net Remoting object that wraps around the DLL and implements the standard interface methods.

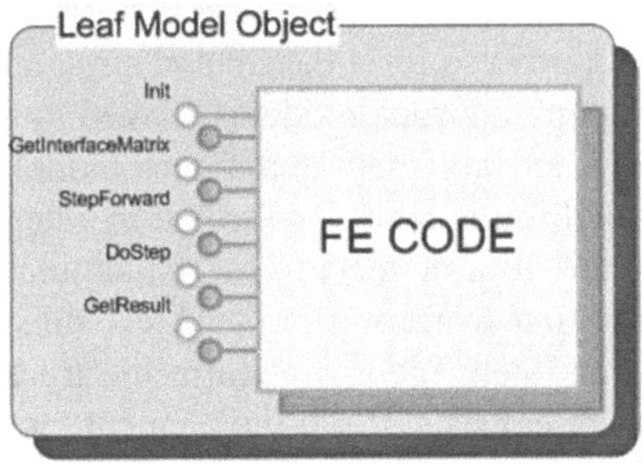

Fig. 10. Wrapping of a legacy FE code

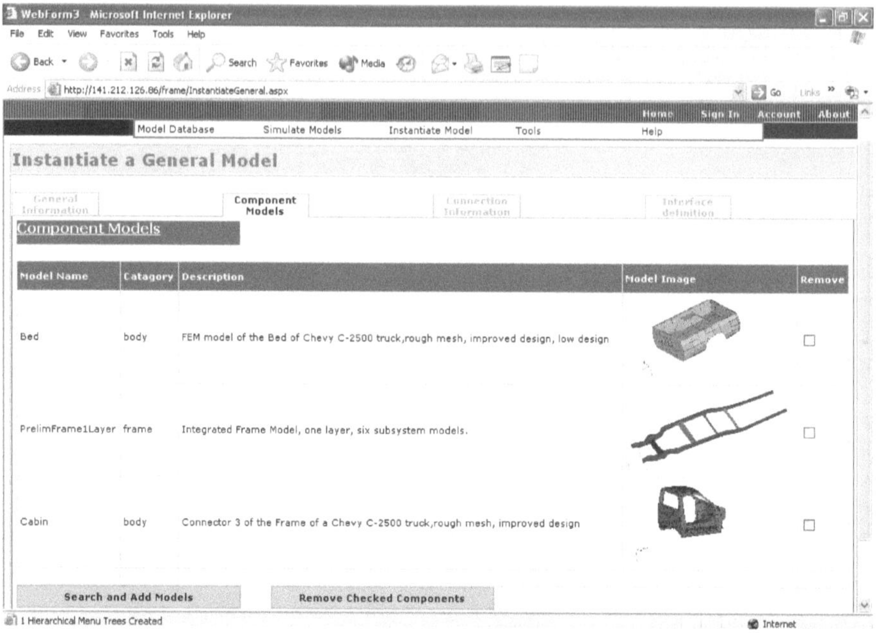

Fig. 11. Snapshot of the demonstration platform

A Web-based user interface has been developed, which is shown in Figure 11.

6 Examples

6.1 Example 1: Gluing a Finite Element Truck Model

The first example is a distributed simulation problem of the truck model shown in Figure 12. The simulation problem has two gluing layers, which simulates the generality of our coupling process for a distributed structural system. In the first layer (Figure 12a), the integrated truck model is formed by assembling a leaf cabin model, a leaf bed model and an integrated frame model. In the second layer (Figure 12b), the integrated frame model is formed by integrating six sub- components: left rail, right rail and four connectors. The simulation system couples the second layer components first to form a higher-level subsystem model, and then couples the first layer subsystem models to form the truck system. As shown in Figure 12, each component consists of a finite element input file and a standalone finite element code (solver).

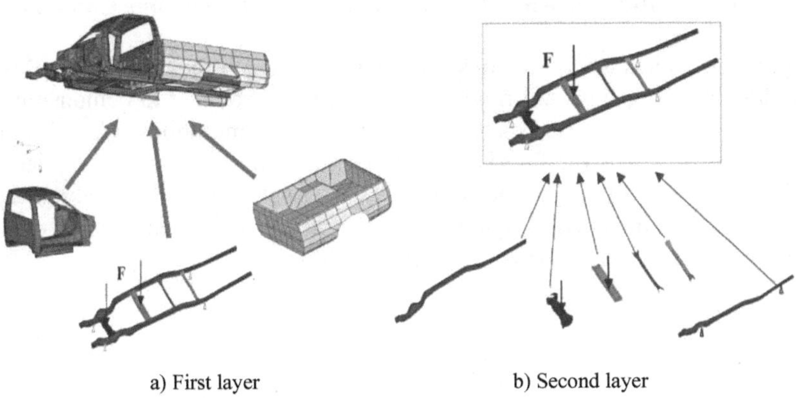

Fig. 12. A two-layer distributed simulation of a truck model

During the gluing process, the models at the two levels communicate by exchanging their interface information at their own level, and the gluing coordinators update the interface variables using the gluing algorithm. Note that each layer has its own coordinator and its own gluing process. Since the problem is linear, there is no need for iteration when updating the interface forces using Equation (3).
Figure 13 shows the results obtained using the gluing system developed compared with results obtained using an "all-at-once" finite element analysis. Here, dynamic loads of f=2000 sin(100πt) N are applied at the middle point of both the connector 1 and 2, along the global z (vertical) direction, and the chassis frame is supported at the four points shown in Figure 12. Figure 13 shows the acceleration at a selected node (node 1193) along the vertical direction. It is clear that the gluing process induced no additional error.

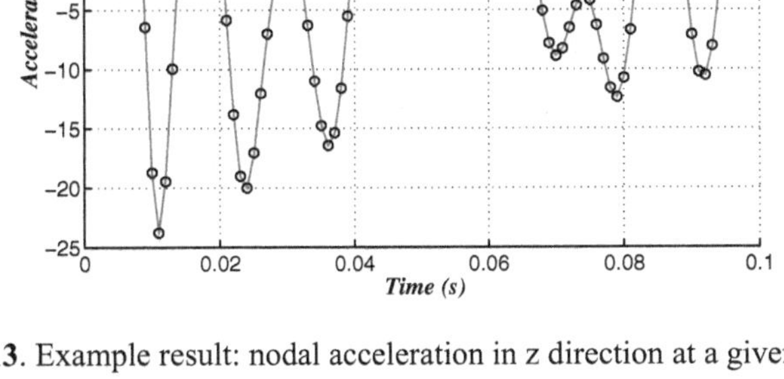

Fig. 13. Example result: nodal acceleration in z direction at a given node

6.2 Example 2: Gluing a 4-Bar Link Mechanism with a Flexible Component

Figure 14 depicts a four-bar link mechanism in which the third bar is flexible while the remaining links are rigid. This example is provided to demonstrate the applicability of the proposed methodology for flexible multibody dynamics problems. Figure 14 shows the four-bar link is separated into two subsystems at joint C, which connects Bodies 2 and 3. The first subsystem includes Body 1 and 2, both of which are modeled as rigid bodies. The second subsystem comprises Body 3, which is modeled as a flexible Euler-Bernoulli beam.

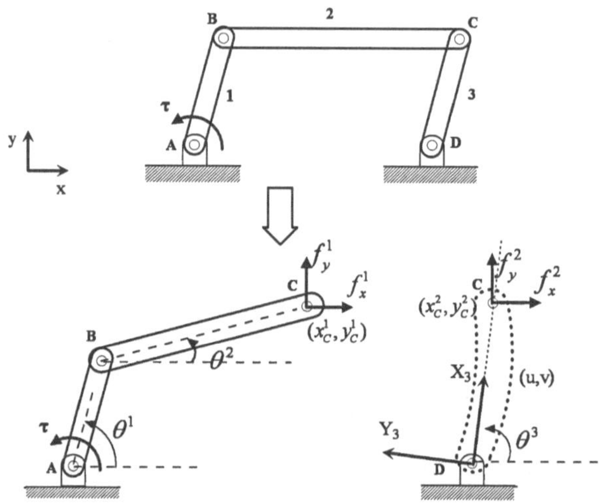

Fig. 14. Gluing simulation of a four-bar link flexible mechanism

Using the floating frame of reference coordinate system shown in Figure 14, the deformation shape of the beam is represented by

$$\begin{Bmatrix} u \\ v \end{Bmatrix} = \mathbf{Sq} = \begin{bmatrix} \sin\left(\frac{\pi x}{2l}\right) & 0 & 0 \\ 0 & \sin\left(\frac{\pi x}{2l}\right) & \sin\left(\frac{2\pi x}{l}\right) \end{bmatrix} \begin{Bmatrix} q^1 \\ q^2 \\ q^3 \end{Bmatrix} \tag{31}$$

The parameters used for each body are:

$$m^1 = 10kg,\, m^2 = 20kg,\, m^3 = 5kg$$

$$l^1 = 1m,\, l^2 = 2m,\, l^3 = 1m$$

where, m^i, (i=1,2,…,3) are the masses of bar 1, 2 and 3 and l^i, (i=1,2,…,3) are the link lengths of the bars. At joint A, a driving torque is applied, defined as

$$\tau = \begin{cases} -1000 * t\ N \cdot m, & t < 0.1s \\ -100\ N \cdot m, & t \geq 0.1s \end{cases}$$

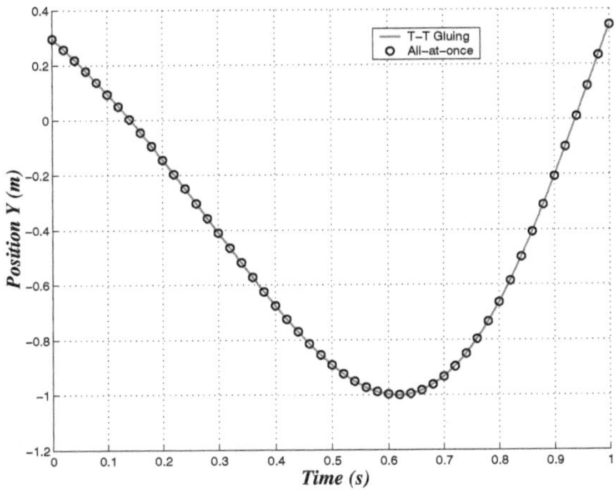

Fig. 15. Displacement along global Y direction of the interface joint C

For the flexible beam, the cross-section is assumed to be circular and uniform, with a radius $r = 0.015m$. The Young's modulus and mass density of the beam are $E = 210\ GPa$ and $\rho = 7.0\ 10^3\ Kg\ m^{-3}$.

The mechanism is driven from an initial position $\theta^1 = \theta^2 = \theta^3 = 0.3\ rad$ with initial speeds $\omega^1 = \omega^3 = -2.0\ rad\ s^{-1}$ and all other initial conditions are zero. The equations of motion of both subsystems are first reduced to ODEs in terms of (θ^1, θ^2) and $(\theta^3, q^1, q^2, q^3)$, respectively. Then both subsystems are solved using the *ode45* solver in Matlab. The error tolerance for the gluing is $\|\mathbf{e}\| \le 10^{-10}$ and the time step size Δt is selected as 10^{-3}. The compatibility condition at joint C is used to update the interface forces with the use of Equation (3). ADAMS/Flex was employed as the all-at-once system benchmark, in which the flexible bar is modeled using 10 beam elements, which are finally reduced to three modal coordinates as used in the gluing simulation. A damping coefficient of 0.1 is applied to all three modes in both simulations. Figs. 15 and 16 compare the results obtained from the T-T method and the ADAMS simulation. Figure 15 compares the displacement along the global Y direction at the cut joint (joint C); Figure 16 compares the velocity at the same joint. Here all the measurements are in the global coordinate system. Good agreement is obtained between the two simulations except for the small oscillation in the velocity as seen in Figure 16, which arises when the mechanism passes through the "singular points". These oscillations are observed in the results from both the T-T method and the ADAMS simulation with a slight difference. Note that the singular point is at the time when all bars lie on the same line when ignoring the deformation.

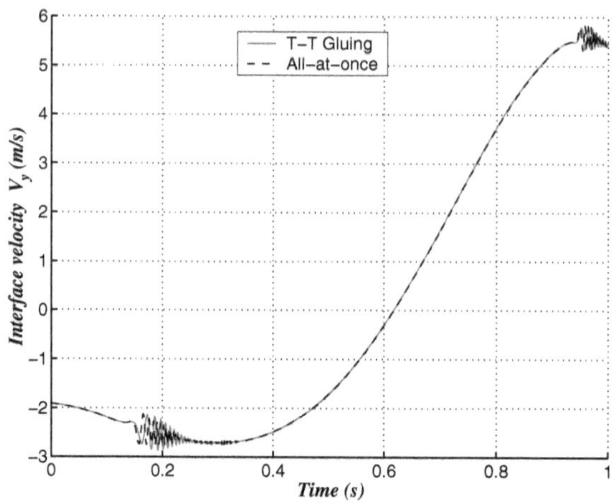

Fig. 16. Velocity along global Y direction of the interface joint C

6.3 Example 3: Gluing with Non-matching Nodes at the Interface

A two dimensional elastic problem is considered to demonstrate the applicability of the gluing algorithm to the non-matching discretization problem at the interface. Figure 17a illustrates a gluing problem with two non-matching interfaces, $\Gamma^1{}_R$ and $\Gamma^2{}_A$, where $\Gamma^1{}_R$ has 5 finite element nodes, while $\Gamma^2{}_A$ has 4 finite element nodes. The problem is to glue the two substructures together for a coupled static analysis. As a reference, an all-at-once system model with the uniform mesh is also used for the comparison purpose, as shown in Figure 17b.

In this example, virtual interfaces $\Gamma^1{}_{VR}$ and $\Gamma^2{}_{VR}$ (refer to Figure 7) are defined with the assumption of $\Gamma^2{}_{VA}=\Gamma^2{}_A$, as shown in Figure 17a. The problem thus becomes to glue the interfaces $\Gamma^1{}_{VR}$ with $\Gamma^2{}_A$, with an numerical interpolation between $\Gamma^1{}_R$ and $\Gamma^1{}_{VR}$. Let $\mathbf{f}^1{}_R$ and $\mathbf{q}^1{}_R$ define the force and displacement vectors at the original interface $\Gamma^1{}_R$, and $\mathbf{p}^1{}_{VR}$ and $\mathbf{v}^1{}_{VR}$ define the force and displacement measures at the virtual interface $\Gamma^1{}_{VR}$. Using an interpolating scheme, we have:

$$\mathbf{f}_R^1 = \mathbf{S}_R^1 \mathbf{p}_{VR}^1 \text{ and } \mathbf{v}_{VR}^1 = \left(\mathbf{S}_R^1\right)^T \mathbf{q}_R^1 \tag{32}$$

where

$$\mathbf{S}_R^1 = \begin{bmatrix} \mathbf{S} & \mathbf{0} \\ \mathbf{0} & \mathbf{S} \end{bmatrix}$$

and **S** is the interpolation matrix associated with a global axis (x or y). To calculate **S**, we first assume a coordinate interpolation **C** between the nodes on $\Gamma^1{}_R$ and the nodes on $\Gamma^1{}_{VR}$,

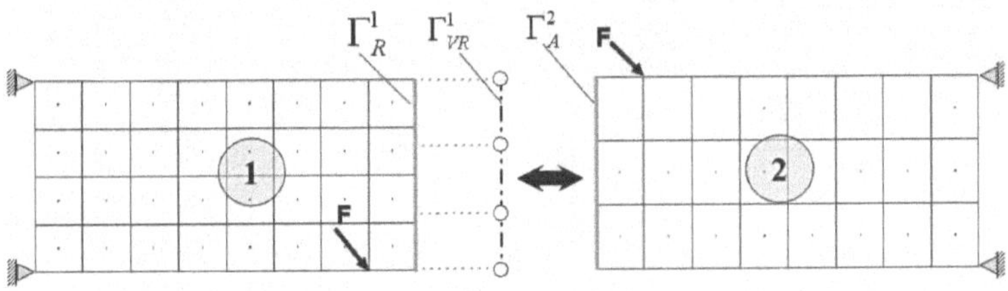

a) Gluing non-matching interfaces

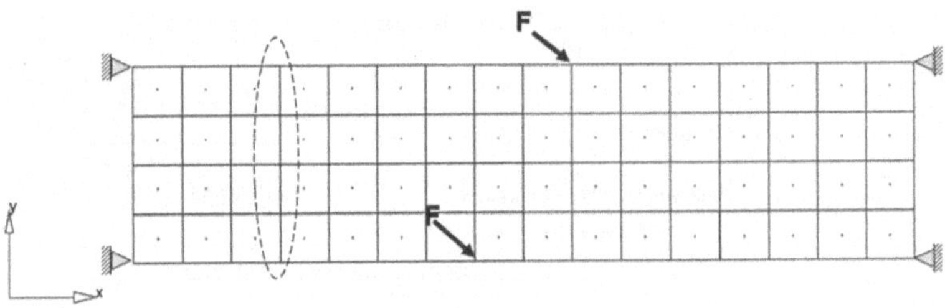

b) Reference all-at-once model

Fig. 17. Two dimensional elastic problems with non-matching meshes in the subsystems

$$\mathbf{C} = \begin{bmatrix} 1 & & & \\ \frac{1}{4} & \frac{3}{4} & & \\ & \frac{1}{2} & \frac{1}{2} & \\ & & \frac{3}{4} & \frac{1}{4} \\ & & & 1 \end{bmatrix} \tag{33}$$

Then **C** is calculated as the transposed pseudo inverse of **C**, namely,

$$\mathbf{S} = \mathbf{C}\left(\mathbf{C}^T\mathbf{C}\right)^{-1} \tag{34}$$

This procedure and the use of Equation (32) are necessary for energy conservation at the interface while enforcing the compatibility condition at the interface for the all nodes.

Figure18 shows the comparison of the nodal displacements calculated using our gluing approach and the all-at-once method. Figure 18a shows the displacements along the x-direction for the five nodes in the column circled in Figure 17b, while Figure 18b shows the relative errors at each node, which shows a maximum value of 0.2%. It is seen that the gluing method can effectively handle the non-matching interface problem.

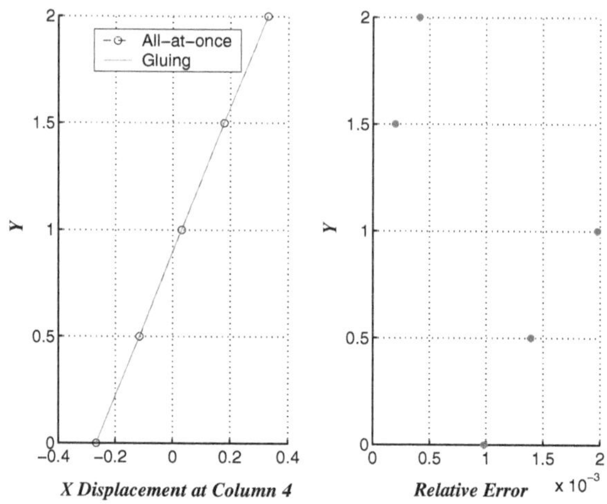

Fig. 18. Comparison of displacements at column 4

7 Conclusions

A general gluing process, the T-T method, is presented in this paper, which can be used to glue, in an effective and accurate way, distributed subsystem models for both structural dynamics and multibody dynamics problems. The formulation of the gluing algorithm is general so it can be extended to deal with a variety of different gluing problems, including linear and non-linear problems. The proposed approach relies only on the interface information exposed by the subsystem models, without requiring internal details of the model. Therefore, each subsystem model can be treated as a black box, regardless of the model and its inherent solution scheme. From the outset of the algorithm development, we considered that subsystem models do not possess knowledge of the gluing algorithm, and that the models may be built using different commercial packages, which usually do not communicate well with each other. These features make the algorithm suitable for practical use in a distributed simulation environment within a real distributed production system.

We have demonstrated that the T-T method can produce exact solutions for linear systems, including static and dynamics problems, without any iteration of the update equations, and with only one-time calculation of the gluing matrix at the beginning of the simulation. With iteration, the gluing algorithm can be used to solve nonlinear multibody dynamics problems, including rigid and flexible multibody dynamics systems. A finite element example and a multibody dynamics example, including flexible members are shown in this paper, along with treatment of interfaces comprising non-matching nodes. Future development will focus on the applicability of the gluing algorithm to a broader class of distributed system simulations.

Acknowledgments

The authors would like to acknowledge the support provided by the U.S. Army Tank-Automotive and Armaments Command (TACOM) through the Automotive Research Center at the University of Michigan under contract DAAE07-98-3-0022.

References

1. Huddi AV, Pidaparti RMV (1995) Distributed finite element structural analysis using the client-server model. Communications in Numerical Methods in Engineering 11:227-233
2. Kumar S, Adeli H (1995) Distributed finite-element analysis on network of workstations-algorithms. Journal of Structural Engineering 121(10):1448-1455
3. Farhat CH, Wilson E (1988) A parallel active column equation solver. Computers and Structures 28:289-304
4. Farhat CH, Roux FX (1994) Implicit parallel processing in structural mechanics. Computational Mechanics Advances 2:1-124
5. Modak S, Sotelino ED (2000) Iterative group implicit algorithm for parallel transient finite element analysis. Int. J. Numer. Methods Eng. 47(4):869-885
6. Tseng FC (2000) Multibody dynamics simulation in network-distributed environments. Ph.D Dissertation, University of Michigan, Ann Arbor, Michigan
7. Craig RR (2000) Coupling of substructures for dynamic analyses: An overview. In Collection of Technical Papers - AIAA/ASME/ASCE/AHS/ASC Structures, Structural Dynamics and Materials Conference, Reston, Virginia, paper AIAA-2000-1573, pp 3-14
8. Tallec, P. L., "Domain decomposition methods in computational mechanics", Computational Mechanics Advances, 1, 1994, 121-220.
9. Kim SS (2002) A subsystem synthesis method for an efficient vehicle multibody dynamics. Multibody System Dynamics 7:189–207
10. Featherstone R (1999) Divide-and-conquer articulated-body algorithm for parallel O(log(n)) calculation of rigid-body dynamics. Part 1: Basic algorithm. International Journal of Robotics Research 18(9):867-875
11. Featherstone R (1999) Divide-and-conquer articulated-body algorithm for parallel O(log(n)) calculation of rigid-body dynamics. Part 2: Trees, loops, and accuracy. International Journal of Robotics Research 18(9):876-892
12. Anderson KS, Duan S (2000) Highly parallelizable low-order dynamics simulation algorithm for multi-rigid-body systems. Journal of Guidance, Control, and Dynamics, 23(2):355-364
13. Duan S, Anderson KS (2000) Parallel implementation of a low order algorithm for dynamics of multibody systems on a distributed memory computing system. Engineering with Computers. 16(2):96-108
14. Sharf I, D'Eleuterio GMT (1992) Parallel simulation dynamics for elastic multibody chains. IEEE Transactions on Robotics and Automation 8:597-606
15. Kübler R, Schiehlen. W (2000) Modular simulation in multibody system dynamics. Multibody System Dynamics 4:107-127

16. Gu B, Asada HH (2001) Co-simulation of algebraically coupled dynamic subsystems. In Proceedings of the American Control Conference (IEEE cat n 01CH37148), Arlington, Virginia, pp 2273-2278
17. Tseng FC, Hulbert GM (2001) A gluing algorithm for network-distributed dynamics simulation. Multibody System Dynamics 6(4):377-396
18. Tseng FC, Ma ZD, Hulbert GM (2003) Efficient numerical solution of constrained multibody dynamics systems. Computer Methods in Applied. Mechanics and Engineering 192:439-472
19. Hulbert GM, Michelena N, Ma ZD, et al. (1999) A case study for network-distributed collaborative design and simulation: Extended life optimization for M1 Abrams tank road arm. Mechanics of Structures and Machines 27(4):423-451
20. Fisette P, Peterkenne JM (1998) Contribution to parallel and vector computation in multibody dynamics. Parallel Computing 24(5-6):717-728
21. Press WH, Teukolsky SA, Vetterling WT, Flannery BP (1992) Numerical recipes: the art of scientific computing, Cambridge University Press

Multiscale Methods for Multibody Systems with Impacts

Werner Schiehlen, Bin Hu and Robert Seifried

Institute B of Mechanics, University of Stuttgart,
Pfaffenwaldring 9, 70550 Stuttgart, Germany
wos@mechb.uni-stuttgart.de

1 Introduction

In the design of mechanical and structural systems, computer simulations are frequently used to check the dynamic response of the system being developed. This form of computer-aided engineering reduces strongly the need to construct and test prototypes. During the last decades modeling and simulation of multibody systems have been throughly investigated in theory and successfully applied in engineering, e.g., in automotive, aerospace and military industry as well as in biomechanics and robotics. The development of the multibody system method is reviewed by Schiehlen [38] for rigid multibody dynamics and Shabana [44] for flexible multibody dynamics. Both authors pointed out that further study should be devoted to modeling and simulation of multibody systems with impact and contact.

For multibody systems with impact, there are mainly two methods known to analyse collisions depending on the duration of contact. The elastic collision approach and the rigid body approach. Both methods will be discussed in this paper based on contributions by Hu and Schiehlen [20] and Schiehlen and Seifried [41].

The first method, see e.g. Bauchau [5], is based on finite contact duration and a finite contact force. In contrast to the rigid body approach, the velocities of the colliding bodies vary continuously. During the contact period, there are two force components active at the point of contact, i.e., the normal contact force F_N and the tangential contact force F_T. The normal contact force follows from a compliant contact model where the local deformations of the colliding bodies in the neighborhood of the contact surface are considered. Usually, the contact force is modeled as a function of the indentation depth. Typical compliant contact models are the Kelvin-Voigt viscoelastic model or the extended Hertzian contact model. In the Kelvin-Voigt viscoelastic model, the contact force is modeled by a linear spring-damper element and results in a linear function of the indentation depth while in the extended Hertzian model the contact force is a nonlinear function of the indentation depth. For modeling the tangential contact force Coulomb's friction law is usually used. It postulates that the friction force for slipping is equal to the normal contact force times the coefficient of kinetic friction and acts in a direction opposing the relative motion. For

J.A.C. Ambrósio (ed.), Advances in Computational Multibody Systems, 95–124.

sticking when the relative motion vanishes, the friction force is smaller than the normal contact force time the coefficient of static friction. Computer simulations using Coulomb's friction encounter numerical difficulties associated with discontinuity of the friction force at zero relative velocity. Consequently, various approximations to Coulomb's law have been recommended. These approximations can be viewed as continuous friction laws that replace the discontinuity at zero relative velocity by a smooth, stiff function of the relative velocity, see e.g. Bauchau [4].

For accurate simulations of the impact responses, the elastic collision approach has to be combined with the dynamics of an elastic body resulting in wave propagation. Since the impact forces as well as the propagating waves have very high frequency components, the simulation with the method of elastodynamics is very time consuming and large computing time prohibits the simulation of the long term impact responses.

The second method for multibody systems with impacts represents rigid body collisions based on a duration of contact approaching zero and the contact force approaching infinite. Thus, the analysis is divided into two intervals: before and after impact. The velocities after impact are obtained by solving a set of momentum equations with a given coefficient of restitution which may be kinematic, kinetic or energetic, defined by Newton, Poisson and Stronge, respectively, see Stronge [46]. The coefficient of restitution represents the kinetic energy loss by wave propagation, viscoelastic effects and plastic deformations during impact. This method was also proposed by Kane [23], and applied to rigid multibody systems by Wehage and Haug [50], and extended to flexible systems by Khulief and Shabana [24]. The rigid collision approach together with the method of rigid multibody systems is very efficient, see Pfeiffer and Glocker [33]. However, it is well known that the coefficient of restitution depends not only on the material parameters but also on the shapes and motion states of the colliding bodies. It is usually difficult to get suitable values of the coefficient of restitution.

For efficient simulation of impact responses with sufficient accuracy, both methods for rigid multibody systems and elastodynamics should be combined. During contact the short time impact responses must be computed using the method of elastodynamics for accuracy and after impact the long term impact dynamics has to be simulated using the method of rigid multibody systems for efficiency.

The first part of this paper deals mainly with an adaptive simulation technique of impact responses during the transition from the short time elastodynamics to the long term rigid body dynamics. For modeling of the elastodynamic phenomena the method of flexible multibody systems with the floating frame of reference formulation is used, and the impact induced elastic deformations are assumed to be small. In the simulation of the transient impact responses the contribution of elastic coordinates to the overall responses are monitored using their response bounds. When these response bounds are small enough, the corresponding elastic coordinates will be deleted. As a consequent, the degrees of freedom of the flexible system will be reduced and the efficiency of the simulation will be improved. Due to the material damping, the impact induced vibrations will decay and the system has then only the rigid body motion. Correspondingly, due to monitoring the response bounds and the adaptive simulation strategy, all elastic coordinates representing the impact induced vibrations will be computed temporarily and only the reference coordinates representing the rigid body motion are simulated in the long run. The application of this adaptive simulation approach is experimentally validated for the longitudinal impact of a rigid body colliding an elastic rod.

Adaptive simulation techniques have been extensively studied in the computations with finite element methods, see e.g. Erhart [11]. For multibody systems, Khulief [26] presented recently an adaptive computational scheme which permits a change in the basis of the modal space in order to regulate the admittance of higher frequencies and to accommodate any change in the kinematic configuration by routinely checking the ratio between the averaged kinetic energy of the low-frequency and the high-frequency motions. Such a kinetic energy index is, however, somewhat empirical and no mathematical grounds were provided in his paper. In contrast to Khulief's approach, the computational scheme in this paper concerns only whether the modal coordinates for deformations associated with elastic vibrations are small enough to be negligible and the mode shapes are not changed during the simulation. Instead of a kinetic energy index, some response bounds for the elastic coordinates are directly used. Therefore, the simulation proves to be more reliable.

The second part of this paper deals mainly with the efficient simulation of impacts in rigid multibody systems using the coefficient of restiution. For rigid body models the kinetic energy loss is measured macromechanically by the coefficient of restitution. This coefficient has to be estimated from experiments and experience but it cannot be computed within the multibody system approach. In this paper results of elastodynamic contact simulations performed additionally on a fast time scale are used to compute the coeffcient of restitution. In the simulations only elastic impacts without plastic deformations or viscoelastic effects are considered. However, the impacts induce waves in the bodies which propagate after separation until they vanish due to material damping. Especially in slender bodies, such as rods, beams, plates and shells a large amount of the initial kinetic energy is transformed into propagating waves. To investigate these high frequency phenomena three different methods are presented. Firstly, the equations of motion for elastodynamics are used and solved by D'Alembert's approach for wave propagation combined with the Hertzian contact law. Secondly, the modal approach for the elastic bodies together with the Hertzian contact law is used. Thirdly, the equations of motion for elastodynamics are solved by discretization using finite elements. Then, these results are compared with each other and with experiments, and they are used to compute the coefficient of restitution. The coefficient is fed back to the multibody system equations and the solution continues on the related slow time scale.

The efficiency and accuracy of the approach presented is shown by numerical and experimental investigations on a slow and fast time scale for the impact of a steel sphere on different objects made of aluminum with approximately the same mass.

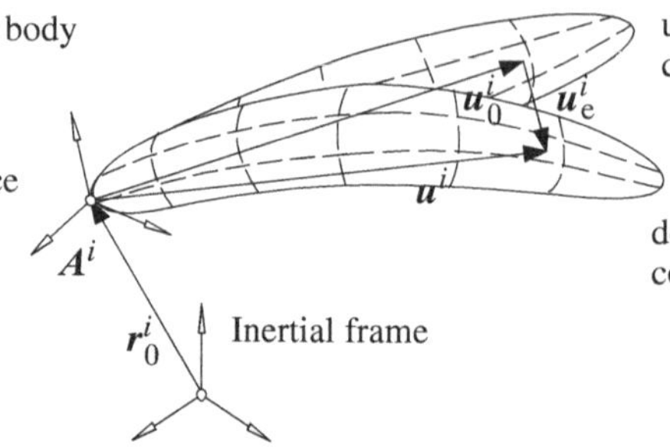

Fig. 1. Description of a flexible multibody system

2 Flexible multibody systems with impact

A flexible multibody system may consist of elastic and rigid bodies which are connected by joints and/or force elements such as springs, dampers and actuators. For modeling flexible multibody systems featuring elastodynamics, there are different approaches available. Well known approaches are the floating frame of reference formulation, the linear theory of elastodynamics, the finite segment method and the absolute nodal coordinate formulation, see Shabana [44]. Among them, the floating frame of reference formulation is currently the most widely used in the computer simulation of flexible systems. Hence, this modeling approach is used in this paper, too, to describe impact responses of flexible multibody systems with impact. In the floating frame of reference formulations, two sets of coordinates are chosen to describe the configuration of the deformable bodies; one set describes the location and orientation of a selected body coordinate system, while the second set describe the deformation of the body with respect to its body coordinate system. Using this description, the global position vector of an arbitrary point on the deformable body i can be written as

$$\mathbf{r}^i = \mathbf{r}_0^i + \mathbf{A}^i(\mathbf{u}_0^i + \mathbf{u}_\mathrm{e}^i) \tag{1}$$

where the vectors appear in this equation are shown in Fig. 1 and $\mathbf{A}^i$ is the transformation matrix that defines the orientation of the body reference frame with respect to the inertial frame. In (1), subscript 0 and e refer to the undeformed position and deformation variable, respectively. The deformation $\mathbf{u}_\mathrm{e}^i$ can further be described by means of a shape matrix $\mathbf{S}^i$ and a set of elastic coordinates $\mathbf{q}^i$,

$$\mathbf{u}_\mathrm{e}^i = \mathbf{S}^i\mathbf{q}^i\,, \tag{2}$$

see Shabana [45]. When the deformation is equal to zero, the kinematic description leads exactly to the modeling in rigid body dynamics. The impact responses usually consist of a gross rigid body motion and small elastic deformations resulting in wave propagation and structural vibrations, respectively. The vector of the generalized coordinates $\mathbf{y}^i$ can be partitioned as

$$\mathbf{y}^i = \begin{bmatrix} \mathbf{y}_\mathrm{r}^i \\ \mathbf{y}_\mathrm{e}^i \end{bmatrix} \tag{3}$$

where subscripts r and e refer to the reference and elastic coordinates, respectively, and the vector of elastic coordinates

$$\mathbf{y}_\mathrm{e}^i = \mathbf{q}^i \,. \tag{4}$$

Using the floating frame of reference, the equation of the deformable body i can be written in the following form

$$\mathbf{M}^i \ddot{\mathbf{y}}^i + \mathbf{k}^i = \mathbf{f}^i + \boldsymbol{\lambda}^i \tag{5}$$

where superscript i refers to the body number, $\mathbf{M}^i$ is the mass matrix, and $\mathbf{k}^i$ is the vector of generalized Coriolis forces including stiffness and damping forces, while $\mathbf{f}^i$ is the vector of externally applied forces, and $\boldsymbol{\lambda}^i$ is the vector of contact forces. Using the coordinate partitioning and performing the linearization for elastic coordinates, the equations of motion of the flexible body i can be written as

$$\begin{bmatrix} \mathbf{M}_\mathrm{rr}^i & \mathbf{M}_\mathrm{re}^i \\ \mathbf{M}_\mathrm{er}^i & \mathbf{M}_\mathrm{ee}^i \end{bmatrix} \begin{bmatrix} \ddot{\mathbf{y}}_\mathrm{r}^i \\ \ddot{\mathbf{y}}_\mathrm{e}^i \end{bmatrix} + \begin{bmatrix} \mathbf{0} & \mathbf{0} \\ \mathbf{0} & \mathbf{C}_\mathrm{ee}^i \end{bmatrix} \begin{bmatrix} \dot{\mathbf{y}}_\mathrm{r}^i \\ \dot{\mathbf{y}}_\mathrm{e}^i \end{bmatrix} + \begin{bmatrix} \mathbf{0} & \mathbf{0} \\ \mathbf{0} & \mathbf{K}_\mathrm{ee}^i \end{bmatrix} \begin{bmatrix} \mathbf{y}_\mathrm{r}^i \\ \mathbf{y}_\mathrm{e}^i \end{bmatrix} = \begin{bmatrix} \mathbf{f}_\mathrm{r}^i \\ \mathbf{f}_\mathrm{e}^i \end{bmatrix} + \begin{bmatrix} \boldsymbol{\lambda}_\mathrm{r}^i \\ \boldsymbol{\lambda}_\mathrm{e}^i \end{bmatrix} + \begin{bmatrix} -\mathbf{k}_\mathrm{r}^i \\ \mathbf{0} \end{bmatrix} \tag{6}$$

where the matrices $\mathbf{M}_\mathrm{ee}^i$, $\mathbf{C}_\mathrm{ee}^i$ and $\mathbf{K}_\mathrm{ee}^i$ are the elastic mass, damping and stiffness matrices of the flexible body i, respectively, and the quantities with index r refer to the rigid body motion.

During contact, the elastic collision approach will be used instead of the rigid collision approach for more accuracy and for more details about contact process. The normal force F_N is assumed to be a function of the indentation depth δ and the tangential force is assumed to be a function of the relative velocity. A general relation between the normal contact force F_N and the indentation depth δ reads

$$F_N = K\delta^n + D\dot{\delta} \,, \tag{7}$$

see Khulief and Shabana [25], Lankarani and Nikravesh [27]. If $n = 1$, the model is the Kelvin-Voigt viscoelastic model and if $n = 3/2$ and $D = 0$, the model describes the Hertzian contact. The parameter K in the Hertzian model depends on the Young's moduli E_1, E_2, Poisson's ratios ν_1, ν_2 and the geometry of the contact surfaces of the colliding bodies, respectively,

$$K = \frac{4 q_k \pi}{3(\frac{1-\nu_1^2}{E_1} + \frac{1-\nu_2^2}{E_2})\sqrt{(A+B)}} \,, \tag{8}$$

where A and B are the curvature parameters of the surface and q_k is a correction faktor depending on the ratio A/B, see Goldsmith [13]. Some experiences show that the Hertzian contact model yields good results, see e.g. Hu, Eberhard and Schiehlen [19]. For the computation of the continuous frictional force, a model described in Bauchau [4] may be used, where the discontinuity at zero relative velocity of the Coulomb's static friction law is replaced by a smooth, stiff function of the relative velocity v_t. The continuous frictional force can be written in the following form

$$F_t = -\mu \mathrm{sign}(v_t)(1 - \mathrm{e}^{-|v_t|/v_*})F_N \tag{9}$$

where v_* is a characteristic relative velocity typically chosen to be small compared to the maximum relative velocity encountered during the simulation, see Bauchau [4].

Because the impact forces have very high frequency components, in addition to the gross overall rigid body motion, the flexible deformations of the colliding bodies must be considered. These flexible deformations are normally small and decay fast compared to the overall motion. The problem is that during the simulation the integration routine has to keep track of these fast but small vibrations which leads to very high computation costs.

After impact, the contact force vanishes. Because of the material damping, the stresses in the colliding bodies approach to zero after some time and the rigid body motions are dominant. Then, the system dynamics can be modeled by the method of rigid multibody systems most efficiently. All elastic coordinates can be assumed to be zero and the equations of motion turn into

$$\mathbf{M}^i_{\mathrm{rr}}\ddot{\mathbf{y}}^i_{\mathrm{r}} = \mathbf{f}_{\mathrm{r}} - \mathbf{k}^i_{\mathrm{r}} \tag{10}$$

When the system can be simulated completely as rigid multibody system (10) depends on the damping of the elastic vibrations. It is well known that the higher frequency elastic vibrations decay faster due to the material damping than the lower ones. During the transition from elastodynamics to rigid body dynamics, the equations of motions for the reference coordinates and for the elastic coordinates given by equation (1) turn into

$$\mathbf{M}^i_{\mathrm{rr}}\ddot{\mathbf{y}}^i_{\mathrm{r}} = \mathbf{f}_{\mathrm{r}} - \mathbf{k}^i_{\mathrm{r}} - \mathbf{M}^i_{\mathrm{re}}\ddot{\mathbf{y}}^i_{\mathrm{e}} \tag{11}$$

$$\mathbf{M}^i_{\mathrm{ee}}\ddot{\mathbf{y}}^i_{\mathrm{e}} + \mathbf{C}^i_{\mathrm{ee}}\dot{\mathbf{y}}^i_{\mathrm{e}} + \mathbf{K}^i_{\mathrm{ee}}\mathbf{y}^i_{\mathrm{e}} = \mathbf{f}^i_{\mathrm{e}} - \mathbf{M}^i_{\mathrm{er}}\ddot{\mathbf{y}}^i_{\mathrm{r}} \tag{12}$$

where the inertial forces $-\mathbf{M}^i_{\mathrm{re}}\ddot{\mathbf{y}}^i_{\mathrm{e}}$ and $-\mathbf{M}^i_{\mathrm{er}}\ddot{\mathbf{y}}^i_{\mathrm{r}}$ represent the time varying coupling between the reference motion and the elastic deformations. The problem is that the large computing time prohibits long time simulation for these high frequency transient vibrations. Therefore, efficient simulations for these elastic vibrations are crucial. The question to answer is when the elastic coordinates are small enough to be negligible so that the method of rigid multibody systems can be used correctly.

2.1 Multi-time scale simulation for transient impact responses

The efficiency of numerical integration methods is strongly depending on the frequency content in the system. If a system consists of fast and slow components, i.e., the eigenvalues are widely spread, the system is characterized as stiff, see Nikravesh [31]. In flexible multibody systems with impact, the contact forces excite all modes of vibration. After the contact force vanishes, the excited higher modes of vibration decay very fast due to the material damping and they exist only over short time intervals. Thus, it will be inefficient to retain them during the entire simulation interval. In order to account for efficiency without loss of the required accuracy, an adaptive simulation approach that provides a capability of changing the number of elastic degrees of freedom during simulation is very useful. The elastic coordinates are considered only during the time interval in which they share a significant portion of the impact responses.

The most common method used for treating the flexible elements in multibody systems is based on the modal superposition. This method allows a truncation of the problem size. The flexible substructure modes may be measured experimentally or determined by solving the flexible substructure eigenvalue problem, see e.g. Ewins [12]. The deformation $\mathbf{u}_e^i$ can be expressed with n_i mode shapes $\varphi_j^i(\mathbf{u}_0)$ and elastic coordinates $q_j^i(t)$ by the modal superposition

$$\mathbf{u}_e^i = \sum_{j=1}^{n_i} \varphi_j^i(\mathbf{u}_0) q_j^i(t) \,. \tag{13}$$

Without loss of generality, it can be assumed that all mode shapes φ_j^i are normalized, usually with the Euclidean norm or the supremum norm to be equal to one. Then, the contribution of the j-th vibration mode of the body i with the mode shape φ_j^i to the deformation $\mathbf{u}_e^i$ is mainly described by the elastic coordinate $q_j^i(t)$. If after an initial time interval the maximum amplitude of the coordinate $q_i^j(t)$ remains always small, then the elastic coordinate q_i^j needs no longer to be considered and the degree of freedom of the elastic vibration is reduced. Usually, the high frequency vibrations are damped more quickly. Correspondingly, the elastic coordinates describing the high frequency vibrations need not to be considered in the simulation after some time. Due to the material damping, all elastic structural vibrations are damped and vanish in time. For the long term impact responses, only the rigid body motion needs to be simulated. But it remains a question when the elastic coordinate $q_i^j(t)$ can be neglected. As a criterion the time instant is used when the amplitude of the elastic coordinate $q_i^j(t)$ begins to be always smaller than a given small amplitude ε. It is well known that the exact computation of the maximum vibration amplitudes is time-consuming. However, an estimation of the amplitude bounds is easy for linear vibrations. The idea for the adaptive simulation is to compute at which instant the maximum amplitude bound is smaller than the given amplitude ε instead of computing the exact instant after that the amplitudes remain always smaller than ε.

For linear vibrations, many amplitude bounds were presented in the literature. Here some results presented in Hu and Eberhard [15] are recalled for application to longitudinal impacts presented in the next section. The equations describing the vibration of an n–degree–of–freedom linear system can be written in the form

$$\mathbf{M}\ddot{\mathbf{y}} + \mathbf{D}\dot{\mathbf{y}} + \mathbf{K}\mathbf{y} = \mathbf{b} \,, \quad \mathbf{y}(0) = \mathbf{y}_0 \,, \quad \dot{\mathbf{y}}(0) = \dot{\mathbf{y}}_0 \,. \tag{14}$$

As usual, it may be assumed that the mass matrix $\mathbf{M}$ and the stiffness matrix $\mathbf{K}$ are symmetric and positive definite, the damping matrix $\mathbf{D}$ is positive semi-definite. The displacement $\mathbf{y}(t)$ is an n–dimensional vector. The initial time is, without loss of generality, chosen to be zero. The vectors $\mathbf{y}_0$ and $\dot{\mathbf{y}}_0$ correspond to the initial displacements and the initial velocities, respectively. According to some investigations about the eigenvalue bounds of system (14), see Nicholson [29, 30] and Hu and Schiehlen [14], the real part of the eigenvalues of the system is smaller than $(-\mu)$, where μ is defined to be

$$\mu = \begin{cases} \lambda_m(\mathbf{D}^*)/2 & \text{for} \quad \lambda_M^2(\mathbf{D}^*) \le 4\lambda_m(\mathbf{K}^*)\,, \\ \min\left\{\frac{1}{2}\left(\lambda_M(\mathbf{D}^*) - \sqrt{\lambda_M^2(\mathbf{D}^*) - 4\lambda_m(\mathbf{K}^*)}\right), \frac{1}{2}\lambda_m(\mathbf{D}^*)\right\} & \text{otherwise}\,. \end{cases} \tag{15}$$

Here λ_m and λ_M denote the minimum and maximum eigenvalue of a matrix. The matrices $\mathbf{D}^*$ and $\mathbf{K}^*$ in Eq. (15) are given by

$$\mathbf{D}^* = \mathbf{M}^{-\frac{1}{2}}\mathbf{D}\mathbf{M}^{-\frac{1}{2}}\,, \quad \mathbf{K}^* = \mathbf{M}^{-\frac{1}{2}}\mathbf{K}\mathbf{M}^{-\frac{1}{2}}\,. \tag{16}$$

For an oscillatory system with all eigenvalues having nonzero imaginary parts, the quantity μ can be set equal to $\lambda_m(\mathbf{D}^*)/2$. Denoting the initial energy

$$E_0 = \frac{1}{2}\dot{\mathbf{y}}^T(0)\mathbf{M}\dot{\mathbf{y}}(0) + \frac{1}{2}\mathbf{y}^T(0)\mathbf{K}\mathbf{y}(0) \tag{17}$$

and another energy quantity

$$E_0^* = E_0 + \mu^2\mathbf{y}_0^T\mathbf{M}\mathbf{y}_0 + \mu\mathbf{y}_0^T\mathbf{M}\dot{\mathbf{y}}_0 - \mu\mathbf{y}_0^T\mathbf{D}\mathbf{y}_0/2 \tag{18}$$

one can give the following response bound for free vibrations

$$|y_i(t)| \le \min\left\{\sqrt{2K_{ii}^{-1}E_0}\,,\ \mathrm{e}^{-\mu t}\sqrt{2K'^{-1}_{ii}E_0^*}\right\}\,, \tag{19}$$

$$|\dot{y}_i(t)| \le \min\left\{\sqrt{2M_{ii}^{-1}E_0}\,,\ \mathrm{e}^{-\mu t}\left(\sqrt{2M_{ii}^{-1}E_0^*} + \mu\sqrt{2K'^{-1}_{ii}E_0^*}\right)\right\}\,, \tag{20}$$

where M_{ii}^{-1} and K'^{-1}_{ii} are the i–th diagonal elements of the matrices $\mathbf{M}^{-1}$ and $\mathbf{K}'^{-1}$, respectively, with the auxiliary matrix

$$\mathbf{K}' = \mathbf{K} - \mu\mathbf{D} + \mu^2\mathbf{M}\,. \tag{21}$$

Using these response bounds, one can compute when the amplitudes of these response bounds instead of the exact responses begin to be smaller than the given small amplitude ε. This estimated instant after that the amplitudes of elastic vibrations remain always to be smaller than ε is conservative. The exact maximum amplitudes of the elastic vibrations after a time instant are usually not available in advance or need too much time to compute, therefore, one use the response bound to estimate when the elastic coordinates begin to be always smaller than ε. It is recommended for the adaptive simulation to choose a corresponding relative amplitude error ε_{r_i} instead of the absolute small amplitude ε and to compute when the term $e^{-\mu t}$ begins to be smaller than the relative error ε_{r_i}.

The response bounds given in (19) and (20) hold true for free vibrations described by the equations of motion (14) with $\mathbf{b} = \mathbf{0}$. Compared with (12), this means that, if the applied force $\mathbf{f}_e^i$ is vanishing, the acceleration of the rigid body motion $\ddot{\mathbf{y}}_r^i$ is negligible or the coupling matrix $\mathbf{M}_{er}^i$ has to be zero. Then, the impact induced elastic vibrations during separation are free. In general, the right side of (12) is not zero and the corresponding elastic vibrations are forced vibrations. For linear forced vibrations, there exist also some response bounds which can be used, though they are somewhat more complicated. More details about the response bounds of linear forced vibrations can be found, for example, in Hu and Eberhard [15]. In the application of the response bounds, if the system (12) can be decoupled into a number independent subsystems, then one can compute the response bounds for every subsystem with lower dimension and get usually sharper response bounds. If all elastic coordinates are decoupled, i.e. they are modal coordinates, then generally the maximal efficiency can be achieved. That is the reason why the mode shapes are preferred.

2.2 Application to longitudinal impacts

In the history of mechanics, longitudinal wave propagation in rods with uniform cross sectional area along the length has been intensively investigated, see Szabó [48]. Vibrations resulting from longitudinal impacts were first examined by Bernoulli, Navier and Poisson. A detailed treatment was provided by St. Venant [47] and later supplemented by Boussinesq [6] and Donnel [10], see Goldsmith [13] and Timoshenko and Goodier [49]. Recently, this classical impact problem has been solved using a computer algebra system, see Hu and Eberhard [18], Hu, Eberhard and Schiehlen [17]. Closed-form solutions for the longitudinal wave are presented. All these theoretical results, however, are based on St. Venant's contact theory developed for perfectly planar contacts, which, up to now, have not been validated experimentally.

Experimental investigations of longitudinal wave propagations in elastic rods were reviewed by Al-Mousawi [1]. For a rigid body colliding with an elastic rod, important results were presented by Crook [7], Ripperger [36], Cunningham and Goldsmith [8], Barton, Volterra and Citron [3], Ramamurti and Ramanamurti [35] and Maekawa, Tanabe and Suzuki [28]. With the advancement of laser techniques, some additional aspects of wave propagation can be observed with Laser-Doppler-Vibrometers nowadays. Recent experimental investigation of a sphere colliding with a rod can be found, for example, in Hu, Eberhard and Schiehlen [19] where some comparisons between the measurements and numerical simulations with different impact theories are discussed. In this paper, the impact responses during the transition from the wave propagation to the rigid body motion is further simulated as an example to show how to use the multi-time scale approach presented.

2.2.1 Impact system and experimental setup

The schematic setup of the experiment is shown in Fig. 2. A steel sphere with radius $R_1 = 1.5\,cm$, mass $m_1 = 0.11\,kg$, Young's modulus $E_1 = 210\,GPa$ and Poisson's

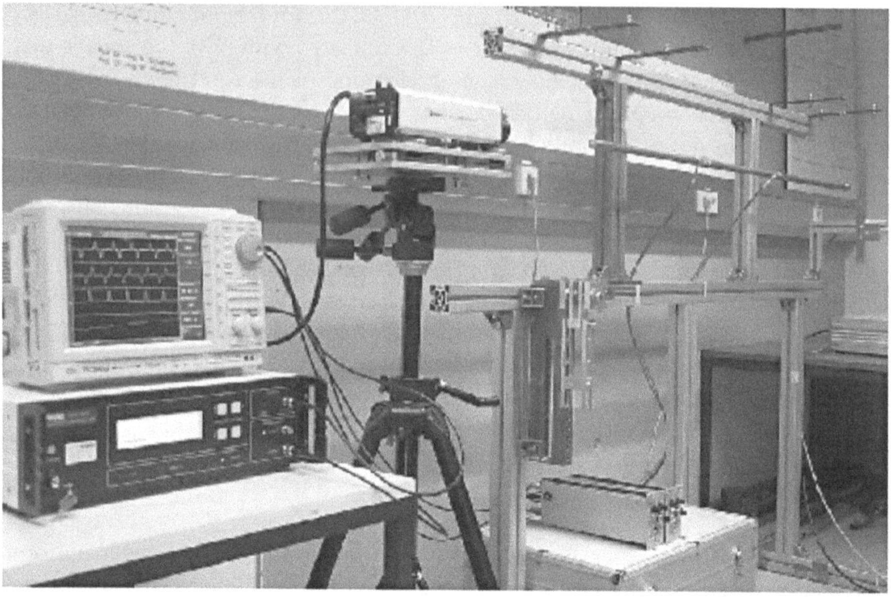

Fig. 2. Overview of the experimental setup and the used instruments

ratio $\nu_1 = 0.30$ is considered as a rigid body and used to strike an aluminum rod with Young's modulus $E = 71.08\,GPa$ and Poisson's ratio $\nu = 0.33$. The striking sphere was suspended in a frame by a thin kevlar wire forming a 'V'. In the experiments it is released at a vertical height h above the axis of the rod and the initial impact velocity of the sphere $v_0 = \sqrt{2gh}$ depends on the height h, where g is the gravitational acceleration.

The used aluminum rod is homogeneous and cylindrical. It has diameter $20\,mm$, length $L = 1\,m$ and density $\rho = 2710\,kg/m^3$. The aluminum rod is suspended by two sets of thin kevlar wires at two locations and positioned horizontally by means of a spirit level. The alignment of the rods and the sphere was also checked by sticking a small piece of adhesive paper backed with carbon paper between the impacting surfaces. Adjustments were then performed until a dense carbon imprint was produced exactly in the middle of the adhesive paper. The rod is suspended at the vertical height $l = 20\,cm$. For the measurement of displacements and velocities, a Laser-Doppler-Vibrometer of type OFV-3000/OFV-302 made by Polytec GmbH was used. The strains were measured with strain gauges of type 3/120XY13 made by Hottinger Baldwin Messtechnik GmbH. The two used DC signal conditioners for the strain signals are of type DMS805B made by Rohrer GmbH with a frequency range up to $1\,MHz$.

2.2.2 Equations of motion

In the modal approach, it is assumed that the displacement of the rod at the position x can be described approximately by a sum of the rigid body motion $x_1(t)$ and N vibrations with the mode shapes $\phi_i(x)$ and the modal coordinates $q_i(t)$, i.e.,

$$r^1(x,t) = x_1(t) + \left(x + u^1(x,t)\right) = x_1(t) + x + \sum_{i=1}^{N(t)} \phi_i(x) q_i(t) \,. \qquad (22)$$

The active degree of freedom of the elastic vibration varies with time so that the impact responses can be simulated adaptively. The mode shapes $\phi_i(x)$ are functions of the spatial coordinate x and satisfy the geometric boundary conditions of systems. For a free rod, they are chosen as

$$\phi_i(x) = \sqrt{2}\cos\left(\frac{i\pi}{L}x\right) \quad \text{for } i = 1, 2, \ldots, N \,. \qquad (23)$$

Compared with (1) and (2), the corresponding vectors and matrices read as

$$\mathbf{r}_0^1(t) = x_1(t)\,, \quad \mathbf{A}^1 = 1\,, \quad \mathbf{u}_0^1 = x\,, \quad \mathbf{S}^1 = [\phi_1(x)\,\phi_2(x)\cdots\phi_N(x)] \,. \qquad (24)$$

The rod is described by one reference coordinate x_1 and N elastic coordinates q_i. The strain of the rod is

$$\varepsilon(x,t) = \sum_{i=1}^{N} \phi_i^{'}(x) q_i(t) = -\frac{\sqrt{2}\pi}{L}\sum_{i=1}^{N} q_i(t)\, i \sin\left(\frac{i\pi}{L}x\right) \qquad (25)$$

where the prime denotes the differentiation with respect to x. The constitutive law of the rod is assumed to be viscoelastic,

$$\sigma(x,t) = E\left[\varepsilon(x,t) + \eta\dot{\varepsilon}(x,t)\right] \,. \qquad (26)$$

where η is a coefficient representing viscosity. According to (25), the stress at the contact end of the rod $x = 0$ remains always to be zero, i.e.

$$\sigma(0,t) = 0 \qquad (27)$$

which is, in fact, not true during contact. The reason for this problem is that the chosen modal shapes (23) do not satisfy the dynamic boundary condition at the contact end

$$EA\sigma(0,t) = F_N(t) \qquad (28)$$

where $F_N(t)$ is the contact force. In the modal approach, it is assumed that the impact force acts on the contact end of the rod like a concentrated force and the stress at the contact end vanishes. The modal shapes which satisfy the dynamic boundary condition (28) for the contact end can hardly be found. According to Eqs. (22) and (23), the motion at the center of mass of the free rod corresponds to the rigid body motion of the rod since

$$r^1(L/2,t) = x_1(t) + L/2\,. \tag{29}$$

The elastic vibrations of the rod are not involved in the motion at the center of mass.

In order to get the equations of motion of the rod, D'Alembert's virtual work principle

$$\int_0^L \rho \ddot{r}^1 \delta r^1 A dx + \int_0^L \sigma \delta\varepsilon A dx = \delta W^a \tag{30}$$

is used. Since the rod is suspended with thin wires and moves with a small amplitude, its rigid body motion can be modeled with a linear damped pendulum, representing the influence of the suspension and the air on the small elastic vibrations, too. Therefore, the virtual work of the applied forces is

$$\delta W^a = F_N \delta r^1(0,t) - mg\frac{x_1}{l}\delta x_1 - d_1 \dot{x}_1 \delta x_1\,. \tag{31}$$

where d_1 is the damping parameter. Using the orthogonality of the mode shapes yields

$$\int_0^L \rho \ddot{r}^1 \delta r^1 A dx = m\ddot{x}_1 \delta x_1 + m\sum_{i=1}^{n} \ddot{q}_i \delta q_i\,, \tag{32}$$

$$\int_0^L \sigma \delta\varepsilon A dx = \frac{EA\pi^2}{L}\sum_i^n i^2\left(q_i + \eta\dot{q}_i\right)\delta q_i\,. \tag{33}$$

Therefore, the equations of motion given by D'Alembert's virtual work principle read as

$$m\ddot{x}_1 + d_1 x_1 + m\frac{g}{l}x_1 = F_N\,, \tag{34}$$

$$\mathbf{M}^1_{ee}\ddot{\mathbf{q}} + \mathbf{C}^1_{ee}\dot{\mathbf{q}} + \mathbf{K}^1_{ee}\mathbf{q} = \mathbf{f}^1_e\,. \tag{35}$$

where the mass matrix, the damping matrix and the stiffness matrix of the rod read as

$$\mathbf{M}^1_{ee} = m_1 \mathbf{diag}\left\{1, 1, \ldots, 1\right\}\,, \tag{36}$$

$$\mathbf{C}^1_{ee} = \frac{\eta EA}{L}\mathbf{diag}\left\{1^2, 2^2, \ldots, n^2\right\}\,, \tag{37}$$

$$\mathbf{K}^1_{ee} = \frac{EA}{L}\mathbf{diag}\left\{1^2, 2^2, \ldots, n^2\right\}\,, \tag{38}$$

respectively, and the generalized force

$$\mathbf{f}^1_e = F_N[\phi_1(0)\ \phi_2(0)\ \ldots\ \phi_n(0)]^T = F_N[\sqrt{2}\ \sqrt{2}\ \ldots\ \sqrt{2}]^T\,. \tag{39}$$

The normal contact force F_N is assumed to be determined by the Hertzian contact law, i.e.,

$$F_N(t) = K\delta^{\frac{3}{2}}(t)\,. \tag{40}$$

The factor K follows from (8) with $A = B = 1/(2R)$ and $q_k = 1/\pi$ where R is the sphere's radius. The indentation depth δ reads as

$$\delta(t) = x_2(t) - r^1(0,t) = x_2(t) - \left(x_1(t) + \sqrt{2}\sum_{i=1}^{n} q_i(t) \right) . \tag{41}$$

This assumption is experimentally validated, see e.g. Hu and Eberhard [16]. In addition to the equations of motion of the elastic rod, one needs also the equation of motion of the sphere

$$m_2 \ddot{x}_2 = -F_N . \tag{42}$$

The elastic vibrations of the sphere are not considered.

During contact the equations of motion of the rod and the sphere as well as the elastic coordinates q_i of the rod are coupled by the contact force. The system is described by a set of nonlinear differential equations. Therefore, one can hardly get the information about the maximum amplitudes of the elastic coordinates and determine when an elastic coordinate can be neglected during contact. In order to describe the wave propagation and to compute the contact force accurately, the degree of freedom of the elastic vibration of the rod $N(t)$ is chosen to be constant during contact and equal to the number $N_0 = 100$. During contact the computation is very time-consuming, but, fortunately, the duration of contact t_c is usually very short.

After impact, i.e $t > t_c$, the rod and the sphere have no contact and the contact force vanishes, i.e., $F_N = 0$. The equations of motion of the rod and the sphere are no longer coupled. Further, the rigid body motion of the rod and the elastic vibrations of the rod are not coupled and all elastic coordinates q_i in (35) are independent since the mode shapes are orthogonal to each other. The vibration equation for i-th mode shape $\phi_i(x)$ reads as

$$m\ddot{q}_i + \frac{\eta EA}{L} i^2 \pi^2 \dot{q}_i + \frac{EA}{L} i^2 \pi^2 q_i = 0 \tag{43}$$

Its natural frequency

$$\omega_i = \sqrt{\frac{EA}{mL}} i\pi = \frac{c}{L} i\pi \tag{44}$$

and the damping ratio

$$\xi_i = \frac{\eta c}{2L} i\pi = \frac{1}{2} \eta \omega_i \tag{45}$$

which is proportional to the eigenfrequency ω_i. Here the wave speed c is introduced as an abbreviation. The damping for the elastic coordinate q_i may be also represented by the constant

$$\mu_i = \xi_i \omega_i . \tag{46}$$

It can be seen that for viscoelastic material higher frequency vibrations are damped more quickly. For the elastic coordinate q_i, one can solve (43) and get exact results for $t \geq t_c$. Here the response bounds for free vibrations discussed before are used as an example and the relative errors ε_{r_i} are chosen to be 0.001. By solving the inequality

$$e^{-\mu_i(t-t_c)} \leq \varepsilon_{r_i}/N_0 \tag{47}$$

one can determine that for

$$t \geq t_c - \frac{1}{\mu_i} \ln \frac{\varepsilon_{r_i}}{N_0} \tag{48}$$

the elastic coordinate q_i needs no longer to be considered. The high frequency vibrations disappear after a short time and only the lower frequency vibrations remain. In the numerical simulation, the damping coefficient $d_1 = 0.1620\,N/m/s$ and the viscosity coefficient $\eta = 3.6914 \times 10^{-8}\,s$ are identified from the measured impact responses. Though the coefficient η is very small, it plays an important role for the decaying of elastic deformations.

2.2.3 Comparisons between Measurements and Simulations

In the experiments the strain signals at two different positions A and B with distance $0.3\,m$ and $0.6\,m$ from the contact end of the rod, respectively, and the velocity and the displacement at the free end of the rod were measured. For the striking velocity $v_0 = 0.28\,m/s$, the comparison between the experimental and simulated results are shown in Fig. 3 for a long time scale of $10\,s$ and in Fig. 4 for a short time scale of $2\,ms$ where the dark lines denote the measured signals and the light lines denote the simulated results. The long term impact behavior shows a discontinuous rigid body motion. After impact the rod changes its velocity suddenly. After one second the strain signals vanish and the rod moves then like a rigid pendulum. Measurement and simulation match perfectly. The short-term behavior shows the phenomenon of the wave propagation. Due to the impact, a wave is induced. It propagates along the rod and is reflected at the both ends of the rod. The wave dispersion can not be seen in Fig. 4. The simulation results for the wave propagation may also be found using St. Venant theory as shown in detail by Hu and Eberhard [16]. The comparison between the measured and simulated results show that the wave propagation can be simulated satisfactorily with the modal approach.

During the transition from the wave propagation to the rigid body motion the impact behavior is characterized by structural vibrations. Due to the material damping, the high frequency elastic vibrations are damped and vanish within one second. Figure 5 shows how the power spectral density (PSD) of the velocity signal at the free end of the rod varies with time. In the computation of the PSD, signals with the same time duration of $40\,ms$, however, at different starting time are considered. In the graphs the PSD with unit m^2/s is shown in a logarithmic scale. It can be seen that at the beginning there exist many elastic vibrations with high frequencies. However, after 0.4 second some high frequency vibrations can not be recognized and at $0.8\,s$ there exist only the fundamental eigenmode. Since many high frequency vibrations have very small amplitudes, their contribution to the impact responses in the time domain can be neglected. In the numerically simulated signals only those lower frequency elastic vibrations are considered. Hence, the measured and simulated results have some differences in the high frequency range.

The comparison between the measured and simulated results during the transition are shown clearly in Fig. 6 with a high resolution at different time. Both results have

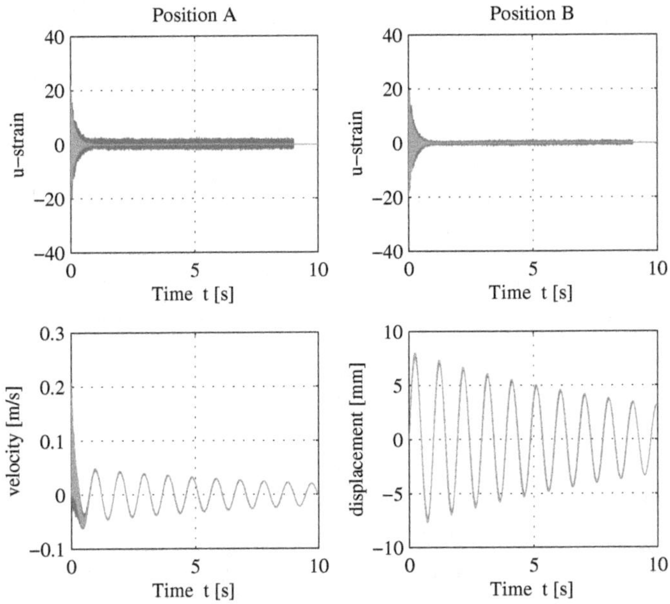

Fig. 3. Comparison between the measured and simulated results for a long time scale

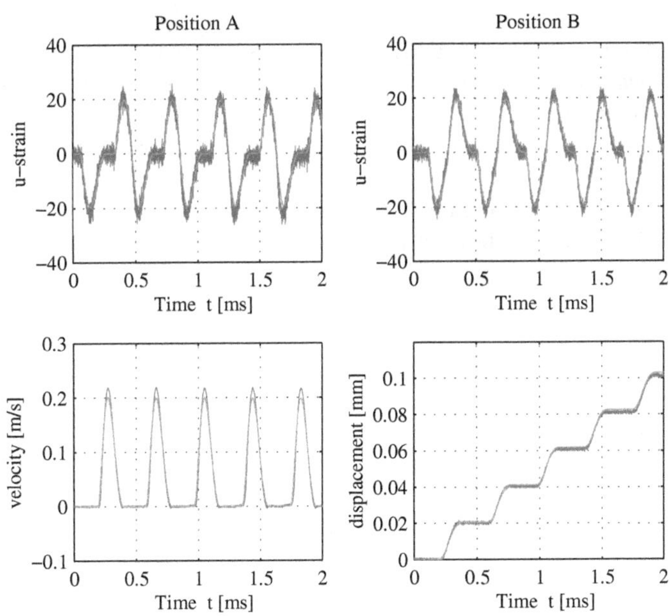

Fig. 4. Comparison between the measured and simulated results for a short time scale

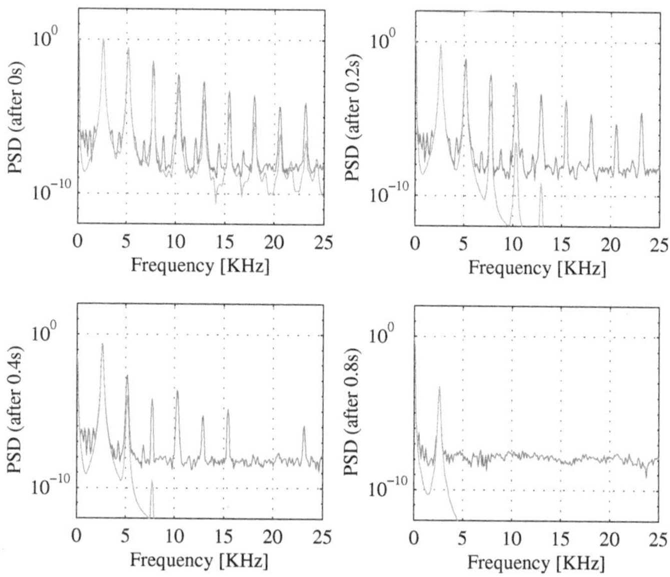

Fig. 5. Comparison between the measured and simulated results in the frequency domain for the rod with the viscous damping

only slight differences. Comparing Fig. 4 and Fig. 6, one can see that the wave form has changed. Due to the material damping, a wave dispersion exists. With a short time scale of millisecond the wave dispersion can not be observed. The wave dispersion makes the motion more uniform. Hence, in the long run there exists only the rigid body motion.

In the multi-time scale simulation, the degree of freedom of the elastic vibration varies with time. Figure 7 shows a time history of the active degrees of freedom $N(t)$. During contact, the degree of freedom N_0 is chosen to be 100. During separation the high frequency vibrations decay very fast. After 0.1 second, one needs only to consider lower frequency vibrations. It can be seen that in less than 0.2 second the degree of freedom $N(t)$ is reduced from 100 to 3. Using the multi-time scale simulation, the efficiency of computation is improved by a factor 100 without affecting the required accuracy.

3 Rigid multibody systems with impact

The method of multibody systems allows the dynamical analysis of machines and structures, see Schiehlen [37, 38, 39]. More recently contact and impact problems featuring unilateral constraints were considered, too, see Pfeiffer and Glocker [33]. Subjected to an impact the multibody system motion is clearly divided into two periods by different sets of initial conditions. The rigid multibody system is represented by its equations as

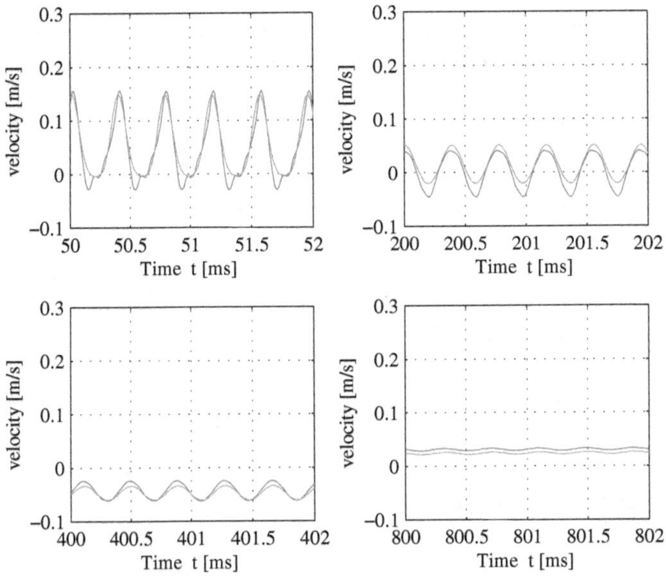

Fig. 6. Comparison between the measured and simulated results during the transition for the rod with the viscous damping

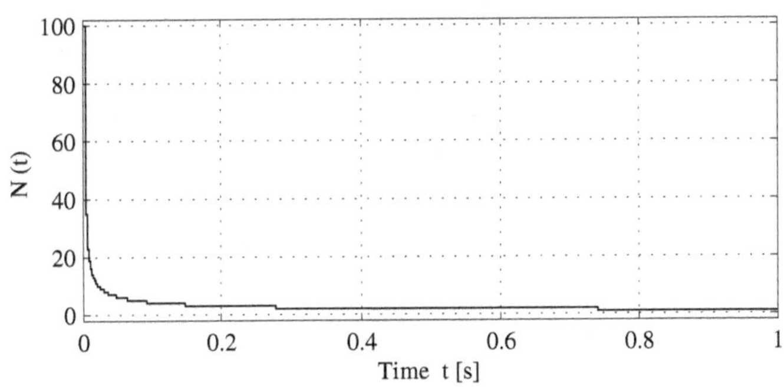

Fig. 7. The time history of the degree of freedom of the elastic vibration

$$\boldsymbol{M}(\boldsymbol{y})\ddot{\boldsymbol{y}} + \boldsymbol{k}(\boldsymbol{y}, \dot{\boldsymbol{y}}) = \boldsymbol{f}(\boldsymbol{y}, \dot{\boldsymbol{y}}) \tag{49}$$

where $\boldsymbol{y}(t)$ is the global position vector featuring f degrees of freedom, $\boldsymbol{M}$ the inertia matrix, $\boldsymbol{k}$ the vector of Coriolis and gyroscopic forces and $\boldsymbol{f}$ the vector of the applied forces, see also (5). The continuous motion of the multibody system during the first period is interrupted by collision. The hypothesis of an infinitely large impact force $\boldsymbol{\lambda}$ leads to the force impulse

$$\triangle \boldsymbol{p} = \lim_{\varepsilon \to 0} \int_{t_1-\varepsilon}^{t_1+\varepsilon} \boldsymbol{\lambda}(\tau)\, d\tau \tag{50}$$

changing the velocities of the motion. Due to the infinitely large impact force all other forces can be neglected during impact. This results in a discontinuous system

$$\boldsymbol{M}(\boldsymbol{y})\ddot{\boldsymbol{y}} + \boldsymbol{k}(\boldsymbol{y},\dot{\boldsymbol{y}}) = \boldsymbol{f}(\boldsymbol{y},\dot{\boldsymbol{y}}) , \quad 0 < t < t_1 - \varepsilon \, , \, t_1 + \varepsilon < t < \infty \, . \tag{51}$$

describing more precisely the motion before and after impact. At the instant of the impact $t = t_1$, the initial velocities of the second period have to be reset, however the displacements remain unchanged. For the instant of impact the impulse equations include the coefficient of restitution which represents the kinetic energy loss during impact. This approach results finally in linear complimentary problems (LCPs) as shown by Pfeiffer and Glocker [33].

3.1 Computation of the coefficient of restitution

For the simulation on the slow time scale using the multibody system approach the coefficient of restitution is required to describe the kinetic energy lost during impact. Therefore, the kinetic coefficient of restitution by Poisson is used. This coefficient is defined as the ratio of linear momentum during the compression and restitution phases of the impact, see e.g. Stronge [46]. The coefficient of restitution cannot be determined within the multibody system approach but must be estimated by costly experiments [13] or experience. However, performing additional elastodynamic contact simulations on a fast time scale the coefficient of restitution can be determined computationally.

Following the assumptions of rigid body impact, the impact duration is infinitesimally small and all forces but the impact force are negligibly small. The linear momentum balance in the central impact line of two colliding bodies during the compression and restitution phases reads as

$$\triangle p_c = m_1(v_1^- - v), \qquad \triangle p_c = m_2(v - v_2^-), \tag{52}$$

$$\triangle p_r = m_1(v - v_1^+), \qquad \triangle p_r = m_2(v_2^+ - v), \tag{53}$$

where v_1^- and v_2^- are the velocities before impact and v_1^+ and v_2^+ are the velocities after impact and m_1 and m_2 are the masses of the colliding bodies. The common velocity of the bodies at the end of the compression phase is v and the linear momentum produced by the impact force during the compression and restitution phase is denoted by $\triangle p_c$ and $\triangle p_r$, respectively. Using the coefficient of restitution e by Poisson, the total linear momentum produced by the impact reads as

$$\triangle p = \triangle p_c + \triangle p_r = \triangle p_c(1 + e). \tag{54}$$

Rearranging Eq. (53) and (54), the coefficient of restitution can be expressed in dependence of the total linear momentum during impact as,

$$e = \frac{(m_1 + m_2)\triangle p}{m_1 m_2 (v_1^- - v_2^-)} - 1. \tag{55}$$

The total linear momentum $\triangle p$ is known from additionally performed elastodynamic contact simulations on the fast time scale as time integral of the computed impact force. Therefore, the coefficient of restitution can be determined by the simulation results on the fast time scale.

The efficiency and accuracy of this multi-time scale approach for rigid body impact is shown on a fast and slow time scale. Therefore, the impact of a steel sphere on different objects of similar mass made of aluminum are investigated numerically and experimentally. The impacted bodies are a slender cylinder called rod in the following due to its longitudinal excitation, a half-circular plate, a compact cylinder called ball in the following due to its similarity to a ball of the same mass and a slender cylinder called beam due to its lateral excitation. The material and geometrical data of the colliding bodies are summarized in Table 1. In Fig. 8 the geometry configuration of the different impact systems is shown.

	Steel Sphere	Aluminum Rod	Aluminum Plate	Aluminum Ball	Aluminum Beam
Radius $[mm]$	15	10	200	36.48	10
Thickness $[mm]$	–	–	5.1	–	–
Length $[mm]$	–	1000	–	73.68	976.5
Young‘s modulus $[GPa]$	210	70.5	72.55	70.5	67.7
Poisson ratio	0.3	0.33	0.33	0.33	0.33
Density [kg/m^3]	7780	2710	2710	2710	2696

Table 1. Geometrical and material data of the colliding bodies

3.2 Elastodynamic contact simulation on the fast time scale

For the elastodynamic contact simulation on the fast time scale three different methods are used in the following. Firstly, the equations of motion for elastodynamics are introduced and solved by D'Alembert's approach for wave propagation combined with the Hertzian contact law. Secondly, the modal approach for the elastic bodies together with the Hertzian contact law is applied. Thirdly, the equations of motion for elastodynamics are solved by discretization using finite elements.

3.2.1 Elastodynamic contact using wave propagation

Using the equation of motion for elastodynamics and solving them by D'Alem-bert's approach for wave propagation is a very time effective method [17, 19] to simulate impacts on the fast time scale. However, this approach is limited to geometrically simple bodies, such as the longitudinal impact of a sphere on a rod as shown in the following. The longitudinal waves in a rod are governed by the partial differential equation

$$\frac{\partial^2 u(x,t)}{\partial t^2} = c^2 \frac{\partial^2 u(x,t)}{\partial x^2} \quad \text{with} \quad c = \sqrt{\frac{E}{\rho}} \tag{56}$$

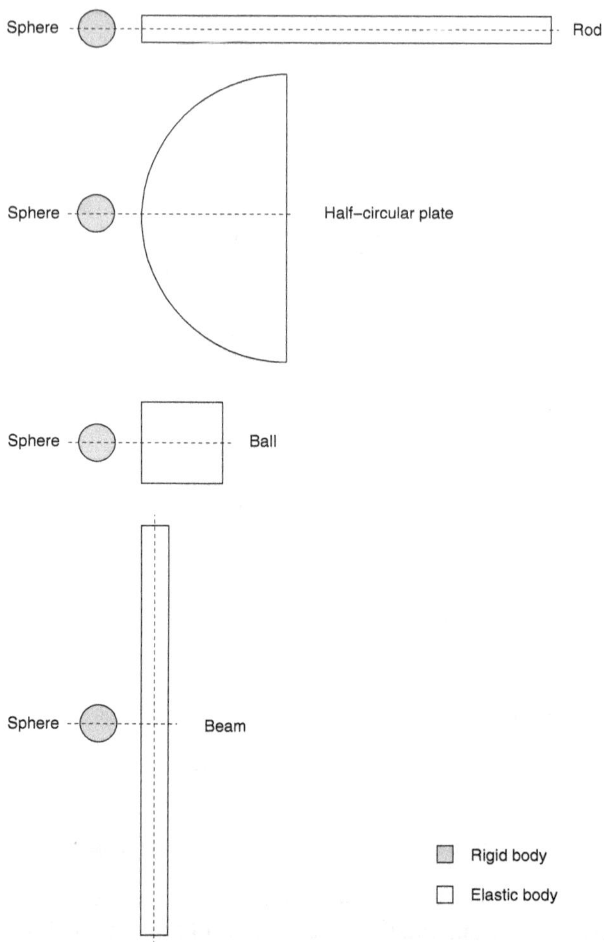

Fig. 8. Colliding bodies of the impact systems

where c represents the wave speed, E, ρ are the rod's Young's modulus and density, respectively. According to D'Alembert's approach the general solution of (56) reads as

$$u(x,t) = f(x - ct) + g(x + ct), \tag{57}$$

where f and g are real functions representing a forward and a backward traveling wave, respectively. Using (56) and (57) with the dynamic boundary condition

$$-EA\frac{\partial u(0,t)}{\partial x} = F(t) \tag{58}$$

for the struck end, the impact of a sphere on a rod is modeled. Thereby the contact force $F(t)$ between the sphere and the rod is described by the Hertzian elastostatic contact law described in (7) i.e.,

$$F(t) = max[0, K\delta(t)^{3/2}] \quad \text{with} \quad \delta(t) = x_s(t) - u(0,t) \tag{59}$$

where $x_s(t)$ describes the sphere's position. The good agreement of this analytical model with experimental results is shown in [19].

3.2.2 Elastodynamic contact using modal approach

Another time efficient method for investigating elastodynamic contact problems is a combination of a modal approach describing the linear elastic body and the Hertzian contact law, as presented in Chapter 2. Using a modal approach, geometrically more complex shaped bodies as well as more complex impact configurations can be modeled efficiently. In the following the modeling of the transverse impact of a sphere on the center of a beam is shown. The sphere is modeled as a rigid body whereas the beam is represented by the modal approach. Therefore, a finite element model using Timoshenko beam elements is created to compute the beam's eigenfrequencies and mode shapes.

Using the modal approach the displacement of a point of the discretized beam is given by the sum of the beam's rigid body motion $x_1(t)$ and the first N eigenmodes given by the previously computed mode shape φ_i and the modal coordinates q_i,

$$u^p(t) = x_1(t) + \sum_{i=1}^{N} \varphi_i^p \, q_i(t) \tag{60}$$

where the index p identifies the nodes of the FE model. For the modeling of impacts on the beam frequencies up to 100kHz are taken into account. The investigated beam is suspended like a pendulum, thus the equations of motion of the beam reads as

$$\begin{aligned} m\ddot{x}_1 + m\frac{g}{l}x_1 &= F(t) \\ \ddot{q}_i + \omega_i^2 \, q_i &= f_i, \qquad i = 1(1)n \end{aligned} \tag{61}$$

where m is the beam's mass, g the gravity, l the length of the pendulum, F the contact force and ω_i the angular frequency of the ith eigenmode. The modal contact force is given by

$$f_i = F \, \varphi_i^c \, , \qquad i = 1(1)n \tag{62}$$

where $p = c$ denotes the point of the contact node. The contact force is computed using the Hertzian contact law according to (59). For the contact between a sphere and a cylinder, it turns out that the contact geometry is given by an ellipse. Following Goldsmith [13] the contact stiffness is given by (8) with

$$A = \frac{1}{2R_s} \quad \text{and} \quad B = \frac{1}{2}(\frac{1}{R_s} + \frac{1}{R}) \tag{63}$$

where R and R_s are the radius of the beam and sphere, respectively. The parameter q_k depends on the ratio A/B and is found in Goldsmith [13] as 0.3414. The comparison of the results of the modal approach with results from a FEM-simulation presented in Daparti [9] shows a good agreement, while the computation time for the modal model is significant lower.

3.2.3 Contact simulation using finite elements

For investigation of elastodynamic impact phenomena on the fast time scale involving bodies with complex geometric shapes resulting in complex contact conditions as well as non-linear material properties, numerical methods such as the FEM have to be used. The the nodal displacement vector $\boldsymbol{U}$ of a fricitonless static contact problem minimizes the potential energy function $\Pi(\boldsymbol{U})$ of the FE system considering the impenetrability condition $\boldsymbol{g}(\boldsymbol{U}) \geq \mathbf{0}$ where the penetrations of the bodies are summarized in $\boldsymbol{g}(\boldsymbol{U})$. Due to its simplicity the penalty method is used in most commercial FE-codes for solving this restricted minimization problem. Other common approaches are the Lagrange multiplier method, the augmented Lagrangean method and hybrid methods, see e.g. Zhong [51].

Using the penalty method the potential energy function $\Pi(\boldsymbol{U})$ is extended by a penalty term of the form $\frac{1}{2}P\boldsymbol{g}^T\boldsymbol{g}$ where P is called the penalty factor. The minimization of the extended potential results in the nodal displacements $\boldsymbol{U}$. The contact force is computed by $\boldsymbol{F} = P\boldsymbol{g}$. For penalty factors P approximating infinity the penetration vanishes and the solution converges to physical reality. However, too high penalty factors result in numerical instabilities. An optimal choice of the penalty factor for static contact is described by Nour-Omid and Wriggers [32].

For impact problems the influence of the penalty factor and the discretization is investigated for the test example of a sphere to rod impact as reported in [43], too. It is shown that the choice of the penalty factor and the discretization of the contact area have a significant influence on the time response of the calculated impact force. Especially the independence of the results from the choice of the penalty factor has to be checked by additional simulations. Also great attention has to be given to the evaluation of the resulting wave propagation in the elastic bodies which requires a small element size.

3.3 Experimental validation of rigid body impacts

For the experimental validation of the simulation results for the impact processes on a slow and a fast time scale the experimental setup described in Sect. 2.2.1 is used. The longitudinal impact on the rod and the ball occurs along the central axis of the bodies and the radial impact on the half-circular plate occurs along the symmetry axis on the curved side of the plate. The transverse impact on the beam occurs also along the symmetry axis of the beam. When rebounding after impact the sphere is set to rest again. Then, the impacted body moves like a free pendulum. The experimental setup for the impact on the beam is shown in Fig. 9.

3.3.1 Experimental validation on the fast time scale

In the following the experimental and simulation results for the different impact systems are presented on the fast time scale. The impact velocity of the steel sphere is approximately $0.3m/s$. The left plot of Fig. 10 shows the measured velocities at the free end of the aluminum rod on the fast time scale. During impact kinetic energy is

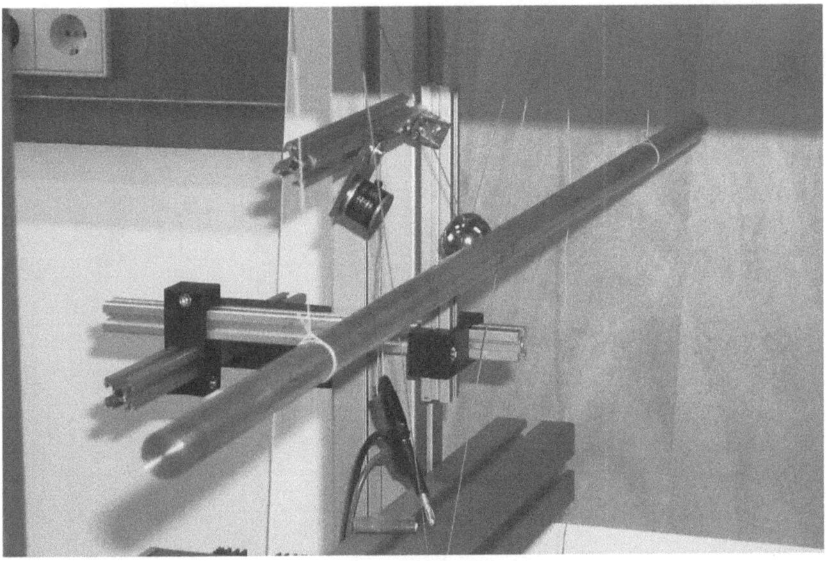

Fig. 9. Experimental setup with the impact on the beam

transformed into strain energy which propagates as wave away from the contact region. This wave is reflected at the free end of the rod. The investigated rod is long so that the contact is dissolved before the reflected wave reaches the struck end. Then, the wave is reflected again. After the wave passes a point of the rod, e.g. the free end, this point remains in rest again until the wave passes through it again. The simulation results of the rod using D'Alembert's approach for wave propagation combined with the Hertzian contact law is indicated in the left plot. A comparison of the described wave phenomena using the modal approach presented in Fig. 4 shows good agreement, too. The right plot of Fig. 10 shows the measured velocity of the half-circular plate. In the half-circular plate the initiated waves are multiply reflected at the boundaries and then the motion is composed of a rigid body motion and an irregular oscillation. In the right plot the results of a FE-simulation of the half-circular plate using ANSYS [2] are indicated, too. This shows the good agreement of FE-simulations and experiments for the fast time scale. For more details see Seifried [42] and Seifried, Hu and Eberhard [43].

In the left plot of Fig. 11 the impact on the aluminum ball is shown. In contrast to the impact on the rod and plate no significant vibrations are observable. The ball moves after impact as rigid body. The simulation of this impact system is performed by a FEM-simulation, too.

Measurements and simulations using a modal approach for the beam agree also very well as shown in the right plot of Fig. 11. The post impact behavior of the beam shows very strong vibration phenomena of a wide frequency range. The bending frequency of the beam is $94Hz$, what is low compared with frequencies of $2500Hz$ for the rod and $4800Hz$ for the plate. The velocity plot of the impacting steel sphere,

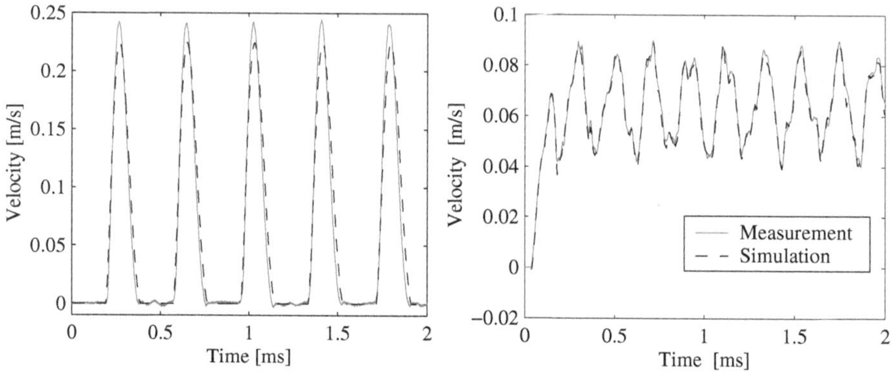

Fig. 10. Aluminum rod and half-circular plate on the fast time scale

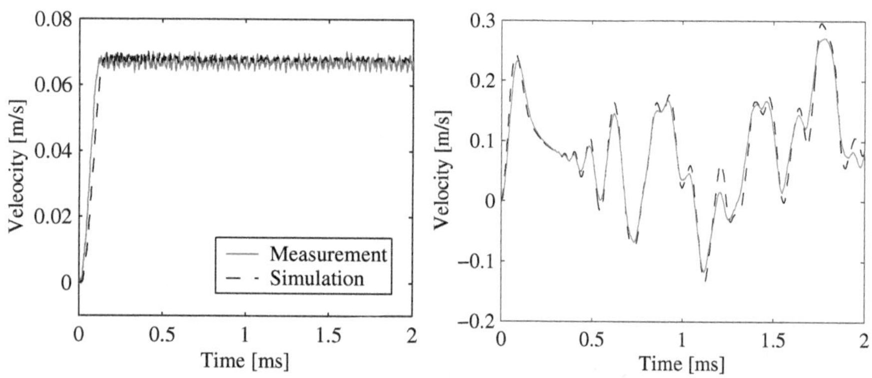

Fig. 11. Aluminum ball and beam on the fast time scale

presented in Fig. 12, shows that there are two successive impacts within $1.5ms$. After the first impact the steel sphere still moves in positive direction with a rigid body velocity of $0.062m/s$. This velocity is larger than the rigid body velocity of the beam with $0.033m/s$. After the first impact the beam bends first forwards, then the beam bends backwards and a second impact occurs. After the second impact the sphere moves in opposite direction. These two successive impacts are also clearly seen in the displacement plot of the sphere and beam as presented in the right plot of Fig. 12. For the simulation of this transverse impact system with two successive impacts the correct identification of the first impact is essential. Variations of the velocities after the first impact might results in very different rebound velocities after the second impact and also the instant of the second impact might vary. Due to several impacts on the aluminum beam plastic deformation and a dynamic hardening of the beam's contact region occurs. This effect can also be observed for the impacts on the other bodies; however its influence on these bodies' motion is much smaller. After several impacts no additional plastic deformation occurs and the material behaves elastic, however due to a deformed contact surface the Hertzian contact law presented in (59) and (63) does not describe the contact accurate enough. Comparison with measurements and first FE-simulations of successive impacts with plasticity and hardening show that in a first approximation a higher Hertzian contact stiffness K must be assumed. Therefore, the value following from (63) is increased by a factor of $1,7$.

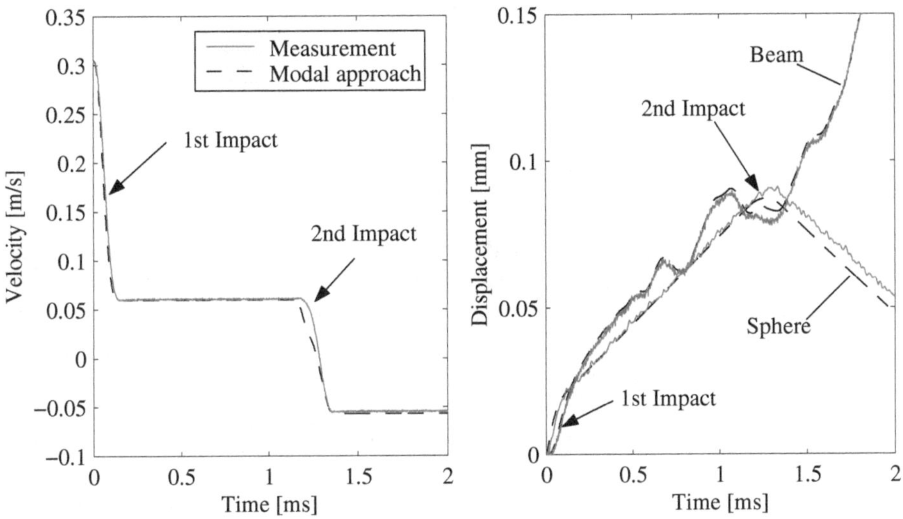

Fig. 12. Velocity of the steel sphere and displacements of the sphere and beam in impact direction

3.3.2 Experimental validation on the slow time scale

With an initial velocity of $0.3m/s$ for the steel sphere a coefficient of restitution of 0.625 is computed for the impact on the aluminum rod. Using this value, the rigid body motion of the bodies can be simulated efficiently using multibody dynamics. The rod's velocity is shown in the left plot of Fig. 13 and compared with experimental results. This shows the good agreement of the rigid body motion. The velocity differs at the beginning due to the strong wave phenomena which cannot be captured by the rigid body model, however the waves decay due to material damping and vanish finally. Compared to Fig. 3 the material damping in Fig. 13 is smaller resulting in a longer decay period. The difference in damping is due to the strain gauges which are not used in the experiment shown in Fig. 13. The right plot of Fig. 13 shows the velocity of the half-circular plate. Using the FEM-results for the impact on the half-circular plate a coefficient of restitution of 0.967 is computed. Hereby much less of the initial kinetic energy is transformed into waves and structural vibrations what is clearly seen in the velocity plot.

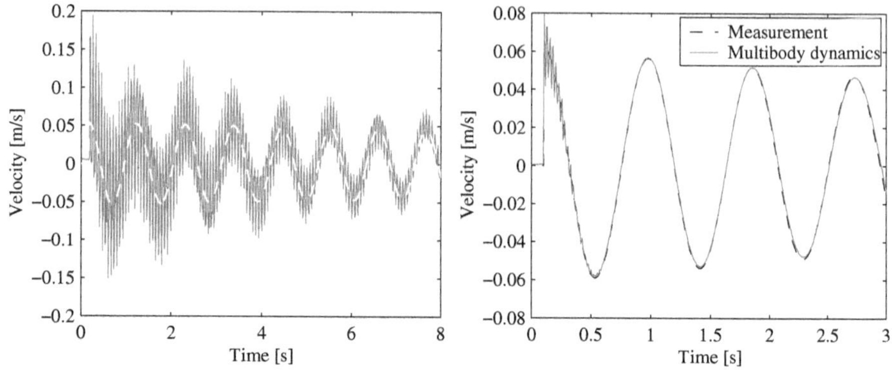

Fig. 13. Aluminum rod and half-circular plate on the slow time scale

The simulation of the impact on the aluminum ball shows that nearly no energy is transformed into wave propagation and therefore the computed coefficient of restitution is very close to 1. Due to the short dimensions of the ball multiple reflected waves travel during impact with high frequency back and forth through the ball and, therefore, the ball behaves nearly rigid as shown in the left plot of Fig. 14. For the impact on the beam a coefficient of restitution of 0.3482 is computed. Most of the initial kinetic energy is transformed into waves and vibrations. The measured velocity of the beam on the slow time scale, presented in the right plot of Fig. 14, shows these strong vibrations. Again, a comparison of the measurement and the simulation show good agreement for the rigid body velocity.

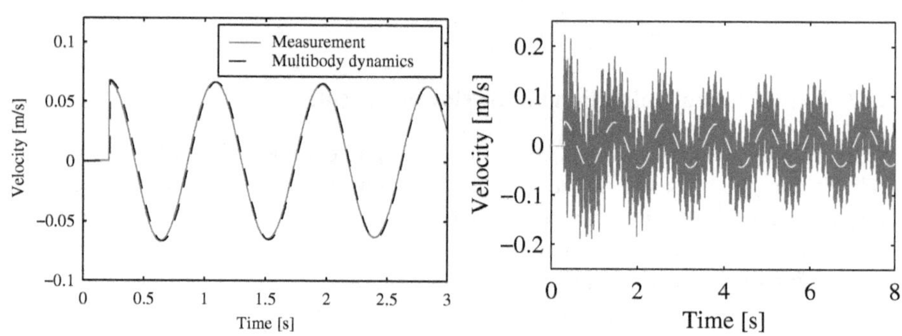

Fig. 14. Aluminum ball and beam on the slow time scale

The coefficient of restitution depends not only on the material parameters, the contact geometry and the body geometry but also on the initial velocity as shown experimentally in Goldsmith [13] and analytically in Hu and Seifried [21]. The influence of the velocity is shown in Fig. 15 in a range of $0.05m/s$ to $0.8m/s$ for the rod, plate and ball. From the simulations for the rod and half-circular plate it is seen that the coefficient of restitution decreases steadily with increasing initial velocity. This indicates an increase of energy transformation from the initial rigid body motion into waves and vibrations with increasing velocity. The computed coefficient of restitution for the ball is close to 1 for the investigated velocity range. For the impact on the ball the transformation of initial kinetic energy into waves and vibrations can be neglected. The computed kinetic energy loss during impact is for all bodies purely due to the initiation of waves in the bodies and is also summarized in Fig. 15. For the impact on the beam first simulations and experiments are performed in the range of $0.25m/s$ to $0.35m/s$ showing values of the coefficient of restitution between 0.30 and 0.38. This indicates a transformation of about 80% of the initial kinetic energy into vibrations of the beam. These results are also indicated in Fig. 15.

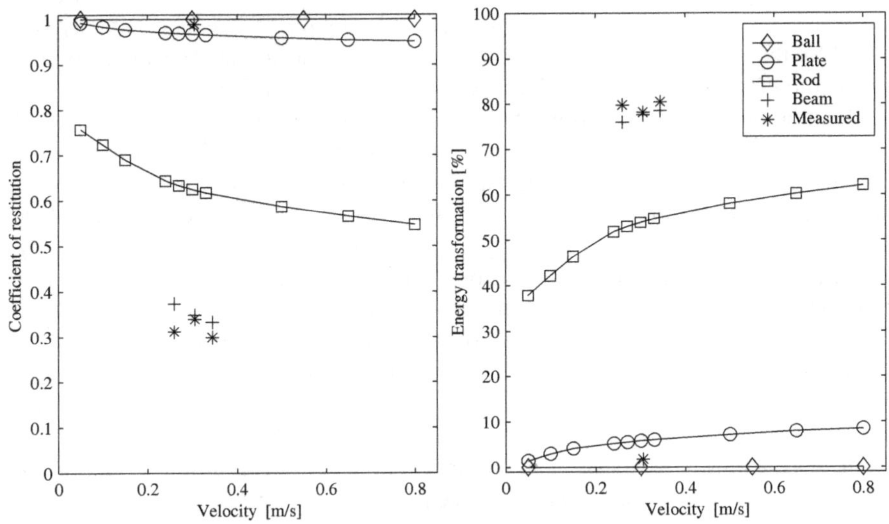

Fig. 15. Computed and some measured coefficients of restitution and energy transformation

A signal analysis of the simulation results shown in Schiehlen and Seifried [40] help to investigate experimental observations such as the tone and noise produced. The impact on the rod produces strong periodic waves which propagate in the rod for a long time before they vanish due to material damping. This produces a clear, harmonical, enduring tone. Due to irregular mode excitation the impact on the plate produces a metal noise while for the impact on the ball only a very short dull noise is observed.

4 Conclusions

A detailed analysis of impacts shows that the impact behavior have three different phenomena: wave propagation, structural vibrations and rigid body motion. For efficient simulation of these different phenomena, different time scales are needed. During contact the short time impact response is computed using the method of flexible multibody systems for accuracy and after separation the long term motion is simulated using the method of rigid multibody systems for efficiency. During the transition from the wave propagation to the rigid body motion, an adaptive simulation method with varying degree of freedom for the elastic vibration is developed. In the simulation, the contribution of each elastic coordinate to the overall motion is monitored using their response bounds. When these response bounds are smaller than a given constant, the corresponding elastic coordinates are deleted. As a consequence, the elastic degrees of freedom are reduced and the efficiency of the simulation is improved. The application of this multi-time scale simulation is experimentally validated for the impact of a rigid body colliding with an elastic rod. The application of this method for more complicated impact systems will be further investigated.

Multibody systems with impact may be analysed directly if the coefficient of restitution is known. Three methods are presented to compute this coefficient on a fast time scale: wave propagation, modal approach and finite elements. The main results include impacts on rods with longitudinal waves, impacts on a plate with plane waves, impacts on a beam with bending vibrations and impacts on a ball shaped body. While the longitudinal waves show strong periodicity, the plane and spatial waves may be highly irregular even for linear structural dynamics models. It is shown that a substantial amount of initial kinetic energy is transformed into waves and vibrations for the impact on the slender rod and beam whereas the ball shaped body behaves like a rigid body. In conclusion, the interaction of multibody dynamics and wave propagation, represents an efficient multiscale tool for the dynamical analysis of mechanical systems with impacts. The required coefficient of restitution is determined out of fast time scale results. The approach allows the consideration of nonlinear material properties, too.

References

1. Al-Mousawi M M (1986) On experimental studies of longitudinal and flexural wave propagations: an annotated bibliography. Appl Mech Rev 39:853–864
2. ANSYS Inc. (1997) ANSYS Theory Reference, Release 5.4. Canonsburg, ANSYS Inc.
3. Barton C S, Volterrra E, Citron S J (1958) On elastic impacts of spheres on long rods. In: Proceedings of the 3rd US National Congress on Applied Mechanics pp. 89–94
4. Bauchau O A (1999) On the modeling of friction and rolling in flexible multi-body systems. Multibody System Dynamics 3:209–239
5. Bauchau O A (2000) Analysis of flexible multibody systems with intermittent contacts. Multibody System Dynamics 4:23–54
6. Boussinesq J (1885) Applications des potentiels á l'étude de l'équilibre et du mouvement des solides élastiques. Gauthier–Villars, Paris

7. Crook A W (1952) A study of some impacts between metal bodies by a piezoeletric method. Proceedings of the Royal Society of London Seriers A 212:377–390
8. Cunningham D M, Goldsmith W (1958) Short impulses produced by longitudinal impact. Proceedings of the Society for Experimental Stress Analysis 16:153–165
9. Daparti M (2003) Numerical and experimental investigation of double impacts on a beam. MA Thesis, DIPL-99, Institute B of Mechanics, University of Stuttgart, Stuttgart
10. Donnell L H (1930) Longitudinal wave transmission and impact. Trans Amer Soc Mech Eng 52:153–167
11. Erhart T, Taenzer L, Diekmann R, Wall, W (2001) Adaptive remeshing issues for fast transient, highly nonlinear processes. In: Proceeding of European Conference on Computation Mechanics, Cracow, Poland
12. Ewins D (2001) Structural dynamics@2000 : current status and future directions. Research Studies Press, Baldock
13. Goldsmith W (1960) Impact: the Theory and Physical Behaviour of Colliding Solids. Edward Arnold Ltd, London
14. Hu B, Schiehlen W (1996) Eigenvalue, frequency response and variance bounds of linear damped systems. Eur. J. Mech. A/Solids 15:617–646
15. Hu B, Eberhard P (1999) Response bounds of linear vibrations. Journal of Applied Mechanics 66:997–1004
16. Hu B, Eberhard P (1999) Experimental and theoretical investigation of a rigid body striking an elastic rod. Report IB-32, Institute B of Mechanics, University of Stuttgart, Stuttgart
17. Hu B, Eberhard P, Schiehlen W (2001) Symbolical impact analysis for a falling conical rod against the rigid ground. Journal of Sound and Vibration 240:41–57
18. Hu B, Eberhard P (2001) Symbolic computation of longitudinal impact waves. Computer Methods in Applied Mechanics and Engineering 190:4805–4815
19. Hu B, Eberhard P, Schiehlen W (2003) Comparison of analytical and experimental results for longitudinal impacts on elastic rods. Journal of Vibration and Control, 9(1):157–174
20. Hu B, Schiehlen W (2003) Multi-time scale simulation for impact systems: From wave propagation to rigid body motion. Archive of Applied Mechanics, 72:885–897
21. Hu B, Seifried R (2002) Impact induced vibrations and the loss of energy. In: Proceedings of the 5th International Conference on Vibration Engineering, Nanjing, China, 18-20 September 2002. Haiyan Hu (ed.) Beijing: China Aviation Industry Press pp. 10–15
22. Johnson K L (1995) Contact Mechanics. Cambridge University Press, Cambridge
23. Kane T (1962) Impulsive motions. ASME J Appl Mech 15:718–732
24. Khulief Y A, Shabana A A (1986) Dynamics analysis of constrained systems of rigid and flexible bodies with intermittent motion. ASME J Mech Transmissions, Automat Design 108:38–44
25. Khulief Y A, Shabana A A (1987) A continuous force model for the impact analysis of flexible multi-body systems. Mech Mach Theory 22:213–224
26. Khulief Y A (2001) Dynamic response calculation of spatial elastic multibody systems with high-frequency excitation. Multibody System Dynamics 5:55–78
27. Lankrani H M, Nikravesh P E (1994) Continuous contact force models for impact analysis in multibody systems. Nonlinear Dynamics 5:193–207
28. Maekawa I, Tanabe Y, Suzuki M (1988) Impact stress in a finite rod. JSME International Journal Series I 31:554–560
29. Nicholson D W (1987) Eigenvalue bounds for linear mechanical systems with nonmodal damping.Mech Res Commun 14:115–122
30. Nicholson D W, Lin B (1996) Stable response of non-classically damped mechanical systems–II. Appl. Mech. Rev. 49:49–54

31. Nikravesh P E (1988) Computer-Aided Analysis of Mechanical Systems. Prentice Hall, Englewood Cliffs
32. Nour-Omid B, Wriggers P (1987) A note on the optimum choice for penalty parameters, Communications in Applied Numerical Methods, 3:581-585
33. Pfeiffer F, Glocker, C (1996) Multibody Dynamics with Unilateral Contacts. Wiley, New York
34. Polytec GmbH (1994) Vibrometer's manual for Polytec Vibrometer Series OFV-3000/OFV-302, OFV501 and OFV502. Manual No. VIB-MAN-9308-e04/01, Polytec GmbH
35. Ramamurti V, Ramanamurti P V (1977) Impact on short length bars. Journal of Sound and Vibration 53:529–543
36. Ripperger E A (1953) The propagation of pulses in cylindrical bars: an experimental study. In: Proceeding of the First Midwestern Conference on Solid Mechanics, Urbana IL, USA, pp. 29–39
37. Schiehlen W (1986) Technische Dynamik. Teubner, Stuttgart
38. Schiehlen W (1997) Multibody System Dynamics: Roots and Perspectives. Multibody System Dynamics 1:149–188
39. Schiehlen W (1999) Unilateral contacts in machine dynamics. In: Pfeiffer F, Glocker C (eds) IUTAM Symp. on Unilateral Multibody Contacts, Series: Solid Mechanics and its Applications 72, Kluwer Academic Publishers, Dordrecht, pp. 287-298
40. Schiehlen W, Seifried R (2002) Multiscale impact models: Interaction between multibody dynamics and wave propagation. In: Proceeding of the IUTAM Symposiom on Nonlinear Stochastic Dynamics, (accepted for publication)
41. Schiehlen W, Seifried R (2003) Multiscale methods for multibody systems. In: J Ambrosio (ed) Proceeding of the ECCOMAS Conference on Advances in Computational Multibody Dynamics 2003, Lisbon, Portugal, 1-4 July 2003, Lisbon, IDMEC, paper MB2003-094
42. Seifried,R (2001) Experimentelle und numerische Untersuchung von gestoßenen Kreisscheiben. Diplomarbeit DIPL-85, Institut B of Mechanics, University of Stuttgart, Stuttgart
43. Seifried R, Hu B, Eberhard P (2003) Numerical and experimental investigation of radial impacts on a half-circular plate. Multibody Systems Dynamics, 9:265–281
44. Shabana A A (1997) Flexible multibody dynamics: review of past and recent developments. Multibody System Dynamics 1:189–222
45. Shabana A A (1998) Dynamics of Multibody Systems. Cambridge University Press, Cambridge
46. Stronge W J (2000) Impact Mechanics. Cambridge University Press, Cambridge
47. St. Venant B D, Flamant M (1883) Résistance vive ou dynamique des solids. Représentation graphique des lois du choc longitudinal. Comptes rendus hebdomadaires des Séances de l'Académie des Sciences 97:127–353.
48. Szabó I (1977) Geschichte der mechanischen Prinzipien und ihrer wichtigsten Anwendungen. Birkhäuser, Basel
49. Timoshenko S P, Goodier J N (1970) Theory of Elasticity. McGraw Hill, New York
50. Wehage R, Haug E J (1982) Dynamic analysis of mechanical systems with intermittent motion. ASME J. Mech. Design 104:778–784
51. Zhong Z-H (1993) Finite Element Procedures for Contact-Impact Problems. Oxford University Press, New York

Unified Modelling Theories for the Dynamics of Multidisciplinary Multibody Systems

John McPhee

Systems Design Engineering, University of Waterloo, Canada
mcphee@real.uwaterloo.ca

1 Introduction

A major goal of research in multibody dynamics is to develop formulations that automatically generate and solve the governing equations of motion for a system of rigid and flexible bodies, given only a description of the system. Much progress has been made in multibody dynamics research over the last few decades and nowadays, there are several commercially-successful computer programs (e.g. Adams, Dads) that automatically analyze the dynamics of multibody mechanical systems.

However, these same programs, and the theoretical formulations on which they are based, are not capable of modelling general multidisciplinary applications in which a multibody system is coupled to other physical domains, e.g. electrical or pneumatic. There are numerous important applications of multidisciplinary multibody systems, including vehicles with active suspensions and traction control, mechatronic systems, and micro-electromechanical systems (MEMS). The design of these multidisciplinary applications would be greatly facilitated by algorithms that could automate their dynamic analysis.

There are two distinct approaches that have been proposed for modelling and simulating the dynamics of multidisciplinary multibody systems. The first is based on coupled simulations, or "co-simulation", in which two separate simulation programs or subroutines are coupled numerically. The advantage of co-simulation is that one can use existing programs that are very well-developed for their particular domain. However, numerical stability problems may arise during a co-simulation [20] and, more importantly, there is no underlying mathematical framework that one could use to generate analytical models of these multidisciplinary applications.

The second approach is to apply a unified systems theory to the dynamic modelling of multidisciplinary multibody systems. This paper focuses on this second approach for several reasons:

- a unified theory leads to analytical models that promote physical insight
- a unified theory can be applied manually or implemented in a computer algorithm
- the computer implementation can be symbolic or numeric

J.A.C. Ambrósio (ed.), Advances in Computational Multibody Systems, 125–154.

- symbolic models are very appropriate for real-time simulation
- symbolic models facilitate design optimization and sensitivity calculations
- symbolic models are easily communicated between colleagues and to students
- a unified theory can be extended to new situations and domains

Two main systems theories dominate the literature: linear graph theory and bond graph theory. Note that linear graph theory is sometimes abbreviated to "graph theory", especially in the literature on electrical circuits. A few authors [17, 22] have also proposed the principle of virtual work as the basis of a third unified theory, but these authors are forced to adopt elements of graph theory in their formulations, e.g. when generating topological equations for electrical subsystems. Thus, virtual work on its own does not constitute a complete and independent systems theory that can be used to automate the dynamic analysis of complex systems.

The goal of this paper is to present the modelling of multidisciplinary multibody systems using bond graph theory and linear graph theory, and to investigate their relative advantages and disadvantages.

2 Representative Problems

The features of modelling with bond graphs and linear graph theory will be demonstrated by means of four example problems, ranging from quite simple to very complex. Three of the examples are multidisciplinary applications, with components from the mechanical and electrical domains, and the last three examples contain multi-dimensional multibody subsystems.

2.1 Condenser Microphone

A simple model of a condenser microphone [3, 4] is shown in Figure 1. A voltage source E_4 is connected in series with a resistor R_1, capacitor C_2, and inductor L_3. The resistor and inductor are modelled by standard linear constitutive equations [10], but the upper plate of the capacitor is free to move, requiring extra consideration. This upper plate represents the mass (m_5), stiffness (k_6), and damping (d_7) of the mechanical portion of the microphone. For completeness, the gravitational force on the upper plate has also been included in the model.

Due to the electrical attraction of the two plates, a voltage v_2 across the capacitor results in an attractive force F_2 between the two plates given by [3]:

$$F_2 = -\frac{1}{2}\frac{dC_2}{dx}v_2^2 \tag{1}$$

where the capacitance C_2 is a function of the plate separation x; it decreases as x increases, i.e. $dC_2/dx < 0$, giving a positive force of attraction in equation (1).

The second constitutive equation for the capacitor is:

$$i_2 = C_2\frac{dv_2}{dt} + \frac{dC_2}{dx}\frac{dx}{dt}v_2 \tag{2}$$

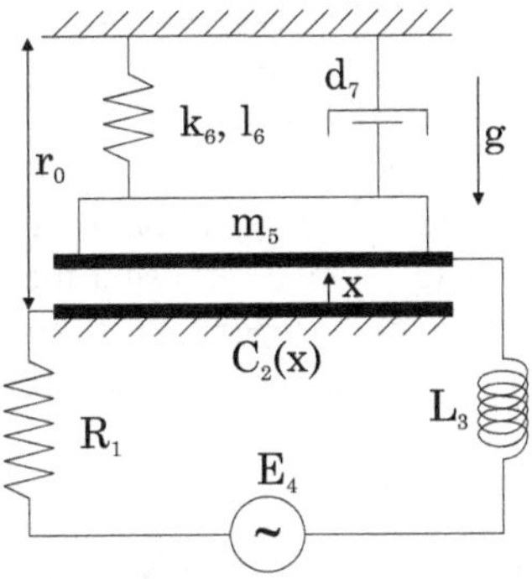

Fig. 1. Simple model of condenser microphone

where i_2 is the current through the capacitor; the second term on the right-hand side is induced by the relative motion of the two plates. Note that there are usually two constitutive equations for transducers, such as this moving-plate capacitor, that couple one physical domain (electrical) to a second (mechanical). It is through these transducers that energy can flow between the two domains.

In this condensor microphone, vibrations of the plate result in an electrical current that can be measured and amplified, if necessary. The goal of this example is to generate the system equations that relate the motion of the plate to the current.

2.2 Inverted double pendulum

This example, taken from Karnopp et al [5], is a planar mechanical multibody system. As shown in Figure 2, it consists of a horizontally-translating mass m, upon which an inverse double pendulum is mounted. The two links of this pendulum have lengths l_1 and l_2, and the masses m_1 and m_2 are assumed to be concentrated at the tips of the links. Gravity acts vertically downwards.

The purpose of this example is to demonstrate some features of multi-dimensional multibody systems, and the application of bond graphs and linear graph theory to these systems.

To automate the dynamic analysis of a multibody system, one must develop a formulation that can express all kinematic quantities in terms of a general set of

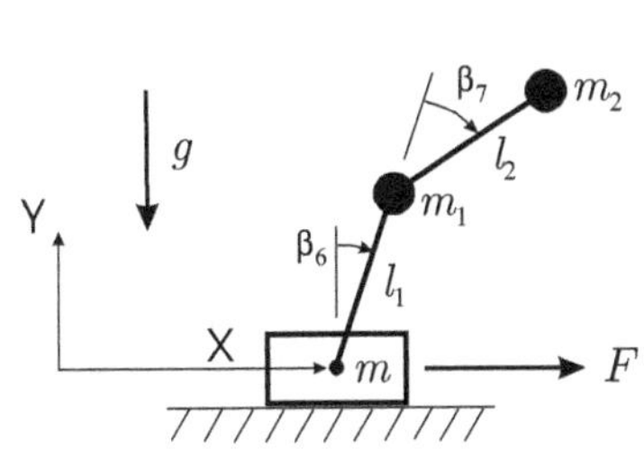

Fig. 2. Inverted double pendulum

coordinates, and generate the dynamic equations in terms of these coordinates. By far, the two most popular sets are absolute (Cartesian) coordinates and relative (joint) coordinates, although other possibilities do exist [23].

Absolute coordinates lead to relatively simple computer implementations, which is why they are used in commercially-successful packages such as Adams, Dads, and Working Model. These coordinates, which represent the position and orientation of every body in the system, will not be independent if there are any holonomic joints in the system. In that case, the n coordinates $\mathbf{q}$ are related by m nonlinear algebraic equations:

$$\mathbf{\Phi}(\mathbf{q}, t) = \mathbf{0} \tag{3}$$

where $n - m = f$, the degrees of freedom of the system. The dynamic equations are easily generated from free-body diagrams of each body, with the constraint reactions represented by Lagrange multipliers $\boldsymbol{\lambda}$:

$$\mathbf{M}\ddot{\mathbf{q}} + \mathbf{\Phi}_{\mathbf{q}}^{T}\boldsymbol{\lambda} = \mathbf{F} \tag{4}$$

where $\mathbf{M}$ is the constant $n \times n$ mass matrix, $\mathbf{\Phi}_{\mathbf{q}}$ is the Jacobian matrix of the constraint equations (3), and $\mathbf{F}$ contains external forces and quadratic velocity terms. Equations (3) and (4) constitute a set of $n + m$ differential-algebraic equations (DAEs) that can be solved for $\mathbf{q}(t)$ and $\boldsymbol{\lambda}(t)$.

Joint coordinates are more difficult to implement in an automated formulation, because one must pay more attention to topological processing [24]. However, the reward is fewer equations to solve than those expressed in absolute coordinates. For open-loop systems, i.e. those having no closed kinematic chains, the joint coordinates are independent and equal in number to the degrees of freedom f. There are no kinematic constraint equations (3) to satisfy, and a minimal set of f ordinary differential equations (ODEs) is obtained for the dynamics:

$$\mathbf{M}\ddot{\mathbf{q}} = \mathbf{F} \tag{5}$$

where the $f \times f$ mass matrix $\mathbf{M}$ is now a function of $\mathbf{q}$.

Since the inverted double pendulum is an example of an open-loop system, it is best modelled by joint coordinates.

2.3 Robot Manipulator

In this example, an experimental two-link robot manipulator is modelled. An overview of the experimental system is shown in Figure 3. It consists of two DC motors, a shoulder motor and an elbow motor, and two links. The interchangeable links may be rigid or flexible; in this example, we model the manipulator with rigid links. The manipulator is supported by air bearings on a large glass surface so as to minimize friction and gravitational effects. The shoulder motor may also be fixed to the glass surface by a vacuum, which is the case considered in this example. A variety of sensors are used to track the motion of the manipulator and a moving payload. These sensors provide feedback to a microcomputer that is responsible for generating control signals to the two motors. More details of the WatFlex experimental manipulator[1] can be found in [21].

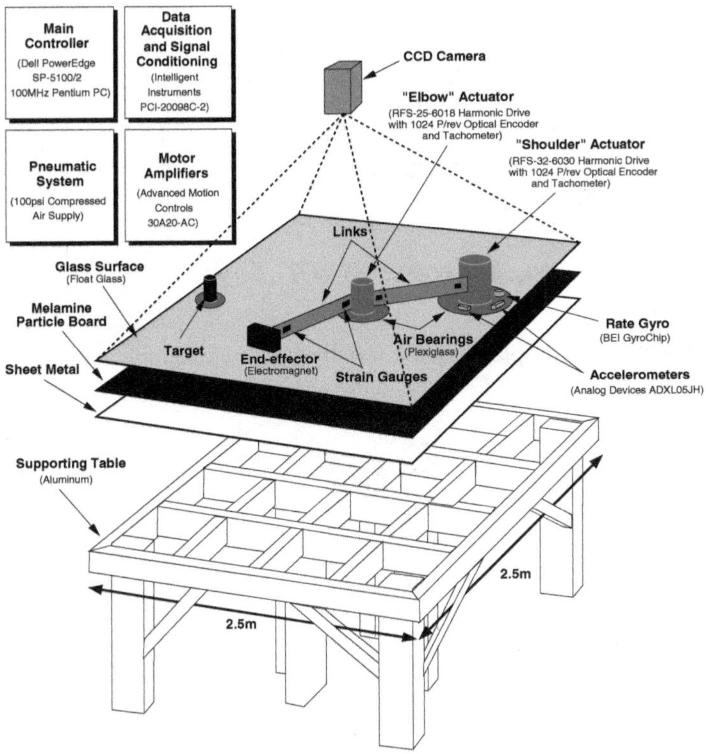

Fig. 3. Watflex robot manipulator

Shown in Figure 4 is a model of the two-link manipulator. We have chosen to include gravity (g) in the model, by re-orienting the robot such that the links rotate about horizontal joint axes. The first joint angle is designated θ_1, while the second is θ_{1-2}. Therefore, the absolute rotation θ_2 of the second link is the sum of the two joint angles.

The objective is to control the robot to follow a prescribed joint trajectory, specifically a rotation of both joints by 90 degrees in 4 seconds at constant angular speeds. These desired joint angles are input to a PD-controller that computes the differences between the desired and actual angles. The controller then supplies the two DC motors with voltages that are proportional to these differences (errors), and the time derivatives of these errors (i.e. the differences between desired and actual angular speeds).

[1] http://real.uwaterloo.ca/˜watflex

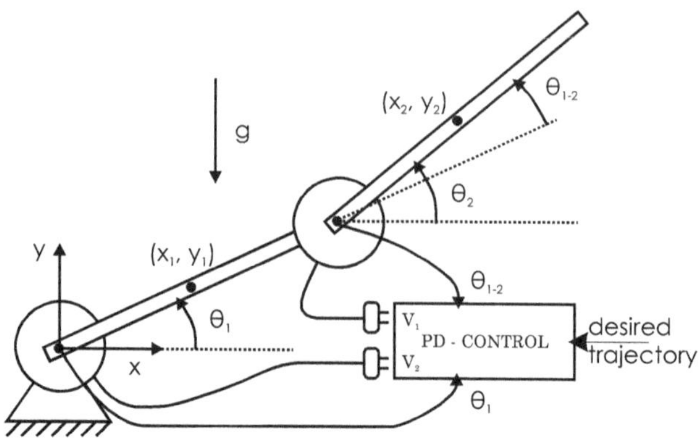

Fig. 4. PD-controlled robot manipulator

Similar to the moving-plate capacitor in the previous example, there are two constitutive equations for each DC motor in the model:

$$v = K_v \frac{d\theta}{dt} + Ri + L\frac{di}{dt} \tag{6}$$

$$T = K_T i - B\frac{d\theta}{dt} \tag{7}$$

where K_v is the voltage (back emf) constant, L and R are the armature inductance and resistance, respectively, K_T is the torque constant, and B is the coefficient of viscous friction in the bearings. The masses and inertias of the stator and rotor are assumed to be lumped with the mechanical component to which they are affixed.

2.4 Parking Gate System

In this fourth example, another electromechanical multibody system is considered. This system is used to raise and lower a flexible barrier, which is typically used to control access to a parking lot. The flexible barrier is 3 m long, and is rigidly connected to link P_2O_2 of the Watt-II six-bar mechanism [8] shown in Figure 5. A spring is used to counter-balance the weights of the moving links and flexible barrier, and an asynchronous 3-phase induction motor is used to drive the input link of the six-bar mechanism. More details of this system may be found in [13].

As shown in Figure 6, the induction motor is in a star-star configuration with a short-circuited rotor. Mutual inductance effects arise within each circuit, and also between the stator and rotor components — which is what converts the electrical currents into a driving torque.

The position of the rotor θ influences these mutual inductance effects, as can be seen from the constitutive equation for the motor torque T:

$$T = \frac{1}{2p}\sum_{j=1}^{p}\sum_{k=1}^{p}\frac{\partial}{\partial\theta}\left(\mathbf{M}_{jk} i_j i_k\right) \tag{8}$$

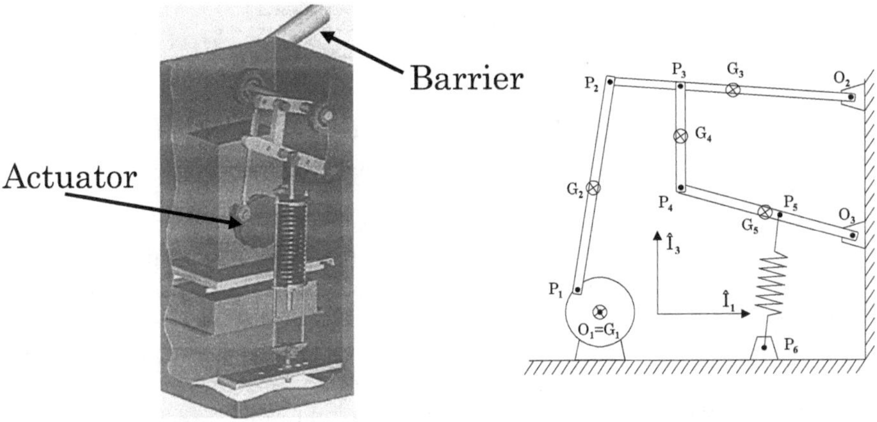

Fig. 5. Parking gate system

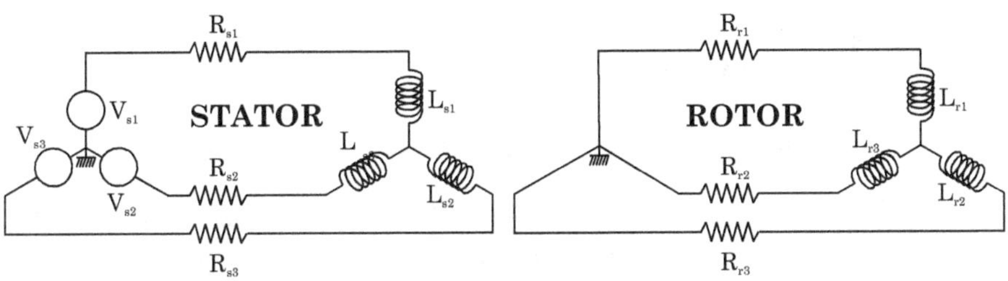

Fig. 6. Equivalent circuit for 3-phase actuator

where $p = 3$ is the number of pairs of poles, i_j is the current in inductor j, the index j ranges over the three inductors (L_{s1}, L_{s2}, L_{s3}) in the stator, k ranges over the three rotor inductors (L_{r1}, L_{r2}, L_{r3}), and $\mathbf{M}_{jk}$ is the entry in the jth row and kth column of the mutual inductance matrix defined by:

$$\mathbf{M} = M_{sr} \begin{pmatrix} \cos(\theta_{em}) & \cos\left(\theta_{em} + \frac{2\pi}{3}\right) & \cos\left(\theta_{em} + \frac{4\pi}{3}\right) \\ \cos\left(\theta_{em} + \frac{4\pi}{3}\right) & \cos(\theta_{em}) & \cos\left(\theta_{em} + \frac{2\pi}{3}\right) \\ \cos\left(\theta_{em} + \frac{2\pi}{3}\right) & \cos\left(\theta_{em} + \frac{4\pi}{3}\right) & \cos(\theta_{em}) \end{pmatrix} \tag{9}$$

where θ_{em} is the electromechanical position given by $\theta_{em} = p\theta$, and M_{sr} is the external mutual inductance between the stator and rotor.

The voltage v_i induced in each stator and rotor inductance is given by the second constitutive equation:

$$v_i = \sum_{j=1}^{p} \mathbf{L}_{ij} \frac{di_j}{dt} - \sum_{k=1}^{p} \frac{d}{dt} (\mathbf{M}_{ik} i_k) \tag{10}$$

where $\mathbf{L}_{ij}$ is the entry in the ith row and jth column of the inductance matrix of the stator or rotor in which inductor i resides:

$$\mathbf{L_s} = \begin{pmatrix} L_s & M_s & M_s \\ M_s & L_s & M_s \\ M_s & M_s & L_s \end{pmatrix} \qquad \mathbf{L_r} = \begin{pmatrix} L_r & M_r & M_r \\ M_r & L_r & M_r \\ M_r & M_r & L_r \end{pmatrix} \tag{11}$$

where L_s and L_r are the stator and rotor self-inductances, respectively, and M_s and M_r are the internal mutual inductances of the stator and rotor. In equation (10), the index j ranges over the inductors that are fixed with respect to the given inductor i, while the index k ranges over the inductors that are in relative motion. Thus, the first summation includes self-inductance terms, while the second summation represents the mutual inductance between stator and rotor components, which depends on the relative angle θ.

The flexible barrier is initially in a horizontal position, with no currents in the stator and rotor. A dynamic analysis and simulation is required to determine the response of the parking gate system to a sinusoidal input voltage (220V, 50Hz) over a period of 2 seconds.

3 Unified Modelling Theories

A perusal through monographs on "system dynamics" [5, 11] reveals that two unified modelling theories have become firmly established over the last few decades.

The first, linear graph theory, was invented by Leonhard Euler in 1736 to solve the famous Königsberg bridge problem [1]. Kirchoff applied graph theory to his analysis of electrical networks in the 1850s, and the generality of the graph theory approach in all physical domains was established by Trent [32] in 1955. The application of linear graph theory to physical system modelling is now well-established [6, 10, 11]. For multidisciplinary applications, the coupling between the different physical domains is not explicitly shown by the graph, but is embedded in the constitutive equations for the coupling elements (transducers).

The second unified modelling theory, bond graph theory, was invented by Henry Paynter [9] as an alternative notation in which energy flows between different physical domains are explicitly represented in the graph. Bond graphs were slow to be accepted by the engineering community, "mostly because of rather hazy mathematical underpinnings" [26]. Birkett and Roe [14], among others, have published a series of papers to address the lack of a combinatorial foundation for bond graphs. Nowadays, bond graphs are very well-known [2, 5, 12] and applied to multidisciplinary problems in both academia and industry.

Some authors [26] have stated that bond graphs are a special case of oriented linear graphs, while others [14] have concluded that bond graphs and linear graphs are distinct special cases of matroids. Regardless of the relationship between bond graph theory and linear graph theory, their representation of and application to multidisciplinary problems is very different in practice, especially for multibody systems.

In the following, a brief overview of bond graphs and linear graph theory is given, and their differences are highlighted through their application to the four example problems.

3.1 Bond Graph Theory

To develop a bond graph or linear graph model of a physical system, it is first decomposed into a finite number of discrete components. In bond graph theory, one distinguishes between "one-port" elements that connect to other components at two locations, and "two-port" elements that have four connection points. A one-port element is represented graphically by a single stroke, or bond, while a two-port element is depicted by two bonds. It is through these bonds that power flows through the system model; the two-port elements can provide an explicit representation of the energy transfer between different physical domains in a multidisciplinary application.

Associated with each bond is an effort (e) and flow (f) variable, the product of which gives power. For electrical components, the effort and flow corresponds to voltage and current, respectively. For mechanical systems, the effort and flow are usually taken as force and velocity, respectively, in what is known as the "force-effort" analogy [12]. In the less popular "force-flow" analogy, forces are flows and velocities are efforts.

By defining these generalized variables, and components with generalized constitutive equations in terms of these variables, one can develop bond graphs that represent multiple physical domains. The set of generalized one-port components includes resistances (R), inertias (I), capacitances (C), effort sources (S_e), and flow sources (S_f). A resistance, through which energy is lost from the system, can model an electrical resistor or a mechanical viscous damper. Similarly, an inertia can model an electrical inductor or a mechanical inertia, in the force-effort analogy. In the force-flow analogy, an inertia represents a mechanical spring. A summary of the bond graphs and constitutive equations for generalized one-port components is shown in Table 1

Table 1. Bond graphs and constitutive equations for generalized one-port components

Component	**Bond graph**	**Equation**
Effort source	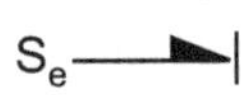	$e = e(t)$
Flow source	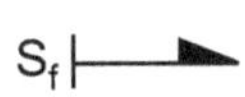	$f = f(t)$
Resistance		$e = e(f)$
Capacitance		$f = C\frac{de}{dt}$
Inertia		$e = I\frac{df}{dt}$

Note that the bond graphs shown in Table 1 have a half-arrow associated with them. This is to identify the direction used to measure positive power flow, a convention established by the modeller for each system. Usually, one tries to predict positive power flows, starting from energy sources and flowing to loads. Thus, the arrow is directed away from the effort and flow sources in Table 1. Since resistances dissipate power, usually in the form of heat, the arrow is directed towards the R component. Finally, power may flow to and from the capacitances and inertias; in Table 1, power flowing into these components (i.e. being stored) is chosen as positive.

Also shown in Table 1 are vertical bars at one end of each bond. This bar is known as a "causal" stroke, and is used to assign causality to a model, i.e. which of the efforts and flows are causes, and which are effects. The convention is that the causal stroke indicates the direction of effort, with flow acting in the reverse direction. Thus, the causal stroke is on the right side of the effort source in Table 1, and on the left side of the flow source. With linear R elements, there is no preferred causality; one may interpret the effort as causing the flow, or vice-versa. With linear inertias, the flow is proportional to the time integral of effort. The causal direction shown in Table 1 for the I element implies that effort is input to the element, and flow is the output. Thus, flow is being calculated from effort, which requires an integration. This is known as "integral causality"; the converse is derivative causality [12].

The assignment of causality to a bond graph model is used to facilitate the computation of the system equations. Causality conflicts can also reveal a fundamental modelling problem. By obtaining integral causality for a bond graph model, the governing equations will take the form of ordinary differential equations that can be readily solved using numerical integration methods. Derivative causality is to be avoided, due to the error-prone nature of computations involving numerical differentiation.

There are two basic types of two-port elements: transformers and gyrators [12]. In a transformer, the efforts in the two bonds are proportional, with the ratio known as the transformer modulus. Since transformers are energy-conserving elements, i.e. the power in one bond equals the power in the other, the flows in the two bonds must be equal to the reciprocal ratio. For gyrators, the effort in one bond is proportional to the flow in the second, and vice-versa for power conservation. The proportionality constant is the gyrator modulus. In modulated transformers and gyrators, the moduli are time-varying inputs.

From the topology of the physical system being modelled, the bond graphs of the discrete components are combined using 0 and 1 junctions, where 0 represents a parallel connection, and 1 represents a series connection. For all components connected by a 0 junction, all efforts are equal and all flows must sum to zero, taking into account the direction assigned to the half-arrows. For all components connected by a 1 junction, all flows are equal and all efforts sum to zero. Thus, the 0 and 1 junctions provide the topological equations for a given system.

By combining the topological equations with the constitutive equations, one has a necessary and sufficient set of equations to generate a complete system model. The bond graph approach is unified, since it can easily handle multidisciplinary applications, and very systematic. Thus, it is quite amenable to computer implementation and several computer programs have been developed using bond graph theory as their basis.

Full details of bond graph theory may be found in [5], [12], and [14]. In the following, the application of bond graph theory to the four examples is presented and discussed.

3.1.1 Bond Graph Model of Microphone

Shown in Figure 7 is a bond graph representation of the condenser microphone shown in Figure 1. A single bond graph provides a unified representation of the electrical and mechanical domains making up this multidisciplinary application. Note the use of the colon (:) notation in the figure to show the parameters associated with a generalized component model. Also, the effort and flow variables are shown explicitly for some of the bonds, with effort above the bond and the flow variable below.

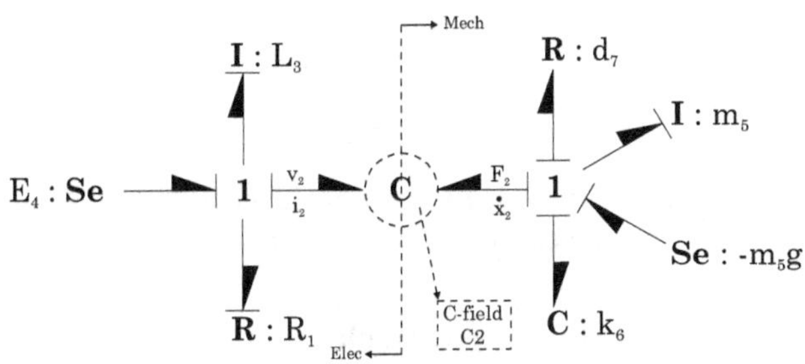

Fig. 7. Bond graph model of condenser microphone

The electrical circuit is modelled on the left using a resistance (R_1), capacitance (C_2), an inertia (L_3), and an effort source for the voltage driver (E_4). These four electrical components are connected in series, i.e. by a 1 junction. The assigned power flow directions are consistent with those shown in Table 1.

The mechanical subsystem is modelled on the right using a resistance for the damper (d_7), a capacitance for the spring (k_6), an inertia for the mass (m_5), and an effort source for the weight. The components in the physical system are connected in parallel; however, we are using the conventional force-effort analogy which requires that the bond graph components be connected by a series 1 junction. This is somewhat counter-intuitive, but it is needed to obtain the correct topological equations for the force-effort analogy. Using the force-flow analogy, parallel and series mechanical connections would be represented by parallel 0 and series 1 junctions, respectively. Regardless of whether one uses a force-effort or force-flow analogy, the bond graph does not bear much resemblance to the physical system.

The bond graphs for the mechanical and electrical domains are explicitly coupled by the capacitance element. In this case, the moving-plate capacitor is modelled as a "C-field" two-port component with the mechanical effort $e_2 = F_2$ and electrical flow $f_2 = i_2$ defined by the constitutive equations (1) and (2), respectively.

The system equations are derived very systematically from the bond graph. One starts with the topological equations for efforts from the 1 junctions:

$$v_4 - v_1 - v_3 - v_2 = 0 \tag{12}$$

$$F_2 + F_5 + F_6 + F_7 - F_g = 0 \tag{13}$$

where v_i and F_i is the voltage or force for component i, and F_g is the weight. The corresponding topological equations for flows are:

$$i_1 = i_3 = i_4 = i_2 \tag{14}$$

$$\dot{x}_2 = \dot{x}_6 = \dot{x}_7 = \dot{x}_g = \dot{x}_5 = \dot{x} \tag{15}$$

Combining the topological equations (12-15) with the constitutive equations for components, and re-arranging, one obtains:

$$C_2'\dot{x}\left(-R_1 i_2 - L_3\frac{di_2}{dt} + E_4(t)\right) + C_2\left(-R_1\frac{di_2}{dt} - L_3\frac{d^2 i_2}{dt^2} + \dot{E}_4\right) = i_2 \tag{16}$$

$$m_5\ddot{x} + d_7\dot{x} + k_6 x + m_5 g = \frac{1}{2}C_2'\left(-R_1 i_2 - L_3\frac{di_2}{dt} + E_4(t)\right)^2 \tag{17}$$

where $C_2' \equiv \frac{dC_2(x)}{dx}$. These two ODEs are equivalent to those derived by hand in [17], and can be solved for the capacitor current $i_2(t)$ and plate separation $x(t)$.

3.1.2 Bond Graph Model of Inverted Double Pendulum

When one switches from one-dimensional systems to multi-dimensional mechanical systems, the bond graph representation becomes considerably more complex. Consider the bond graph model of the inverted double pendulum, taken from [5] and shown in Figure 8. This bond graph requires some explanation since it bears little resemblance to the physical system in Figure 2.

The bond graph model consists mainly of one-port elements, plus two multi-port modulated transformers (MTFs). Although the system has only 3 degrees of freedom, it is modelled by 7 absolute speeds: $\dot{x}$ for the sliding mass, and $\dot{x}, \dot{y}, \dot{\theta}$ for each of the two links. Associated with each of these speeds is an inertia in the bond graph, e.g. the moment of inertia J_1 corresponding to $\dot{\theta}_1$. Note that $\theta_1 = \beta_6$, the coordinate for the first revolute joint in Figure 2, and $\theta_2 = \beta_6 + \beta_7$. Gravity and the applied force F are modelled as effort sources in the $-y$ and x directions, respectively.

Clearly, there are a lot of bonds making up the model of this relatively simple mechanical system. This is due to the fact that bond graphs were designed to operate on scalar variables. Hence, there is one bond for each of the 7 absolute speeds. One can combine the speeds associated with a given body into a single "multibond graph" [31] to obtain a simpler graph representation. However, the speeds are still represented by a column matrix of scalar variables, and not as a frame-invariant tensor, e.g. a Gibbs vector. Thus, the modeller must keep track of local reference frames and introduce manually-derived rotation transformations into the graph, as needed.

Note also that there are no components to explicitly represent the joints that connect the rigid bodies and constrain their absolute speeds. Thus, the 4 kinematic constraint equations (3) that express the interdependency of the 7 absolute coordinates must also be derived manually, with 2 equations obtained from each revolute joint.

The constraint forces that arise in the joints are converted into their corresponding Lagrange multipliers by the modulated transformers. Thus, these MTF elements

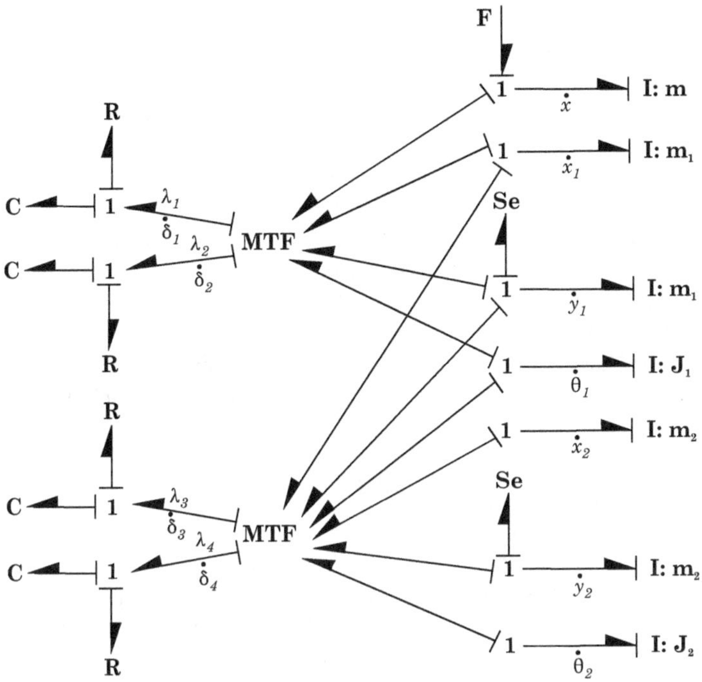

Fig. 8. Bond graph model of inverted double pendulum [5]

represent the rows of the Jacobian matrices (transposed) from equation (4); the top MTF in the bond graph corresponds to the Jacobian of the two kinematic constraints from the lower revolute joint in Figure 2, while the bottom MTF represents the Jacobian of the two constraint equations from the upper revolute joint.

Bond graphs were designed to operate on speeds, so that power is obtained from the product of efforts and flows. However, multibody system equations are generally expressed in terms of displacements, which are needed to calculate kinematic constraint equations, their Jacobians, and kinematic transformations between frames. These displacements are not represented as effort or flow variables in the bond graph, and must be obtained from numerical integration of the absolute speeds.

The use of absolute coordinates results in large systems of DAEs corresponding to equations (3) and (4). To avoid having to solve DAEs, resistance and capacitance elements are added to the bond graph to model damping and stiffness in the joints [33]. The translational speeds $\dot{\delta}$ across the joint, which are zero for an ideal joint model, are the complementary variables to the Lagrange multipliers. By modelling the joints in this fashion, the constraint equations are only needed to compute the Jacobians for the MTF elements, and the DAEs are converted to stiff ODEs. For the inverted double pendulum modelled with 7 absolute speeds, the 11 DAEs are reduced to 7 ODEs. Numerical integration of the latter may require special solvers for stiff systems, and will not guarantee that the 4 constraint equations are satisfied.

Tiernego and Bos [31] use multibond graphs to analyze the dynamics of open-loop multibody systems and, by applying well-known velocity transformations from absolute speeds to joint speeds [18], they are able to generate a minimal set of ODEs corresponding to equation (5). However, their procedure cannot handle the closed kinematic chains that typically arise in mechanical systems, and it seems that much of their analysis (e.g. derivation of velocity transformations) must also be performed manually before the bond graph is constructed.

Favre and Scavarda [15] present an extension of the work by Tiernego and Bos to systems with closed kinematic chains, but it seems that manual derivations of kinematic transformations are still required (no equations are presented for their two examples). The current state of bond graph modelling of multibody systems is best summarized by Karnopp et al [5]: "It is true that low-order, linear systems can be simulated with virtually no effort from the user. But complex nonlinear systems do require significant user input".

3.1.3 Bond Graph Model of Robot

Consider the bond graph representation, shown in Figure 9, of the two-link PD-controlled robot manipulator from Figure 4. The bond graph is quite complex, making it difficult to interpret, but it does constitute a single unified representation of this multidisciplinary application.

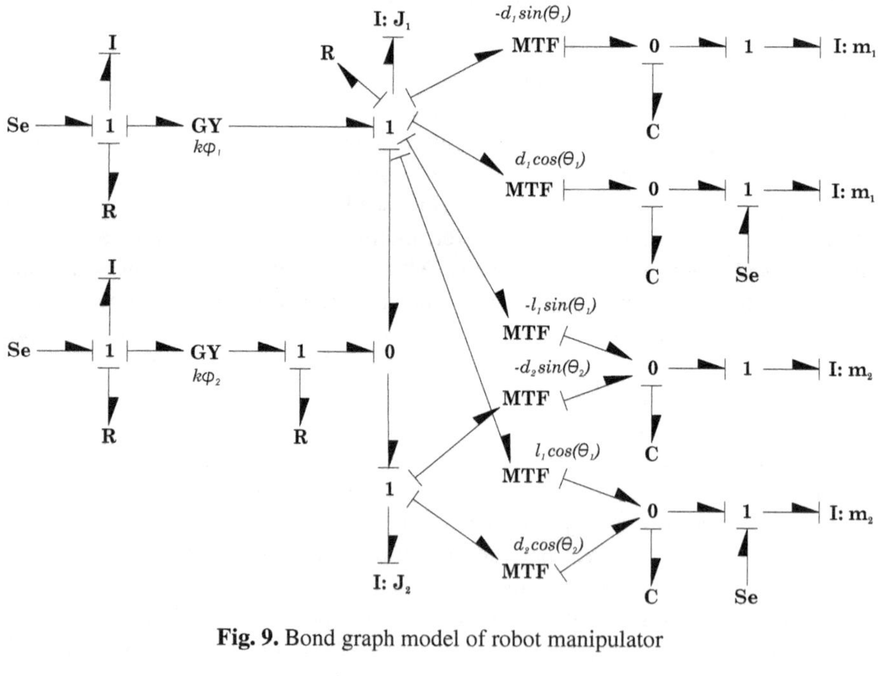

Fig. 9. Bond graph model of robot manipulator

The two DC-motors are modelled on the left of the bond graph. The applied voltage (coming from the PD-controller) is represented as an effort source, while the I and R elements correspond to the armature inductances and resistances, respectively. These elements are connected in series (1 junction) with a gyrator (GY), that couples the electrical domain to the mechanical domain. Substituting the component equations into the sum of efforts for the 1 junction gives the constitutive equation (6) for the DC-motor model.

The mechanical portion of this bond graph is somewhat different from that shown in Figure 8 for the inverted double pendulum. The mechanical subsystem is still represented by absolute speeds, with inertias corresponding to each of these 6 speeds. The joints are again modelled by stiffness (C) elements, but damping across the joint is not included. The modulated transformers are now used to convert between the Cartesian $\dot{x}, \dot{y}$ speeds and the absolute rotational speeds $\dot{\theta}_1$ and $\dot{\theta}_2$, i.e. velocity transformations. These transformations are manually derived and shown in the bond graph as parameters for the MTF two-port elements. The absolute rotational speeds are easily converted into joint angular speeds using 0 and 1 junctions. The rotational damping in the DC motor, seen in equation (7), is modelled by a resistance acting on the joint speed.

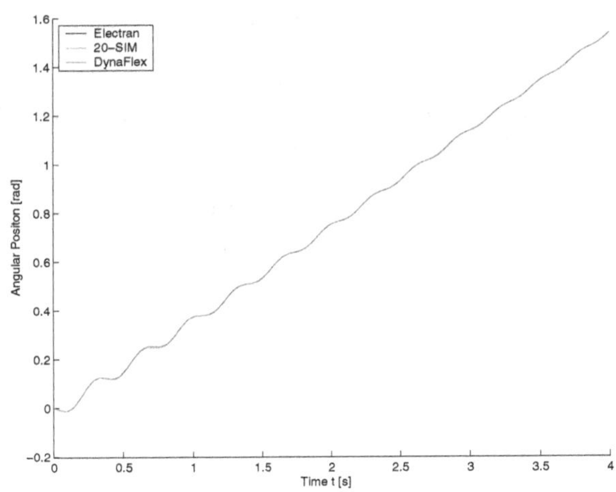

Fig. 10. Angular displacement of first joint versus time

To evaluate the performance of the PD-controller, the dynamic response of the robot is simulated by numerically integrating the 6 ODEs for the absolute coordinates. This was accomplished using the commercial software package 20-sim[2], with which the system equations can be generated from the bond graph model and numerically integrated. The 20-sim results for the angular displacement and speed of the first joint are shown in Figures 10 and 11, respectively. Also shown are the re-

[2] www.20sim.com

sults from two other software packages: DynaFlex, based on linear graph theory, and Robotran/Electran [13, 16], based on the virtual work principle and Kirchoff's laws.

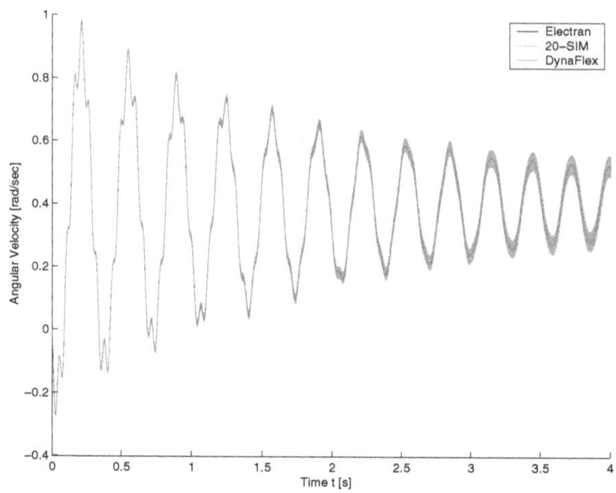

Fig. 11. Angular speed of first joint versus time

Because the robot is started from rest and immediately commanded to move at a constant speed of 90/4 deg/s (0.39 rad/s), large voltages are supplied by the PD controller to accelerate the system. This sets up an oscillation in the system response that is still evident at the end of the simulation. The plot of joint speed shows that the response is slowly converging to 0.39 rad/s, and that the bond graph results from 20-sim are in good agreement with those from the two other programs. However, additional high-frequency oscillations are visible in 20-sim velocity plot, resulting from the use of stiff springs to model the revolute joints. In contrast, the Electran and DynaFlex programs make direct use of rigid joint models, and their responses do not exhibit this high-frequency oscillation. As a result, they are able to simulate this problem faster than 20-sim.

3.1.4 Bond Graph Model of Gate System

Shown in Figure 12 is a bond graph model, created by Sass et al [27], of the 3-phase induction motor used to actuate the parking gate system in Figure 5. The inductive effects and the electromechanical coupling are modelled by a mixed IC-field for which the corresponding inductance matrices $\mathbf{L_s}$, $\mathbf{L_r}$, and $\mathbf{M}$ are defined by equations (9) and (11). The star-star connection shown in Figure 6 is modelled by the two 0 junctions that force the sum of the three currents to be zero. This star-star connection results in two constraints that can be seen in the bond graph as two derivative causalities for the I part of the IC-field.

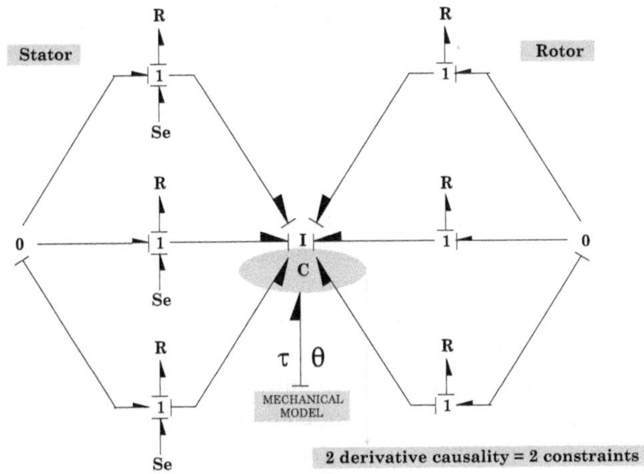

Fig. 12. Bond graph model of 3-phase actuator

A multibond graph of the 6-bar mechanism is shown in Figure 13. Using the approach of Favre and Scavarda [15] for this mechanism, every vector is assumed to be resolved in the inertial reference frame, and the center of mass is taken as the point of reference for each body. Thus, the bond graph takes the shape of a diamond for each body, as can be seen in the figure. This bond graph contains 5 moving bodies (for simplicity, the flexible beam is not shown in the graph) and the loop constraints are imposed by means of the zero-velocity flow source ($S_f = 0$). This method for opening and closing kinematic chains is only valid for revolute joints connected to the ground. Closing the loops elsewhere would require the manual calculation of the relative velocities of the two points connected by the revolute joint. Other joint types, e.g. prismatic or universal, would be even more problematic.

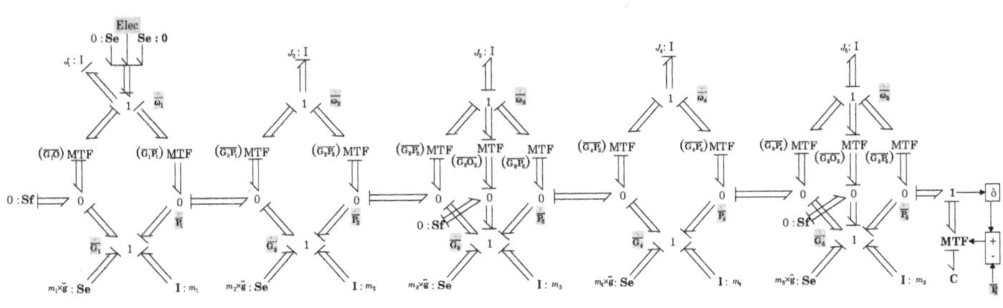

Fig. 13. Bond graph model of parking gate mechanism

One can see that the causality assignment and derivation of equations is not straightforward for a multidisciplinary system of this complexity. Furthermore, the manual derivation of kinematic transformations is very tedious for a system containing several bodies, especially if some of them are flexible. As a result of these difficulties, a numerical simulation of the system response using bond graphs and the 20-sim software was abandoned.

3.2 Linear Graph Theory

Linear graph theory is a branch of mathematics devoted to the study of system topology. It has been combined with the characteristics of physical components to obtain a unified systems theory for modelling multidisciplinary applications. The term "graph-theoretic modelling" (GTM) is often used to denote this systems theory. In a nutshell, a system model is obtained by combining topological relationships from linear graph theory with the constitutive equations for individual components. This systems theory is very methodical and well-suited to computer implementation.

To model a physical system, individual components are identified and their constitutive equations are determined. In general, these constitutive relationships are obtained from experimental measurements of the component's "through" and "across" variables; through variables (τ) are measured by an instrument in series with the component, while across variables (α) are obtained from an instrument in parallel. For electrical systems, the through and across variables are current and voltage, respectively. For mechanical systems, force and displacement (or its derivatives) play the role of through and across variables, respectively. Note that through and across variables may be tensors of any order, including scalars and vectors.

One can see that the through and across variables correspond (approximately) to the flow and effort variables, respectively, from bond graph theory. However, the definition of through and across variables as experimental measurements naturally results in the force-flow analogy for mechanical systems. Furthermore, mechanical displacements are represented explicitly, and not as numerical integrals of velocities. Finally, the GTM approach is not restricted to scalar variables.

Once the constitutive equations are determined, the component models are combined in the topology defined by the structure of the physical system. A linear graph, consisting of lines (edges) and circles (nodes or vertices), is used to represent the system topology. The edges represent the individual components, whereas nodes represent the points of their interconnection. From this graph, linear topological equations are systematically obtained in terms of the through and across variables for all components. The system model is simply the combination of these topological equations with the individual constitutive equations.

Graph-theoretic modelling has been applied to a wide variety of disciplines and multidisciplinary applications, including electrical and mechanical systems [6, 10], electromechanical multibody systems [28], and electrohydraulic multibody systems [25]. The essential features of graph-theoretic modelling are presented in the following by means of the four example problems.

3.2.1 Linear Graph Model of Microphone

For the condenser microphone from Figure 1, a linear graph representation is shown in Figure 14. Edges R_1, C_2, L_3, and E_4 represent the resistor, capacitor, inductor, and voltage source, respectively. Note that the linear graph resembles the physical system, which is an advantage when it comes to modelling using this approach. Directions are assigned to each edge to establish a positive convention for measuring the through and across variables, similar to setting the polarity on a measuring instrument. The constitutive equations for electrical components are expressed in terms of the scalar variables, current (i) and voltage (v). For the purpose of this example, we assume standard linear relationships for these components, e.g. $v_1 = R_1 i_1$ and $v_3 = L_3 \frac{di_3}{dt}$, but the constitutive equations may be highly nonlinear in general.

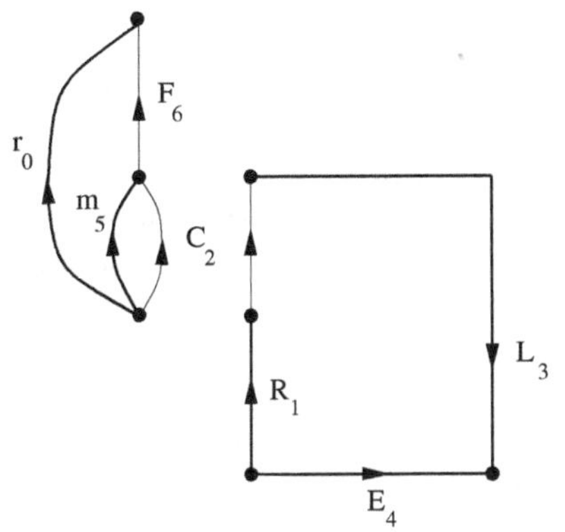

Fig. 14. Linear graph model of condenser microphone

Also shown in Figure 14 is the linear graph of the mechanical part of the condenser microphone. The edge m_5 represents the inertia and weight of the moving mass; the edge begins at a ground-fixed (inertial) reference node and terminates at the center of mass. Its constitutive equation is given by the combination of gravity with the d'Alembert form of Newton's Second Law: $\mathbf{F}_5 = m_5\mathbf{g} - m_5\boldsymbol{a}_5$, where the vector force $\mathbf{F}_5$ depends on gravity $\mathbf{g} = -g\hat{\boldsymbol{\imath}}$, the vector acceleration $\boldsymbol{a}_5 = \ddot{x}\,\hat{\boldsymbol{\imath}}$, and the upwards unit vector $\hat{\boldsymbol{\imath}}$ (parallel to x). The edge F_6 represents the combined effects of the spring and damper components (these could easily be split into separate edges for the spring and damper, if desired). Its vector constitutive equation is $\mathbf{F}_6 = -k_6(|\mathbf{r}_6| - l_6)\hat{\boldsymbol{r}}_6 - d_7(\mathbf{v}_6 \cdot \hat{\boldsymbol{r}}_6)\hat{\boldsymbol{r}}_6$, where l_6 is the undeformed spring length, k_6 and d_7 are the stiffness and damping coefficients, $\mathbf{v}_6$ is the relative velocity of the endpoints, and $\hat{\boldsymbol{r}}_6 = \mathbf{r}_6/|\mathbf{r}_6|$ is the unit vector parallel to the component. Finally, the edge r_0 locates the point where the spring-damper is attached to the ground: $\mathbf{r}_0 = l_0\,\hat{\boldsymbol{\imath}}$.

Note that the graph consists of two parts, one for each physical domain. This is always the case for linear graph models of multidisciplinary applications. The parts of the graph are not coupled explicitly, but by the constitutive equations for transducer components, which have an edge in each of the coupled domains. In this example, the electrical and mechanical domains are coupled by the moving-plate capacitor, which is characterized by two constitutive equations: the scalar equation

(2) for the electrical edge and the vector equation $\mathbf{F}_2 = -F_2\hat{\imath}$ for the mechanical domain, where the attractive force F_2 is defined by equation (1) and acts in the $-X$ direction.

For each of the two parts of the linear graph, mechanical and electrical, we can generate sets of topological equations that relate the through and across variables. This can be done manually by inspection of the graph, or by applying matrix operations to an "incidence matrix" that encapsulates the topology of the physical system. For a linear graph with e edges and v vertices, entry I_{jk} of the $e \times v$ incidence matrix $\mathbf{I}$ is [0, -1, or +1] if edge k is [not incident upon, incident and away from, or incident and towards] the vertex j.

The Vertex Postulate [6] then allows us to write:

$$\mathbf{I}\boldsymbol{\tau} = 0 \tag{18}$$

where $\boldsymbol{\tau}$ is the column matrix of through variables for all edges. For electrical systems, the Vertex Postulate reduces to Kirchoff's Current Law at every node. For mechanical systems, the Vertex Postulate gives v equations for dynamic equilibrium.

Starting from the Vertex Postulate, two very useful sets of topological equations, the "cutset" and "circuit" equations, can be systematically derived by selecting a tree and applying elementary matrix operations to $\mathbf{I}$. For electrical networks, the circuit equations correspond to Kirchoff's Voltage Law around a closed circuit, while the cutset equations are linear combinations of the vertex equations for all the nodes in a given subgraph.

A tree is a set of $v - 1$ edges ("branches") that connects all of the vertices but does not contain any closed loops. A very attractive feature of linear graph theory is that *by selecting a tree, one can control the primary variables appearing in the final system*: they are the across variables $\boldsymbol{\alpha}_b$ for branch elements, and the through variables $\boldsymbol{\tau}_c$ for cotree elements ("chords"). This is accomplished by:

- re-writing the cutset equations as the chord transformations $\boldsymbol{\tau}_b = -\mathbf{A}_c\boldsymbol{\tau}_c$, where $\boldsymbol{\tau}_b$ are the branch through variables and $\mathbf{A}_c$ is obtained from elementary row operations on $\mathbf{I}$
- by re-writing the circuit equations as the branch transformations $\boldsymbol{\alpha}_c = -\mathbf{B}_b\boldsymbol{\alpha}_b$, where $\boldsymbol{\alpha}_c$ are the cotree across variables

The Principle of Orthogonality, which represents a very generalized energy conservation principle, guarantees that $\mathbf{B}_b = -\mathbf{A}_c^T$.

By selecting edges R_1, L_3, and E_4 into the tree for the electrical sub-graph in Figure 14, one gets the chord transformations:

$$\begin{Bmatrix} i_1 \\ i_3 \\ i_4 \end{Bmatrix} = \begin{bmatrix} 1 \\ 1 \\ -1 \end{bmatrix} i_2 \tag{19}$$

and the single branch transformation:

$$v_2 = -\begin{bmatrix} 1 \\ 1 \\ -1 \end{bmatrix}^T \begin{Bmatrix} v_1 \\ v_3 \\ v_4 \end{Bmatrix} \tag{20}$$

Assuming that there is one constitutive equation for each of the v elements, substituting the branch and chord transformations into these constitutive equations will result in v system equations in terms of the v primary variables.

It is possible to reduce the equations to an even smaller set by exploiting the linear nature of the electrical constitutive equations. One approach is to generate one equation for each capacitor and inductor, and to use the remaining constitutive equations and branch/chord transformations to express all other variables in terms of the capacitor voltages and inductor currents. This approach was successfully implemented by Muegge [7]. For the electrical portion of the linear graph shown in Figure 14, one would get two first-order ODEs in terms of v_2 and i_3.

Another approach is to express all variables in terms of the currents associated with chords, or the voltages associated with branches. The former is called the current formulation, while the latter is named the voltage formulation; both were implemented in the Maple symbolic programming language by Scherrer and McPhee [28]. By selecting the tree appropriately, one can significantly reduce the final number of system equations.

For the example shown in Figure 14 with the capacitor C_2 selected into the cotree, the current formulation will give a single second-order ODE in terms of the corresponding current i_2.

This is accomplished by substituting the chord transformations in (19) into the constitutive equations for the branches, giving:

$$\begin{aligned} v_1 &= R_1 i_2 \\ v_3 &= L_3 \frac{di_2}{dt} \\ v_4 &= E_4(t) \end{aligned}$$

where $E_4(t)$ is the prescribed voltage source. Substituting these constitutive equations into the branch transformation (20) gives:

$$v_2 = -R_1 i_2 - L_3 \frac{di_2}{dt} + E_4(t) \tag{21}$$

which expresses the capacitor voltage in terms of its current. Substituting this equation into the electrical constitutive equation (2), the single ODE for the electrical domain is obtained:

$$C_2' \dot{x} \left(-R_1 i_2 - L_3 \frac{di_2}{dt} + E_4(t) \right) + C_2 \frac{d}{dt} \left(-R_1 i_2 - L_3 \frac{di_2}{dt} + E_4(t) \right) = i_2 \tag{22}$$

where $C_2' \equiv dC_2/dx$ and the primary electrical variable is the cotree current i_2.

For the mechanical domain, the fixed vector r_0 and mass m_5 are selected into the tree shown as bold edges in Figure 14, resulting in only one primary across variable $r_5 = x$. This is an independent coordinate for the 1-dof mechanical subsystem, and a single dynamic equation is obtained from the cutset equation for the mass, projected onto its motion space defined by $\hat{\imath}$:

$$(\mathbf{F}_5 + \mathbf{F}_2 - \mathbf{F}_6 = 0) \cdot \hat{\imath} \tag{23}$$

Substituting the mechanical constitutive equations into this expression, evaluating, and re-arranging,

$$-m_5\ddot{x} - m_5 g + \frac{1}{2}\frac{dC_2}{dx}v_2^2 + d_7\dot{x}_6 + k_6(x_6 - l_6) = 0 \tag{24}$$

Assuming that the spring is unstretched at $x = 0$, which implies that $l_6 = r_0$, one gets the branch transformations:

$$x_2 = x$$
$$x_6 = r_0 - x, \qquad \dot{x}_6 = -\dot{x}$$

which shows that the spring-damper shortens as x increases. Substituting these branch transformations and the capacitor voltage (21) into equation (24), one gets the single ODE for the mechanical domain:

$$m_5\ddot{x} + d_7\dot{x} + k_6 x + m_5 g = \frac{1}{2}\frac{dC_2}{dx}\left(-R_1 i_2 - L_3\frac{di_2}{dt} + E_4(t)\right)^2 \tag{25}$$

Together, equations (22) and (25) can be solved for the primary variables $i_2(t)$ and $x(t)$. Equations (22) and (25) are equivalent to those derived by hand in [17].

3.2.2 Linear Graph Model of Inverted Double Pendulum

The same basic concepts apply when one models a multi-dimensional mechanical (multibody) system using linear graph theory: the system model is obtained by combining the constitutive equations for individual components with the linear cutset and circuit equations resulting from their connectivity. Again, the selection of a tree determines the primary variables appearing in the system equations. The cutset and circuit equations retain a simple form because linear graph theory allows the use of vector modelling variables. However, the constitutive equations for some components will be nonlinear due to the finite rotations of bodies in the system. Furthermore, the physical interpretation of nodes and edges must be generalized.

To illustrate, consider the inverted double pendulum shown in Figure 2, and its linear graph representation in Figure 15. For the sake of clarity, the three bodies are superimposed on the graph with dashed lines. One can see that the topology of the physical system is closely mirrored by the structure of the linear graph.

Each node in the linear graph represents the position and orientation of a body-fixed reference frame, while the edges represent transformations between frames corresponding to physical components. For each element, there are now two sets of through and across variables: translational and rotational. Thus, there will be two sets of cutset and circuit equations, since these variables cannot be added together. Although the incidence matrix is the same for each, selecting different trees can be used to create different cutset and circuit equations for translation and rotation. This can be used to reduce the system equations to a set that is smaller in number than those generated by conventional multibody formalisms [23].

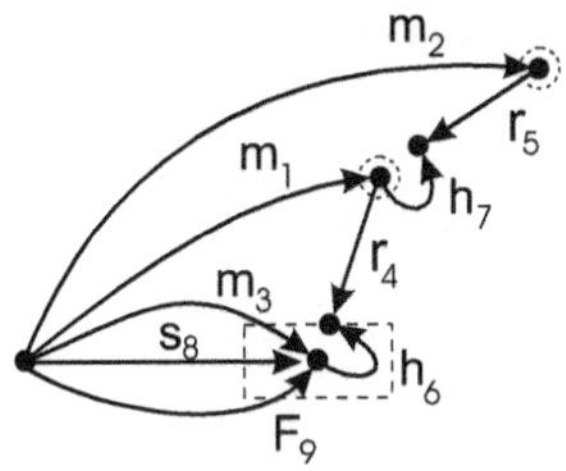

Fig. 15. Linear graph model of inverted double pendulum

In Figure 15, the edges m_1, m_2, and m_3 represent both the translational and rotational inertia of the three rigid bodies (slider and two links). These bodies are connected by the revolute joints h_6 and h_7, and by the prismatic joint s_8 between the slider and the ground. The "rigid arm" elements r_4 and r_5 define the position and orientation, relative to the center of mass frames on the bodies, of the body-fixed frames that define the connection points of these joints. Finally, the external force on the slider is modelled by the force element F_9, originating at the inertial frame (node) and terminating at the slider.

The constitutive equations for the multi-dimensional translation of rigid bodies (and spring-dampers) are the same as that shown in the previous section. However, a second equation relates the d'Alembert torque on the body to its rotational inertia. This equation corresponds to Euler's equations for rotational motion.

For the rigid-arm elements, e.g. r_4, the tip node does not rotate relative to the tail (center of mass) node; hence, the angular velocity (e.g. ω_4) is zero. However, the translational velocity of the rigid-arm is a nonlinear function of the angular velocity of the body on which it resides, e.g. $\mathbf{v}_4 = \boldsymbol{\omega}_1 \times \mathbf{r}_4$, which is a well-known result from rigid body kinematics. For the ideal joints, one always finds that the motion allowed by a joint, e.g. $\mathbf{r}_8 = x\hat{\imath}$ where $\mathbf{r}_8$ is the translational displacement of the slider along X, is orthogonal to the reaction forces and torques that arise in the joint, e.g. $\mathbf{F}_8 = F_8\,\hat{\jmath}$ and $\mathbf{T}_8 = T_8\hat{k}$ where $\hat{\jmath}$ and $\hat{k}$ are unit vectors parallel to Y and Z, respectively. This is a result of the fact that ideal joints do no work, and can be used to eliminate joint reactions in the system dynamic equations.

The topological equations remain linear regardless of the nonlinearities in the constitutive equations. Furthermore, the selection of trees can again be used to define the primary variables $\mathbf{q}$ and $\boldsymbol{\lambda}$ in the final system equations, where the "branch coordinates" $\mathbf{q}$ are the unknown across variables for elements (branches) in the tree.

By selecting into the tree those components with known across variables, i.e. r_4 and r_5, the number n of branch coordinates (and system equations) is reduced. If the tree is completed by m_1, m_2, and m_3, then the final equations are in terms of the absolute coordinates for the three bodies. In that case, the system equations take the form of the DAEs (3) and (4).

If joints h_6, h_7, and s_8 are selected into the tree in place of m_1, m_2, and m_3, then one obtains equations in the joint coordinates β_6, β_7, and x; for this open-loop system, the governing equations would then take the form of the ODEs shown in

(5). Thus, linear graph theory provides a unification of traditional absolute and joint coordinate formulations.

Any joints left in the cotree, e.g. h_6 in the absolute coordinate formulation, will provide the reaction loads appearing in the Lagrange multipliers $\boldsymbol{\lambda}$. Furthermore, these cotree joints will also provide one kinematic constraint equation for each reaction load. These m constraint equations (3) express the relationships between the branch coordinates, which will not be independent if there are joints in the cotree; this is always the case for a system with closed kinematic chains. These constraint equations are always found by projecting the circuit equations for the cotree joints onto their reaction spaces.

To demonstrate, consider a mixed-coordinate formulation that results from selecting m_1, m_2, r_4, r_5, and s_8 into the tree for Figure 15. The corresponding branch coordinates $\mathbf{q} = [x_1, y_1, \theta_1, x_2, y_2, \theta_2, x]^T$ are identical to those used in the previous bond graph model of this system. (Note that the joint speed $\dot{x}$ is the same as the absolute speed used in the bond graph model, which neglected the other two absolute speeds for the sliding mass in an ad hoc manner). The two revolute joints left in the cotree, h_6 and h_7, each provide 2 Lagrange multipliers and 2 constraint equations to the system DAEs, for a total of $m = 4$ constraints. For example, one can generate the translational circuit equation for joint h_6:

$$\mathbf{r}_1 + \mathbf{r}_4 - \mathbf{r}_6 - \mathbf{r}_8 = 0 \tag{26}$$

where $\mathbf{r}_6 = 0$ and $\mathbf{r}_8 = x\hat{\imath}$ from the constitutive equations for revolute and prismatic joints, respectively. For the rigid body, $\mathbf{r}_1 = x_1\hat{\imath} + y_1\hat{\jmath}$, and $\mathbf{r}_4 = -l_1 \sin\theta_1 \hat{\imath} - l_1 \cos\theta_1 \hat{\jmath}$ for the body-fixed vector.

Note that the circuit equations represent the zero summation of displacement vectors around a closed kinematic chain. The reaction space for h_6 is spanned by unit vectors $\hat{\imath}$ and $\hat{\jmath}$, since the revolute joint prevents translations along these two axes (but allows rotation along $\hat{k}$, which defines the joint motion space). Projecting the vector circuit equation (26) onto these two unit vectors, and substituting all constitutive equations, results in the kinematic constraint equations:

$$x_1 - l_1 \sin\theta_1 - x = 0 \tag{27}$$

$$y_1 - l_1 \cos\theta_1 = 0 \tag{28}$$

which are easily verified by hand. These equations, which are very systematically generated in terms of $\mathbf{q}$, correspond to two of the four rows in the constraint equations (3) for this example. The other two equations are obtained in exactly the same manner from the cotree joint h_7.

To obtain the dynamic equations of the system, the cutset equations for each branch are projected onto the motion space for that branch. This is why, in the previous example, the cutset equation for the capacitor mass m_5 was projected onto the unit vector $\hat{\imath}$ defining its motion space.

Note that one can also generate the dynamic equations by combining linear graph theory with analytical mechanics, e.g. the principle of virtual work. This approach is very useful for incorporating flexible bodies into the multibody system model

[30]. It has been implemented using the Maple symbolic programming language into a multibody dynamics program called DynaFlex[3], which can reduce the DAEs to ODEs by means of symbolic coordinate partitioning. DynaFlex can be also be used to model electromechanical multibody systems, as shown in the next section.

3.2.3 Linear Graph Model of Robot

For the PD-controlled robot manipulator in Figure 4, the linear graph representation is shown in Figure 16.

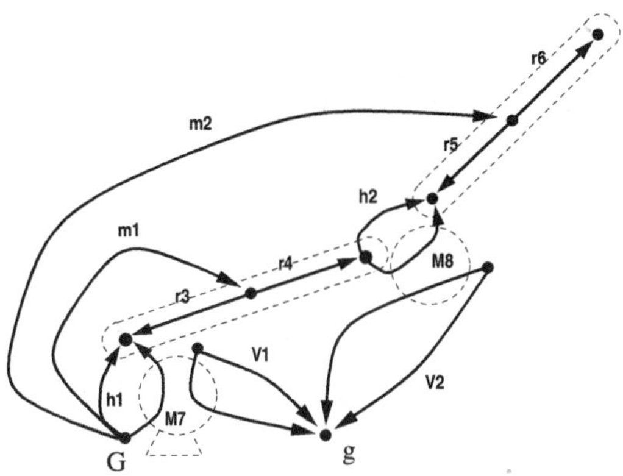

Fig. 16. Linear graph model of robot manipulator

The mechanical subgraph is obtained by applying the systematic procedure described in the previous section:

- one node is added for each body-fixed reference frame, e.g. a center of mass frame, joint connection point, or force application point;
- these nodes are connected by edges corresponding to different physical components (rigid body m, revolute joint h, rigid arm r).

In a similar manner, the electrical subgraph is directly obtained by drawing an edge for each physical component in the electrical circuit, on a one-to-one basis (e.g. voltage source V). As described previously, the two physical domains are coupled by the DC-motor transducer elements (M), which have an edge associated with each domain. The constitutive equations for the electrical and mechanical edges are defined by equations (6) and (7), respectively.

Note that many of the contributing terms in these transducer constitutive equations could be modelled by separate elements in the electrical and mechanical subgraphs, as was done in the bond graph model; here, they are combined into this "subsystem" representation of the DC-motor, for modelling convenience [29].

[3] http://real.uwaterloo.ca/~dynaflex/

To generate a minimal set of system equations, all rigid arms and revolute joints were selected into the tree of the mechanical subgraph. For the electrical subsystem, the voltage sources were selected into the tree and a current formulation was applied. The system equations were automatically generated in symbolic form by DynaFlex; these ODEs are explicit functions of the joint angles θ_1 and θ_{1-2} and the two motor currents i_7 and i_8. These 4 ODEs were exported to Matlab and solved using a standard numerical integrator. As shown in Figures 10 and 11, the DynaFlex results are in exact agreement with those from the independent software package Electran.

3.2.4 Linear Graph Model of Gate System

Figure 17 depicts the linear graph representation of the parking gate system shown in Figures 5 and 6. The mechanical subgraph consists of components discussed in previous sections, plus three new components: a weld joint (w), a flexible body (fb), and an induction motor (im). The weld joint, as its name suggests, simply locks two reference frames together.

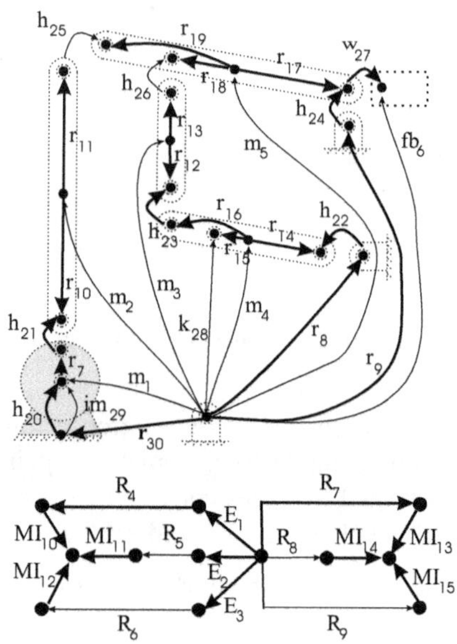

[htbp]

Fig. 17. Linear graph model of parking gate

The flexible body element represents the flexible parking barrier. Details regarding the constitutive equations for the flexible body can be found in [30]. To summarize, a Rayleigh beam model is used in conjunction with polynomial shape functions that represent the axial and torsional deformations, as well as bending about two lateral axes.

For this planar mechanism, only the in-plane bending was considered; 5 elastic variables were used to model this bending deflection. The value of 5 was obtained by progressively adding more deformation variables until the simulation results converged. It was also found that including axial deformations in the flexible beam model had no effect on the simulation results.

The mechanical domain is coupled to the electrical domain by the induction motor. The constitutive equation for the mechanical edge (im_{29}) of this transducer is defined by equation (8).

The graph for the electrical domain is almost identical to the circuit shown in Figure 6, the only difference being that the inductors have been replaced by mutual inductance components. Note that the induction motor transducer has multiple edges in the electrical subgraph. The constitutive equation for each one of these *MI* edges is defined by equation (10).

By selecting an appropriate tree (shown in bold in Figure 17) and using a current formulation, DynaFlex was used to generate 14 ordinary differential equations in symbolic form: 5 for the rigid multibody system in terms of joint coordinates β_{20} – β_{24}, 5 for the deformation variables of the beam, and 4 for the electrical system in terms of the cotree resistor currents i_5, i_6, i_8 and i_9. These ODEs were exported to Matlab and solved by a standard numerical integrator.

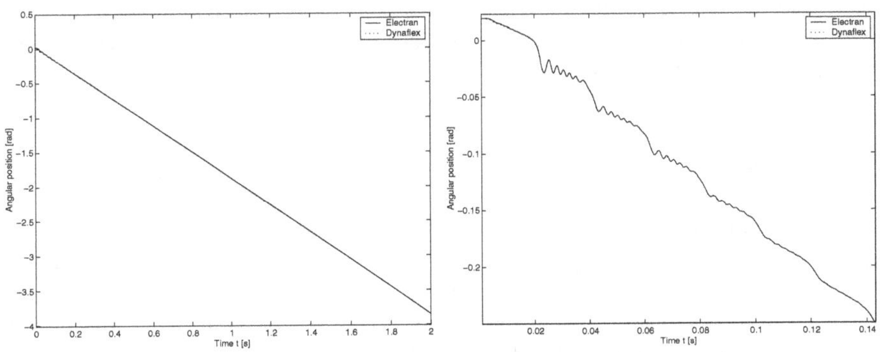

Fig. 18. Angular displacement of input link

Shown in Figures 18 and 19 are the numerical results from DynaFlex and Electran for the angular displacement of the input link and the current through a rotor inductor, respectively. The plots on the right of these figures are over a shorter time scale, in order to show the slight differences between the DynaFlex and Electran results that are not visible in the plots on the left. These results have been obtained using a 1:53 gear ratio between the motor and the 6-bar mechanism.

The results from the two software packages are nearly identical, with only slight differences in the currents. There are two possible sources of this difference:

- DynaFlex uses a Rayleigh beam model, while the Electran results were generated with a flexible beam model that consisted of several rigid bodies connected by rotational springs.

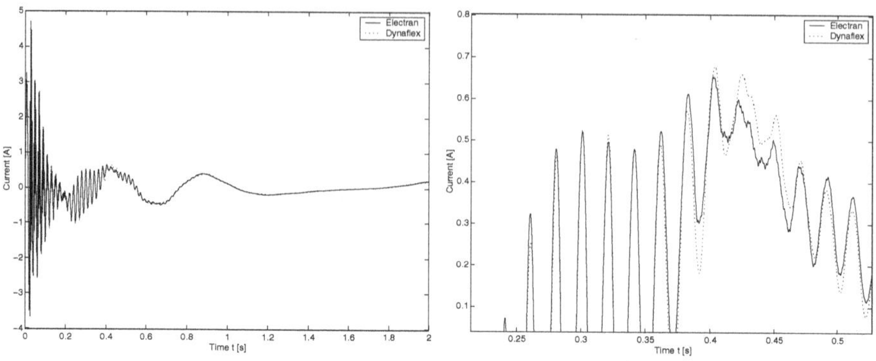

Fig. 19. Current through rotor inductor L_{r2} (MI_{15})

- the Electran results were generated with a different set of joint coordinates than those used by DynaFlex.

Note that the input link rotor oscillations shown in Figure 18 are damped. The only dissipative elements in this model are the electrical resistance; changing the resistance value has an effect on the damping of these mechanical oscillations. This illustrates the tight interaction between the electrical and mechanical subsystems in this multidisciplinary application.

4 Conclusions

The two most prominent unified theories for modelling multidisciplinary systems — linear graph theory and bond graph theory — have been examined in detail. Although similar in their decomposition of a system into a collection of discrete component models that are combined using topological equations, they are quite different in their graphical representation, the variables that they use, and the range of problems for which they are suited.

A bond graph model provides a single, unified representation of a multidisciplinary application, in which the coupling between physical domains is represented explicitly by a transducer element in the graph. Associated with each element are effort and flow variables, the product of which is the power flowing through the graph. Bond graphs are very powerful for modelling systems governed by scalar variables, e.g. electrical networks and 1-dimensional mechanical systems, and a number of software packages are available for simulating bond graph models. However, they were not designed to use tensor variables or mechanical displacements, and are therefore not very well-suited to the modelling of planar or spatial mechanical systems. Furthermore, the popular use of the force-effort analogy leads to a discrepancy between the multibody system topology and the junctions used to model this topology, the use of absolute coordinates in bond graph models leads to relatively large systems of DAEs, and systems with several bodies require many tedious manual calculations of kinematic transformations. Further research is needed to address these issues in bond graph modelling of multibody systems.

A linear graph model also provides a single, unified representation of a multidisciplinary application. However, the coupling between domains is not as evident in the graph, which consists of separate parts for each domain; this coupling is embedded in the constitutive equations for transducer elements, which have an edge in each of the two connected domains. Graph-theoretic modelling is supported by very few software products, but it is very well-suited to the modelling of multi-dimensional multidisciplinary applications. This can be attributed to the use of mechanical displacements as across variables, the ability to use tensors as through and across variables, and the existence of models for a variety of mechanical joints and rigid or flexible bodies. A unique and powerful feature of a graph-theoretic model is that, by selecting a tree, one can control the variables that appear in the final system equations. For multibody systems, these variables can include absolute or joint coordinates, or some combination of the two.

5 Acknowledgements

Special thanks goes to Laurent Sass and Chad Schmitke for their work on modelling and simulating the examples, and for many fruitful discussions. The author's research is funded by a Premier's Research Excellence Award (Province of Ontario), Waterloo Maple Inc., and the Natural Sciences and Engineering Research Council of Canada.

References

1. Biggs N, Lloyd E, Wilson R (1976) Graph Theory: 1736-1936. Oxford University Press, Oxford
2. Breedveld P, Dauphin-Tanguy G (1992) Bond Graphs for Engineers. Elsevier Science, Amsterdam
3. Crandall S, Karnopp D, Kurtz E, Pridmore-Brown D (1968) Dynamics of Mechanical and Electromechanical Systems. McGraw-Hill, New York
4. Gayford M (1971) Electroacoustics – Microphones, Earphones and Loudspeakers. Elsevier Publishing Company, New York
5. Karnopp D, Margolis D, Rosenberg R (2000) System Dynamics: Modeling and Simulation of Mechatronic Systems. John Wiley and Sons, New York
6. Koenig H, Tokad Y, Kesavan H (1967) Analysis of Discrete Physical Systems. McGraw-Hill, New York
7. Muegge B (1996) Graph-Theoretic Modelling and Simulation of Planar Mechatronic Systems. University of Waterloo, Canada
8. Norton R (2001) Design of Machinery, 2nd ed. McGraw-Hill, New York
9. Paynter H (1961) Analysis and Design of Engineering Systems. MIT Press, Cambridge, Massachusetts
10. Roe P (1966) Networks and Systems. Addison-Wesley, Reading, Massachusetts
11. Rowell D, Wormley D (1997) System Dynamics: An Introduction. Prentice Hall, New Jersey

12. Thoma J (1975) Introduction to Bond Graphs and their Applications. Pergamon Press, Oxford
13. Sass L, Fisette P, Grenier D (2002) Modelling of mechatronic systems — study of the actuation of a swinging barrier. In: Proceedings of The 8th Mechatronics Forum International Conference, University of Twente, Enschede, Netherlands
14. Birkett S, Roe P (1990) The Mathematical Foundations of Bond Graphs - III. Matroid Theory. J Frank Inst 327(1):87–108
15. Favre W, Scavarda S (1998) Bond Graph Representation of Multibody Systems with Kinematic Loops. J Frank Inst 335B(4):643–660
16. Fisette P, Postiau T, Sass L, Samin J-C (2002) Fully Symbolic Generation of Complex Multibody Models. Mech Struct Mach 30(1):31–82
17. Hadwich V, Pfeiffer F (1995) Principle of Virtual Work in Mechanical and Electromechanical Systems. Arch Appl Mech 65:390–400
18. Jerkovsky W (1978) The structure of multibody dynamics equations. J Guid Control 1(3):173–182
19. Karnopp D (1997) Understanding Multibody Dynamics Using Bond Graph Representations. J Frank Inst 334B(4):631–642
20. Kübler R, Schiehlen W (2000) Modular Simulation in Multibody System Dynamics. Multibody Sys Dyn 4:107–127
21. Lovekin D, Heppler G, McPhee J (2000) Design and Analysis of a Facility for Free-floating Flexible Manipulators. Trans CSME 24(2):375–390
22. Maisser P, Enge O, Freudenberg H, Kielau G (1997) Electromechanical Interactions in Multibody Systems Containing Electromechanical Drives. Multibody Sys Dyn 1:281–302
23. McPhee J (1998) Automatic Generation of Motion Equations for Planar Mechanical Systems Using the New Set of "Branch Co-ordinates". Mech Mach Theory 33(6):805–823
24. McPhee J (1996) On the use of linear graph theory in multibody system dynamics. Nonlin Dyn 9:73–90
25. Papadopoulos E, Gonthier Y (2002) On the Development of a Real-Time Simulator Engine for a Hydraulic Forestry Machine. Int Journal Fluid Power 3(1):55–65
26. Perelson A, Oster G (1976) Bond Graphs and Linear Graphs. J Frank Inst 302(2):159–185
27. Sass L, McPhee J, Schimitke C, Fisette P, Grenier D (2004) A Comparison of Different Methods for Modelling Electromechanical Multibody Systems. Multibody Sys Dyn 12(3):209–250
28. Scherrer M, McPhee J (2003) Dynamic Modelling of Electromechanical Multibody Systems. Multibody Sys Dyn 9:87-115
29. Schmitke C, McPhee J (2003) A Procedure for Modeling Multibody Systems Using Subsystem Models. Int J Multiscale Comp Eng 1(2):139-159
30. Shi P, McPhee J (2000) Dynamics of Flexible Multibody Systems Using Virtual Work and Linear Graph Theory. Multibody Sys Dyn 4:355–381
31. Tiernego M, Bos A (1985) Modelling the Dynamics and Kinematics of Mechanical Systems With Multibond Graphs. J Frank Inst 319(1/2):37–50
32. Trent H (1955) Isomorphisms between linear graphs and lumped physical systems. J Acoust Soc Am 27:500–527
33. Zeid A, Chung C-H (1992) Bond Graph Modeling of Multibody Systems: A Library of Three-Dimensional Joints. J Frank Inst 329(4):605–636

A Biomechanical Multibody Model with a Detailed Locomotion Muscle Apparatus

Jorge A.C. Ambrósio and Miguel P.T. Silva

Institute of Mechanical Engineering, Instituto Superior Técnico
Av. Rovisco Pais 1, 1049-001 Lisboa, Portugal
{jorge,pcms}@dem.ist.utl.pt

The ability of the animals to repeat the same movement or posture by recruiting different muscles or by using different muscle activation patterns is well known. From the mathematical point of view the solution of this problem and consequent determination of the recruited set of muscles and associated forces involves the solution of an optimization problem, in which the intrinsic objectives used by the central nervous system to recruit the referred set of muscles are represented by means of proper physiological cost functions. The objective of this work is to present a multibody dynamics based methodology to model the human body and the relevant features of the locomotion apparatus required for gait. For this purpose, a whole-body biomechanical model is used within the framework of an inverse dynamic analysis formulation with fully Cartesian coordinates, to calculate the individual muscle forces in the locomotion apparatus, the net moments of force at the joints of the upper body and the joint reaction forces developed between the anatomical segments of the biomechanical model when performing the specified task. Myoactuators representing the most relevant muscles of the locomotion apparatus are introduced using a Hill-type muscle model. Different cost functions are used to represent the objectives of the central nervous system when developing a particular task for the gait cycle. Sequential quadratic optimization tools are used to resolve the force-sharing problem arising from having a number of unknowns, associated to the individual muscle forces, higher than the number of available equations of motion, representing the dynamics of the anatomical segments that represent the human body. The methodologies proposed here are applied to a normal cadence gait cycle. In the process the most suitable cost functions for the specific task under analysis are identified and the quality of the results produced for this type of indeterminate problems is discussed.

1 Introduction

Biomechanical models based on multibody dynamics are used in a wide range of applications where the human motion is characterized by large displacements. In fact almost all models used for impact biomechanics, in vehicle passive safety applications, in sports sciences or in gait analysis rely on the use of multibody dy-

J.A.C. Ambrósio (ed.), Advances in Computational Multibody Systems, 155–184.

namics approaches to obtain the kinematics of the human body represented by the model, the reaction forces that develop between the different anatomical segments, the muscle forces that are required to develop specific motions or even to design protective equipment for sportsmen or for vehicle occupants or to devise new strategies for human-machine interaction. Among the possible applications of the biomechanical models this work focus on those used for the gait analysis emphasizing the requirements to construct them.

The biomechanical models applied on the study of the human locomotion require that the major anatomical segments of the lower part of the human body are represented. The upper body may be represented by lumping some of the anatomical segments or by having a more or less detailed representation of the major segments. In any case, the multibody description of each segment requires that one or more rigid bodies are associated to it. The anatomical joints are represented either by kinematic joints or by contact joints in the multibody model, depending in the objectives of the analysis. The ligaments and other passive tissues required to provide stability or stiffness to the anatomical joints are typically represented as spring-damper elements with linear or nonlinear characteristics. The muscles of the locomotion apparatus need also to be represented in the models being possible to use different strategies to include them in the biomechanical model. In particular, the use of a detailed description of the muscles in the locomotion apparatus poses a problem known as 'redundant problem in biomechanics'[1], which requires the calculation of the redundant forces produced by the muscle apparatus. Finally, the use of multibody biomechanical models for gait analysis also requires a comprehensive description of the contact between the different segments of the models and external objects, such as the ground.

It is the purpose of this work to present a multibody based methodology that together with the use of optimization procedures, allows for the calculation of the redundant muscle forces, generated in a particular muscle apparatus of the human body. The proposed methodology uses a multibody formulation with natural coordinates where rigid bodies and kinematic joints are modeled using the Cartesian coordinates of a set of anatomical points and unit vectors [2,3]. Using this general-purpose methodology, a whole body biomechanical model is constructed using rigid bodies interconnected by revolute and universal joints. The biofidelity of the model is improved using the subject's anthropometric link lengths together with biomechanical information regarding the physical characteristics of the anatomical segments. This information is collected from a general database and scaled for the subject dimensions and total body mass [4,5]. The motion of the subject is acquired together with all the externally applied forces.

Two different types of actuators are used to drive the biomechanical model through the acquired motion: joint actuators, that drive the degrees-of-freedom of the biomechanical model associated with joints that are not crossed by muscles, and muscle actuators that drive the degrees-of-freedom of the joints crossed by the muscles forming the muscle apparatus under analysis. It should be noted that when the aim of the analysis is to calculate exclusively reaction forces and net moments-of-force in a particular joint or set of joints, then only joint actuators need to be introduced and the solution to the inverse dynamics problem is unique and non redundant [6]. When it is required to evaluate the muscle forces, muscle actuators need to be introduced, which represents a mechanical system with a redundant nature, i.e., the system has more unknowns than equations.

Optimization tools are used to resolve this indeterminate problem. These tools consider that muscle forces are generated according to the minimization of some performance criteria and that any optimal solution obtained for the muscle forces must satisfy the equations of motion of the biomechanical system. These performance criteria are analytical expressions that represent the decisions taken by the central nervous system when executing of the prescribed task. In the present work, the performance criteria used to calculate the muscle force generated during a normal cadence walking cycle are the sum of the average individual muscle stress raised to a power of three and the sum of the squares of the individual muscle force [7]. In particular, regarding the first performance criteria, some studies show that the minimization of the total muscle stresses is strongly related with the maximization of the muscle endurance [8].

To each muscle actuator a muscle model is associated that simulates its activation-contraction dynamics [9,10,11]. In the present work, a Hill type muscle model is applied, being the force produced by the muscle contractile element calculated as a function of the muscle activation, maximum isometric peak force, muscle length and muscle rate of shortening. Using this constitutive law, the equations of motion of the biomechanical system and the performance criteria used in de optimization procedure are expressed in terms of muscle activations instead of muscle forces. For a detailed description of the methods used the interested reader is referred to the work presented in reference [11].

2 Multibody Formulation

A multibody formulation, using natural or fully Cartesian coordinates, is applied to the study and analysis of the human body movement. With this formulation, the rigid bodies are constructed by using the Cartesian coordinates of a set of points and unit vectors. These points and unit vectors, usually located at the joints and extremities of the model anatomic components, are used not only to define the kinematic structure of the rigid bodies, but also, when shared by different rigid bodies, to define in natural way simple kinematic joints such as the spherical and revolute joint that represent hip and knee.

A vector of generalized coordinates is constructed with the Cartesian coordinates of the points and vectors used in the definition of the mechanical system [2]:

$$\mathbf{q} = \{q_1 \; q_2 \; q_3 \cdots q_n\}^T \tag{1}$$

where $n=3(np+nv)$ is the total number of natural coordinates and np and nv are, respectively, the total number of points and unit vectors of the model.

2.1 Kinematic Constraints

Rigid body constraints are the most common type since the number of natural coordinates that defines a rigid body is always higher than the number of its degree-of-freedom. Consequently, not all the coordinates are independent and some kinematic constraint equations need to be added to express such dependences. Rigid body constraints represent physical properties of rigid bodies such as the constant distance between two points, the constant angle between two segments or the constant length of a vector, which are shown as a scalar product given by:

$$\Phi^{(SP,1)}(\mathbf{q},t) = \mathbf{v}^T\mathbf{u} - L_v L_u \cos\left(\langle \mathbf{v},\mathbf{u}\rangle(t)\right) = 0 \qquad (2)$$

where **v** and **u** are two generic vectors used in the definition of rigid bodies, L_v and L_u are the respective norms and <**v**,**u**>(t) is the angle between them.

Equation (2) is also applied to the definition of kinematic joints and driving constraints. In the case of driving constraints, used to prescribe the motion of the system over time, the angles are functions of time. Depending on vectors **v** and **u**, Equation (2) has different physical meanings. Considering that $\mathbf{r}_i$, $\mathbf{r}_j$, $\mathbf{r}_k$ and $\mathbf{r}_l$ are the Cartesian coordinates of points i, j, k and l and that **a** and **b** are unit vectors, the most relevant kinematic constraints involving the scalar product and their respective physical meanings are presented in Table 1. Note also that the most common kinematic constraints are obtained in this formulation either by scalar products or by sharing points and vectors between bodies.

Table 1. Physical meanings of the scalar product constraint

Constraint Description	v	u	L_v	L_u	<v,u>	Representation
Constant distance between points i and j.	$(\mathbf{r}_j-\mathbf{r}_i)$	$(\mathbf{r}_j-\mathbf{r}_i)$	L_{ij}	L_{ij}	0	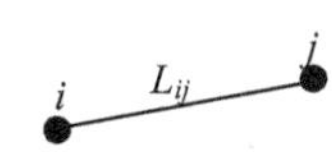
Unit module vector.	**a**	**a**	1	1	0	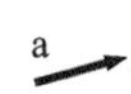
Constant angle between unit vectors **a** and **b**.	**a**	**b**	1	1	α	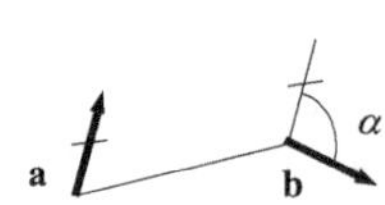
Constant angle between segment $\mathbf{r}_{ij}$ and unit vector **a**.	$(\mathbf{r}_j-\mathbf{r}_i)$	**a**	L_{ij}	1	β	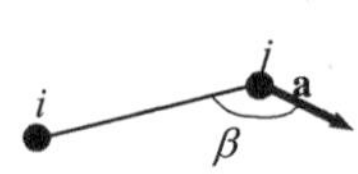
Constant angle between segments $\mathbf{r}_{ij}$ and $\mathbf{r}_{kl}$.	$(\mathbf{r}_j-\mathbf{r}_i)$	$(\mathbf{r}_l-\mathbf{r}_k)$	L_{ij}	L_{kl}	γ	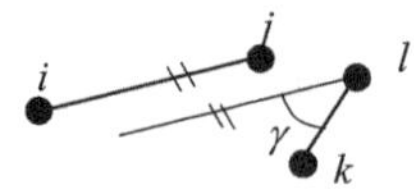
Rotational driver around revolute joint located in point i	$(\mathbf{r}_j-\mathbf{r}_i)$	$(\mathbf{r}_k-\mathbf{r}_i)$	L_{ij}	L_{ik}	ϕ=f(t)	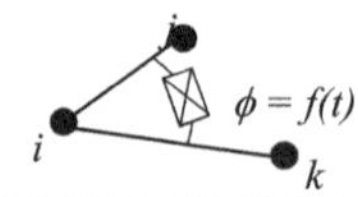

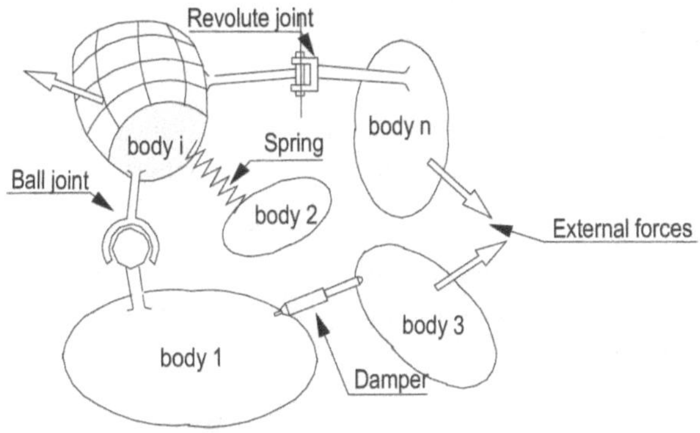

Fig. 1. General multibody system.

The constraint equations associated with the mechanical system are assembled in a single vector and written as:

$$\mathbf{\Phi}(\mathbf{q},t)=\mathbf{0} \tag{3}$$

It should be noted that, not all the constraints assembled in vector $\mathbf{\Phi}$ are independent. To prevent instabilities, numerical methods capable of dealing with redundant constraints are used in the solution of the kinematic and dynamic analyses[12].

There are other types of kinematic constraints, such as the linear combination constraint or the cross product constraints that are also used when modeling with natural coordinates [2]. All kinematic constraints mentioned have a quadratic or a linear dependency on the coordinates. Their contribution to the Jacobian matrix of the constraints is either linear or constant.

2.2 Equations of Motion

The equations of motion of a constrained multibody system acted upon by external applied forces, such as the one presented in Figure 1, are given by:

$$\mathbf{M}\ddot{\mathbf{q}}+\mathbf{\Phi}_{\mathbf{q}}^{T}\boldsymbol{\lambda}=\mathbf{g} \tag{4}$$

where $\mathbf{M}$ is the global mass matrix of the system, $\mathbf{\Phi}_{\mathbf{q}}$ the Jacobian matrix of the constraints, $\ddot{\mathbf{q}}$ the vector of natural accelerations, $\mathbf{g}$ the generalized force vector and the vector of Lagrange multipliers [2,13,14]. Due to the presence of redundant constraints, multiple solutions of Equation (4) can be found. In order to calculate a single solution, the minimum norm condition is applied [2], assuring that the vector of Lagrange multipliers is orthogonal to the null space of $\mathbf{\Phi}_{\mathbf{q}}{}^{T}$.

When performing an inverse dynamic analysis, the Lagrange multipliers vector, that represents the reaction forces and the driving forces and moments, associated to muscle forces and net moments of force at the joints of a biomechanical model, is the only unknown of Equation (4), being all other quantities calculated from kinematic data and force measuring devices or given as modeling data.

3 The Biomechanical Model

The biomechanical model, presented in Figure 2, is a three-dimensional model of the human body [3,5,6,12]. It is described using the general multibody formulation with natural coordinates presented before and it has a kinematic structure made of thirty-three rigid bodies, interconnected by revolute and universal joints, in such a way that sixteen anatomical segments are identified. A complete description of these segments and the corresponding rigid bodies used in their definition is presented in reference [11].

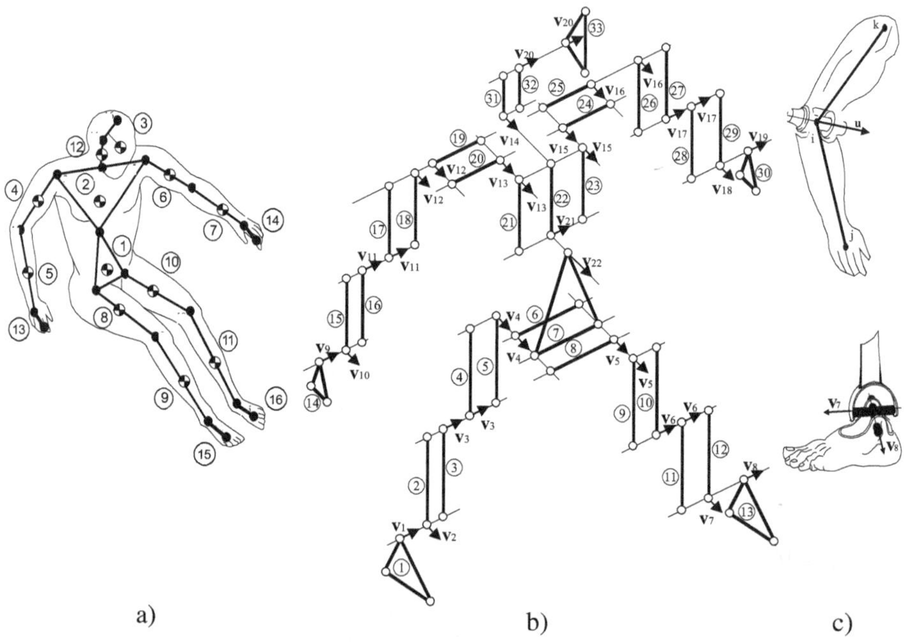

Fig. 2. The biomechanical model: a) the sixteen anatomical segments; b) the kinematic structure; c) kinematic joints for the elbow and ankle

A set of physical characteristics, obtained from anthropometric measures of the human body, is associated to each anatomical segment of the biomechanical model. The most important properties of the anatomical segments, in what multibody modeling is concerned, are its mass, principal moments of inertia, lengths, and distance of its center of mass to its proximal joint. The properties mentioned before are obtained from the literature for the 50th-percentile human male [4,5,15]. However, in order to improve the biofidelity of the data with respect to the anthropometrics of the subject modeled in any particular application, these physical properties are scaled using non-dimensional scaling factors [4,5].

3.1 Description of the Kinematic Structure of the Biomechanical Model

The 33 rigid bodies are constructed using an underlying kinematic structure made of 25 points and 23 unit vectors, accounting for a total number of 141 natural coordinates. Two types of rigid bodies are used: rigid bodies defined by two points and one unit vector and rigid bodies defined by three points and one unit vector. Two types of kinematic joints are used in the biomechanical model: revolute and universal joints. Since most joints are defined naturally by sharing points and vectors between rigid bodies, the most common kinematic constraints that arise are of the rigid body type. A total number of 97 non-redundant kinematic constraints are introduced in this form. Consequently, the model has 44 degrees of freedom that correspond to 38 rotations about 26 revolute joints and 6 universal joints, plus 6 degrees-of-freedom that are associated with free body rotations and translations of the base body, which coincides with the pelvis in the model.

3.2 Input Data for the Biomechanical Model

The main objective, when performing the inverse dynamic analysis of the biomechanical system, is to calculate the reaction and driving forces that the model has to develop at the joints and in the actuators to perform a predetermined movement. The kinematic information required for the analysis consists in the trajectories of a set of points, located at the joints and extremities of the subject under analysis [16] that are acquired using three-dimensional motion reconstruction techniques [18-21]. The trajectories of points are filtered in order to reduce the noise levels introduced in the motion reconstruction procedure [15,19] and their consistency with the kinematic structure of the biomechanical model is enforced [16,17].

Velocity and acceleration curves are calculated, for each point using the derivatives of the constraint equations (3). These curves are also used to calculate joint direction unit vectors, average link lengths and the history curves of each degree-of-freedom of the model. These history curves are used later to define the joint actuators that drive the model during the analysis period.

Another important set of input data consists in the externally applied forces over the biomechanical model. In dynamic analyses of biomechanical systems, these forces are associated to impact forces, seat reaction forces, pressure-distributed forces or, in the present case of gait analysis, to the ground reaction forces at the feet.

The overall layout of the gait lab used to collect the kinematic and dynamic data for the present gait analysis is presented in Figure 3. An apparatus consisting in four video cameras and three force plates is used. The video cameras and force plates are synchronized to collect data with a sampling frequency of 60 Hz.

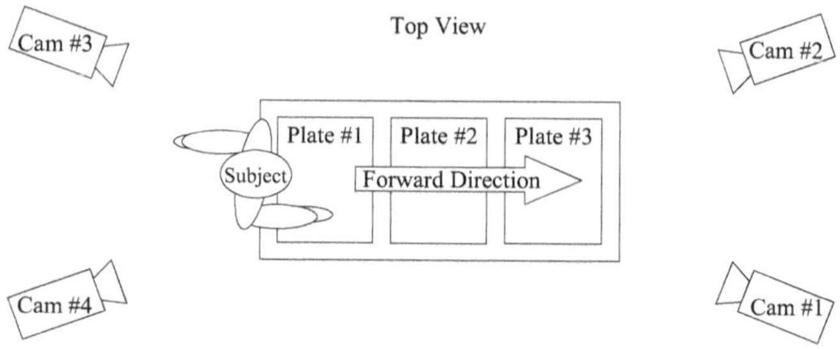

Fig. 3. Overall layout of the lab apparatus.

3.3 Joint Rotational Actuators

To drive the biomechanical model throughout the inverse dynamic analysis joint actuators, such as the one shown in Figure 4 for the knee joint, are specified. For each degree-of-freedom, a joint actuator equation is added to the equations of motion of the system. Joint actuator equations are kinematic constraints of scalar product type presented in Equation (2), in which the angle between the two vectors is a function of time that describes the motion of each kinematic joint. These equations are added to the system kinematic equation, so that the number of non-redundant constraint equations becomes equal to the number of natural coordinates that describe the model. The results obtained, by solving Equation (4), include the Lagrange multipliers of the joint actuators that represent the net moments-of-force of the muscles crossing those joints. The inverse dynamics problem, as stated here, is totally determined, i.e., Equation (4) has a unique solution.

The results obtained using this type of analysis provides valuable information regarding the reaction forces at the joints and the overall behavior of the muscles during the execution of a task. However, these results do not provide any kind of information regarding specific muscle forces or muscle activation patterns.

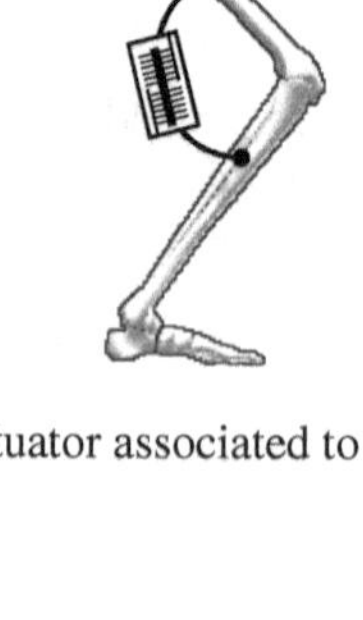

Fig. 4. Joint actuator associated to the knee joint.

4 Redundant Muscle Forces by Optimization Techniques

In complex biomechanical systems such as the human body, nearly every joint is crossed by several muscles or muscle groups. This means that different muscle activation patterns can generate forces that produce the same net moments-of-force at the joints and, as result, the same posture or movement. It is the central nervous system that, depending on the task being performed and objectives to be achieved, selects and activates the muscles that best fulfill some physiological criteria.

Muscle force distribution problem results, from a mathematical point of view, from the fact that the number of load-transmitting elements at a joint usually exceeds the number of available equilibrium equations and consequently a unique solution for the distribution of those forces can not be obtained [7]. Optimization techniques are applied to resolve the indeterminate problem by choosing from an infinite set of solutions the one that minimizes one or more cost functions. The cost functions are mathematical expressions that intend to represent physiological principles used by the central nervous system to select the muscles recruited for a given activity.

4.1 Muscle Actuators in Multibody Systems

Muscles are introduced in the equations of motion of the multibody system as point-to-point kinematic driver actuators, also designated by myoactuators. Two of the muscles of the lower extremity muscle apparatus are shown Figure 5 to illustrate different complexities in their path. The *semimembranosus* is a two-point muscle, the origin and insertion points, and the *tensor fasciae latea* is a muscle defined by multiple points for an accurate characterization of its curvature.

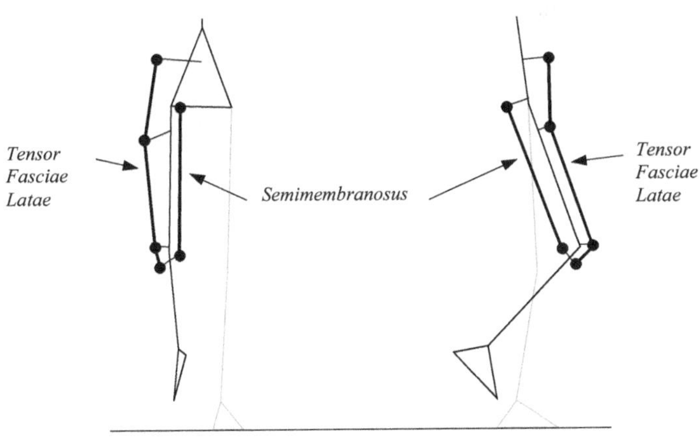

Fig. 5. Muscle actuators defined with two or more points.

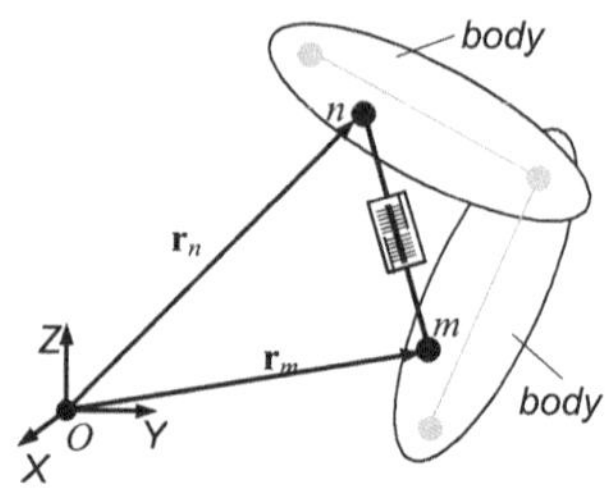

Fig. 6. Muscle actuator defined between points n and m of rigid bodies i and j.

A constraint equation that specifies the muscle action during the analysis period is associated to each muscle actuator. These kinematic relations constrain the distance between two generic points of different rigid bodies to change in agreement with a specified length history previously calculated from the motion reconstruction. Considering a two-point muscle actuator, with an origin located in point n of rigid body i, and an insertion located in point m of rigid body j, as depicted by Figure 6, the mathematical expression used to define the constraint is

$$\mathbf{\Phi}^{(MA,1)}(\mathbf{q},t) = \left(\mathbf{r}_m - \mathbf{r}_n\right)^T \left(\mathbf{r}_m - \mathbf{r}_n\right) - L_{nm}^2(t) = 0 \tag{5}$$

where $\mathbf{r}_m$ and $\mathbf{r}_n$ are respectively the global position vectors of the origin and insertion points and $L_{nm}(t)$ is the muscle total length, calculated for each time step of the analysis. The two generic points n and m of rigid bodies i and j that are used to define the muscle actuator do not belong to the set of basic points and unit vectors used in the construction of the rigid bodies to which they are attached. It is necessary that the quantities involved in Equation (5) are expressed in terms of the generalized set of natural coordinates.

Let vectors $\mathbf{q}_i$ and $\mathbf{q}_j$ define the generalized natural coordinates of rigid bodies i and j. The coordinates of points n and m are now rewritten as

$$\begin{aligned} \mathbf{r}_n &= \mathbf{C}_i^n \mathbf{V}_i \mathbf{q}_i \\ \mathbf{r}_m &= \mathbf{C}_j^m \mathbf{V}_j \mathbf{q}_j \end{aligned} \tag{6}$$

where $\mathbf{C}_i^n$ and $\mathbf{C}_j^m$ are constant transformation matrices that relate the Cartesian coordinates of a generic point with the generalized natural coordinates of the elementary rigid body [2]. $\mathbf{V}_i$ and $\mathbf{V}_j$ are transformation matrices relating the generalized coordinates of the elementary rigid body with the generalized coordinates of the rigid bodies used to describe the biomechanical model [2,16].

Substituting Equation (6) in the muscle actuator constraint Equation (5) leads to the following result for the muscle actuator constraint equation:

$$\mathbf{\Phi}^{(MA,1)}(\mathbf{q},t) = \left(\mathbf{C}_j^m \mathbf{V}_j \mathbf{q}_j - \mathbf{C}_i^n \mathbf{V}_i \mathbf{q}_i\right)^T \left(\mathbf{C}_j^m \mathbf{V}_j \mathbf{q}_j - \mathbf{C}_i^n \mathbf{V}_i \mathbf{q}_i\right) - L_{nm}^2(t) = 0 \tag{7}$$

where all quantities are expressed in terms of the natural coordinates of rigid bodies i and j. Note that Equation (7) presents a quadratic dependency on the natural coordinates, which means that the contributions of this constraint to the Jacobian matrix of the constraints are linear.

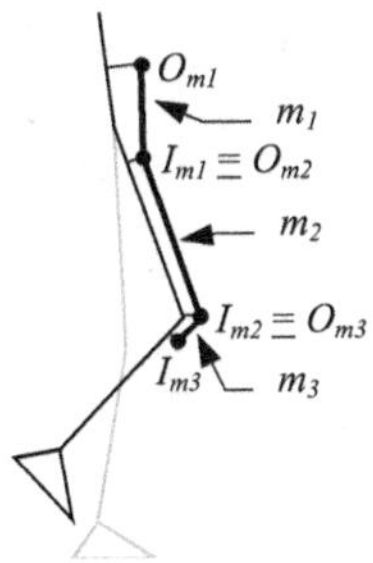

Fig. 7. Describing more complex muscle actuators.

A Lagrange multiplier is associated to each muscle actuator of the locomotion apparatus. The physical dimension of this multiplier, used in the context of the actuators described before, is a force per unit of length. In order to obtain muscle forces or muscle activations, these multipliers must be multiplied by proper scalar factors, which are closely related with the type of muscle model adopted.

Muscle actuators defined with more than two points are introduced in the Jacobian matrix of the constraints as a sum of several two-point muscle actuators. Consider, for example, the muscle *tensor fasciae latea* presented in Figure 5. This muscle is described using three two-point muscle actuators, labeled respectively m_1, m_2 and m_3 in Figure 7. The Lagrange multipliers λ_{m1}, λ_{m2} and λ_{m3} are associated to muscle actuators m_1, m_2 and m_3, respectively.

With the information presented before, the term $\mathbf{\Phi_q}^T\boldsymbol{\lambda}$ of Equation (4) is assembled for muscle actuators m_1, m_2 and m_3, and written as:

$$\begin{array}{c} \\ \mathbf{q}_3 \\ \mathbf{q}_4 \\ \mathbf{q}_5 \\ \mathbf{q}_6 \\ \mathbf{q}_7 \end{array} \begin{array}{c} \begin{array}{ccc} m_1 & m_2 & m_3 \end{array} \\ \begin{bmatrix} \partial\Phi^{m1}/\partial\mathbf{q}_3 & \mathbf{0} & \mathbf{0} \\ \partial\Phi^{m1}/\partial\mathbf{q}_4 & \partial\Phi^{m2}/\partial\mathbf{q}_4 & \mathbf{0} \\ \mathbf{0} & \partial\Phi^{m2}/\partial\mathbf{q}_5 & \partial\Phi^{m3}/\partial\mathbf{q}_5 \\ \mathbf{0} & \mathbf{0} & \mathbf{0} \\ \mathbf{0} & \mathbf{0} & \partial\Phi^{m3}/\partial\mathbf{q}_5 \end{bmatrix} \end{array} \begin{Bmatrix} \lambda_{m1} \\ \lambda_{m2} \\ \lambda_{m3} \end{Bmatrix} \tag{8}$$

where $\mathbf{q}_3$ to $\mathbf{q}_7$ indicate the rows of the Jacobian matrix and represent the set of natural coordinates defining the rigid bodies interconnected by the muscle, and $\bullet\mathbf{\Phi}^{mi}/\bullet\mathbf{q}_j$ are the partial derivatives of muscle actuator equation m_i with respect to the natural coordinate $\mathbf{q}_j$.

Consider that any muscle must have a constant force per unit of length from its origin to its insertion. For muscles defined with more than two points the Lagrange multipliers associated with each segment must be equal. In the muscle *tensor fasciae latea* the Lagrange multipliers must be

$$\lambda_{m1} = \lambda_{m2} = \lambda_{m3} = \lambda_{TFL} \tag{9}$$

Substituting Equation (9) in Equation (8) lead the three actuators associated with muscles m_1, m_2 and m_3 to add up to form a single actuator equation expressing the kinematics of the muscle with more complex path, written as

$$\begin{array}{c} \\ \mathbf{q}_3 \\ \mathbf{q}_4 \\ \mathbf{q}_5 \\ \mathbf{q}_6 \\ \mathbf{q}_7 \end{array} \begin{array}{c} m_1 + m_2 + m_3 \\ \left[\begin{array}{c} \partial\Phi^{m1}/\partial\mathbf{q}_3 + \mathbf{0} + \mathbf{0} \\ \partial\Phi^{m1}/\partial\mathbf{q}_4 + \partial\Phi^{m2}/\partial\mathbf{q}_4 + \mathbf{0} \\ \mathbf{0} + \partial\Phi^{m2}/\partial\mathbf{q}_5 + \partial\Phi^{m3}/\partial\mathbf{q}_5 \\ \mathbf{0} + \mathbf{0} + \mathbf{0} \\ \mathbf{0} + \mathbf{0} + \partial\Phi^{m3}/\partial\mathbf{q}_5 \end{array} \right] \end{array} \{\lambda_{TFL}\} \tag{10}$$

The muscle driver actuator equations are introduced in the Jacobian matrix of the constraints together with the kinematic constraint equations defining the motion of the biomechanical model and kinematic joints. Therefore, the muscle and reaction forces of the biomechanical model are all evaluated at the same time.

Joint rotational actuators, introduced in last section, are removed from all the joints crossed by muscle actuators. However in the remaining joints, where muscle actuators are not used, joint rotational actuators are still maintained allowing for the muscle actuators to be introduced only in the anatomical segments of the biomechanical model under analysis.

4.2 Dynamics of Muscle Tissue

The dynamics of muscle tissue can be divided into activation dynamics and muscle contraction dynamics [9], as schematically indicated in Figure 8. The activation dynamics generates a muscle tissue state that transforms the neural excitation produced by the central nervous system, into activation of the contractile apparatus. The activation dynamics, although not implemented in this work, describes the time lag between neural signal and the corresponding muscle activation [11].

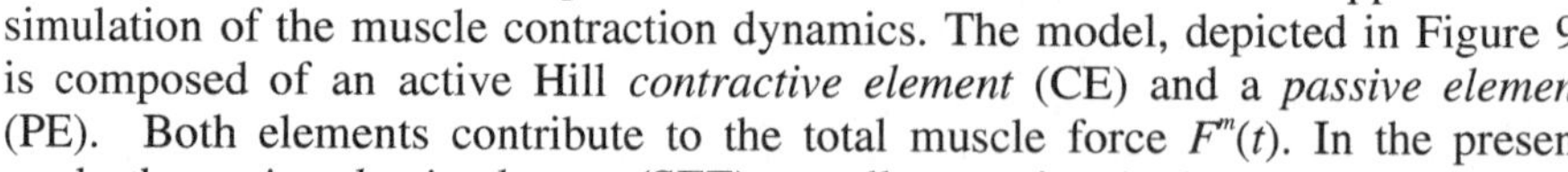

Fig. 8. Dynamics of muscle tissue.

The muscle contraction dynamics requires that a mathematical model of the muscle is introduced. In the present work the Hill muscle model is applied to the simulation of the muscle contraction dynamics. The model, depicted in Figure 9, is composed of an active Hill *contractive element* (CE) and a *passive element* (PE). Both elements contribute to the total muscle force $F^m(t)$. In the present work, the *series elastic element* (SEE), usually associated with cross-bridge stiffness, is not included in the model since it can be neglected in coordination studies not involving short-tendon actuators [9].

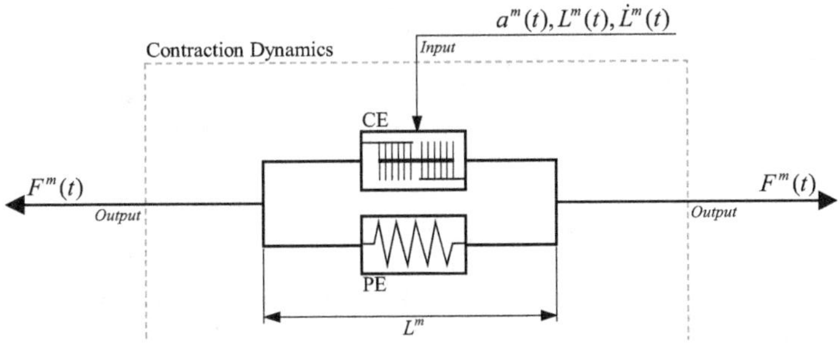

Fig. 9. Contraction dynamics using a Hill-type muscle model.

In the Hill muscle model, the contractile properties of the muscle tissue are controlled by its current length $l^m(t)$, rate of length change $\dot{l}^m(t)$ and activation $a^m(t)$. The force produced by the active Hill *contractile element*, for muscle *m*, is

$$F_{CE}^m(a^m(t), l^m(t), \dot{l}^m(t)) = \frac{F_l^m(l^m(t)) F_{\dot{l}}^m(\dot{l}^m(t))}{F_0^m} a^m(t) \tag{11}$$

where F_0^m is the maximum isometric force and $F_l^m(l^m(t))$ and $F_{\dot{l}}^m(\dot{l}^m(t))$ are two functions that represent the muscle force-length and force-velocity dependency, respectively [9,11]. These two functions are approximated analytically by [11]

$$F_l^m(l^m(t)) = F_0^m e^{-\left[\left[-\frac{9}{4}\left(\frac{l^m(t)}{l_0^m} - \frac{19}{20}\right)\right]^4 - \frac{1}{4}\left[-\frac{9}{4}\left(\frac{l^m(t)}{l_0^m} - \frac{19}{20}\right)\right]^2\right]} \tag{12}$$

and

$$F_{\dot{l}}^m(\dot{l}^m(t)) = \begin{cases} 0 & -\dot{l}_0^m > \dot{l}^m(t) \\ -\dfrac{F_0^m}{\arctan(5)} \arctan\left(-5 \dfrac{\dot{l}^m(t)}{\dot{l}_0^m}\right) + F_0^m & 0.2\dot{l}_0^m \ge \dot{l}^m(t) \ge -\dot{l}_0^m \\ \dfrac{\pi F_0^m}{4\arctan(5)} + F_0^m & \dot{l}^m(t) \ge 0.2\dot{l}_0^m \end{cases} \tag{13}$$

where l_0^m is the muscle resting length and $\dot{l}_0^m$ is the maximum contractile velocity above which the muscle cannot produce force. The passive element is independent of the activation and it only starts to produce force when stretched beyond its resting length l_0^m. The force produced by the passive element is approximated by [11]:

$$F_{PE}^m(l^m(t)) = \begin{cases} 0 & l_0^m > l^m(t) \\ 8\dfrac{F_0^m}{l_0^{m3}}\left(l^m - l_0^m\right)^3 & 1.63 l_0^m \ge l^m(t) \ge l_0^m \\ 2F_0^m & l^m(t) \ge 1.63 l_0^m \end{cases} \tag{14}$$

Equation (14) shows that the force produced by the passive element is only a function of the muscle length, being its value completely determined during the total time of the analysis. Since the force produced by the passive element is not an unknown it is treated here as an external force, which is directly applied to the rigid bodies interconnected by the muscle.

The forces produced by the contractile element are the only unknown forces. In order to calculate these forces, a muscle actuator equation is associated to each contractile element. This association is accomplished multiplying each actuator equation by a proper scalar factor, so that the Lagrange multiplier associated to the actuator, represents muscle force or muscle activation. The factors for the muscle force and activation are respectively

$$C_\lambda^m = \frac{1}{2l^m} \quad ; \quad C_\lambda^m = \frac{F_l^m F_i^m}{2F_0^m l^m} \tag{15}$$

Note that if the Lagrange multiplier represents muscle activation, the associated muscle force is calculated using Equation (10).

4.3 Muscle Database

A muscle locomotion apparatus, with thirty-five muscle actuators, is used to simulate the right lower extremity intermuscular coordination. The muscle apparatus and a brief description of each muscle action [22] are presented in Table 2. The physiological information regarding the muscle definition is obtained from the literature [23,24] and compiled in a muscle database. This information consists in the maximum isometric force, resting length, attachment points, wrap-around bodies and the local coordinates of the origin, insertion and via points. The whole muscle apparatus is presented in Figure 10.

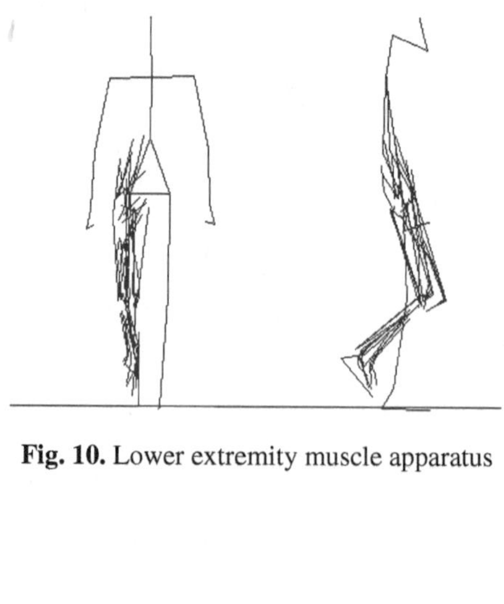

Fig. 10. Lower extremity muscle apparatus

Table 2. List and description of the lower extremity muscle apparatus

Nr	Muscle Name	Muscle Action
1	Adductor Brevis	Adducts and flexes and helps to laterally rotate the thigh.
2	Adductor Longus	Adducts and flexes the thigh; helps to laterally rotate the hip.
3	Adductor Magnus	Thigh adductor; superior horizontal fibers also help to flex the thigh, while vertical fibers help extend the thigh.
4	Biceps Femoris (long head)	Flexes the knee, and rotates the tibia laterally; long head extends the hip joint.
5	Biceps Femoris (short head)	Flexes the knee, and rotates the tibia laterally; long head extends the hip joint.
6	Extensor Digitorum Longus	Extend toes 2 – 5 and dorsiflexes ankle.
7	Extensor Hallucis Longus	Extends great toe and dorsiflexes ankle.
8	Flexor Digitorum Longus	Flexes toes 2 – 5; also helps in plantar flexion of ankle.
9	Flexor Hallucis Longus	Flexes great toe, helps to supinate ankle, and is a very weak plantar flexor of ankle.
10	Gastrocnemius (lateral head)	Powerful plantar flexor of ankle.
11	Gastrocnemius (med. head)	Powerful plantar flexor of ankle.
12	Gemellus (inf. and superior)	Rotates the thigh laterally; also helps abduct the flexed thigh.
13	Gluteus Maximus	Major extensor of hip joint; rotate laterally the hip; superior fibers abduct the hip; inferior fibers tighten the iliotibial band.
14	Gluteus Medius	Major abductor of thigh; anterior fibers help to rotate hip medially; posterior fibers help to rotate hip laterally
15	Gluteus Minimus	Abducts and medially rotates the hip joint.
16	Gracilis	Flexes the knee, adducts the thigh, helps to medially rotate the tibia on femur.
17	Iliacus	Flex the torso and thigh with respect to each other.
18	Pectineus	Adducts the thigh and flexes the hip joint.
19	Peroneus Brevis	Everts foot and plantar flexes ankle.
20	Peroneus Longus	Everts foot and plantar flexes ankle; helps to support the transverse arch of the foot.
21	Peroneus Tertius	Dorsiflexes, everts and abducts foot.
22	Piriformis	Lateral rotator of the hip joint; helps abduct the hip if it is flexed.
23	Psoas	Flex the torso and thigh with respect to each other.
24	Quadratus Femoris	Rotates the hip laterally; also helps adduct the hip.
25	Rectus Femoris	Extends the knee.

Nr	Muscle Name	Muscle Action
26	Sartorius	Flexes and laterally rotates the hip joint and flexes the knee.
27	Semimembranosus	Extends the thigh, flexes the knee, and also rotates the tibia medially, especially when the knee is flexed.
28	Semitendinosus	Extends the thigh and flexes the knee, and also rotates the tibia medially, especially when the knee is flexed.
29	Soleus	Powerful plantar flexor of ankle.
30	Tensor Fasciae Lata	Helps stabilize and steady the hip and knee joints by putting tension on the iliotibial band of fascia.
31	Tibialis Anterior	Dorsiflexor of ankle and invertor of foot.
32	Tibialis Posterior	Principal invertor of foot; adducts foot, plantar flexes ankle, helps to supinate foot.
33	Vastus Intermedius	Extends the knee.
34	Vastus Lateralis	Extends the knee.
35	Vastus Medialis	Extends the knee.

4.4 Static Optimization: Cost Functions and Constraint Equations

The solution of the inverse dynamics problem, with muscle actuators instead of joint actuators, introduces indeterminacy in the equations of motion of the biomechanical system, since it involves more unknowns than available equations of motion. Indeterminate systems present an infinite set of possible solutions, being the aim of optimization techniques to find, from all the possible solutions, the one that minimizes a prescribed objective function, subjected to a certain number of restrictions or constraints. Mathematically, the optimization problems is stated as:

$$\text{minimize } \mathcal{F}_0(u_i)$$

$$\text{subject to:}\begin{cases} f_j(u_i)=0 & j=1,\dots,n_{ec} \\ f_j(u_i)\geq 0 & j=(n_{ec}+1),\dots,n_{tc} \\ u_i^{lower}\leq u_i \leq u_i^{upper} & i=1,\dots,n_{sv} \end{cases} \tag{16}$$

where u_i are the state variables bounded respectively by u_{lower} and u_{upper}, $\mathcal{F}_0(u_i)$ is the objective or cost function to minimize and $f_j(u_i)$ are constraint equations that restrain the state variables. In Equation (16), n_{st} represents the total number of state variables and n_{tc} the total number of constraint equations in which n_{ec} are of the equality type.

The minimization of cost functions simulates the physiological criteria adopted by the central nervous system when deciding which muscles to recruit as well as the level of activation that produce the adequate motion or posture for a specific task. Many cost functions have been used by researchers in the study of the redundant problem in biomechanics [1,7,8]. The selection of the most appropriate criterion to use in the optimization process resides upon several important aspects such as the type of motion under analysis, the objectives to achieve or the presence of any type of pathology. For instance if the subject under analysis suffers from severe pain in the knee joint then the central nervous system is certainly concerned in minimizing the pain in the referred joint. In this case a possible criterion could be the minimization of the reaction forces at the knee joint. In a normal gait example, the central nervous system focus in maximizing the comfort, i.e., the muscle endurance and therefore minimizing the muscle fatigue. Here, possible criteria would be the minimization of the total muscle stress or the minimization of the total muscle force, which are closely related with the muscular fatigue [1,8].

A cost function must reflect the inherent physical activity or pathology and to include relevant physiological characteristics and functional properties, such as the maximum isometric force or the electromyographic activity [29]. From the computational point of view, a cost function must be numerically stable and fast to evaluate. Some of the most commonly used cost functions are presented hereafter:

i) Sum of the square of the individual muscle forces:

$$\mathcal{F}_0 = \sum_{m=1}^{n_{ma}} \left(F_{CE}^m\right)^2 \tag{17}$$

When applied to the study of human locomotion, this cost function is considered to fulfill the objective of energy minimization. This cost-function does not include any physiological or functional capabilities [29].

ii) Sum of the cube of the average individual muscle stress:

$$\mathcal{F}_0 = \sum_{m=1}^{n_{ma}} \left(\sigma_{CE}^m\right)^3 \tag{18}$$

This cost function was introduced by Crowninshield and Brand [8] and it is based on a quantitative force-endurance relationship and on experimental results. It includes physiological information, namely the value of the physiological cross sectional area of each muscle and it is reported to predict co-activation of muscle groups in a more physiologically realistic manner [29].

iii) Sum of the square of the normalized muscular forces:

$$\mathcal{F}_0 = \sum_{m=1}^{n_{ma}} \left(\frac{F_{CE}^m}{F_0^m}\right)^2 \tag{19}$$

This cost function it is similar to the first one but including physiological information, namely the maximum isometric force that each muscle is able to produce.

iv) Sum of the square of the muscle forces related with the maximum instantaneous moment that each muscle can exert:

$$\mathcal{F}_0 = \sum_{m=1}^{n_{ma}} \left(\frac{F_{CE}^m}{M^m}\right)^2 \tag{20}$$

This cost function has the important particularity of including physiological as well as functional information [29]. The calculation of the maximum instantaneous moment that the muscle is capable to generate not only requires the use of the maximum isometric force for that muscle but also its instantaneous moment arm.

v) "Soft saturation" cost function:

$$\mathcal{F}_0 = \sum_{m=1}^{n_{ma}} \sqrt{1 - \left(\frac{F_{CE}^m}{F_0^m} \right)^2} \tag{21}$$

This cost function includes physiological information regarding the maximum isometric force that each muscle is able to generate. It produces a more realistic synergistic function of the muscle in particular when activation and co-activation of muscles is concerned.

Cost functions can also be sum of the instantaneous muscle power or the sum of the square of the total reaction forces at the joints [7,29]. In the present work the principles of minimization of the sum of the square of the muscle forces [7] and of the sum of the cube of the individual muscle stresses [8] are used in applications involving human locomotion. Substituting Equation (11) in Equations (17) and (18), results for each one of this functions are

$$\begin{aligned} \mathcal{F}_0 &= \sum_{m=1}^{n_{ma}} \left(F_{CE}^m\right)^2 = \sum_{m=1}^{n_{ma}} \left(\frac{F_L^m F_{\dot{L}}^m}{F_0^m} a^m \right)^2 \\ \mathcal{F}_0 &= \sum_{m=1}^{n_{ma}} \left(\sigma_{CE}^m\right)^3 = \sum_{m=1}^{n_{ma}} \left(\bar{\sigma} \frac{F_L^m F_{\dot{L}}^m}{F_0^{m^2}} a^m \right)^3 \end{aligned} \tag{22}$$

where n_{ma} are the number of muscle actuators and $\bar{\sigma}$ is the specific muscle strength with a constant value of 31.39 N/cm^2 [23,24]. Note that only state variables associated with muscle actuators are used to evaluate the cost functions, although the complete set of state variables also include the Lagrange multipliers associated with the rest of the kinematic constraints. Using these cost functions, the state variables associated with muscle actuators represent muscle activations and for that reason are bounded to assume values between *0* and *1*. No bounds are specified for all other state variables.

In inverse dynamic analyses the equations of motion of the biomechanical system are the constraint equations that the state variables must fulfill. Therefore, all the constraint equations are of equality type and their number is equal to the number of equations of motion, i.e., equal to the number of natural coordinates defining the system. A vector containing all optimization constraints is defined as:

$$\mathbf{f} = \left\{ \begin{array}{c} f_1 \\ \vdots \\ f_{n_{tc}} \end{array} \right\} = \mathbf{\Phi}_\mathbf{q} \boldsymbol{\lambda} + \left(\mathbf{M}\ddot{\mathbf{q}} - \mathbf{g} \right) = \mathbf{0} \tag{23}$$

Equation (23) represents a set of linear equations on the state variables given by the Lagrange multipliers vector. This means that the gradients associated to these equations can be obtained analytically, i.e., the gradient of vector **f** in order to the state variables is given by the Jacobian matrix of the constraints already calculated. Analytically, this is:

$$\nabla_\lambda \mathbf{f} = \frac{\partial \mathbf{f}}{\partial \boldsymbol{\lambda}} = \boldsymbol{\Phi}_\mathbf{q} \tag{24}$$

This important result reveals that there is no need to calculate the sensitivities by finite differences or by any other numerical method, because these are readily available, in analytical form, and can be used directly in the optimization process.

Three optimization packages are tested in the solution of the optimization of the redundant muscle forces: DOT 5.0 - Design Optimization Tools [25], DNCONG from IMSL Math Library [26] and MMA – Method of Moving Asymptotes [27]. The first two methods use successive quadratic programming algorithms, while the third one uses the globally convergent method of moving asymptotes with inner and outer iterations. The analytical gradients of the cost function and the optimization constraint equations are supplied to the three optimization routines.

5 Application Case

The methodology and the biomechanical model described before are applied to the gait analysis of a subject with a normal cadence stride period. The subject under analysis is a 25-year-old male with a height of 1.70 m and a total body mass of 70 kg. The subject is wearing running shoes. The trial starts at the time step just before the right heel contact with the floor, and continues until the subsequent occurrence of the same foot. During the stride period, the subject has to walk over three force plates that measure the ground reaction forces for both feet [5,6]. A total number of 66 frames are recorded with a sampling frequency of 60 Hz. The trial has a total duration of 1.083s that corresponds to a walking cadence of approximately 111 steps per minute. This frequency is within the expected value reported in the literature for normal cadence stride periods [28].

An inverse dynamic analysis is performed to calculate the net moments of force developed by the joint actuators at the joints of the biomechanical model. The net moments of force for the joint actuators of the right ankle, knee and hip joints are presented in Figure 11. These results are within the expected values reported in the literature for a normal cadence stride period [28].

The muscle forces developed in the right leg during the stride period are calculated using the optimization procedure described before. For this purpose, muscle actuators associated with the muscles described in Table 2 are introduced in the biomechanical model while the joint actuators driving the joints crossed by the muscle apparatus are removed. The results obtained with two optimization packages, are presented in Figure 12 for the muscle *Gluteus Minimus*, an abductor of the hip and for *Soleus*, a powerful plantar flexor of the ankle.

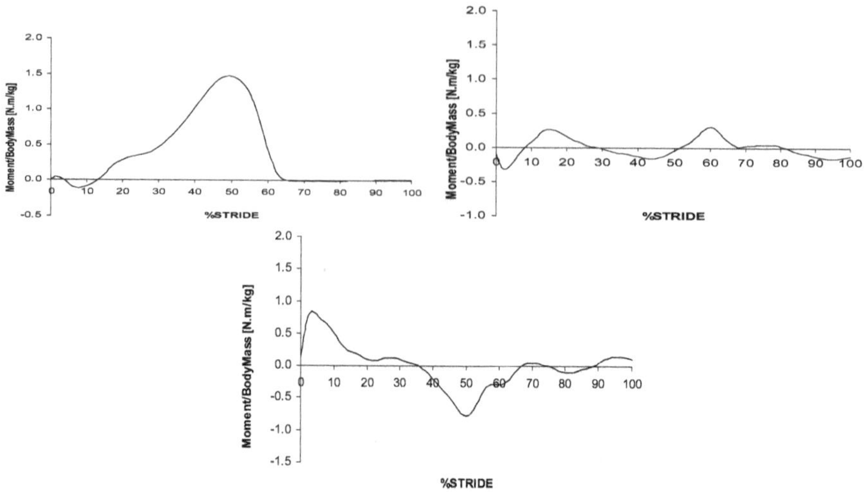

Fig. 11. Net moment-of-force (scaled by the body mass) for the ankle, knee and hip joints.

The results present a good correlation between the two optimization routines DOT and DNCONG. The forces present similar behaviors although with different force levels. The net moment-of-force produced by all muscles crossing a specified joint is equal to the net moment-of-force calculated for the joint actuators of that joint in the determinate inverse dynamics problem. This means that the results presented in Figure 12 are two possible solutions obtained by different optimization packages and that both solutions fulfill the equations of motion of the system. In terms of CPU time, the optimization package DOT about 160 times longer than DNCONG to optimize the 66 time steps.

The muscle activations represented in Figure 13 cannot be directly translated into muscle forces, as these are also related to the contraction velocity and to the muscle length through equation (11). The forces of the muscular apparatus are displayed in Figure 14 for some the muscles used in the biomechanical model.

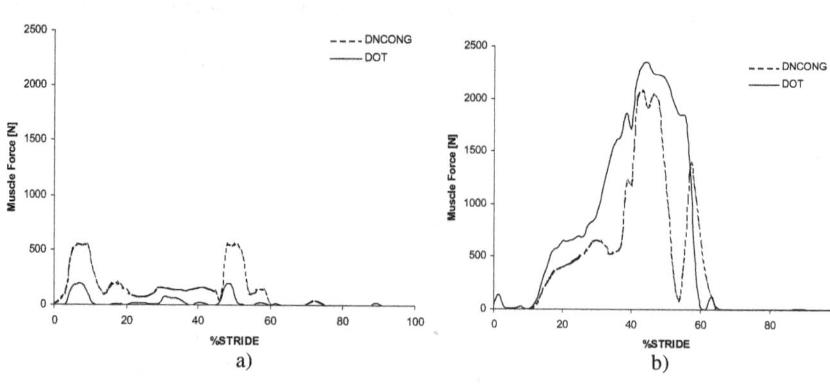

a) b)

Fig. 12. Muscle Forces: a) *Gluteus Minimus*, b) *Soleus*

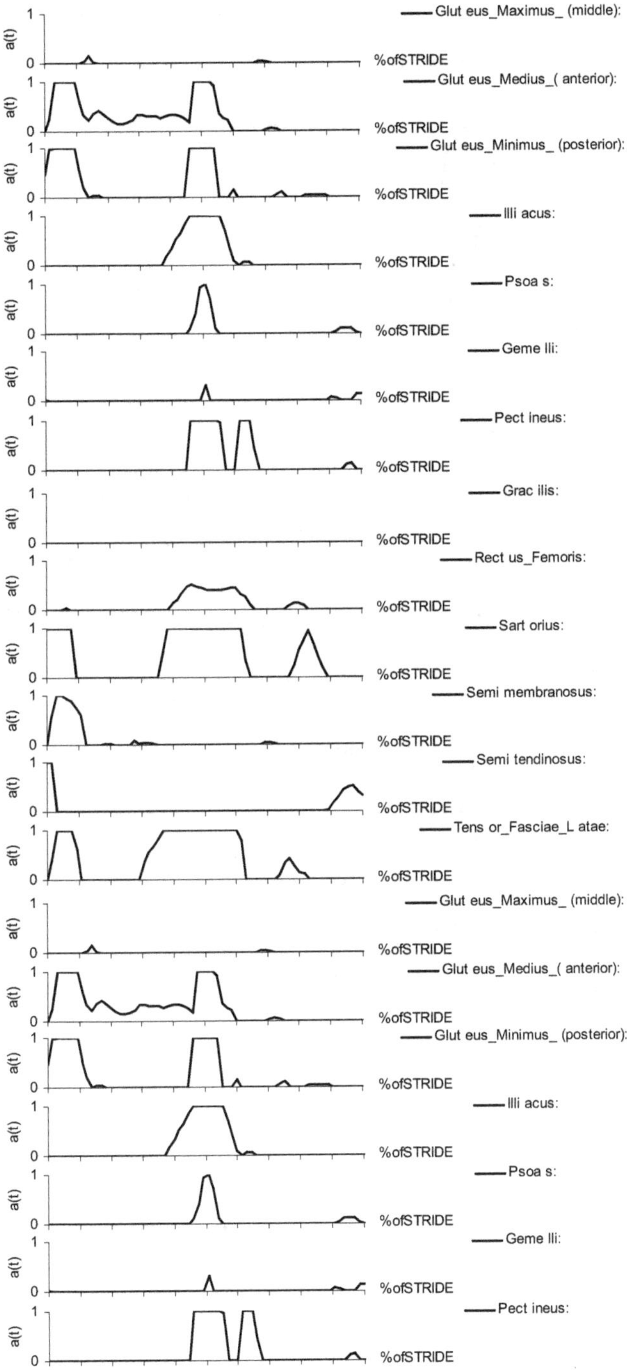
a(t)
1
0
%ofSTRIDE
Glut eus_Maximus_ (middle):
Glut eus_Medius_(anterior):
Glut eus_Minimus_ (posterior):
Illi acus:
Psoa s:
Geme lli:
Pect ineus:
Grac ilis:
Rect us_Femoris:
Sart orius:
Semi membranosus:
Semi tendinosus:
Tens or_Fasciae_L atae:
Glut eus_Maximus_ (middle):
Glut eus_Medius_(anterior):
Glut eus_Minimus_ (posterior):
Illi acus:
Psoa s:
Geme lli:
Pect ineus:

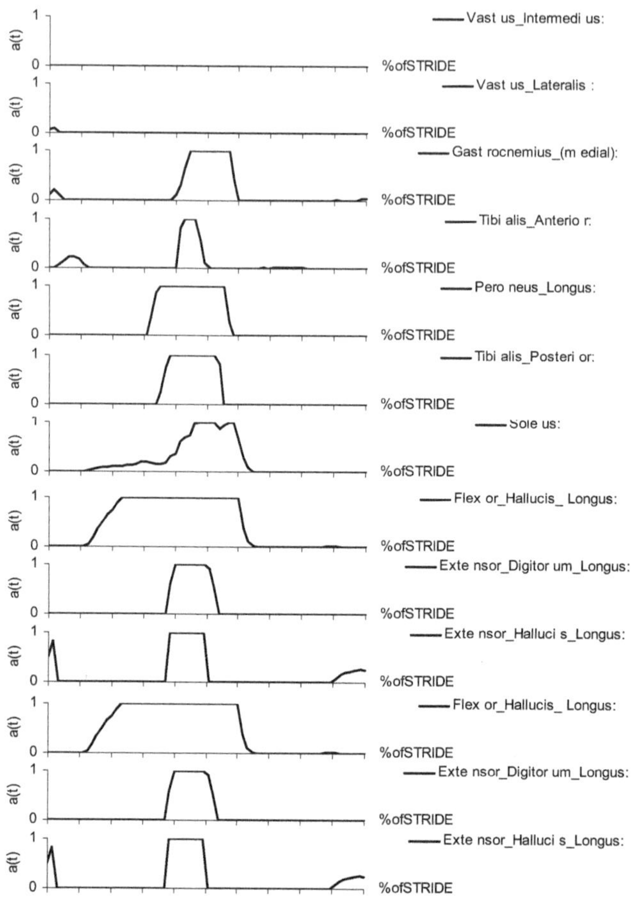

Fig. 13. Activation patterns for the muscles of the locomotion apparatus (right leg).

Comparing the results obtained for the normal cadence gait with those obtained by other authors, and in particular with those provided by Crowninshield and Brand [8], it is observed that the activation patterns shown in Figure 13 are similar to the EMG activation patterns reported for many of the muscles included in the model. Also the muscle forces obtained for many of the muscles described in Figure 14 are similar to those reported by Crowninshield and Brand [8]. But the opposite is also true, i.e., there is a significative number of muscles for which there is no correlation for the forces calculated in this work and those reported in the literature. The reason for not having a full agreement of all the muscle forces has to do with the detail of the muscle system used in the model and with the optimization procedures used. Not only the number of muscles used by Crowninshield and Brand [8] is smaller than the number of muscles used to represent the locomotor apparatus in this work but also all muscles have a linear line of action in the reference while in this work they are modeled as curved muscles, when appropriate.

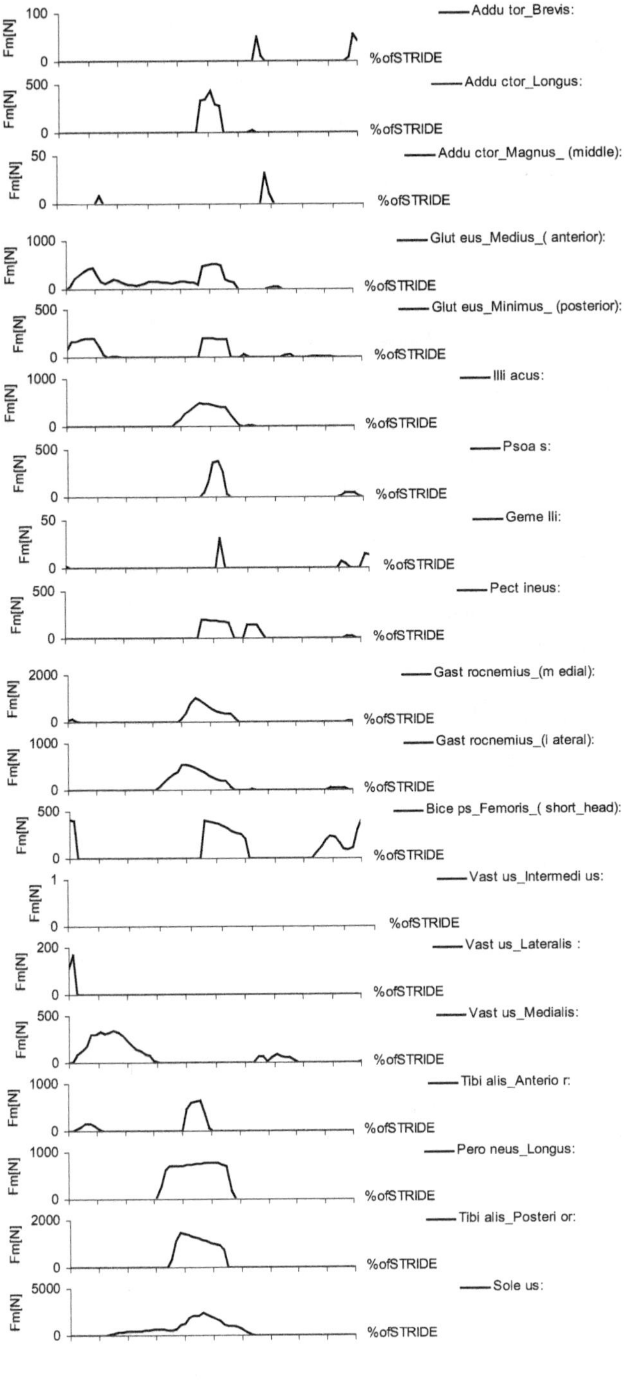
Fm[N]
%ofSTRIDE
Addu tor_Brevis:
Addu ctor_Longus:
Addu ctor_Magnus_ (middle):
Glut eus_Medius_(anterior):
Glut eus_Minimus_ (posterior):
Illi acus:
Psoa s:
Geme lli:
Pect ineus:
Gast rocnemius_(m edial):
Gast rocnemius_(l ateral):
Bice ps_Femoris_(short_head):
Vast us_Intermedi us:
Vast us_Lateralis :
Vast us_Medialis:
Tibi alis_Anterio r:
Pero neus_Longus:
Tibi alis_Posteri or:
Sole us:

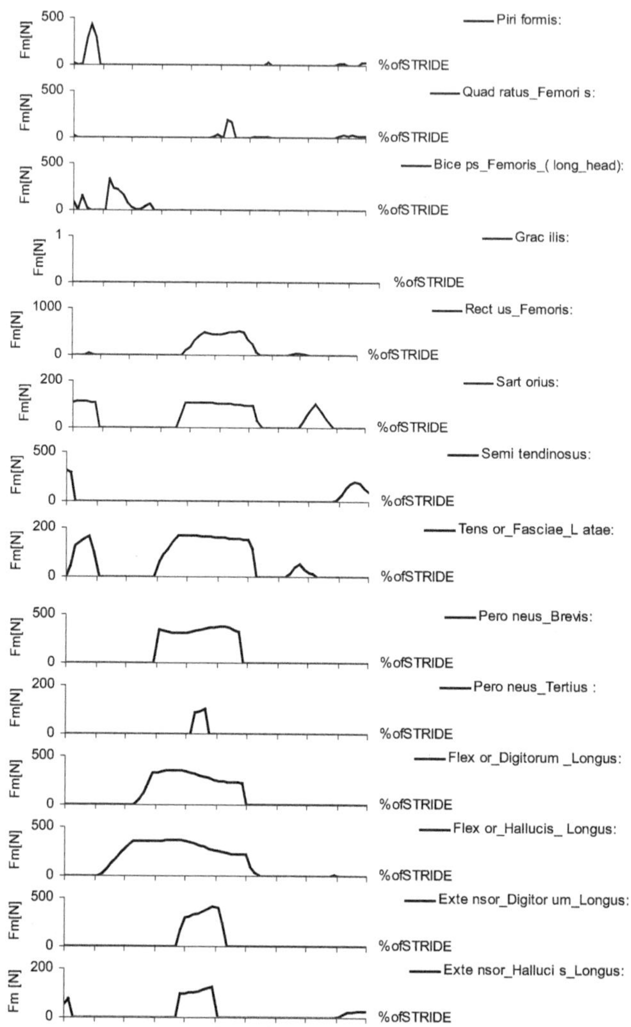

Fig. 14. Muscle force patterns for the muscles of the locomotion apparatus.

The analysis of the existing literature shows that the data provided is not sufficient to perform a quantitative comparison between published results and those reported here. In some cases only the temporal patterns of registered muscle activity are provided [7]. In other cases, the muscle forces are presented but only in qualitative terms [23], or in terms of their activation patterns [30]. However, in works where qualitative and quantitative information is provided, it is observed that the comparison of the results still may be difficult and misleading. To demonstrate this, the results obtained for the individual muscle forces are compared, with those of Patriarco *et al.* [31] and Crowninshield *et al.* [8] and presented in Figure 15 for the *Iliacus*, *Soleus*, *Adductor Brevis* and *Biceps Femoris* (SH).

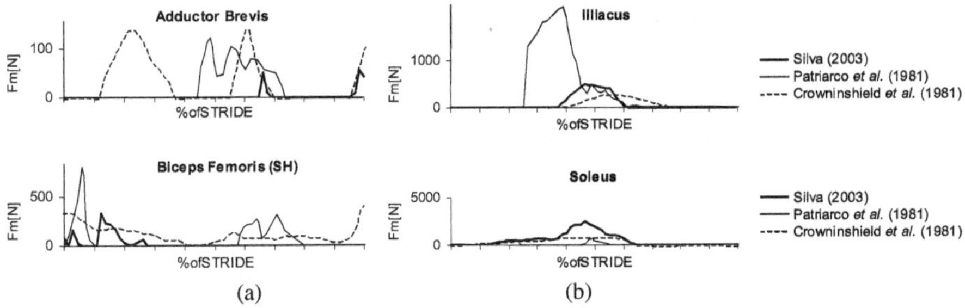

Fig. 15. Comparison of results of individual muscle forces obtained by different authors during a normal gait stride period. (a) No correlation is observed between the results of the different works. (b) A minimum correlation is observed between different works.

The analysis of the muscle forces presented in Figure 15 shows that there are several muscles where a minimum correlation can be obtained between the compared works, as in Figure 15 (b), but there are also some muscles where no correlation is found at all, even between the results presented by the different authors, as observed in Figure 15 (a). The difficulty in validating results against published data was already diagnosed in the literature. In its work, Patriarco *et al.* [8] refers that comparisons between different approaches and results of different investigators are difficult. Pedersen *et al.* [32], also refer that the solutions of the redundant muscle problem, predicted by different investigators, differ considerably for a variety of reasons, in which are included the use of different mathematical approaches, biomechanical models, input data, anthropometric models, acquisition devices, human subjects, muscle models, optimization techniques or objective functions.

A form of getting a better understanding of the quality of the results for the redundant muscle forces is to consider the physiological function of each muscle group and this action on the observed motion. With such purpose, the muscles are grouped considering their most important function. Eight different muscle groups are assembled: the hip adductors, composed by the *adductor brevis*, *adductor longus*, *adductor magnus*, *pectineus* and *quadratus femoris*; the hip abductors, composed by the *gluteus minimus*, *gluteus medius*, *gemelli* and *piriformis*; the ilipsoas, composed by the *iliacus* and *psoas*; the *quadriceps femoris*, composed by the *vastus medialis*, *vastus intermedius*, *vastus lateralis* and *rectus femoris*; the hamstrings, composed by the *semitendinosus*, *semimembranosus*, *biceps femoris*, *sartorius* and *gracilis*; the *triceps surae*, composed by the *soleus* and the two heads of the *gastrocnemius*; the ankle plantar flexors (without the *triceps surae* group), composed by the *tibialis posterior*, *peroneus brevis*, *peroneus longus*, *flexor digitorum longus* and *flexor hallucis longus*; and the ankle dorsiflexors, composed by the *tibialis anterior*, *peroneus tertius*, *extensor digitorum longus* and *extensor hallucis longus*. The force developed by each one of these muscle groups as a function of the percentage of stride is represented in Figure 16.

Six important events occurring during the stride period are highlighted, considering their relevance for the physiological analysis that follows. These events are:

the initial (right) heel-strike (HS) at 0% of stride, the opposite (left) toe-off (OTO) that occurs roughly after 15% of stride, the moment when the left foot passes a point immediately below the right hip joint (LFRH) at approximately 30% of stride, the opposite (left) heel-strike (OHS) that occurs at 50% of stride, the moment when the right foot passes a point immediately below the left hip joint (RFLH) approximately at 80% of stride, and the consecutive right HS at 100% of the stride period.

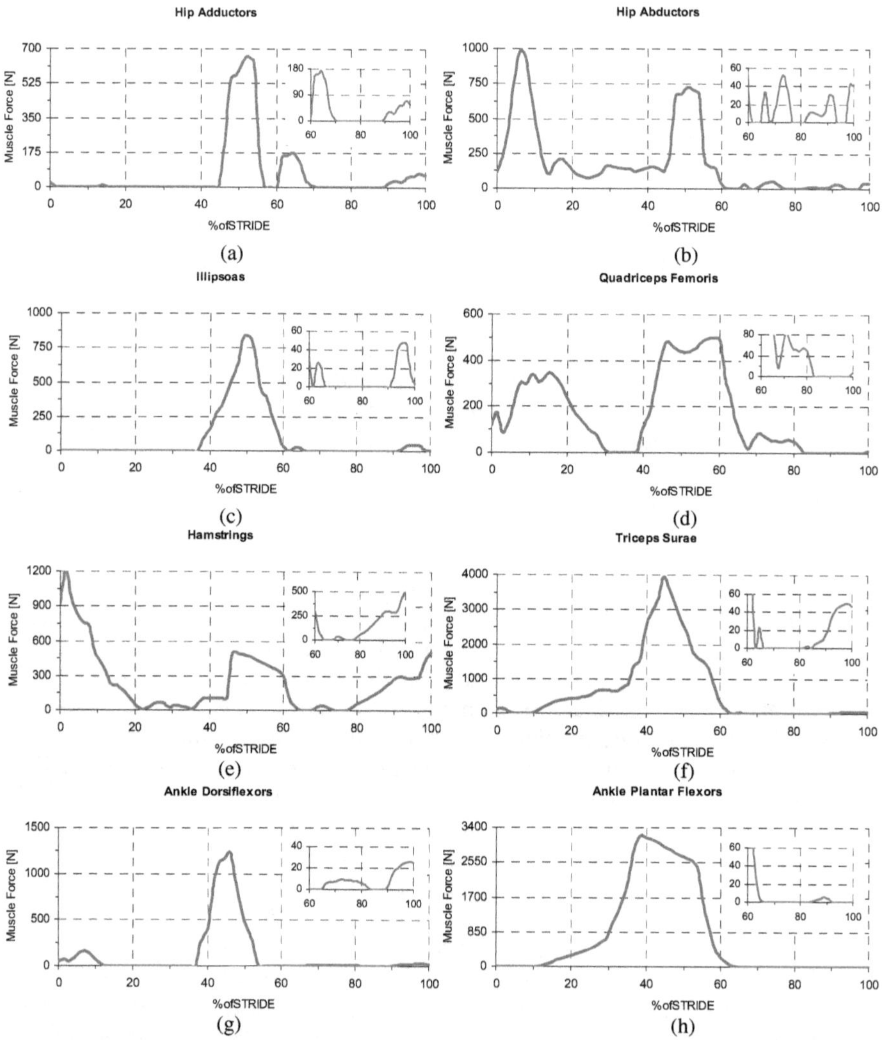

Fig. 16. Muscle forces arrange by functional muscle groups. (a) Hip adductors; (b) Hip abductors; (c) Ilipsoas; (d) Quadriceps femoris; (e) Hamstrings; (f) Triceps surae; (g) Ankle dorsiflexors; (h) Ankle plantar flexors.

The results presented in Figure 16 allow distinguishing two major and distinct phases during the stride period. The *stance phase* that starts after right heel-strike (HS), at 0% of stride, and ends with right toe-off (TO), shortly after 60% of stride. The *swing phase* that starts after right TO and ends with the succeeding right HS at 100% of stride. The stance phase is characterized by high levels of muscle activation and consequently by high levels of muscle forces, which develop to support the body weight and to thrust the body forward. Conversely the swing phase is characterized by small levels of muscle activity and consequently by considerably small muscle forces. In this latter phase, the central nervous system takes advantage of the pendulum-like motion of the leg, using the momentum of the body to reduce muscular activity.

6 Conclusions

This work presents a methodology that allows for the calculation of the net moments-of-force and reactions at the joints, and also for the calculation of the muscle forces developed in a specific muscle apparatus of a subject describing a prescribed motion. The subject is simulated using a whole body biomechanical model constructed with rigid bodies interconnected by revolute and universal joints. To improve the biofidelity of the model, the most important characteristics and dimensions of its rigid bodies are scaled from biomechanical data contained in a database with the most important physiological characteristics and properties of the principal anatomical segments of the human body. A general multibody formulation using natural or fully Cartesian coordinates is applied to the definition of rigid bodies and kinematic joints. With this formulation, the equations of motion of the biomechanical system are assembled and solved in a systematic manner.

The model is driven through the prescribed motion using kinematic driver actuators of two types: joint driver actuators and muscle driver actuators. A Lagrange multiplier is associated to each driver actuator that, depending on the actuator type, represents, for the case of a joint actuator, the net moment-of-force produced by all the muscles that cross the specified joint, or, in the case of the muscle actuator, the muscle force associated with a specified muscle or muscle complex. Depending on the objectives of the analysis, the two types of actuators can be used independently or cooperatively. It has been demonstrated that the use of muscle actuators usually introduces indeterminacy in the equations of motion of the biomechanical system. When this occurs, optimization techniques are the tools that allow for the selection of the solution that minimizes a specific cost-function. This cost function represents the physiological criteria used during the trial by the central nervous system of the subject under analysis.

It was shown in this work that the use of different optimization procedures lead to different sets of results, but still fulfilling the constraint equations and the joint moments-of-force obtained in the determinant inverse dynamic analysis. It is noticeable that some subsets of muscle forces compare better with the reported work by some authors while other muscle forces subsets are comparable with the results provided by other authors, all for cases of normal gait patterns. The same type of relations is identified when comparing the muscle activations obtained with the EMG patterns reported in the literature. This confirms that there is a sensitivity of the results obtained to the type of optimization procedure adopted and to the detail of the muscle apparatus implemented in the model.

From a quantitative point of view, the comparison of results with those reported by other investigators is difficult and eventually misleading. If there are cases of individual muscle forces for which a minimum correlation can be found between all works, there are other muscles for which no correlation can be obtained at all, even when comparing the results among the other referred investigators. However, when the results obtained in this work, for the solution of the force-sharing problem and prediction of the individual muscle forces, are grouped in terms of their action and after that they are analyzed from a physiological point of view and compared with a similar procedure obtained from the literature, a strong agreement is observed in all the phases of the stride period and for every muscle group considered in the study.

Acknowledgements

The work developed in this article was supported by Fundação para a Ciência e Tecnologia through the project PRAXIS/P/EME 14040/98, entitled *Human Locomotion Biomechanics Using Advanced Mathematical Models and Optimization Procedures* and through PhD scholarship PRAXIS/4/4.1/BD/2851. The authors want to gratefully acknowledge the valuable inputs and discussions with Dr. Matthew Kaplan.

References

1. Yamaguchi GT, Moran DW, Si J (1995) A computationally efficient method for solving the redundant problem in biomechanics. J. Biomech 28(8): 999-1005
2. Jalon J G, Bayo E (1994) Kinematic and Dynamic Simulation of Mechanical Systems – The Real-Time Challenge. Springer-Verlag, Berlin, Germany
3. Silva M, Ambrosio J and Pereira M (1997) Biomechanical Model with Joint Resistance for Impact Simulation. Multibody System Dynamics 1(1): 65-84
4. Laananen D, Bolokbasi A, Coltman J (1983) Computer simulation of an aircraft seat and occupant in a crash environment – Volume I: technical report, US Dept of Transp., Federal Aviation Administration, Report n DOT/FAA/CT-82/33-I
5. Ambrósio J, Silva M, Abrantes J (1999) Inverse Dynamic Analysis of Human Gait Using Consistent Data. In Proc of the IV Int. Symp. on Computer Methods in Biomechanics and Biomedical Engng, October13-16, Lisbon, Portugal
6. Silva M, Ambrósio J (2004) Sensitivity of the Results Produced by the Inverse Dynamic Analysis of a Human Stride to Perturbed Input Data. Gait and Posture 19(1): 35-49

7. Collins J (1995) The redundant nature of locomotor optimization laws, J. Biomech. 28(3): 251-267
8. Crowninshield R, Brand R (1981) Physiologically Based Criterion of Muscle Force Prediction in Locomotion. J. Biomech. 14(11): 793-801
9. Zajac F (1989) Muscle and tendon: properties, models, scaling, and application to biomechanics and motor control. Critical Reviews in Biomedical Engineering 17(4): 359-411
10. Hatze H (1984) Quantitative Analysis, Synthesis and Optimization of Human Motion. Human Movement Science 3: 5-25
11. Silva M (2003) Human Motion Analysis Using Multibody Dynamics and Optimization Tools. Ph.D. Dissertation, Instituto Superior Técnico, Technical University of Lisbon, Lisbon, Portugal
12. Silva M, Ambrósio J, Pereira M (1997) A multibody approach to the vehicle and occupant integrated simulation. Int. J. of Crashworthiness 2(1): 73-90
13. Nikravesh P (1988) Computer-Aided Analysis of Mechanical Systems. Prentice Hall, Englewood-Cliffs, New Jersey
14. Haug E (1989) Computer Aided Kinematics and Dynamics of Mechanical Systems. Allyn and Bacon, Boston, Massachusetts
15. Winter D (1990) Biomechanics And Motor Control Of Human Movement, 2nd Ed., John Wiley & Sons, Toronto, Canada
16. Silva M, Ambrósio J (2002) Kinematic Data Consistency in the Inverse Dynamic Analysis of Biomechanical Systems. Multibody System Dynamics 8(2): 219-239
17. Celigüeta J (1996) Multibody Simulation of Human Body Motion in Sports. In Abrantes J (ed.) Proceedings of the XIV International Symposium on Biomechanics in Sports, June 25-29, Funchal, Portugal, pp 81-94
18. Addel-Aziz Y, Karara H (1971) Direct Linear Transformation from Comparator Coordinates into Object Space Coordinates in Close-Range Photogrammetry. In Proc. of the Symposium on Close-range Photogrammetry, Falls Church, Virginia, pp. 1-18.
19. Ambrósio J, Silva M, Lopes G (1999) Reconstrução do Movimento Humano e Dinâmica Inversa Utilizando Ferramentas Numéricas Baseadas em Sistemas Multicorpo. In Proceedings of the IV Congreso de Métodos Numéricos en Ingeniería, Sevilha, 7-10 June
20. Nigg B, Herzog W (1999) Biomechanics of the Musculo-skeletal System, John Wiley & Sons, New York, New York
21. Allard P, Stokes I, Blanchi J (1995) Three-Dimensional Analysis of Human Movement. Human Kinetics, Champaign, Illinois
22. Richardson M (2001) Lower Extremity Muscle Atlas, in internet address http://www.rad.washington.edu/atlas2/ , University of Washington - Department of Radiology,Washington
23. Yamaguchi G (2001) Dynamic Modeling of Musculoskeletal Motion. Kluwer Academic Publishers, Boston, Massachussetts
24. Carhart M (2000) Biomechanical Analysis of Compensatory Steping: Implications for paraplegics Standing Via FNS., Ph.D. Dissertation, Department of Bioengineering, Arizona State University, Tempe, Arizona
25. Vanderplaats R&D (1999) DOT – Design Optimization Tools – USERS MANUAL – Version 5.0, Colorado Springs, Colorado
26. V. Numerics (1995) IMSL FORTRAN Numerical Libraries – Version 5.0, Microsoft Corp.

27. Svanberg K (1999) The MMA for Modeling and Solving Optimization Problems. In Proceedings of the 3rd World Congress of Structural and Multidisciplinary Optimization, May 17-21, New York
28. Winter D (1991) The biomechanics and motor control of human gait: Normal, Elderly and Pathological, 2nd Ed. University of Waterloo Press, Waterloo, Canada
29. Tsirakos D, Baltzopoulos V, Bartlett R (1997) Inverse Optimization: Functional and Physiological Considerations Related to the Force-Sharing Problem. Critical Reviews in Biomedical Engineering 25(4-5): 371-407
30. Anderson F, Pandy M (2001) Static and Dynamic Optimization Solutions for Gait are Practically Equivalent. J. Biomech. 34: 153-161
31. Patriarco A., Mann R, Simon S, Mansour J (1981) An Evaluation of the Approaches of Optimization Models in the Prediction of Muscle Forces During Gait. J. Biomech. 14(8): 513-525
32. Pedersen D, Brand R, Cheng C, Arora J (1987) Direct Comparison of Muscle Force Predictions Using Linear and Nonlinear Programming. ASME Journal of Biomechanical Engineering 109: 192-199
33. Palastanga N, Field D, Soames R (2002) Anatomy and Human Movement - Structure and Function. Elsevier, Amsterdam, The Netherlands

Understanding Mean-Axis Conditions as Floating Reference Frames

Parviz E. Nikravesh

Department of Aerospace and Mechanical Engineering, University of Arizona, Tucson, AZ 85721, USA
pen@email.arizona.edu

The use of the so-called mean axis conditions as a floating reference frame for moving deformable bodies is discussed. This reference frame is compared against the commonly used nodal-fixed frame. The discussion on various issues associated with these frames is facilitated through simple schematic visualizations. The equations of motion for a moving deformable body are analyzed in both nodal and modal spaces. Issues associated with the extraction of the free-free and constrained modes are also discussed. It is shown that the mean-axis conditions can be employed effectively to represent a floating reference frame for a moving deformable body.

1 Introduction

An interesting topic in multibody dynamics is how to attach a reference frame to a moving deformable body. In structural finite element analysis, it is a simple task to define a reference frame since, in most cases, the structure is assumed not to have any rigid body motion. However, when a deformable body is allowed to translate and rotate in space, the issue of attaching a reference frame to the body deserves special attention. It is a common practice to define six simple conditions to eliminate six of the nodal deflections from the equations of motion. These conditions attach the structure to the reference frame or vise versa. We refer to this type of frame as the nodal-fixed axes. Another less known and sometimes misunderstood reference frame is the mean-axis conditions. The mean axes, which are based on the minimization of the deformation kinetic energy, impose six conditions on all of the nodes impartially unlike the nodal-fixed conditions. Although there are other types of reference frames for moving deformable bodies, those frames are not discussed in this paper; e.g., using the instantaneous principal axes of the deformable body or using the nonmoving inertial frame.

It has been common among researchers to use the nodal-fixed axes as the moving reference frame for deformable bodies. This is due to the fact that since the nodal-fixed frame is commonly used in the structural finite-element models, it is natural to use the same type of frame in a multibody system. Furthermore, since the frame is specifically fixed to some of the nodes, it is easy to locate the frame if

J.A.C. Ambrósio (ed.), Advances in Computational Multibody Systems, 185–203.

the positions of the nodes are known or vise-versa. In contrast, to someone unfamiliar with the mean-axis frames, it would be difficult to visualize how a frame could follow a deformable body without being fixed to any specific nodes. It is the objective of this document to illustrate that the mean-axis conditions provide a more meaningful moving frame compared to the nodal-fixed conditions. Furthermore, the resulting equations of motion from the mean axes contain terms that are more meaningful than those derived from the nodal-fixed equations.

In this paper we first discuss the equations of motion for a deformable body. These equations are stated without proof in several forms. It is assumed that the reader has some familiarity with these equations. In order to concentrate on the concepts and not to be distracted by the complex form of these equations, we assume that a finite element node exhibits only *translational* degrees-of-freedom. Elimination of the *rotational* degrees-of-freedom does not make the discussion any less general. The rotational degrees-of-freedom can be added to these formulations if necessary. Furthermore, some of the equations have been stated in semi-abstract forms. Interested readers who may want to formulate these equations into a computer program should be aware of the hidden details.

Following the presentation of the equations of motion, conditions for defining a nodal-fixed or a mean-axis reference frames are discussed. Comparison between the two types of frames is made via a simple example. Then, the transformation of the equations of motion the nodal to modal space is considered. Issues associated with the use of free-free and constrained modes are discussed. Finally, some general conclusions are drawn.

It is important that the reader to become familiar with the notation first. Therefore, it is recommended that the reader pay special attention to the following section on notation.

2 Notation

In this paper matrix notation is used in order to keep the attention on concepts without much loss of details. The reader should find the notation very effective in multibody formulations, especially when deformable bodies are involved. The following nomenclature is used:

Reference frames:
- x-y-z Inertial
- ξ–η–ζ Body attached

Vectors and arrays:
- Bold-face, lower-case characters
  - Roman (regular) (contains x-y-z components)
  - Italic (contains ξ–η–ζ components)

Matrices:
- Bold-face, upper-case characters
  - Roman (regular) (described in x-y-z frame)
  - Italic (described in ξ–η–ζ frame)

Right superscripts:

a	a node (or degree-of-freedom) defining reference axes
b	a boundary node (or degree-of-freedom)
m	a master node (or degree-of-freedom)
s	a slave node (or degree-of-freedom)
u	an unconstrained node (or degree-of-freedom)

Over-scores:

~	(tilde) transforms a 3-vector to a skew-symmetric matrix
^	(hat) stacks vertically 3-vectors or 3×3 skew-symmetric matrices
–	(bar) repeats a 3×3 matrix to form a block-diagonal matrix

Strike-through:

–	(bar) denotes an entity in modal space

The following examples should clarify the notation. Assume that $\mathbf{b}^i$ is a 3-vector and $\tilde{\mathbf{b}}^i$ is a 3×3 skew-symmetric matrix for $i=1, ..., n$. Let $\mathbf{I}$ be a 3×3 identity matrix and $\mathbf{A}$ to be a 3×3 rotational transformation matrix. Then the following stack and block-diagonal matrices can be constructed:

$$\mathbf{b} = \begin{Bmatrix} \mathbf{b}^1 \\ \Box \\ \mathbf{b}^n \end{Bmatrix}, \quad \hat{\tilde{\mathbf{b}}} = \begin{bmatrix} \tilde{\mathbf{b}}^1 \\ \Box \\ \tilde{\mathbf{b}}^n \end{bmatrix}, \quad \hat{\mathbf{I}} = \begin{bmatrix} \mathbf{I} \\ \Box \\ \mathbf{I} \end{bmatrix}, \quad \bar{\mathbf{I}} = \begin{bmatrix} \mathbf{I} & \cdots & \mathbf{0} \\ \Box & \ddots & \Box \\ \mathbf{0} & \cdots & \mathbf{I} \end{bmatrix}, \quad \bar{\mathbf{A}} = \begin{bmatrix} \mathbf{A} & \cdots & \mathbf{0} \\ \Box & \ddots & \Box \\ \mathbf{0} & \cdots & \mathbf{A} \end{bmatrix}$$

3 Equations of Motion for a Deformable Body

In this section the equations of motion for a deformable body are stated. In order to simplify the form of the equations, without loss of generality, it is assumed that a finite element node exhibits only three *translational* degrees-of-freedom.

3.1 Kinematics

Assume that a deformable body is described by n nodes that are allowed only translational deflections; i.e., $n_{dof}=3\times n$. A body-reference frame denoted as $\xi-\eta-\zeta$ is defined to move with the body without stating at this point how the frame is attached to the body; i.e., no boundary conditions are defined between the frame and the nodes. For a typical node i, as shown in Figure 1, the translational deflection vector is denoted by a 3-vector $\boldsymbol{\delta}^i$. This node is positioned from the origin of the body-frame in its undeformed state by vector $\mathbf{s}^i$ and in the deformed state by vector $\mathbf{b}^i=\mathbf{s}^i+\boldsymbol{\delta}^i$. For all the nodes, the arrays of nodal deflections, the undeformed positions, and the deformed positions are defined as

$$\boldsymbol{\delta} = \begin{Bmatrix} \boldsymbol{\delta}^1 \\ \Box \\ \boldsymbol{\delta}^n \end{Bmatrix} \quad \mathbf{s} = \begin{Bmatrix} \mathbf{s}^1 \\ \Box \\ \mathbf{s}^n \end{Bmatrix} \quad \mathbf{b} = \begin{Bmatrix} \mathbf{b}^1 \\ \Box \\ \mathbf{b}^n \end{Bmatrix} \tag{1}$$

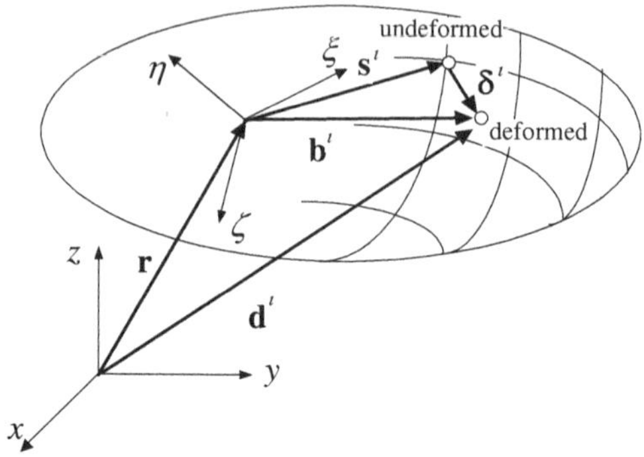

Fig. 1. A deformable body and its body frame positioned in an inertial frame.

In addition to the ξ–η–ζ frame, we also introduce an inertial x-y-z frame where between the two frames a 3×3 rotational transformation matrix $\mathbf{A}$ is defined. The elements of this matrix; i.e., the nine direction cosines, can be determined and computed in term of any desired rotational coordinates that we define between the two frames. Components of any vector in ξ–η–ζ frame can be transformed into the x-y-z frame using this matrix; e.g., $\boldsymbol{\delta}^i = \mathbf{A}\,\boldsymbol{\delta}^i$, $\mathbf{s}^i = \mathbf{A}\,\mathbf{s}^i$, or $\boldsymbol{\delta} = \overline{\mathbf{A}}^{\mathrm{T}}\,\boldsymbol{\delta}$. Note that a vector or an array in *italic* indicates that the components are in ξ–η–ζ frame.

The absolute position of a typical node i, as shown in Figure 1, is expressed as

$$\mathbf{d}^i = \mathbf{r} + \mathbf{b}^i = \mathbf{r} + \mathbf{s}^i + \boldsymbol{\delta}^i \tag{2}$$

The absolute velocity and acceleration of this node are expressed as

$$\dot{\mathbf{d}}^i = \dot{\mathbf{r}} - \tilde{\mathbf{b}}^i\,\boldsymbol{\omega} + \dot{\boldsymbol{\delta}}^i \tag{3}$$

$$\ddot{\mathbf{d}}^i = \ddot{\mathbf{r}} - \tilde{\mathbf{b}}^i\,\dot{\boldsymbol{\omega}} + \ddot{\boldsymbol{\delta}}^i + \mathbf{w}^i \tag{4}$$

where $\boldsymbol{\omega}$ and $\dot{\boldsymbol{\omega}}$ are the angular velocity and acceleration vectors of the body frame, and all the quadratic velocity terms are grouped together as

$$\mathbf{w}^i = \tilde{\boldsymbol{\omega}}\tilde{\boldsymbol{\omega}}\mathbf{b}^i + 2\tilde{\boldsymbol{\omega}}\dot{\boldsymbol{\delta}}^i \tag{5}$$

If arranged in stack form, the position, velocity, and acceleration of all the nodes can be written as

$$\mathbf{d} = \hat{\mathbf{I}}\mathbf{r} + \mathbf{b} \tag{6}$$

$$\dot{\mathbf{d}} = \hat{\mathbf{I}}\dot{\mathbf{r}} - \hat{\tilde{\mathbf{b}}}\boldsymbol{\omega} + \dot{\boldsymbol{\delta}} = \begin{bmatrix} \hat{\mathbf{I}} & -\hat{\tilde{\mathbf{b}}} & \bar{\mathbf{I}} \end{bmatrix} \begin{Bmatrix} \dot{\mathbf{r}} \\ \boldsymbol{\omega} \\ \dot{\boldsymbol{\delta}} \end{Bmatrix} \tag{7}$$

$$\ddot{\mathbf{d}} = \hat{\mathbf{I}}\ddot{\mathbf{r}} - \hat{\tilde{\mathbf{b}}}\dot{\boldsymbol{\omega}} + \ddot{\boldsymbol{\delta}} + \mathbf{w} = \begin{bmatrix} \hat{\mathbf{I}} & -\hat{\tilde{\mathbf{b}}} & \bar{\mathbf{I}} \end{bmatrix} \begin{Bmatrix} \ddot{\mathbf{r}} \\ \dot{\boldsymbol{\omega}} \\ \ddot{\boldsymbol{\delta}} \end{Bmatrix} + \mathbf{w} \tag{8}$$

where

$$\mathbf{w} = \tilde{\bar{\boldsymbol{\omega}}}\tilde{\bar{\boldsymbol{\omega}}}\,\mathbf{b} + 2\tilde{\bar{\boldsymbol{\omega}}}\dot{\boldsymbol{\delta}} \tag{9}$$

3.2 Kinetics

For a deformable body, the mass and stiffness matrices, $\boldsymbol{M}$ and $\boldsymbol{K}$, are initially constructed by any finite-element package in the body-frame ξ–η–ζ. It should be noted that the stiffness matrix $\boldsymbol{K}$ has a rank deficiency of six since we have not yet imposed any reference conditions. These two matrices can be transformed into the x-y-z frame as[1]

$$\mathbf{M} = \bar{\mathbf{A}}\,\boldsymbol{M}\,\bar{\mathbf{A}}^{\mathrm{T}}, \quad \mathbf{K} = \bar{\mathbf{A}}\,\boldsymbol{K}\,\bar{\mathbf{A}}^{\mathrm{T}} \tag{10}$$

In order to better understand the development of the equations of motion for a moving deformable body, we start with the standard equations of motion for a structure. We first assume that the ξ–η–ζ is a non-moving frame. The equations of motion for this *structure* are written as

$$\boldsymbol{M}\,\ddot{\boldsymbol{\delta}} = \boldsymbol{f} - \boldsymbol{K}\,\boldsymbol{\delta} \tag{11}$$

where $\boldsymbol{f}$ is the array of external forces. We are reminded that in this equation all the entities are described in terms of their components in the ξ–η–ζ frame.

Using the block-diagonal transformation matrix $\bar{\mathbf{A}}$, Equation (11) is expressed in terms of the x-y-z components of its entities as

$$\mathbf{M}\ddot{\boldsymbol{\delta}} = \mathbf{f} - \mathbf{K}\boldsymbol{\delta} \tag{12}$$

Now we allow the ξ–η–ζ frame to move relative to the x-y-z frame. The equations of motion can now be stated as

$$\mathbf{M}\ddot{\mathbf{d}} = \mathbf{f} - \mathbf{K}\boldsymbol{\delta} \tag{13}$$

where $\ddot{\mathbf{d}}$ contains the absolute nodal accelerations. Note that all entities in this equation are described in the x-y-z frame. Also note in Equation (12) rigid-body motion of the deformable body is not allowed but in Equation (13) the body is allowed to translate and rotate in space while it deforms.

We now transform Equation (13) into other useful forms. We substitute Equation (8) into Equation (13) to obtain

$$\mathbf{M}\begin{bmatrix} \hat{\mathbf{I}} & -\hat{\tilde{\mathbf{b}}} & \bar{\mathbf{I}} \end{bmatrix}\begin{Bmatrix} \ddot{\mathbf{r}} \\ \dot{\boldsymbol{\omega}} \\ \ddot{\boldsymbol{\delta}} \end{Bmatrix} = \mathbf{g} \tag{14}$$

---

[1] If the mass matrix is constructed properly, then $\mathbf{M} = \boldsymbol{M}$. Further discussion on the characteristics of the mass matrix is outside the scope of this paper.

where

$$\mathbf{g} = \mathbf{f} - \mathbf{M}\mathbf{w} - \mathbf{K}\boldsymbol{\delta} \tag{15}$$

If Equation (14) is pre-multiplied by the transpose of the coefficient matrix of Equation (8), we get

$$\begin{bmatrix} \hat{\mathbf{I}}^T\mathbf{M}\hat{\mathbf{I}} & -\hat{\mathbf{I}}^T\mathbf{M}\hat{\tilde{\mathbf{b}}} & \hat{\mathbf{I}}^T\mathbf{M} \\ -\hat{\tilde{\mathbf{b}}}^T\mathbf{M}\hat{\mathbf{I}} & \hat{\tilde{\mathbf{b}}}^T\mathbf{M}\hat{\tilde{\mathbf{b}}} & -\hat{\tilde{\mathbf{b}}}^T\mathbf{M} \\ \mathbf{M}\hat{\mathbf{I}} & -\mathbf{M}\hat{\tilde{\mathbf{b}}} & \mathbf{M} \end{bmatrix} \begin{Bmatrix} \ddot{\mathbf{r}} \\ \dot{\boldsymbol{\omega}} \\ \ddot{\boldsymbol{\delta}} \end{Bmatrix} = \begin{Bmatrix} \hat{\mathbf{I}}^T\mathbf{g} \\ -\hat{\tilde{\mathbf{b}}}^T\mathbf{g} \\ \mathbf{g} \end{Bmatrix} \tag{16}$$

We define the total mass of the body as m and the rotational inertia matrix of the body as $\mathbf{J}$; i.e.,

$$\hat{\mathbf{I}}^T\mathbf{M}\hat{\mathbf{I}} = m\mathbf{I}\,,\ \hat{\tilde{\mathbf{b}}}^T\mathbf{M}\hat{\tilde{\mathbf{b}}} = \mathbf{J} \tag{17}$$

Equation (16) can be expressed as

$$\begin{bmatrix} m\mathbf{I} & -\hat{\mathbf{I}}^T\mathbf{M}\hat{\tilde{\mathbf{b}}} & \hat{\mathbf{I}}^T\mathbf{M} \\ -\hat{\tilde{\mathbf{b}}}^T\mathbf{M}\hat{\mathbf{I}} & \mathbf{J} & -\hat{\tilde{\mathbf{b}}}^T\mathbf{M} \\ \mathbf{M}\hat{\mathbf{I}} & -\mathbf{M}\hat{\tilde{\mathbf{b}}} & \mathbf{M} \end{bmatrix} \begin{Bmatrix} \ddot{\mathbf{r}} \\ \dot{\boldsymbol{\omega}} \\ \ddot{\boldsymbol{\delta}} \end{Bmatrix} = \begin{Bmatrix} \hat{\mathbf{I}}^T\mathbf{g} \\ -\hat{\tilde{\mathbf{b}}}^T\mathbf{g} \\ \mathbf{g} \end{Bmatrix} \tag{18}$$

So far we have described the equations of motion for a moving deformable body in three forms. The first form is given in Equation (13) where there are as many equations as the number of unknown absolute accelerations. The second form is provided by Equation (14) where we have six more unknown accelerations than the number of equations. Equation (18) is the third form of the equations of motion where the number of equations and the number of unknowns are the same. From these three forms, only Equation (13) is solvable for the accelerations—in Equation (14) we have more unknowns than the number of equations and the coefficient matrix in Equation (18) has a rank deficiency of six. Equations (14) and (18) are not solvable since we have not yet defined reference conditions.

For practical purposes, the array of nodal deflections in Equations (14) and (18) is normally described in terms of the body-fixed components. Therefore these equations are written as

$$\mathbf{M}\begin{bmatrix} \hat{\mathbf{I}} & -\hat{\tilde{\mathbf{b}}} & \bar{\mathbf{A}} \end{bmatrix} \begin{Bmatrix} \ddot{\mathbf{r}} \\ \dot{\boldsymbol{\omega}} \\ \ddot{\boldsymbol{\delta}} \end{Bmatrix} = \mathbf{g} \tag{19}$$

and

$$\begin{bmatrix} m\mathbf{I} & -\hat{\mathbf{I}}^T\mathbf{M}\hat{\tilde{\mathbf{b}}} & \hat{\mathbf{I}}^T\mathbf{M}\bar{\mathbf{A}} \\ -\hat{\tilde{\mathbf{b}}}^T\mathbf{M}\hat{\mathbf{I}} & \mathbf{J} & -\hat{\tilde{\mathbf{b}}}^T\mathbf{M}\bar{\mathbf{A}} \\ \bar{\mathbf{A}}^T\mathbf{M}\hat{\mathbf{I}} & -\bar{\mathbf{A}}^T\mathbf{M}\hat{\tilde{\mathbf{b}}} & \bar{\mathbf{A}}^T M\bar{\mathbf{A}} \end{bmatrix} \begin{Bmatrix} \ddot{\mathbf{r}} \\ \dot{\boldsymbol{\omega}} \\ \ddot{\boldsymbol{\delta}} \end{Bmatrix} = \begin{Bmatrix} \hat{\mathbf{I}}^T\mathbf{g} \\ -\hat{\tilde{\mathbf{b}}}^T\mathbf{g} \\ g \end{Bmatrix} \tag{20}$$

These equations may further be transformed to other forms. For example, the angular acceleration vector can be expressed with respect to the body frame. This or any further transformations will be left to the reader.

Several simplifications are made on the right-hand side of Equation (18). For example, the expansion of the terms in $\hat{\mathbf{I}}^{\mathrm{T}}\mathbf{g}$ and $\hat{\mathbf{b}}^{\mathrm{T}}\mathbf{g}$ reveals that $\hat{\mathbf{I}}^{\mathrm{T}}\mathbf{K}\boldsymbol{\delta}=\mathbf{0}$ and $\hat{\mathbf{b}}^{\mathrm{T}}\mathbf{K}\boldsymbol{\delta}=\mathbf{0}$ [2]. Such simplifications are not shown in these and the following equations.

In the following section we will discuss two choices of reference frames for moving deformable bodies. For this and other purposes, we may split the nodes or the deformation degrees-of-freedom into different subsets. For example, if we split the nodes or the deformation degrees-of-freedom) into the *boundary* and the *unconstrained* (or free) nodes, we denote them with superscripts "*b*" and "*u*" respectively. Based on this categorization, all nodal entities will also be split into subsets. For example, the mass matrix will be split as:

$$\mathbf{M}=\begin{bmatrix}\mathbf{M}^{bb} & \mathbf{M}^{bu}\\ \mathbf{M}^{ub} & \mathbf{M}^{uu}\end{bmatrix}=\begin{bmatrix}\mathbf{M}^{b,}\\ \mathbf{M}^{u,}\end{bmatrix}=\begin{bmatrix}\mathbf{M}^{,b} & \mathbf{M}^{,u}\end{bmatrix}$$

4 Reference Axis Conditions

Attaching a reference frame to a deformable body requires defining six conditions. This reduces the number of nodal deflection degrees-of-freedom to $n_{dof}=3\times n-6$. The most common procedure, as it is normally done in structural analysis with finite element method, is to set six of the nodal deflections to zero. We refer to this procedure as the *nodal-fixed* method. Another method that is not widely used is to define six equations called *mean-axis* conditions. Other methods such as using the instantaneous principal axes of the deformable body [1, 2] or using the inertial reference frame instead of the body-fixed frame [3] are not discussed in this paper.

4.1 Nodal-Fixed Frame

In structural finite-element analysis, when nodal rotational degrees-of-freedom are not considered, the frame is normally attached to three nodes of the deformable body. We assume that the origin of the frame is always at node *o*, that node *j* always remains along the ξ–axis, while node *k* remains on the ξ–η plane. This yields six conditions as:

$$\begin{aligned}\delta^{o}_{(\xi)}=\delta^{o}_{(\eta)}=\delta^{o}_{(\zeta)}&=0\\ \delta^{j}_{(\eta)}=\delta^{j}_{(\zeta)}&=0\\ \delta^{k}_{(\zeta)}&=0\end{aligned}\tag{21}$$

[2] These are the sum of internal (structural) forces and moments respectively.

With these six conditions, knowing where the nodes are at any instant, we can locate the ξ–η–ζ frame, and vise-versa. In order to incorporate these six conditions into the equations of motion, we split the array of nodal displacement and any associated entities into the *axis* and *unconstrained* sets. These sets are denoted by superscripts "*a*" and "*u*" respectively as:

$$\delta = \begin{Bmatrix} \delta^a \\ \delta^u \end{Bmatrix} = \begin{Bmatrix} \mathbf{0} \\ \delta^u \end{Bmatrix}, \quad \ddot{\delta} = \begin{Bmatrix} \ddot{\delta}^a \\ \ddot{\delta}^u \end{Bmatrix} = \begin{Bmatrix} \mathbf{0} \\ \ddot{\delta}^u \end{Bmatrix} \tag{22}$$

With this partitioning of the nodal deflections, Equation (19) becomes:

$$\begin{bmatrix} \mathbf{M}\hat{\mathbf{I}} & -\mathbf{M}\hat{\tilde{\mathbf{b}}} & \bar{\mathbf{A}}\mathcal{M}^{\cdot u} \end{bmatrix} \begin{Bmatrix} \ddot{\mathbf{r}} \\ \dot{\boldsymbol{\omega}} \\ \ddot{\delta}^u \end{Bmatrix} = (\mathbf{f} - \bar{\mathbf{A}}\mathcal{M}^{\cdot u}\ddot{\delta}^u - \mathbf{M}^{\cdot u}\mathbf{w}^u) \tag{23}$$

where we have as many equations in as the number of accelerations. It would also be possible to incorporate the nodal-fixed conditions into a form similar to that of Equation (18).

4.2 Mean-Axis Conditions

The mean-axis conditions are six constraints that enforce the body-frame to follow the motion of the nodes in such a way that the kinetic energy associated with the deformation stays at a minimum. In order to derive these conditions, the deformation kinetic energy is expressed as

$$T = \tfrac{1}{2}\dot{\boldsymbol{\delta}}^{\mathrm{T}}\mathbf{M}\dot{\boldsymbol{\delta}} = \tfrac{1}{2}(\dot{\mathbf{d}} - \hat{\mathbf{I}}\dot{\mathbf{r}} + \hat{\tilde{\mathbf{b}}}\boldsymbol{\omega})^{\mathrm{T}}\mathbf{M}(\dot{\mathbf{d}} - \hat{\mathbf{I}}\dot{\mathbf{r}} + \hat{\tilde{\mathbf{b}}}\boldsymbol{\omega}) \tag{24}$$

where we have used Equation (8). The partial derivatives of the kinetic energy with respect to the translational and rotational velocity vectors of the reference frame are

$$T_{\dot{r}} = (\dot{\mathbf{d}} - \hat{\mathbf{I}}\dot{\mathbf{r}} + \hat{\tilde{\mathbf{b}}}\boldsymbol{\omega})^{\mathrm{T}}\mathbf{M}\hat{\mathbf{I}} = -\dot{\boldsymbol{\delta}}^{\mathrm{T}}\mathbf{M}\hat{\mathbf{I}}$$
$$T_{\omega} = (\dot{\mathbf{d}} - \hat{\mathbf{I}}\dot{\mathbf{r}} + \hat{\tilde{\mathbf{b}}}\boldsymbol{\omega})^{\mathrm{T}}\mathbf{M}\hat{\tilde{\mathbf{b}}} = \dot{\boldsymbol{\delta}}^{\mathrm{T}}\mathbf{M}\hat{\tilde{\mathbf{b}}}$$

The *translational* and *rotational* mean axis conditions are obtained by setting these partial derivatives equal to zero[3] [4]:

$$\hat{\mathbf{I}}^{\mathrm{T}}\mathbf{M}\dot{\boldsymbol{\delta}} = \mathbf{0} \qquad (\hat{\mathbf{I}}^{\mathrm{T}}\mathcal{M}\dot{\delta} = \mathbf{0}) \tag{25}$$

$$\hat{\tilde{\mathbf{b}}}^{\mathrm{T}}\mathbf{M}\dot{\boldsymbol{\delta}} = \mathbf{0} \qquad (\hat{\tilde{b}}^{\mathrm{T}}\mathcal{M}\dot{\delta} = \mathbf{0}) \tag{26}$$

[3] If nodal rotational degrees-of-freedom are included, the mean axis conditions find a slightly different form than that of Equations 22 and 23.

The time derivatives of these equations, following some simplifications, are[4]

$$\hat{\mathbf{I}}^{\mathrm{T}}\,\mathbf{M}\ddot{\boldsymbol{\delta}} = \mathbf{0} \qquad (\hat{\mathbf{I}}^{\mathrm{T}}\,\boldsymbol{M}\ddot{\boldsymbol{\delta}} = \mathbf{0}) \tag{27}$$

$$\hat{\hat{\mathbf{b}}}^{\mathrm{T}}\,\mathbf{M}\ddot{\boldsymbol{\delta}} = \mathbf{0} \qquad (\hat{\hat{\mathbf{b}}}^{\mathrm{T}}\,\boldsymbol{M}\ddot{\boldsymbol{\delta}} = \mathbf{0}) \tag{28}$$

Other identities can be derived based on Equations (25) and (26). For example, the integral of Equation (25) is

$$\hat{\mathbf{I}}^{\mathrm{T}}\,\mathbf{M}\boldsymbol{\delta} = \mathbf{0} \qquad (\hat{\mathbf{I}}^{\mathrm{T}}\,\boldsymbol{M}\boldsymbol{\delta} = \mathbf{0}) \tag{29}$$

If the origin of the frame in the undeformed state of the body is positioned at the mass center; i.e., $\hat{\mathbf{I}}^{\mathrm{T}}\mathbf{M}\mathbf{s} = \mathbf{0}$, then Equation (29) becomes

$$\hat{\mathbf{I}}^{\mathrm{T}}\mathbf{M}\mathbf{b} = \mathbf{0} \qquad (\hat{\mathbf{I}}^{\mathrm{T}}\boldsymbol{M}\boldsymbol{b} = \mathbf{0}) \tag{30}$$

This means that the origin of the reference frame stays at the instantaneous mass center of the body. Furthermore, if the mass matrix is properly constructed, it can be shown that Equation (39) yields

$$\hat{\mathbf{I}}^{\mathrm{T}}\,\mathbf{M}\hat{\hat{\mathbf{b}}} = \hat{\hat{\mathbf{b}}}^{\mathrm{T}}\,\mathbf{M}\hat{\mathbf{I}} = \mathbf{0} \qquad (\hat{\mathbf{I}}^{\mathrm{T}}\,\boldsymbol{M}\hat{\hat{\boldsymbol{b}}} = \hat{\hat{\boldsymbol{b}}}^{\mathrm{T}}\boldsymbol{M}\hat{\mathbf{I}} = \mathbf{0}) \tag{31}$$

Note that all entities in Equations (25)-(31) can be described either in their x-y-z or ξ–η–ζ components.

4.3 Interpretation of the Mean Axes

The best way to understand how the mean-axis conditions enforce a reference frame to follow the nodes of a deformable body is through a simple example and visualization. Consider a simple planar deformable rod that is modeled by seven nodes as shown in its undeformed state in Figure 2(a). Under some applied forces the rod finds the undeformed shape shown in Figure 2(b). Now we observe the deflection of the nodes from two different reference frames.

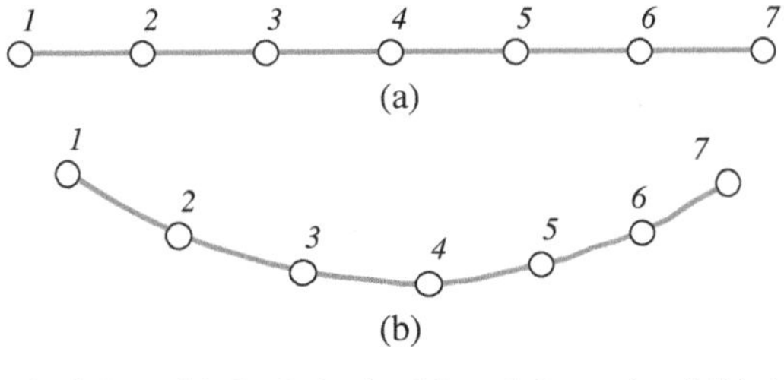

Fig. 2. A simple deformable body in the (a) undeformed and (b) deformed states

---

[4] The right-hand side of Equation 25 may contain a quadratic velocity term if the mass matrix is not properly constructed.

The first reference frame that we consider is the conventional nodal-fixed axes. As shown in Figure 3(a), it is assumed that the origin of this frame is fixed to node *1* (setting two of the deformation degrees-of-freedom to zeros) and the ξ-axis remains on node *2* (setting one of the deformation degrees-of-freedom to zero). With this description of the body-frame, the deformation of the nodes are viewed as shown in Figure 3(b). Note that to a viewer attached to this frame it would appear that node *7* has the largest deflection, that node *2* has a slight deflection only along the ξ-axis, and that node *1* has no deflection at all.

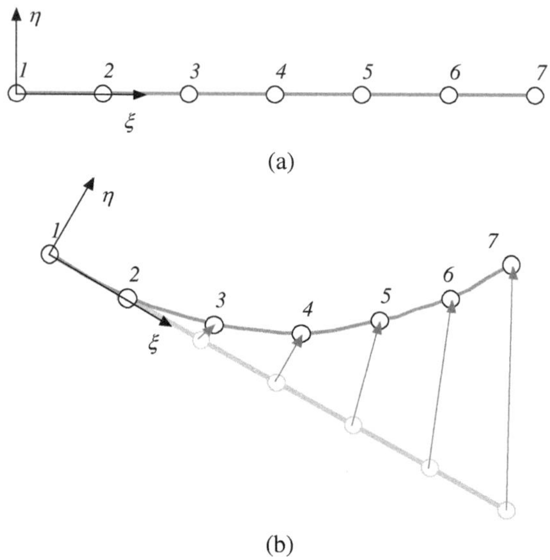

Fig. 3. Deflection of a rod as viewed from a nodal-fixed frame

The next reference frame that we consider is the mean-axis conditions. As shown in Figure 4(a), the origin of the frame is initially positioned at the undeformed mass center of the rod that coincides with node *4*. The axes of the frame can initially be rotated in any desired orientation since the rotational mean-axis condition does not exist at the coordinate level. In our example, the ξ-axis is positioned initially along the rod axis to the right. In the deformed position, as shown in Figure 4(b), the origin of the frame has remained at the instantaneous mass center and its orientation has followed the condition of Equation (26). Since this condition is on the velocities, it cannot be visualized as easily as the translational condition. Note that to a viewer attached to this frame every node would appear to have some deflection, and the deflections appear to be more uniformly distributed compared to the deflections viewed from the nodal-fixed frame.

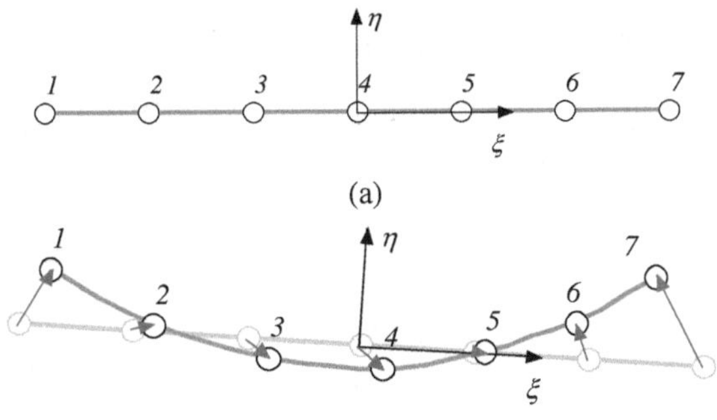

Fig. 4. Deflection of a rod as viewed from a mean axes frame

From the results shown in Figs. 3 and 4 we observe that the nodal deflections of a deformable body would not appear the same from two different body frames. In addition, the position and orientation of the two frames are not the same with respect to the body nor with respect to a nonmoving inertial frame. Probably the most important difference is that the largest deflection as observed from a mean-axis frame would, in general, be smaller than the largest deflection observed from a nodal-fixed frame. This may be an important issue when we consider the finite-element approximations associated with the assumption of *small deflections*.

4.4 Equations of Motion with Mean-Axes

Substituting Equations(27), (28), and (31) into Equation (18) yields:

$$\begin{bmatrix} m\mathbf{I} & \mathbf{0} & \mathbf{0} \\ \mathbf{0} & \mathbf{J} & \mathbf{0} \\ \mathbf{M}\hat{\mathbf{I}} & -\mathbf{M}\hat{\tilde{\mathbf{b}}} & \mathbf{M} \end{bmatrix} \begin{Bmatrix} \ddot{\mathbf{r}} \\ \dot{\boldsymbol{\omega}} \\ \ddot{\boldsymbol{\delta}} \end{Bmatrix} = \begin{Bmatrix} \hat{\mathbf{I}}^{\mathrm{T}}\mathbf{g} \\ -\hat{\tilde{\mathbf{b}}}^{\mathrm{T}}\mathbf{g} \\ \mathbf{g} \end{Bmatrix} \tag{32}$$

The coefficient matrix in this equation is non-singular. Therefore, if all the forces are known, this equation can be solved for the accelerations.

Since the mean-axis conditions at the acceleration level are not explicitly present in Equation(32), there is no guarantee that the resultant values for $\ddot{\boldsymbol{\delta}}$ would satisfy the acceleration constraints; i.e., these constraints may slightly be violated. Although the amount of violation could be small, it is possible that during forward integration of the equations of motion, this error may cause numerical instability. In order to eliminate this particular error, Equations (27) and (28) can be incorporated into Equation (32) (or Equation (18)) with the aid of Lagrange multipliers to obtain

$$\left[\begin{array}{cccc} m\mathbf{I} & \mathbf{0} & \mathbf{0} & \mathbf{0} \\ \mathbf{0} & \mathbf{J} & \mathbf{0} & \mathbf{0} \\ \mathbf{M}\hat{\mathbf{I}} & -\mathbf{M}\hat{\hat{\mathbf{b}}} & \mathbf{M} & \mathbf{D}_{m-a}^{\mathrm{T}} \\ \hline \mathbf{0} & \mathbf{0} & \mathbf{D}_{m-a} & \mathbf{0} \end{array}\right] \left\{\begin{array}{c} \ddot{\mathbf{r}} \\ \dot{\boldsymbol{\omega}} \\ \ddot{\boldsymbol{\delta}} \\ \hline \boldsymbol{\lambda}_{m-a} \end{array}\right\} = \left\{\begin{array}{c} \hat{\mathbf{I}}^{\mathrm{T}}\mathbf{g} \\ -\hat{\hat{\mathbf{b}}}^{\mathrm{T}}\mathbf{g} \\ \mathbf{g} \\ \hline \mathbf{0} \end{array}\right\} \tag{33}$$

where

$$\mathbf{D}_{m-a} = \begin{bmatrix} \hat{\mathbf{I}}^{\mathrm{T}}\mathbf{M} \\ \hat{\hat{\mathbf{b}}}^{\mathrm{T}}\mathbf{M} \end{bmatrix}$$

and $\boldsymbol{\lambda}_{m-a}$ contains six Lagrange multipliers. The solution of Equation (33) guarantees that the mean-axis conditions at the acceleration level are always satisfied.

5 Nodal Reduction of Equations of Motion

The number of nodes and consequently the number of degrees-of-freedom associated with a moving deformable body may be too large for any realistic numerical simulation within the context of multibody modeling. Therefore, different processes have been developed in the past to reduce the number of degrees-of-freedom in such systems.

The reduction process may be performed in the nodal space. In the so-called static or Guyan condensation, it is assumed that some of the nodes of a finite element model can be kept and the rest can be deleted [5]. For this purpose, we use superscripts "*m*" and "*s*" for *master* (kept) and *slave* (deleted) nodes. The structural equations of motion are re-written as

$$\begin{bmatrix} \boldsymbol{M}^{mm} & \boldsymbol{M}^{ms} \\ \boldsymbol{M}^{sm} & \boldsymbol{M}^{ss} \end{bmatrix} \begin{Bmatrix} \ddot{\boldsymbol{\delta}}^{m} \\ \ddot{\boldsymbol{\delta}}^{s} \end{Bmatrix} + \begin{bmatrix} \boldsymbol{K}^{mm} & \boldsymbol{K}^{ms} \\ \boldsymbol{K}^{sm} & \boldsymbol{K}^{ss} \end{bmatrix} \begin{Bmatrix} \boldsymbol{\delta}^{m} \\ \boldsymbol{\delta}^{s} \end{Bmatrix} = \begin{Bmatrix} \boldsymbol{f}^{m} \\ \boldsymbol{f}^{s} \end{Bmatrix} \tag{34}$$

If we assume that the inertia of the entire structure is allocated to the master nodes and no external forces are applied to the slave nodes, then the condensed mass and stiffness matrices are found as:

$$\boldsymbol{M}_{G} = \boldsymbol{M}^{mm} + \boldsymbol{M}^{ms}\boldsymbol{G} + \boldsymbol{G}^{\mathrm{T}}\boldsymbol{M}^{sm} + \boldsymbol{G}^{\mathrm{T}}\boldsymbol{M}^{ss}\boldsymbol{G} \tag{35}$$

$$\boldsymbol{K}_{G} = \boldsymbol{K}^{mm} + \boldsymbol{K}^{ms}\boldsymbol{G} \tag{36}$$

where the static condensation matrix is defined as

$$\boldsymbol{G} = -\boldsymbol{K}^{ss\text{-}1}\boldsymbol{K}^{sm} \tag{37}$$

The condensed matrices of Equations (35) and (36) can be used instead of $\boldsymbol{M}$ and $\boldsymbol{K}$ in Equations (13), (14), and (18) in order to reduce the number of equations. Following that we can enforce either the nodal-fixed or the mean-axis conditions[5].

The nodal condensation is one of the simplest methods to reduce the number of degrees-of-freedom. However, we must note that this condensation process alters the modal characteristics of a deformable body; i.e., the reduced number of natural frequencies is not a subset of the original body. This feature could be a reason for not using this reduction method in certain applications.

6 Modal Transformation

The equations of motion can be transformed from nodal to modal coordinates. The mass and stiffness matrices yield matrices of mode shapes and modal frequencies denoted as Ψ and Λ. The transformation between the modal and nodal coordinates is expressed as:

$$\delta = \Psi \mathbf{z} \tag{38}$$

The modal mass and stiffness matrices are computed as:

$$\not{\boldsymbol{M}} = \Psi^{\mathrm{T}} \boldsymbol{M} \Psi \,, \quad \not{\boldsymbol{K}} = \Psi^{\mathrm{T}} \boldsymbol{K} \Psi \tag{39}$$

In order to follow the form of the equations of motion that we have discussed so far, for transformation purposes we define another modal matrix as

$$\mathrm{\Psi} = \bar{\mathbf{A}} \Psi \tag{40}$$

Therefore the transformation of Equation (38) becomes

$$\boldsymbol{\delta} = \Psi \mathbf{z} \tag{41}$$

As it will be discussed in the upcoming sections, the characteristics of matrix Y depend on the lack or presence of boundary conditions on the nodes.

6.1 Free-Free Modes

Without applying any boundary conditions on $\boldsymbol{M}$ and $\boldsymbol{K}$, the corresponding matrices of mode shapes and modal frequencies are expressed as:

$$\begin{bmatrix} \hat{\mathbf{I}} & -\hat{\tilde{\mathbf{s}}} & \Psi_f \end{bmatrix}, \quad diag\,(\mathbf{0},\ \mathbf{0},\ \Lambda_f) \tag{42}$$

where $\hat{\mathbf{I}}$ and $-\hat{\tilde{\mathbf{s}}}$ represent the translational and rotational rigid-body mode shapes, each being an $n_{dof}\times 3$ matrix, and Ψ_f is an $n_{dof}\times(n_{dof} -6)$ matrix represent-

[5] Some of the simplifications in the mean-axis conditions may no longer be applicable when we use a condensed mass matrix.

ing the deformation mode shapes. The diagonal matrix of eigen-values contains six zero eigen-values associated with the rigid-body modes in addition to the deformation eigen-values denoted by $\boldsymbol{\Lambda}_f$. The subscript *"f"* emphasizes that these entities correspond to a *free-free* structure. The corresponding transformations are

$$\boldsymbol{\delta} = \boldsymbol{\Psi}_f \mathbf{z}\,, \quad \mathcal{M}_f = \boldsymbol{\Psi}_f^{\mathrm{T}} \boldsymbol{M} \boldsymbol{\Psi}_f\,, \quad \mathcal{K}_f = \boldsymbol{\Psi}_f^{\mathrm{T}} \boldsymbol{K} \boldsymbol{\Psi}_f \tag{43}$$

The transformations of Equation (43) are substituted into Equation (32) to obtain:

$$\begin{bmatrix} m\mathbf{I} & \mathbf{0} & \mathbf{0} \\ \mathbf{0} & \mathbf{J} & \mathbf{0} \\ \boldsymbol{\Psi}_f^{\mathrm{T}}\mathbf{M}\hat{\mathbf{I}} & -\boldsymbol{\Psi}_f^{\mathrm{T}}\mathbf{M}\hat{\hat{\mathbf{b}}} & \mathcal{M}_f \end{bmatrix} \begin{Bmatrix} \ddot{\mathbf{r}} \\ \dot{\boldsymbol{\omega}} \\ \ddot{\mathbf{z}} \end{Bmatrix} = \begin{Bmatrix} \hat{\mathbf{I}}^{\mathrm{T}}\mathbf{g} \\ -\hat{\hat{\mathbf{b}}}^{\mathrm{T}}\mathbf{g} \\ \boldsymbol{\Psi}_f^{\mathrm{T}}\mathbf{g} \end{Bmatrix} \tag{44}$$

The characteristics of matrix $\boldsymbol{\Psi}_f$ guarantee that the mean-axis conditions are automatically satisfied. Therefore, there is no need to substitute Equation (43) into Equation (33)—such a substitution will result into six redundant equations. Furthermore, it is not of any use to incorporate the free-free modal data into the nodal-fixed equations of motion.

6.2 Constrained Modes

In structural analysis, the nodes that are connected to the ground are considered as the boundary nodes. Therefore, we denote these nodes with superscript "b" and the remaining nodes with superscript "u". These boundary nodes also include the removed degrees-of-freedom that define the reference axes in the nodal-fixed conditions. For these nodes in a nodal-fixed frame we set $\boldsymbol{\delta}^b = \ddot{\boldsymbol{\delta}}^b = \mathbf{0}$. The corresponding sub-matrices $\boldsymbol{M}^{uu}$ and $\boldsymbol{K}^{uu}$ are used to obtain the modal matrices $\boldsymbol{\Psi}_c$ and $\boldsymbol{\Lambda}_c$. The subscript *"c"* emphasizes that these entities correspond to a *constrained* structure. The corresponding transformation equations are expressed as:

$$\boldsymbol{\delta}^u = \boldsymbol{\Psi}_c \mathbf{z}\,, \quad \mathcal{M}_c = \boldsymbol{\Psi}_c^{\mathrm{T}} \boldsymbol{M}^{uu} \boldsymbol{\Psi}_c\,, \quad \mathcal{K}_c = \boldsymbol{\Psi}_c^{\mathrm{T}} \boldsymbol{K}^{uu} \boldsymbol{\Psi}_c \tag{45}$$

The transformations of Equation (45) can be used to transform the nodal-fixed equations of motion to modal space. However, if we use the mean-axis conditions, the boundary conditions can no longer be described as $\boldsymbol{\delta}^b = \ddot{\boldsymbol{\delta}}^b = \mathbf{0}$ and, therefore, the modal data must be obtained differently. This is the subject of discussion in the following sub-section.

6.3 Constrained Modes and Mean-Axes

In order to understand how the mode shapes are viewed from different reference frames, we consider the schematic presentation of the bending of a planar rod. The fixed-fixed rod is constrained to the ground at nodes *1* and *10* as shown in Figure 5. Two possible reference frames are defined for the rod—a nodal-fixed frame and a mean-axes frame. The nodal-fixed frame is attached to nodes *1* and *10*; i.e., to the ground, where the mean-axes frame moves with the mass center as depicted schematically. The objective of this example is to understand how the mode shapes are viewed from different frames. We will look at two bending modes.

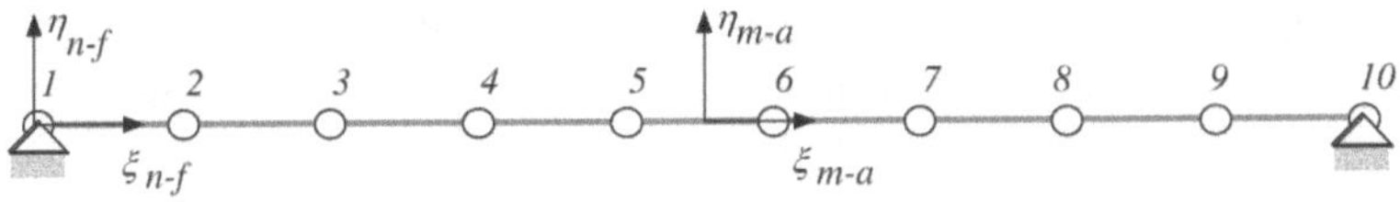

Fig. 5. A fixed-fixed constrained rod carrying a nodal-fixed frame and a mean-axes frame

The first bending mode of this rod is shown in Figure 6(a). In this figure the position and orientation of the two frames are also shown. Obviously, since the nodal-fixed frame is attached to the ground, it has no movement, where the mean-axes frame has moved with the mass center of the first bending mode. There is no rotation of the mean-axes frame since this bending mode is symmetric with respect to the η-axis. This bending mode as viewed from the nodal-fixed frame is shown in Figure 6(b). However, the same bending mode appears totally different to the mean-axis frame as shown in Figure 6(c) where the constrained nodes *1* and *10* appear to be displaced with respect to the frame.

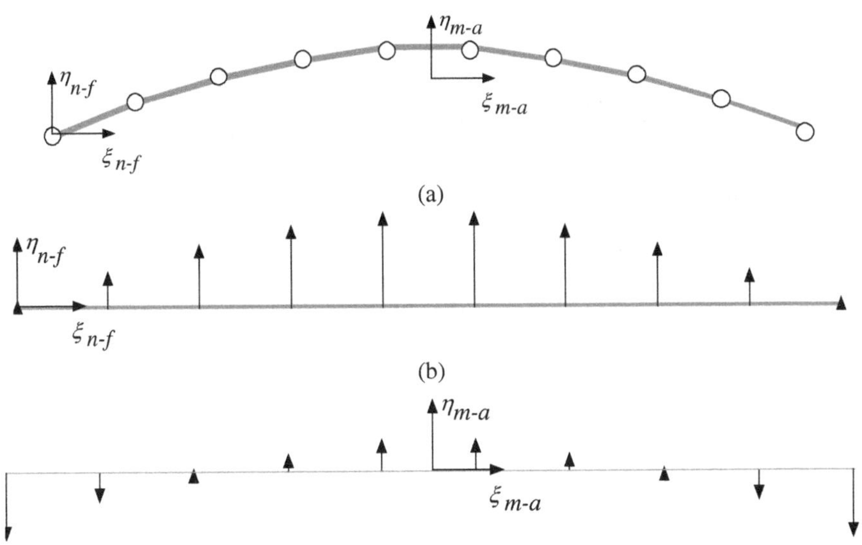

Fig. 6. The first bending mode of the rod: (a) the actual bending; and the bending mode as viewed from (b) the nodal-fixed frame and (c) the mean-axes frame

The second bending mode of this rod and the two reference frames are shown in Figure 7(a). Similar to the first bending mode, obviously, there is no displacement associated with the nodal-fixed frame. The figure schematically shows that the mean-axes frame has not translated but it has undergone a rotation. Since the mass center of the beam, as far as this mode is concerned, has remained at the same location, there is no translational displacement of this frame. In order to understand the cause of the rotation of this frame, we must consider Equation (26) and rewrite it for small displacements as:

$$\hat{\hat{\mathbf{b}}}^{\mathrm{T}}\,\mathbf{M}\,(\Delta\boldsymbol{\delta}) = \mathbf{0}$$

Therefore, in order to satisfy this condition, the frame must undergo the necessary rotation. This bending mode as viewed from the nodal-fixed and the mean-axes frames are shown in Figs. 7(b) and (c).

We should note that although the mode shapes have different appearances in the two frames, the natural frequencies are not effected by the choice of reference frames. It is possible to transform the mode shapes from one frame to another, if all the mode shapes are available in one of the frames.

The constrained mode shapes, when we use the mean-axes, can be computed directly from the mass and stiffness matrices. We consider the equations of motion for a free deformable body from Equation (32) or (33). We split the nodes (or nodal degrees-of-freedom) into the boundary and unconstrained sets to get

$$\left[\begin{array}{ccccc|c} m\mathbf{I} & \mathbf{0} & \mathbf{0} & \mathbf{0} & \mathbf{0} \\ \mathbf{0} & \mathbf{J} & \mathbf{0} & \mathbf{0} & \mathbf{0} \\ \mathbf{M}^{b\cdot}\hat{\mathbf{I}} & -\mathbf{M}^{b\cdot}\hat{\hat{\mathbf{b}}}^{b} & \mathbf{M}^{bb} & \mathbf{M}^{bu} & \mathbf{D}^{b\ \mathrm{T}}_{m-a} \\ \mathbf{M}^{u\cdot}\hat{\mathbf{I}} & -\mathbf{M}^{u\cdot}\hat{\hat{\mathbf{b}}}^{u} & \mathbf{M}^{ub} & \mathbf{M}^{uu} & \mathbf{D}^{u\ \mathrm{T}}_{m-a} \\ \hline \mathbf{0} & \mathbf{0} & \mathbf{D}^{b}_{m-a} & \mathbf{D}^{u}_{m-a} & \mathbf{0} \end{array}\right] \left\{\begin{array}{c} \ddot{\mathbf{r}} \\ \dot{\boldsymbol{\omega}} \\ \ddot{\boldsymbol{\delta}}^{b} \\ \ddot{\boldsymbol{\delta}}^{u} \\ \hline \boldsymbol{\lambda}_{m-a} \end{array}\right\} = \left\{\begin{array}{c} \hat{\mathbf{I}}^{\mathrm{T}}\mathbf{g} \\ -\hat{\hat{\mathbf{b}}}^{\mathrm{T}}\mathbf{g} \\ \mathbf{g}^{b} \\ \mathbf{g}^{u} \\ \hline \mathbf{0} \end{array}\right\} \tag{46}$$

Now the boundary nodes can be constrained as

$$\ddot{\mathbf{d}}^{b} = \left[\hat{\mathbf{I}} \quad -\hat{\hat{\mathbf{b}}}^{b} \quad \bar{\mathbf{I}}\right] \left\{\begin{array}{c} \ddot{\mathbf{r}} \\ \dot{\boldsymbol{\omega}} \\ \ddot{\boldsymbol{\delta}}^{b} \end{array}\right\} + \mathbf{w}^{b} = \mathbf{0} \tag{47}$$

With the aid of Lagrange multipliers, the equations of motion become:

$$\left[\begin{array}{cccc|cc} m\mathbf{I} & \mathbf{0} & \mathbf{0} & \mathbf{0} & \mathbf{0} & \hat{\mathbf{I}}^{\mathrm{T}} \\ \mathbf{0} & \mathbf{J} & \mathbf{0} & \mathbf{0} & \mathbf{0} & -\hat{\hat{\mathbf{b}}}^{b\mathrm{T}} \\ \mathbf{M}^{b\cdot}\hat{\mathbf{I}} & -\mathbf{M}^{b\cdot}\hat{\hat{\mathbf{b}}}^{b} & \mathbf{M}^{bb} & \mathbf{M}^{bu} & \mathbf{D}^{b\ \mathrm{T}}_{m-a} & \bar{\mathbf{I}} \\ \mathbf{M}^{u\cdot}\hat{\mathbf{I}} & -\mathbf{M}^{u\cdot}\hat{\hat{\mathbf{b}}}^{u} & \mathbf{M}^{ub} & \mathbf{M}^{uu} & \mathbf{D}^{u\ \mathrm{T}}_{m-a} & \mathbf{0} \\ \hline \mathbf{0} & \mathbf{0} & \mathbf{D}^{b}_{m-a} & \mathbf{D}^{u}_{m-a} & \mathbf{0} & \mathbf{0} \\ \hat{\mathbf{I}} & -\hat{\hat{\mathbf{b}}}^{b} & \bar{\mathbf{I}} & \mathbf{0} & \mathbf{0} & \mathbf{0} \end{array}\right] \left\{\begin{array}{c} \ddot{\mathbf{r}} \\ \dot{\boldsymbol{\omega}} \\ \ddot{\boldsymbol{\delta}}^{b} \\ \ddot{\boldsymbol{\delta}}^{u} \\ \hline \boldsymbol{\lambda}_{m-a} \\ \boldsymbol{\lambda} \end{array}\right\} = \left\{\begin{array}{c} \hat{\mathbf{I}}^{\mathrm{T}}\mathbf{g} \\ -\hat{\hat{\mathbf{b}}}^{\mathrm{T}}\mathbf{g} \\ \mathbf{g}^{b} \\ \mathbf{g}^{u} \\ \hline \mathbf{0} \\ -\mathbf{w}^{b} \end{array}\right\} \tag{48}$$

From this equation we need to extract mass and stiffness matrices for modal analysis.

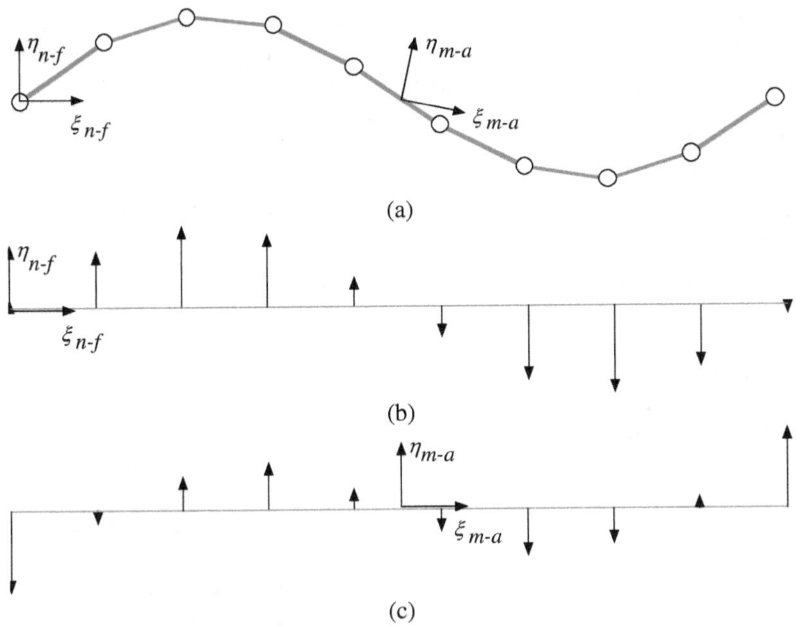

Fig. 7. The second bending mode: (a) the actual bending; and the bending mode as viewed from (b) the nodal-fixed frame and (c) the mean-axes frame

The right-hand-side array of Equation (48) can be expanded as

$$\left\{\begin{array}{c}\hat{\mathbf{I}}^{\mathrm{T}}\mathbf{g}\\ -\hat{\hat{\mathbf{b}}}^{\mathrm{T}}\mathbf{g}\\ \mathbf{g}^{b}\\ \mathbf{g}^{u}\\ \hline \mathbf{0}\\ -\mathbf{w}^{b}\end{array}\right\} \Rightarrow \left\{\begin{array}{c}\hat{\mathbf{I}}^{\mathrm{T}}(\mathbf{f}-\mathbf{M}\mathbf{w}-\mathbf{K}\boldsymbol{\delta})\\ -\hat{\hat{\mathbf{b}}}^{\mathrm{T}}(\mathbf{f}-\mathbf{M}\mathbf{w}-\mathbf{K}\boldsymbol{\delta})\\ \mathbf{f}^{b}-\mathbf{M}^{\cdot b}\mathbf{w}^{b}-\mathbf{K}^{\cdot b}\boldsymbol{\delta}^{b}\\ \mathbf{f}^{u}-\mathbf{M}^{\cdot u}\mathbf{w}^{u}-\mathbf{K}^{\cdot u}\boldsymbol{\delta}^{u}\\ \hline \mathbf{0}\\ -\mathbf{w}^{b}\end{array}\right\}$$

For modal extraction, we eliminate the applied forces and the velocity dependent terms. Furthermore, we observe that $\hat{\mathbf{I}}^{\mathrm{T}}\mathbf{K}=\mathbf{0}$ and $\hat{\hat{\mathbf{b}}}^{\mathrm{T}}\mathbf{K}=\mathbf{0}$[6]. Hence we get:

$$\left\{\begin{array}{c}\hat{\mathbf{I}}^{\mathrm{T}}\mathbf{g}\\ -\hat{\hat{\mathbf{b}}}^{\mathrm{T}}\mathbf{g}\\ \mathbf{g}^{b}\\ \mathbf{g}^{u}\\ \hline \mathbf{0}\\ -\mathbf{w}^{b}\end{array}\right\} \Rightarrow \left\{\begin{array}{c}-\hat{\mathbf{I}}^{\mathrm{T}}\mathbf{K}\boldsymbol{\delta}\\ \hat{\hat{\mathbf{b}}}^{\mathrm{T}}\mathbf{K}\boldsymbol{\delta}\\ -\mathbf{K}^{\cdot b}\boldsymbol{\delta}^{b}\\ -\mathbf{K}^{\cdot u}\boldsymbol{\delta}^{u}\\ \hline \mathbf{0}\\ \mathbf{0}\end{array}\right\} \Rightarrow \left\{\begin{array}{c}\mathbf{0}\\ \mathbf{0}\\ -\mathbf{K}^{\cdot b}\boldsymbol{\delta}^{b}\\ -\mathbf{K}^{\cdot u}\boldsymbol{\delta}^{u}\\ \hline \mathbf{0}\\ \mathbf{0}\end{array}\right\} \Rightarrow -\begin{bmatrix}\mathbf{0}&\mathbf{0}&\mathbf{0}&\mathbf{0}&\mathbf{0}&\mathbf{0}\\ \mathbf{0}&\mathbf{0}&\mathbf{0}&\mathbf{0}&\mathbf{0}&\mathbf{0}\\ \mathbf{0}&\mathbf{0}&\mathbf{K}^{bb}&\mathbf{K}^{bu}&\mathbf{0}&\mathbf{0}\\ \mathbf{0}&\mathbf{0}&\mathbf{K}^{ub}&\mathbf{K}^{uu}&\mathbf{0}&\mathbf{0}\\ \mathbf{0}&\mathbf{0}&\mathbf{0}&\mathbf{0}&\mathbf{0}&\mathbf{0}\\ \mathbf{0}&\mathbf{0}&\mathbf{0}&\mathbf{0}&\mathbf{0}&\mathbf{0}\end{bmatrix}\left\{\begin{array}{c}\mathbf{0}\\ \mathbf{0}\\ \boldsymbol{\delta}^{b}\\ \boldsymbol{\delta}^{u}\\ \hline \mathbf{0}\\ \mathbf{0}\end{array}\right\} \tag{49}$$

[6] These two terms represent the sum of internal (structural) forces and moments, respectively, to be zero.

Since modal extraction is performed at the undeformed state of the body, in the mass matrix of Equation (48), the **b** array is replaced by the **s** array. Therefore, the mass and stiffness matrices for modal extraction are:

$$\left[\begin{array}{cccc|cc} m\mathbf{I} & \mathbf{0} & \mathbf{0} & \mathbf{0} & \mathbf{0} & \hat{\mathbf{I}}^{\mathrm{T}} \\ \mathbf{0} & \mathbf{J} & \mathbf{0} & \mathbf{0} & \mathbf{0} & -\hat{\tilde{\mathbf{s}}}^{b\mathrm{T}} \\ \mathbf{M}^{b,}\hat{\mathbf{I}} & -\mathbf{M}^{b,}\hat{\tilde{\mathbf{s}}}^{b} & \mathbf{M}^{bb} & \mathbf{M}^{bu} & \mathbf{D}_{m-a}^{b\ \mathrm{T}} & \bar{\mathbf{I}} \\ \mathbf{M}^{u,}\hat{\mathbf{I}} & -\mathbf{M}^{u,}\hat{\tilde{\mathbf{s}}}^{u} & \mathbf{M}^{ub} & \mathbf{M}^{uu} & \mathbf{D}_{m-a}^{u\ \mathrm{T}} & \mathbf{0} \\ \hline \mathbf{0} & \mathbf{0} & \mathbf{D}_{m-a}^{b} & \mathbf{D}_{m-a}^{u} & \mathbf{0} & \mathbf{0} \\ \hat{\mathbf{I}} & -\hat{\tilde{\mathbf{s}}}^{b} & \bar{\mathbf{I}} & \mathbf{0} & \mathbf{0} & \mathbf{0} \end{array}\right], \quad \begin{bmatrix} \mathbf{0} & \mathbf{0} & \mathbf{0} & \mathbf{0} & \mathbf{0} & \mathbf{0} \\ \mathbf{0} & \mathbf{0} & \mathbf{0} & \mathbf{0} & \mathbf{0} & \mathbf{0} \\ \mathbf{0} & \mathbf{0} & \mathbf{K}^{bb} & \mathbf{K}^{bu} & \mathbf{0} & \mathbf{0} \\ \mathbf{0} & \mathbf{0} & \mathbf{K}^{ub} & \mathbf{K}^{uu} & \mathbf{0} & \mathbf{0} \\ \mathbf{0} & \mathbf{0} & \mathbf{0} & \mathbf{0} & \mathbf{0} & \mathbf{0} \\ \mathbf{0} & \mathbf{0} & \mathbf{0} & \mathbf{0} & \mathbf{0} & \mathbf{0} \end{bmatrix} \tag{50}$$

Note that in matrix $\mathbf{D}_{m\text{-}a}$, the **b** array must also be replaced by the **s** array. Furthermore, for extracting modal data, all entities associated with deformation in both matrices must be transformed to their components in the body-frame.

An eigen-analysis with the mass and stiffness matrices from Equation (50) would reveal several zero eigen-values associated with the mean-axes and the boundary constraints. The deformation eigen-values and the corresponding eigen-vectors represent the natural frequencies and the mode-shapes for the constrained structure as viewed from the mean-axes reference frame. These are the mode-shapes that we should use in the transformations of Equation (45).

7 Conclusions

In this document two methods for defining a moving reference frame for a deformable body are discussed: the nodal-fixed frame which is quite conventional, and the mean-axis frame which is not well understood. The main purpose of this document is to describe the use of these two reference frames, to provide a better understanding of the mean-axis conditions, and to compare some advantages and disadvantages of each frame.

These two reference frames each impose six conditions on the deformation degrees-of-freedom of a body. These conditions allow the corresponding reference frame to follow the deformable body without constraining its deformation. The six conditions for the nodal-fixed frame are so simple that they can be implemented in the equations of motion explicitly in order to reduce the number of equations of motion and the corresponding variables. In contrast, the mean-axis conditions increase the number of equations of motion by at least six.

The deflection of a deformable body appears different to each of these frames. In a nodal-fixed frame, the boundary nodes will appear not to have any displacements and, therefore, some of the nodes will appear to have large displacements. However, since a mean-axes frame treats all of the nodes according to their inertia, nodal displacements would appear more uniformly from this frame. In general, the largest displacement as viewed from a mean-axis frame would be smaller than the largest displacement viewed from a nodal-fixed frame. This could be an important factor in error reduction in some applications if we consider the "small displacement" assumption in linear finite element modeling.

Transformation of the equations of motion from nodal to modal space has been traditionally performed with the nodal-fixed frame. It has been the general belief that the mean-axes frame can only be employed for free-free structures. This document shows that if we interpret the deformation mode shapes properly, the mean-axes frames can be used whether the structure is constrained or not.

Although the formulation of a moving deformable body is not extended in this document to multibody systems, the use of mean-axis conditions is highly recommended for such formulations. These conditions provide more natural moving frames for deformable bodies that undergo large spatial translation and rotation, while interconnected to other rigid or deformable components. Whereas the nodal-fixed frame is more suitable for structures that do not exhibit large translational and rotational displacements.

References

1. Nikravesh PE (2003) Model Reduction Techniques in Flexible Multibody Dynamics. In Schiehlen W, Valasek M (eds.) Virtual Nonlinear Multibody Systems, Kluwer Academic Publishers, Dordrecht, The Netherlands, pp 83-102
2. Nikravesh PE, Lin Y-S (2003) Body Reference Frames in Deformable Multibody Systems. Int. J. for Multiscale Engineering 1 (2-3):201-217
3. Shabana AA, Yakoub RY (2001) Three-Dimensional Absolute Nodal Coordinate Formulation for Beam Elements: Theory. ASME J. of Mechanical Design 123(4):606-613
4. Agrawal OP, Shabana AA (1986) Application of Deformable-Body Mean Axis to Flexible Multibody System Dynamics. Computer methods in Applied Mechanics and Engineering 56:217-245
5. Guyan RJ (1965) Reduction of Stiffness and Mass matrices. AIAA Journal 3(2):380

Robust analysis of flexible multibody systems and joint clearances in an energy conserving framework

J.C. García Orden and J.M. Goicolea

E.T.S. Ingenieros de Caminos. Universidad Politécnica de Madrid.
jcgarcia@mecanica.upm.es

1 Introduction

Flexible multibody systems (MBS) appear in a number of mechanical applications, in which the model must consider the deformation of some or all of the bodies. A classical method for considering flexibility has been the floating frame technique (25), generally limited to small strains. A more general approach based on inertial coordinates may be formulated by nonlinear finite element methods (27; 28; 11), which are versatile and computationally efficient. Furthermore, employing energy-momentum time integration algorithms they prove to be extremely stable for nonlinear stiff behaviours which frequently arise in such systems.

We present briefly the overall dynamic formulation, but we focus mainly on the formulation of constraints for joints with or without clearances.

Integration algorithms that conserve both momentum and energy have been proposed in (29; 26; 16; 15), attracting considerable attention in the last few years. One of the main benefits of their robustness is their ability to perform stable long-term simulations in nonlinear systems.

The approach followed here, in contrast to other energy-momentum formulations (29; 16; 8; 5; 9; 23), differs in two key aspects: 1) a rotation-free parametrisation for rigid bodies, based on inertial cartesian coordinates of body points, forming a dependent set which is subject to constraints; 2) a penalty formulation for constraints. As will be explained below, this allows for a simple, efficient and robust numerical implementation.

Penalty methods are associated to a non-exact fulfilment of constraints, and in order to ensure sufficient numerical accuracy, large enough penalty parameters must be employed. These may lead to a stiff behaviour and further difficulties in the numerical solution of the problem. Their effect on the system is analogous to introducing very stiff elements between the constrained degrees of freedom. As a consequence, a penalty approach introduces high-frequency components in addition to the already existing ones in flexible multibody systems, due to wave propagation in the deformable bodies. This causes severe numerical difficulties for most time integration algorithms (8), being the main drawback of penalty formulations. However,

J.A.C. Ambrósio (ed.), Advances in Computational Multibody Systems, 205–237.

the energy-momentum algorithm employed here performs exceedingly well, allowing a stable integration even with large time-steps, which is the key to the success of the method. This may be seen in the applications presented here (section 7) and in previous work (12; 21; 13).

The parametrisation for rigid bodies employed is based on inertial cartesian coordinates of body points, following ideas originally proposed by (10) ("natural coordinates"). As a consequence, issues related to integration of rotation parameters are of no concern here. Rotation-free parametrisation for flexible systems has been also used recently by (30; 24). An important remark is that in order to use cartesian coordinates of points for rigid bodies a dependent set of such points must be employed, leading to associated rigid-body constraints. However, these constraints are of the constant distance type (quadratic), and pose no special numerical drawbacks. On the contrary, this choice has some convenient features such as producing a constant mass matrix.

In summary, using cartesian coordinates of points and a penalty approach has two main advantages: firstly it allows a simple formulation of the numerical algorithms, and secondly it leads to a problem formulated as a set of ordinary differential equations (ODE's) with the same number of equations as variables in the model. This is not the case of the Lagrange multipliers method, which leads to a set of differential-algebraic equations (DAE's) with additional variables (Lagrange multipliers).

Regarding conservation properties, the use of a penalty approach has a useful advantage in that the constraint forces may be derived from a potential function, the *constraint energy*. The *total energy* of the system, defined to include this constraint energy as well, is conserved by construction of the time integration algorithm. It must be noted that the *true energy* (i.e. that corresponding to rigid bodies and elastic bodies) does not exactly coincide with this total energy, the difference being the said constraint energy, which is being continuously interchanged with that of the rest of the system; however, using large penalty factors, the violation of the constraints and their associated energy may be maintained sufficiently small for most practical purposes.

Special attention is paid to study of real joints incorporating smooth clearances. The basic phenomenon is the contact between the surfaces defining the joint, which has particular features in the context of multibody dynamics: high number of intermitent contacts and very different space and time scales.

The numerical models described here have been coded in a stand-alone computer program using object-oriented programming (C++). This code structure has allowed an efficient and simple framework for including different objects such as rigid bodies, constraints and finite elements within a single system.

In the rest of this chapter, we first discuss briefly the equations of motion for constrained flexible multibody systems. We then describe in more detail the rotation-free formulation for rigid bodies. Following, the energy-momentum time integration scheme is established. Attention is then focused on the modelling of joints, either perfect or with clearances. Finally, some representative numerical applications are shown.

2 Parametrisation and General Equations of Motion

Following we introduce the notation and discuss briefly the equations of motion, first for rigid bodies, identified with subscript $(\cdot)_{\mathrm{R}}$; next, elastic bodies, identified as $(\cdot)_{\mathrm{E}}$; and finally, the joint equations for the complete flexible MBS.

2.1 Constrained System of Rigid Bodies

Let us consider a rigid body $\mathcal{B}_{\mathrm{R}}$ with mass M_R, being X and x respectively the convected (i.e. material) and spatial coordinates in an inertial frame for an arbitrary point of the body. The kinematics of $\mathcal{B}_{\mathrm{R}}$ is defined in terms of the position of the centre of mass (C.O.M.) x_G and the rotation $\boldsymbol{\Lambda} \in \mathrm{SO}(3) = \{\boldsymbol{\Lambda} : \mathbb{R}^3 \mapsto \mathbb{R}^3;\ \boldsymbol{\Lambda}\boldsymbol{\Lambda}^{\mathrm{T}} = 1;\ \det(\boldsymbol{\Lambda}) = +1\}$ as follows:

$$x = x_G + \boldsymbol{\Lambda}(X - X_G). \tag{1}$$

The spatial angular velocity $\boldsymbol{\omega}$ is the axial vector of the skew-symmetric tensor $\widehat{\boldsymbol{\omega}} = \dot{\boldsymbol{\Lambda}}\boldsymbol{\Lambda}^{\mathrm{T}}$. The convected central inertia tensor is constant, defined as $\mathbb{J}_G = \int_{\mathcal{B}_{\mathrm{R}}} \left[(X - X_G)^2\mathbf{1} - (X - X_G) \otimes (X - X_G)\right] \rho\,\mathrm{d}V$, where $\mathbf{1}$ is the unit 2nd order tensor, and $\otimes$ is the dyadic product. The non-constant spatial inertia tensor takes the form $\mathbb{I}_G = \boldsymbol{\Lambda}\mathbb{J}_G\boldsymbol{\Lambda}^{\mathrm{T}}$.

Let us consider a rigid body subject to q holonomic constraints $\boldsymbol{\Phi}(\mathbf{q}_\Phi) = \mathbf{0}$, in terms of p coordinates $\mathbf{q}_\Phi$ ($\mathbb{R}^p \ni \mathbf{q}_\Phi \mapsto \boldsymbol{\Phi}(\mathbf{q}_\Phi) \in \mathbb{R}^q$). The dynamics may be formulated in weak terms based on Lagrange multipliers $\boldsymbol{\lambda} \in \mathbb{R}^q$ as:[1]

$$\begin{aligned} \delta x_G \cdot M_{\mathrm{R}}\ddot{x}_G + \delta\boldsymbol{\theta} \cdot \frac{\mathrm{d}}{\mathrm{d}t}(\mathbb{I}_G\boldsymbol{\omega}) + \frac{1}{2}\,\delta\mathbf{q}_\Phi^{\mathrm{T}} \cdot \left[(\mathrm{D}\boldsymbol{\Phi})^{\mathrm{T}} \cdot \boldsymbol{\lambda}\right] \\ + \frac{1}{2}\,\delta\boldsymbol{\lambda}^{\mathrm{T}} \cdot \boldsymbol{\Phi} - \delta x_G \cdot f_{\mathrm{R}} - \delta\boldsymbol{\theta} \cdot m_{\mathrm{R}} = 0 \qquad \forall(\delta x_G, \delta\boldsymbol{\theta}, \delta\boldsymbol{\lambda}) \end{aligned} \tag{2}$$

where f_{R} and m_{R} are the force and torque resultants at G, and D is the derivative operator ($\mathrm{D}\boldsymbol{\Phi} \stackrel{\mathrm{def}}{=} \partial\boldsymbol{\Phi}/\partial\mathbf{q}$). The term $(\mathrm{D}\boldsymbol{\Phi})^{\mathrm{T}} \cdot \boldsymbol{\lambda}$ represents the constraint forces (f_Φ), in terms of the Lagrange multipliers $\boldsymbol{\lambda}$. The variations δx_G are arbitrary, and $\delta\boldsymbol{\theta}$ is the axial vector of $\widehat{\delta\boldsymbol{\theta}} = (\delta\boldsymbol{\Lambda})\boldsymbol{\Lambda}^{\mathrm{T}}$. The variations $\delta\mathbf{q}_\Phi$ in (2) are defined by the following kinematic relationship:

$$\delta\mathbf{q}_\Phi = \delta x_G + \delta\boldsymbol{\theta} \wedge (\mathbf{q}_\Phi - x_G), \tag{3}$$

where it has been assumed, with no loss of generality, that the constraints are applied to a single point, $\mathbf{q}_\Phi$.

In this work we avoid the parametrisation of rotations $\boldsymbol{\Lambda}$, instead we choose to formulate the rigid body configuration in terms of the cartesian inertial coordinates

[1] In this equation and hereafter we shall employ preferably the notation of upright boldface ($\mathbf{M}, \mathbf{q}$) for matrix quantities and slanted boldface ($\boldsymbol{x}, \boldsymbol{f}$) for tensor or vector quantities. Dots $(\cdot)$ are employed for matrix products as well as for dot products between vectors.

of a set of m points in $\mathcal{B}_{\mathrm{R}}$, $\mathbf{q}_{\mathrm{R}} \in \mathbb{R}^{3m}$. It will be shown in section 3 that, under this assumption, the kinetic energy can always be expressed as:

$$T_{\mathrm{R}} = \frac{1}{2}\dot{\mathbf{q}}_{\mathrm{R}}^{\mathrm{T}} \cdot (\mathbf{M}_{\mathrm{R}} \cdot \dot{\mathbf{q}}_{\mathrm{R}}), \tag{4}$$

with $\mathbf{M}_{\mathrm{R}}$ being the mass matrix (section 3.2). This matrix will be constant provided that $m \geq 4$ and the selected points are non-coplanar and non-aligned in groups of three.

We now establish a set of coordinates $\mathbf{q}_{\mathrm{R}}$ which comprises basic points for definition of the rigid body, as well as any extra points which may be necessary for formulation of constraints ($\mathbf{q}_{\Phi}$), point loads $\mathbf{Q}_{\mathrm{R}}$, etc. In terms of $\mathbf{q}_{\mathrm{R}}$, the weak form of the equations is given by:

$$\delta\mathbf{q}_{\mathrm{R}}^{\mathrm{T}} \cdot (\mathbf{M}_{\mathrm{R}} \cdot \ddot{\mathbf{q}}_{\mathrm{R}}) + \frac{1}{2}\delta\mathbf{q}_{\mathrm{R}}^{\mathrm{T}} \cdot \left[(\mathrm{D}\boldsymbol{\Phi})^{\mathrm{T}} \cdot \boldsymbol{\lambda}\right] + \frac{1}{2}\delta\boldsymbol{\lambda}^{\mathrm{T}} \cdot \boldsymbol{\Phi} - \delta\mathbf{q}_{\mathrm{R}}^{\mathrm{T}} \cdot \mathbf{Q}_{\mathrm{R}} = 0$$
$$\forall\delta\mathbf{q}_{\mathrm{R}} \in \mathcal{V}^{3m-q},\ \forall\delta\boldsymbol{\lambda} \tag{5}$$

where $\mathcal{V}^{3m-q}$ is a $(3m-q)$-dimensional space of the variations of $\mathbf{q}_{\mathrm{R}}$ compatible with the constraints $\boldsymbol{\Phi}$.

The above equation (5) involves consideration of $\boldsymbol{\lambda}$ as additional variables, subject to the constraints $\boldsymbol{\Phi}$, resulting in a DAE system. Here we employ a different approach, penalty methods. This may be defined by considering the Lagrange multipliers and the associated constraint forces to be:

$$\boldsymbol{\lambda} = \boldsymbol{\alpha} \cdot \boldsymbol{\Phi}\,; \qquad \mathbf{f}_{\Phi} = -(\mathrm{D}\boldsymbol{\Phi})^{\mathrm{T}} \cdot \boldsymbol{\lambda} = -(\mathrm{D}\boldsymbol{\Phi})^{\mathrm{T}} \cdot (\boldsymbol{\alpha} \cdot \boldsymbol{\Phi})\,, \tag{6}$$

where $\boldsymbol{\alpha}$ is a $(q \times q)$ constant penalty matrix, positive definite and symmetric. This leads also to $\delta\boldsymbol{\lambda} = \boldsymbol{\alpha} \cdot \mathrm{D}\boldsymbol{\Phi} \cdot \delta\mathbf{q}_{\mathrm{R}}$, where the variations $\delta\mathbf{q}_{\mathrm{R}}$ need no longer be compatible with the constraints $\boldsymbol{\Phi}$. Introducing these variations in (5) and considering that now $\delta\mathbf{q}_{\mathrm{R}}$ are arbitrary, one obtains the following set of differential equations (ODE's):

$$\mathbf{M}_{\mathrm{R}} \cdot \ddot{\mathbf{q}}_{\mathrm{R}} + \underbrace{(\mathrm{D}\boldsymbol{\Phi})^{\mathrm{T}} \cdot (\boldsymbol{\alpha} \cdot \boldsymbol{\Phi})}_{-\mathbf{f}_{\Phi\mathrm{R}}} - \mathbf{Q}_{\mathrm{R}} = \mathbf{0}\,. \tag{7}$$

A further alternative for treatment of constraints is the *augmented Lagrangian* method, which may be understood to start from a penalty scheme followed by an iteration to obtain the multipliers with the desired accuracy:

$$\mathbf{f}_{\Phi} = -(\mathrm{D}\boldsymbol{\Phi})^{\mathrm{T}} \cdot (\boldsymbol{\alpha} \cdot \boldsymbol{\Phi}) + (\mathrm{D}\boldsymbol{\Phi})^{\mathrm{T}} \cdot \boldsymbol{\lambda}^{*}\,; \quad \boldsymbol{\lambda}_{i+1}^{*} = \boldsymbol{\lambda}_{i}^{*} + \boldsymbol{\alpha} \cdot \boldsymbol{\Phi}_{i+1}\,, \tag{8}$$

where vector $\boldsymbol{\lambda}^{\star}$ converges to the exact Lagrange multiplier vector $\boldsymbol{\lambda}$ as $\boldsymbol{\Phi} \to \mathbf{0}$.

The enforcement of constraints is the key issue in the numerical representation of joints, and will be addressed again with more detail in section 5.1.

With the above approach, a multibody system formed by several rigid bodies $\{\mathcal{B}_{\mathrm{R}_i},\ i = 1,\ldots,N\}$ is assembled by simply collecting in $\mathbf{q}_{\mathrm{R}}$ all the variables of each body.

2.2 Deformable Bodies

Let us consider an elastic body is designed by $\mathcal{B}_{\mathrm{E}}$. A general motion is defined by a smooth differentiable mapping $\varphi : X \mapsto x$. The deformation gradient F defines the derivate of this mapping $F = \mathrm{D}\varphi = \partial x / \partial X$ (with the abuse of notation involved in the use of x instead of $\varphi(X)$).

We consider hyperelastic materials in which a stored energy density function exists of the type $W(C)$, with $C = F^{\mathrm{T}}F$, from which the symmetric Piola-Kirchhoff stress tensor may be obtained as $S = 2\,\mathrm{D}W$. The strain energy of the elastic body will be $\Pi = \int_{\mathcal{B}_{\mathrm{E}_0}} W\,\mathrm{d}V_0$. Introducing the notation for an L_2-inner product, $\langle \cdot, \cdot \rangle_{\Box} = \int_{\Box}(\cdot)(\cdot)\,\mathrm{d}\Box$, the variation of Π for an arbitrary virtual displacement is given by $\delta\Pi = \langle FS, \mathrm{D}(\delta x)\rangle_{\mathcal{B}_{\mathrm{E}_0}}$. The weak form of the equations of motion for a single elastic body $\mathcal{B}_{\mathrm{E}}$ with boundary $\partial\mathcal{B}_{\mathrm{E}}$ is:

$$\langle \rho\ddot{x}, \delta x\rangle_{\mathcal{B}_{\mathrm{E}}} - \langle F\,S, \mathrm{D}(\delta x)\rangle_{\mathcal{B}_{\mathrm{E}_0}} - \langle f, \delta x\rangle_{\mathcal{B}_{\mathrm{E}}} - \langle t, \delta x\rangle_{\partial\mathcal{B}_{\mathrm{E},t}} = 0, \qquad \forall \delta x \in \mathcal{W}, \quad (9)$$

where f and t are respectively the volumetric loads and traction forces at the boundary. In this case, $\mathcal{W}$ is the infinite-dimensional space of the variations of x compatible with the existing constraints (if any).

Further details on the treatment of elastic bodies are given in (19; 12) and will be omitted here for brevity.

2.3 Global Flexible Multibody Equations

For a multibody system composed of both rigid ($\mathcal{B}_{\mathrm{R}}$) and elastic ($\mathcal{B}_{\mathrm{E}}$) bodies, the weak form of the dynamic equations may be obtained gathering the terms from rigid bodies (5) and elastic bodies (9):

$$\begin{aligned}\langle \rho\ddot{x}, \delta x\rangle_{\mathcal{B}_{\mathrm{E}}} - \langle F\,S, \mathrm{D}(\delta x)\rangle_{\mathcal{B}_{\mathrm{E}_0}} &+ \delta\mathbf{q}_{\mathrm{R}}^{\mathrm{T}} \cdot (\mathbf{M}_{\mathrm{R}} \cdot \ddot{\mathbf{q}}_{\mathrm{R}}) \\ &+ \delta\mathbf{q}_{\mathrm{R}}^{\mathrm{T}} \cdot \left[(\mathrm{D}\boldsymbol{\Phi})^{\mathrm{T}} \cdot (\boldsymbol{\alpha} \cdot \boldsymbol{\Phi})\right] - \langle f, \delta x\rangle_{\mathcal{B}_{\mathrm{E}}} \\ &- \langle t, \delta x\rangle_{\partial\mathcal{B}_{\mathrm{E},t}} - \delta\mathbf{q}_{\mathrm{R}}^{\mathrm{T}} \cdot \mathbf{Q}_{\mathrm{R}} = 0, \quad \forall \delta\mathbf{q}_{\mathrm{R}}, \forall \delta x \in \mathcal{W}. \quad (10)\end{aligned}$$

In the above equation constraints are considered by a penalty method.

2.4 Finite Element Approximation

The elastic body $\mathcal{B}_{\mathrm{E}}$ continuum is discretized in space through standard finite element techniques, such that both material (X) and spatial (x) coordinates of an arbitrary point within an element Ω^e are interpolated from the nodal coordinates through shape functions N_i:

$$x = \sum_{i=1}^{n} N_i x_i^e = \underbrace{\left(N_1\mathbf{I}_3 \middle| N_2\mathbf{I}_3 \middle| \dots \middle| N_n\mathbf{I}_3\right)}_{\mathbf{N}} \begin{Bmatrix} x_1^e \\ \hline x_2^e \\ \hline \dots \\ \hline x_n^e \end{Bmatrix} = \mathbf{N} \cdot \mathbf{q}_{\mathrm{E}}^e \quad (11)$$

$$X = \mathbf{N} \cdot \mathbf{q}_{\mathrm{E}_0}^e, \quad (12)$$

where $\mathbf{I}_3$ represent unit (3×3) matrices. Additionally, in isoparametric elements, the displacement (u) is interpolated in a similar fashion such that $u = \mathbf{N} \cdot (\mathbf{q}_{\mathrm{E}}^e - \mathbf{q}_{\mathrm{E}_0}^e)$, and as a consequence $\delta x = \mathbf{N} \cdot \delta \mathbf{q}_{\mathrm{E}}^e$. The contribution of an element Ω^e to the discrete weak formulation (10) can be written in terms of $\delta \mathbf{q}_{\mathrm{E}}^e \in \mathcal{W}^n$, defined in a finite-dimensional space, as:

$$(\delta \mathbf{q}_{\mathrm{E}}^e)^{\mathrm{T}} \cdot \Big(\underbrace{\int_{\Omega^e} \rho \mathbf{N}^{\mathrm{T}} \cdot \mathbf{N} \, \mathrm{d}V}_{\mathbf{M}_{\mathrm{E}}^e} \ddot{\mathbf{q}}_{\mathrm{E}}^e - \underbrace{\int_{\Omega_0^e} \mathbf{B} S^e \, \mathrm{d}V_0}_{\mathbf{f}_{\mathrm{int}}^e} - \mathbf{Q}_{\mathrm{E}}^e \Big) \tag{13}$$

Details about the computation of the $\mathbf{f}_{\mathrm{int}}^e$ term (internal forces) are given in (12; 19) and will be omitted here. Assembly of the element matrices $\mathbf{M}_{\mathrm{E}}^e$, $\mathbf{f}_{\mathrm{int}}^e$ and $\mathbf{Q}_{\mathrm{E}}^e$ allows the calculation of global matrices $\mathbf{M}_{\mathrm{E}}$, $\mathbf{f}_{\mathrm{int}}$, $\mathbf{Q}_{\mathrm{E}}$. Additionally, a force vector $\mathbf{f}_{\Phi_{\mathrm{E}}}$ similar to $\mathbf{f}_{\Phi_{\mathrm{R}}}$ is introduced for constraints acting within the deformable bodies.

Finally, assembling a new vector $\mathbf{q} \in \mathbb{R}^{3m+n}$ which merges $\mathbf{q}_{\mathrm{R}}$ and $\mathbf{q}_{\mathrm{E}}$, whose variations $\delta \mathbf{q}$ are arbitrary, global dynamics will be expressed by the following set of ordinary differential equations,written in partitioned form:

$$\underbrace{\begin{pmatrix} \mathbf{M}_{\mathrm{R}} & \mathbf{0} \\ \mathbf{0} & \mathbf{M}_{\mathrm{E}} \end{pmatrix}}_{\mathbf{M}} \underbrace{\begin{Bmatrix} \ddot{\mathbf{q}}_{\mathrm{R}} \\ \ddot{\mathbf{q}}_{\mathrm{E}} \end{Bmatrix}}_{\ddot{\mathbf{q}}} = \underbrace{\begin{Bmatrix} \mathbf{f}_{\Phi_{\mathrm{R}}} \\ \mathbf{f}_{\Phi_{\mathrm{E}}} \end{Bmatrix} + \begin{Bmatrix} \mathbf{f}_{\Phi_{\mathrm{RE1}}} \\ \mathbf{f}_{\Phi_{\mathrm{RE2}}} \end{Bmatrix}}_{\mathbf{f}_{\Phi}} + \underbrace{\begin{Bmatrix} \mathbf{0} \\ \mathbf{f}_{\mathrm{int}} \end{Bmatrix} + \begin{Bmatrix} \mathbf{Q}_{\mathrm{R}} \\ \mathbf{Q}_{E} \end{Bmatrix}}_{\mathbf{Q}} . \tag{14}$$

In this expression $\mathbf{f}_{\Phi_{\mathrm{RE}i}}$ are coupling terms from constraints between the rigid and the elastic sub-systems.

It must be remarked that in the above equation (14) the global mass matrix $\mathbf{M}$ will always be constant, due to the choice of inertial cartesian coordinates made both for rigid and elastic bodies. The applied force vector $\mathbf{Q} = \mathbf{Q}(\mathbf{q}, \dot{\mathbf{q}}, t)$ arises from external or internal forces (imposed loads, internal forces of elastic bodies, springs, dashpots, etc).

3 Rotation-free Formulation of Rigid Bodies

The position of a rigid body $\mathcal{B}_{\mathrm{R}}$ is defined by a collection of m points $\{\mathbf{q}_k, \; k = 1, \ldots m\}$ constrained to remain at constant relative distances, through $(3m-6)$ constraint equations. The parametrisation in terms of this set of dependent coordinates arouses two issues: the constant-distance constraints and the consistent formulation of the mass matrix. The approach taken for these two aspects follows the ideas proposed originally in (10).

3.1 Constant Distance Constraint

Given two points (a and b) defined by their cartesian coordinates $x_a = (x_a, y_a, z_a), x_b = (x_b, y_b, z_b)$, the constant-distance constraint equation is:

$$\Phi = (x_b - x_a)^2 + (y_b - y_a)^2 + (z_b - z_a)^2 - l_{ab}^2 \, . \tag{15}$$

The penalty potential is $V_\Phi = \frac{1}{2}\alpha\Phi^2$, and the force vector, considering $\mathbf{q}^{\mathrm{T}} = \left(x_a^{\mathrm{T}} | x_b^{\mathrm{T}} \right)$ is:

$$\mathbf{f}_\Phi = \begin{Bmatrix} x_a - x_b \\ y_a - y_b \\ z_a - z_b \\ x_b - x_a \\ y_b - y_a \\ z_b - z_a \end{Bmatrix} 2\alpha \left((x_b - x_a)^2 + (y_b - y_a)^2 + (z_b - z_a)^2 - l_{ab}^2 \right) \, .$$

Remark 1. Constraint equation (15) is quadratic and as a consequence results in a linear Jacobian, which is particularly convenient from a computational point of view.

3.2 Consistent Mass Matrix

For a rigid body $\mathcal{B}_{\mathrm{R}}$ in 3D, it can be shown that the minimum number of points in order to obtain a constant mass matrix is 4, being non-coplanar and non-aligned in groups of three. Let these points be $\{x_0, x_1, x_2, x_3\}$, where $O \equiv x_0$ may be taken as a reference point and $t_j = (x_j - x_0)$, $(j = 1, \ldots 3)$ are non-unit vectors for reference directions. These define a convected frame $S' \equiv \{O, t_j\}$, which in general may be non-orthonormal. The position of a generic point $P \in \mathcal{B}_{\mathrm{R}}$ defined by vector x may be expressed as:

$$x = x_0 + X_1' t_1 + X_2' t_2 + X_3' t_3 \, ,$$

being $\{X_j'\}$ constants defining relative adimensional coordinates. In compact matrix form, this may be written as:

$$x = \mathbf{C} \cdot \mathbf{q}_{\mathrm{R}}, \qquad \text{with } \mathbf{C} = \left(X_0' \mathbf{I}_3 | X_1' \mathbf{I}_3 | X_2' \mathbf{I}_3 | X_3' \mathbf{I}_3 \right) \, , \tag{16}$$

being $\mathbf{q}_{\mathrm{R}}$ the 12×1 column matrix built with the cartesian coordinates of the 4 points and $X_0' = 1 - \sum_{j=1}^{3} X_j'$.

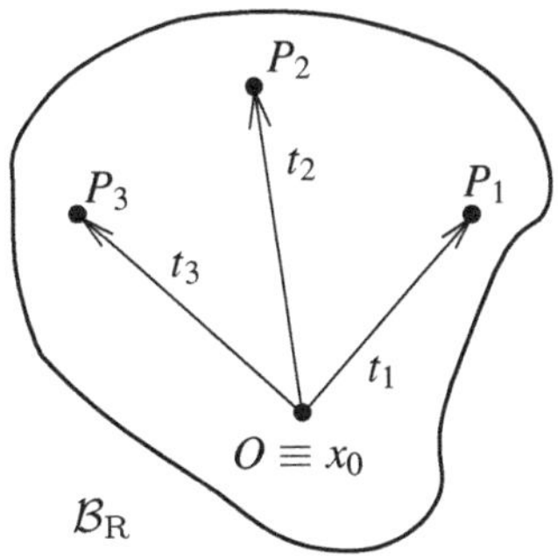

Fig. 1. 3-D rigid body

Differentiating (16) one obtains $\ddot{x} = \mathbf{C} \cdot \ddot{\mathbf{q}}_{\mathrm{R}}$ and $\delta x = \mathbf{C} \cdot \delta \mathbf{q}_{\mathrm{R}}$, and the virtual work of the inertia forces for a compatible displacement $\delta \mathbf{q}_{\mathrm{R}}$ in the body $\mathcal{B}_{\mathrm{R}}$ leads to the following expression:

$$\delta W_{\mathrm{iner}} = \int_{\mathcal{B}_{\mathrm{R}}} \rho(\delta x^{\mathrm{T}} \cdot \ddot{x})\,\mathrm{d}V = \delta \mathbf{q}^{\mathrm{T}} \cdot \underbrace{\left[\int_{\mathcal{B}_{\mathrm{R}}} \rho \mathbf{C}^{\mathrm{T}} \cdot \mathbf{C}\,\mathrm{d}V\right]}_{\mathbf{M}_{\mathrm{R}}} \cdot \ddot{\mathbf{q}}_{\mathrm{R}}\,, \tag{17}$$

where the mass matrix $\mathbf{M}_{\mathrm{R}}$ can be identified. The most convenient procedure to calculate the integral in (17) is in terms of the relative position vector X of an arbitrary point,

$$X = \mathbf{T}X', \qquad \text{with } \mathbf{T} = \left(t_1 | t_2 | t_3\right), \qquad \text{and } X' = \left(X'_1\ X'_2\ X'_3\right)\,,$$

and in terms of the planar inertia tensor. This may be defined at O as $\mathbb{P}_O = I_O I - \mathbb{I}_O$, where I_O is the polar moment of inertia and $\mathbb{I}_O$ is the inertia tensor. The expression in the convected non-orthonormal frame S' is $\mathbb{P}'_O = \mathbf{T}^{-1}\mathbb{P}_O\mathbf{T}^{-\mathrm{T}}$. Further, the relative position vector of the C.O.M. in this same frame is $X'_G = \mathbf{T}^{-1}X_G$. In terms of these, the mass matrix obtained in (17) may be expressed as:

$$\mathbf{M} = \left(\begin{array}{c|c|c|c} m_{00}\mathbf{I}_3 & m_{01}\mathbf{I}_3 & m_{02}\mathbf{I}_3 & m_{03}\mathbf{I}_3 \\ \hline m_{01}\mathbf{I}_3 & & & \\ m_{02}\mathbf{I}_3 & & \widetilde{\mathbb{P}}'_O & \\ m_{03}\mathbf{I}_3 & & & \end{array}\right),$$

where the notation $\widetilde{\mathbb{P}}'_O$ is used to define an *inflated* matrix, i.e.

$$\widetilde{\mathbb{P}}'_O = \left(\begin{array}{c|c|c} \mathbb{P}'_{11}\mathbf{I}_3 & \mathbb{P}'_{12}\mathbf{I}_3 & \mathbb{P}'_{13}\mathbf{I}_3 \\ \hline \mathbb{P}'_{21}\mathbf{I}_3 & \mathbb{P}'_{22}\mathbf{I}_3 & \mathbb{P}'_{23}\mathbf{I}_3 \\ \hline \mathbb{P}'_{31}\mathbf{I}_3 & \mathbb{P}'_{32}\mathbf{I}_3 & \mathbb{P}'_{33}\mathbf{I}_3 \end{array}\right).$$

The parameters m_{0i} take the following values:

$$m_{00} = \int_{\mathcal{B}_{\mathrm{R}}} X'_0 X'_0 \rho\,\mathrm{d}V = M(2X'_{G0} - 1) + \sum_{i=1}^{3}\sum_{j=1}^{3} \mathbb{P}'_{ij}\,,$$

$$m_{0i} = \int_{\mathcal{B}_{\mathrm{R}}} X'_0 X'_i \rho\,\mathrm{d}V = MX'_{Gi} - \sum_{j=1}^{3} \mathbb{P}'_{ij}\,,$$

where M is the total mass of the rigid body and $X'_{G0} = 1 - \sum_{i=1}^{3} X'_{Gi}$.

These expressions are greatly simplified for $O \equiv G$, and for directions t_i which coincide with principal inertia directions. Mass matrices for two and one dimensional rigid bodies can be calculated in a similar fashion.

4 Conservative Time Integration

Following (15), a conservative time integration for mechanical systems may be formulated in terms of the discrete derivative of the Hamiltonian. A Hamiltonian canonical system $(P, \mathbf{J}, H)$ is defined by evolution equations expressed as:

$$\dot{z} = \mathbf{J}\,\mathrm{D}H(z) \text{ with } \quad \mathbf{J} = \begin{pmatrix} \mathbf{0}_n & \mathbf{I}_n \\ -\mathbf{I}_n & \mathbf{0}_n \end{pmatrix},$$

where $z = (q, p) \in P$ (generalised coordinates and moments, being $P \in \mathbb{R}^{2n}$ the phase space). $\mathbf{0}_n$ and $\mathbf{I}_n$ are the null and $(n \times n)$ unit matrices respectively. The skew-symmetric operator $\mathbf{J}$ introduces a symplectic structure, and $H : P \rightarrow \mathbb{R}$ is the Hamiltonian function.

The discrete derivative of a function $f : P \rightarrow \mathbb{R}$, denoted by $\mathsf{D}f(z_n, z_{n+1})$, with $z_n, z_{n+1} \in P$, is defined as a function of the continuous derivative $\mathrm{D}f$ by:

$$\begin{aligned} \mathsf{D}f(z_n, z_{n+1}) &= \mathrm{D}f(z_{n+\frac{1}{2}}) \\ &+ \frac{f(z_{n+1}) - f(z_n) - \mathrm{D}f(z_{n+\frac{1}{2}}) \cdot (z_{n+1} - z_n)}{||z_{n+1} - z_n||^2} (z_{n+1} - z_n), \end{aligned}$$

being $z_{n+\frac{1}{2}} \overset{\text{def}}{=} \frac{1}{2}(z_n + z_{n+1})$. This calculation could be simplified if f can be expressed in terms of a set of elemental functions $\pi = \pi_i(z), i = 1...p$ at most quadratic, with the relation:

$$\mathsf{D}f(z_n, z_{n+1}) \overset{\text{def}}{=} \mathsf{D}\tilde{f}(\pi_n, \pi_{n+1}) \cdot \mathrm{D}\pi(z_{n+\frac{1}{2}}) \quad \text{with} \quad \mathrm{f}(z) = \tilde{\mathrm{f}}(\pi)$$

where $\tilde{f}(\pi)$ is called the *reduced function* associated to f.

Applying the discrete derivative to system $(P, \mathbf{J}, H)$, we obtain a discrete Hamiltonian system, given by the finite difference equation:

$$z_{n+1} - z_n = \Delta t\, \mathbf{J}\left[\mathsf{D}H(z_n, z_{n+1})\right]; \qquad z_n, z_{n+1} \in P \tag{18}$$

For an autonomous system in which H has the meaning of the total energy and is conserved, it is possible to show (15) that the discrete system (18) algorithmically inherits this property, thus achieving exact energy conservation.

Before applying the discrete derivative directly to a simple example, let us analyse the invariance of other first integrals in a Hamiltonian system, such as angular momentum. The conservation of this quantity holds if the Hamiltonian is invariant under the action of the rotation group $\mathrm{SO}(3)$. In this case, the Hamiltonian system is said to possess *symmetry*, and it is possible to define another discrete derivative (called G-equivariant discrete derivative and labelled D^G), such that the discrete system given by:

$$z_{n+1} - z_n = \Delta t\, \mathbf{J}\left[\mathsf{D}^G H(z_n, z_{n+1})\right]; \qquad z_n, z_{n+1} \in P \tag{19}$$

algorithmically conserves both energy and angular momentum. If H can be expressed in terms of a set of *invariant* and independent functions π at most quadratic, the G-equivariant discrete derivative can be calculated with an expression which is formally identical to that previously employed for the standard discrete derivative:

$$\mathsf{D}^G H(z_n, z_{n+1}) \stackrel{\text{def}}{=} \mathsf{D}\tilde{H}(\pi_n, \pi_{n+1}) \cdot \mathrm{D}\pi(z_{n+\frac{1}{2}}), \quad \text{with} \quad H(z) = \tilde{H}(\pi) .$$

Example 1. Particle under a central force Let us suppose a simple mechanical system composed by a single particle with mass m subjected to a central force, with coordinates $q \in \mathbb{R}^3$. The force $F(q) = F(q)q/q$ derives from a potential $V(q) = \hat{V}(q) = \int F(q)dq$.

The mechanical system so defined has a Hamiltonian function

$$H = \frac{1}{2m}p^2 + \hat{V}(q) ,$$

which is independent of time and invariant under the action of rotations. These properties introduce symmetries in the Hamiltonian structure: i.e. total energy ($H = E = T + V$) and angular momentum are conserved throughout the motion. Besides, inspection of the Hamiltonian function reveals that the following quadratic invariants exist: $\{\pi_1 = q^2, \pi_2 = q \cdot p, \pi_3 = p^2\}$. An energy-momentum conserving algorithm is obtained by application of the G-equivariant discrete derivative (19), arriving at:

$$\begin{aligned} p_{n+1} - p_n &= -\Delta t\, \mathsf{D}^G V(q_n, q_{n+1}) \\ &= -\Delta t \frac{V(q_{n+1}) - V(q_n)}{q_{n+1} - q_n} \frac{q_{n+\frac{1}{2}}}{\overline{q}_{n+\frac{1}{2}}} \end{aligned} \tag{20}$$

$$q_{n+1} - q_n = \Delta t\, \frac{1}{2}(\dot{q}_n + \dot{q}_{n+1}) = \Delta t\, \dot{q}_{n+\frac{1}{2}} \tag{21}$$

where $\overline{(\cdot)}_{n+\frac{1}{2}} \stackrel{\text{def}}{=} \frac{1}{2}\left[(\cdot)_n + (\cdot)_{n+1}\right]$

Remark 2. It is important to note that, in general, $\overline{q}_{n+\frac{1}{2}} = \frac{1}{2}(q_n + q_{n+1}) \neq q_{n+\frac{1}{2}} = ||q_{n+\frac{1}{2}}||$.

Remark 3. The energy-momentum method, as suggested by expressions (20) and (21), can be understood as a modified implicit midpoint rule, just by altering the force term in right-hand side of momentum balance equation in order to obtain (20), which no longer has the meaning of the force at the mid-step but rather of a specific discrete expression to obtain algorithmic conservation.

5 Conservative Formulation of Perfect Joints

A joint is understood as perfect as long as it is permanent, smooth (there is no friction) and has no clearances. This section outlines the formulation of such perfect joints in the context of the conservative time integration described in section 4.

In the numerical model, a perfect joint is represented by one or more equality constraints, which are formulated in terms of the generalised coordinates that determine the configuration of the connected bodies. Following this idea, a general approach to the conservative formulation of constraints is presented first; in a second stage, these ideas will be extended to the formulation of practical perfect joints.

5.1 Conservative formulation of constraints

If we denote by $\mathbf{q}_\Phi \in \mathbb{R}^p$ the vector that contains the p generalised coordinates associated to the connected bodies, a constraint Φ is a function defined as:

$$\Phi : (\mathbb{R}^p \times \mathbb{R}^p \times \mathbb{R}) \ni (\mathbf{q}, \dot{\mathbf{q}}, t) \longmapsto \Phi(\mathbf{q}, \dot{\mathbf{q}}, t) \in \mathbb{R}$$

However, for most practical perfect joints, the associated constraints neither depend on the generalised velocities nor time, which is related to the fact that they are permanent and smooth. As a consequence, we shall study at this point only *holonomic* and *escleronomic* constraints, which take the simpler form:

$$\Phi : \mathbb{R}^p \ni \mathbf{q} \longmapsto \Phi(\mathbf{q}) \in \mathbb{R} \tag{22}$$

Enforcement of a set of this type of constraints, grouped into a single vector $\boldsymbol{\Phi}(\mathbf{q}) \in \mathbb{R}^q$, may be accomplished by several methods. From the different alternatives, we choose the *penalty method* as the basic methodology for the constraint enforcement, based on the ideas outlined in the following remarks.

Remark 4. The Lagrange multipliers method provides an exact enforcement of the constraints. However, the penalty method and the augmented Lagrangian methods have a useful advantage from a numerical point of view: they lead to an ODE system in terms of configuration parameters, without additional algebraic constraints or additional mixed variables representing multipliers.

Remark 5. In the penalty method the constraint force (6) derives from a *constraint potential*:

$$V_\Phi = \frac{1}{2}\boldsymbol{\Phi}^{\mathrm{T}} \cdot (\boldsymbol{\alpha} \cdot \boldsymbol{\Phi}) \,, \qquad \mathbf{f}_\Phi = -\mathrm{D}V_\Phi \,;$$

this fact has relevance in the conservative formulation of constraints in the context of the energy-momentum method.

Remark 6. The large penalty parameters $\boldsymbol{\alpha}$ may introduce severe numerical difficulties which are, in fact, the main drawback of the penalty method. However, this problem is overcome by the energy-momentum formulation (13).

Note that constraints defined by (22) do not dissipate energy in any compatible displacement (satisfying $\Phi = 0$) which justifies the interest of developing a discrete conservative formulation which inherits this important property.

In the context of this section, is useful to distinguish between *scalar* and *vector* constraints. This will be related with the cartesian coordinates of selected points that are used to define the system configuration.

- In a *scalar* constraint between two points, the corresponding equation is a function of a single scalar variable (e.g. the modulus of the relative distance). This is the case of the constant-distance constraint described in section 3.1.
- In a *vector* constraint between two or more points, the expression is a general function of the cartesian coordinates of the constrained points. One example can be the equation that forces four points to lie in the same plane.

5.1.1 Scalar constraints

Let us consider a closed two-particle system with a configuration defined by position vectors (x_1, x_2). The global coordinate vector is $\mathbf{q}$, which collects the six components from both. The relative position is given by $r = x_2 - x_1$, with modulus $r = |r|$. Assume an internal constraint exists defined by a function $\Phi : \mathbb{R} \ni r \longmapsto \Phi(r) \in \mathbb{R}$, with constraint energy $V_\Phi = (1/2)\alpha\Phi^2$. This is a Hamiltonian system with symmetry, and total energy and momentum are constant.

Equations (20) and (21) developed for a single particle under a central force can be directly applied to this case, giving the algorithmic equation:

$$\mathbf{M}(\dot{\mathbf{q}}_{n+1} - \dot{\mathbf{q}}_n) = -\Delta t\, \mathrm{D}^G V_\Phi(\mathbf{q}_n, \mathbf{q}_{n+1}) = -\Delta t\, \alpha \frac{\Phi_{n+1}^2 - \Phi_n^2}{r_{n+1}^2 - r_n^2} \begin{Bmatrix} r_{n+\frac{1}{2}} \\ -r_{n+\frac{1}{2}} \end{Bmatrix} \tag{23}$$

$$\mathbf{q}_{n+1} - \mathbf{q}_n = \Delta t \frac{1}{2}(\dot{\mathbf{q}}_{n+1} + \dot{\mathbf{q}}_n) = \Delta t \dot{\mathbf{q}}_{n+1/2}$$

where $r_{n+\frac{1}{2}} \stackrel{\text{def}}{=} \frac{1}{2}(r_{n+1} + r_n)$ and α is the penalty factor.

It is important to verify that expression $(23)_1$ is well-behaved in the limit $\Delta r = (r_{n+1} - r_n) \to 0$. This is due to the fact that the most computationally-efficient predictor value in the first iteration, when integrating from t_n to t_{n+1}, is $r_{n+1}^0 = r_n$. In this limit case, the expression of the constraint force is given by:

$$(\mathbf{f}_\Phi)_{\Delta r \to 0} = -\alpha \frac{\Phi_n \Phi_n'}{r_n} \begin{Bmatrix} r_n \\ -r_n \end{Bmatrix}, \quad \text{where} \quad \Phi' \stackrel{\text{def}}{=} \frac{\mathrm{d}\Phi}{\mathrm{d}r}. \tag{24}$$

Expressions 23 define by themselves an implicit time marching scheme that demands a built-in iterative procedure. This iterative process implies the linearision of (23), leading to the calculation of a consistent tangent matrix. The contribution $(\mathbf{K}_\Phi)$ to the global tangent matrix at t_{n+1} results from the exact linearision of the constraint force vector, and is given by:

$$\mathbf{K}_\Phi = \frac{\partial \mathbf{f}_\Phi}{\partial \mathbf{q}_{n+1}} = \begin{pmatrix} \mathbf{A} & -\mathbf{A} \\ -\mathbf{A} & \mathbf{A} \end{pmatrix}, \tag{25}$$

with:

$$\mathbf{A} \stackrel{\text{def}}{=} -\left[2\frac{V'_{\Phi_{n+1}} - \sigma r_{n+1}}{r_{n+1}(r_{n+1}^2 - r_n^2)} (r_{n+\frac{1}{2}} \otimes r_{n+1}) + \frac{1}{2}\sigma 1 \right]$$

and

$$\sigma \stackrel{\text{def}}{=} \frac{V_{\Phi_{n+1}} - V_{\Phi_n}}{\frac{1}{2}(r_{n+1}^2 - r_n^2)} .$$

It is important to remark that (25) results in a non-symmetric matrix.

In the limit $\Delta r = (r_{n+1} - r_n) \to 0$, the contribution to the consistent tangent matrix is symmetric and given by:

$$(\mathbf{K}_\Phi)_{\Delta r \to 0} = \begin{pmatrix} \mathbf{B} & -\mathbf{B} \\ -\mathbf{B} & \mathbf{B} \end{pmatrix} \tag{26}$$

with

$$\mathbf{B} \stackrel{\text{def}}{=} -\left[\frac{(V''_\Phi)_n - \sigma^*}{2r_n^2}(r_n \otimes r_n) + \frac{1}{2}\sigma^* 1\right],$$

and σ^* given by:

$$\sigma^* = \frac{\alpha \Phi_n \Phi'_n}{r_n}.$$

Example 2. Constant distance constraints, essential in the definition of rigid bodies, are a particular case of scalar constraints. They can be expressed as $\Phi(r) = r - r_0$, being r_0 the initial distance. With this notation, and applying expression (23), the constraint force results:

$$\mathbf{f}_\Phi = -\alpha\left(1 - \frac{r_0}{\overline{r}_{n+\frac{1}{2}}}\right)\begin{Bmatrix} r_{n+\frac{1}{2}} \\ -r_{n+\frac{1}{2}} \end{Bmatrix},$$

which is well-behaved in the limit $\Delta r \to 0$ if $r_n \neq 0$. The consistent linearision of this force, leading to a contribution to the global tangent matrix, can be found in (21).

5.1.2 Vector constraints

In this case, we consider N points defined by their position vectors $x_1, .., x_N$. We define a new vector $\mathbf{q} \in \mathbb{R}^{3N}$, collecting all coordinates of every position vector. A vector constraint is defined by a general function $\Phi : \mathbb{R}^{3N} \ni \mathbf{q} \longmapsto \Phi(\mathbf{q}) \in \mathbb{R}$.

Let us assume that each point corresponds to a particle, and the only applied forces are due to the enforcement of the constraint. It can be shown (19) that this results in a Hamiltonian system without symmetry, and only total energy is conserved along the movement.

A convenient way of considering this type of constraints is to start from a modified midpoint rule given by:

$$\mathbf{M}(\dot{\mathbf{q}}_{n+1} - \dot{\mathbf{q}}_n) = -\Delta t \sigma (\mathrm{D}\Phi)_{n+\beta}\,; \quad \mathbf{q}_{n+1} - \mathbf{q}_n = \Delta t \dot{\mathbf{q}}_{n+\frac{1}{2}}\,,$$

where $(\cdot)_{n+\beta}$ expresses evaluation at point $\mathbf{q}_{n+\beta} \stackrel{\text{def}}{=} \mathbf{q}_n + \beta(\mathbf{q}_{n+1} - \mathbf{q}_n)$, and σ is a scalar which will be calculated for exact energy conservation.

The kinetic energy variation is in this case:

$$
\begin{aligned}
T_{n+1} - T_n &= \frac{1}{2}\dot{\mathbf{q}}_{n+1}^{\mathrm{T}} \cdot (\mathbf{M}\,\dot{\mathbf{q}}_{n+1}) - \frac{1}{2}\dot{\mathbf{q}}_n^{\mathrm{T}} \cdot (\mathbf{M}\,\dot{\mathbf{q}}_n) \\
&= \frac{1}{2}(\dot{\mathbf{q}}_{n+1} + \dot{\mathbf{q}}_n)^{\mathrm{T}} \cdot \mathbf{M}\,(\dot{\mathbf{q}}_{n+1} - \dot{\mathbf{q}}_n) \\
&= \frac{1}{\Delta t}(\mathbf{q}_{n+1} - \mathbf{q}_n)^{\mathrm{T}}(-\Delta t)\sigma(\mathrm{D}\Phi)_{n+\beta} \\
&= -\sigma(\mathrm{D}\Phi)_{n+\beta}^{\mathrm{T}} \cdot (\mathbf{q}_{n+1} - \mathbf{q}_n)\,.
\end{aligned} \tag{27}
$$

Additionally, note that the mean-value theorem shows that, Φ being continuous with continuous derivative, there exists a value $\beta \in (0, 1)$ such that the following expression holds:

$$
\psi(\beta) = (\mathrm{D}\Phi)_{n+\beta}^{\mathrm{T}} \cdot (\mathbf{q}_{n+1} - \mathbf{q}_n) - (\Phi_{n+1} - \Phi_n) = 0\,, \tag{28}
$$

Equation (27) shows that exact energy conservation implies two conditions:

1. When the constraint is exactly satisfied ($\Phi_n = \Phi_{n+1} = 0$), the kinetic energy variation of the constrained points should be zero, since $(V_\Phi)_n = (V_\Phi)_{n+1} = 0$. As a consequence, expression (27) has to verify:

$$
(\mathrm{D}\Phi)_{n+\beta}^{\mathrm{T}} \cdot (\mathbf{q}_{n+1} - \mathbf{q}_n) = 0\,, \tag{29}
$$

where existence of β is guaranteed by (28).

2. When the constraint is not satisfied exactly ($\Phi_n \neq 0, \Phi_{n+1} \neq 0$), the increment of kinetic energy should be equal to the decrement in potential energy:

$$
T_{n+1} - T_n = (V_\Phi)_n - (V_\Phi)_{n+1} = -\frac{1}{2}\alpha(\Phi_{n+1}^2 - \Phi_n^2)\,;
$$

allowing, with expressions (27) and (28), the calculation of σ:

$$
\sigma = \frac{1}{2}\alpha\,\frac{\Phi_{n+1}^2 - \Phi_n^2}{(\mathrm{D}\Phi)_{n+\beta}^{\mathrm{T}} \cdot (\mathbf{q}_{n+1} - \mathbf{q}_n)} = \alpha\overline{\Phi}_{n+\frac{1}{2}}\,,
$$

where, again, $\overline{(\cdot)}_{n+\frac{1}{2}} \stackrel{\mathrm{def}}{=} \frac{1}{2}\left[(\cdot)_n + (\cdot)_{n+1}\right]$.

Summarising, the energy-momentum algorithm takes the general form:

$$
\begin{aligned}
\mathbf{M}(\dot{\mathbf{q}}_{n+1} - \dot{\mathbf{q}}_n) &= -\Delta t\,\alpha\overline{\Phi}_{n+\frac{1}{2}}\,(\mathrm{D}\Phi)_{n+\beta}\,; \\
\mathbf{q}_{n+1} - \mathbf{q}_n &= h\dot{\mathbf{q}}_{n+\frac{1}{2}}\,,
\end{aligned} \tag{30}
$$

where β is the solution of the scalar equation:

$$
\Psi(\beta) = 0 = (\mathrm{D}\Phi)_{n+\beta}^{\mathrm{T}} \cdot (\mathbf{q}_{n+1} - \mathbf{q}_n) - \Phi_{n+1} + \Phi_n\,. \tag{31}
$$

Remark 7. Equation (31) must be solved at every iteration within each time step, it is non-linear and may be solved for β with different numerical methods. One possible choice is to use the following Newton-Raphson iterative scheme:

$$\beta_{k+1} = \beta_k - \frac{1}{\mathrm{D}\Psi(\beta_k)}\Psi(\beta_k)\,, \tag{32}$$

but other choices (quasi-newton, bisection, etc.) could be appropriate as well. This issue will be addressed again related to the conservative formulation of clearances in section 6.

Remark 8. If the constraint is at most quadratic, it is straightforward to see that $\beta = 1/2$, and equation (31) does not need to be solved.

The linearisation of the constraint force given in (30) leads to the following non-symmetric contribution to the consistent tangent matrix:

$$\mathbf{K}_{\Phi} = -\frac{\alpha}{2}\left[\left((\mathrm{D}\Phi)_{n+\beta}\otimes(\mathrm{D}\Phi)_{n+1}\right) + 2\beta\overline{\Phi}_{n+\frac{1}{2}}(\mathrm{D}^2\Phi)_{n+\beta}\right] \tag{33}$$

In practice, it is important to verify that the force constraint is well-behaved when $\Delta\mathbf{q} = (\mathbf{q}_{n+1} - \mathbf{q}_n) \to 0$, for the same reasons exposed earlier for scalar constraints. In this case, the contribution to the tangent matrix is symmetric and given by:

$$(\mathbf{K}_{\Phi})_{\Delta\mathbf{q}\to\mathbf{0}} = -\frac{\alpha}{2}\left[\left((\mathrm{D}\Phi)_n\otimes(\mathrm{D}\Phi)_n\right) + \Phi_n(\mathrm{D}^2\Phi)_n\right]$$

Remark 9. The functional dependency of β with $\mathbf{q}_{n+1}$ has been neglected in the linearisation leading to the tangent matrix contribution given by (33). This approximation significantly simplifies the formulation and is justified by several numerical applications (19), (13).

5.2 Conservative formulation of perfect joints

The most representative types are the spherical, revolute, cylindrical, prismatic and planar joints. Each of these can be modelled with one or more basic holonomic constraints, with the conservative formulation developed in the previous section. This set of constraints is what we call a *basic constraint library*. With the type of coordinates employed throughout this work, the basic library may be expressed in the following form:

1. *Constant distance* (l_{12}) between two points defined by position vectors (x_1, x_2). Calling $x_{ij} \stackrel{\text{def}}{=} x_j - x_i$, this constraint is given by: $\Phi = |x_{12}| - l_{12}$
2. *Constant relative position vector* between two points defined by position vectors (x_1, x_2), given by: $\mathbf{\Phi} = x_{12} - (x_{12})_0$. If this constraint expresses the coincidence between two points of different bodies ($x_{12} = 0$), is instead preferred to share the point, avoiding the explicit definition of a constraint.
3. *Alignment of three points* defined by position vectors (x_1, x_2, x_3), given by: $\mathbf{\Phi} = x_{13} \wedge x_{12}$
4. *Constant angle* α_0 between two directions defined by four points with position vectors (x_1, x_2, x_3, x_4). A cross product is employed for $0° \leq \alpha_0 < 45°$, and a dot product for $45° \leq \alpha_0 \leq 90°$:

$$\Phi = \frac{|x_{12} \wedge x_{34}|}{|x_{12}||x_{34}|} - \sin(\alpha_0) \qquad 0^\circ \leq \alpha_0 < 45^\circ$$
$$\Phi = \frac{x_{12} \cdot x_{34}}{|x_{12}||x_{34}|} - \cos(\alpha_0) \qquad 45^\circ \leq \alpha_0 \leq 90^\circ$$

5. *Four coplanar points* with position vectors (x_1, x_2, x_3, x_4); given by: $\Phi = x_{14} \cdot (x_{12} \wedge x_{13}) = 0$

Example 3. A *revolute joint* between two bodies A and B can be formulated in terms of two basic constraints. Assuming that the revolute axis is defined by points labelled as 1 and 2 in body A, and there are other two points labelled 3 and 4 in body B, these constraints are expressed as:

$$x_{13} = 0 \quad , \qquad \cos(\widehat{x_{12}, x_{34}}) - \cos(\alpha_0) = 0 \tag{34}$$

Details about the definition of other types of joints in terms of this basic constraint library can be found in (21).

Despite the simplicity of this approach, it is possible to obtain more efficient formulations taking advantage of particular features of specific types of joints. This is the case of a revolute joint between two rigid bodies, which is analysed in the following example.

Example 4. A revolute joint connects two rigid bodies (see Figure 2), each one parametrised by the cartesian inertial coordinates of 4 selected points, with one of the points in each body necessarily located at the joint (see Figure 3). If the joint is perfect (no clearance) both bodies can share this point, and only a total number of 7 points are needed to define the two-body system. The revolute axis is defined with one vector e located at the point defining the joint.

In order to obtain the constraint function associated with this particular joint, we define two different body (convected) reference frame systems $\{S_A, S_B\}$. System S_A is attached to body A at point 2 and defined by the three independent vectors $\{t_1, t_2, t_3\}$:

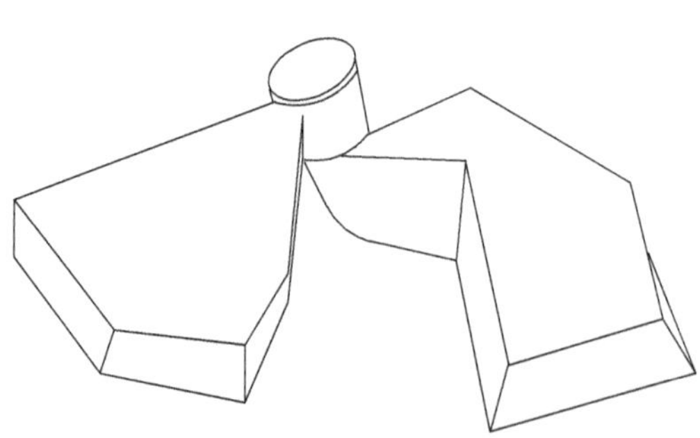

Fig. 2. Perfect three-dimensional revolute joint

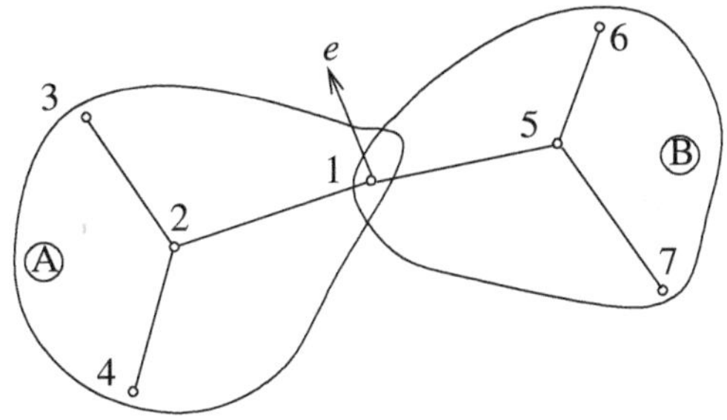

Fig. 3. Numerical model for a perfect revolute joint.

$$t_1 = x_1 - x_2 \quad , \qquad t_2 = x_3 - x_2 \quad , \qquad t_3 = x_4 - x_2 \quad , \tag{35}$$

and system S_B is attached to body B at point 5 and defined by vectors $\{n_1, n_2, n_3\}$:

$$n_1 = x_1 - x_5 \quad , \qquad n_2 = x_6 - x_5 \quad , \qquad n_3 = x_7 - x_5 \tag{36}$$

Note that these convected systems are not necessary orthonormal. The constraint function for this joint states that vector e, which defines the revolute axis, is the same expressed in both convected reference systems $\{S_A, S_B\}$:

$$e = X'_1 t_1 + X'_2 t_2 + X'_3 t_3 = Y'_1 n_1 + Y'_2 n_2 + Y'_3 n_3 \tag{37}$$

Observe that equation (37) does not constrain any movement of the revolute axis in space, which effectively follows the movement of the two connecting bodies. In terms of an unknown vector $\mathbf{q} \in \mathbb{R}^{21}$, which contains the cartesian inertial coordinates of the 7 points in a column-matrix format, the vectorial equation (37) can be stated in a more compact form:

$$\mathbf{\Phi} = \mathbf{C}\mathbf{q} \qquad \text{with} \quad \mathbf{q}^{\mathrm{T}} = \{ x_1^T | x_2^{\mathrm{T}} | \ldots | x_7^{\mathrm{T}} \} \tag{38}$$

where the (3×21) constant matrix $\mathbf{C}$ has the expression:

$$\mathbf{C} = \left((X'_1 - Y'_1)\mathbf{I}_3 | X'_0 \mathbf{I}_3 | X'_2 \mathbf{I}_3 | X'_3 \mathbf{I}_3 | Y'_0 \mathbf{I}_3 | Y'_2 \mathbf{I}_3 | Y'_3 \mathbf{I}_3 \right) \quad ,$$

with: $X'_0 = -\sum_{i=1}^{3} X'_i$, $Y'_0 = -\sum_{i=1}^{3} Y'_i$, and being $\mathbf{I}_3$ the (3×3) identity matrix.

Remark 10. Note the simplicity of the constraint equation (38), which is linear in the unknown vector $\mathbf{q}$, unlike the second equation (34), which is quadratic. The conservative formulation of any of them does not need the β parameter calculation, but the contribution to the total tangent matrix associated to the linear expression (38) is constant.

6 Conservative Formulation of Real Joints with Clearances

Real joints incorporate complex effects, such as clearances and friction, that may significantly affect the overall performance of a a mechanism.

In this section we will focus on *clearances*, that typically appear due to geometrical imperfections during the manufacturing process or are caused by wear. The basic phenomenon is the contact between the surfaces defining the joint. Nevertheless, contact problems in the context of multibody dynamics have special features (22): high number of intermittent contacts, very different time scales (rigid body motion and contact characteristic times are orders of magnitude apart) and very different space scales (clearances are much smaller than the typical size of the components). From the numerical point of view, this scenario typically leads to a stiff ordinary differential equation system, which is very demanding for the time integration scheme.

In this context, it is desirable to study methodologies that stress robustness and efficiency in time scale simulations typical in multibody systems. This time scale is much larger than that required to study the contact phenomena in detail, but this level of high detail is not generally an issue in multibody simulations. The challenge is to be able to model clearances, which involve a very high number of contact/impact events during typical rigid body time scales, in a robust and efficient manner, without demanding great detail for local effects of the individual contacts.

This section presents a methodology that has three key ingredients directly inspired by these special features of intermittent contacts in multibody simulations:

- It takes advantage of the analytical definition of the surfaces involved in the contact events, which reduces the number of variables involved in the problems;
- It does not use the gap as a primary variable of the formulation. This implies that the costly calculation of the minimal distance between the contacting bodies is avoided, speeding up the computations, and
- It sets the overall formulation in the context of an energy-conserving scheme, which has proved to have clear advantage in terms of stability in dynamic contact applications (1; 2; 18). Nevertheless, unconditionally energy-decaying methods are a promising alternative which has been proposed in recent years (3; 4; 7; 6), and should be considered for future developments.

Note that the introduction of both friction and material damping in the model have an stabilising effect from the numerical point of view. Based on this, only the most limiting situation will be considered, which is the case of a smooth clearance made of materials such that the joint impacts are perfectly elastic. Note also that the conservative formulation of such type of joints is clearly justified, since in this case there are no energy dissipation due to the work of the joint forces.

6.1 Clearance formulation

Whereas a perfect joint is formulated in terms of an equality relationship between the generalised coordinates of the bodies (e.g. expression (38) for a revolute joint), a smooth clearance is usually described through an inequality. This inequality is often stated in terms of the gap between the contact surfaces within the clearance, in the form of the Kuhn-Tucker complementary conditions (31):

$$g \geq 0\,, \quad p \leq 0\,, \quad pg = 0 \tag{39}$$

where g is the gap defined as the minimum distance between the contacting bodies (negative when there is penetration) and p the contact pressure (no adhesive stresses are allowed in the contact interface).

In contrast with the previous approach, we present here a methodology that is based on an equality constraint and does not explicitly use the concept of contact gap. This idea will be introduced first with a special type of joint clearance, referred as the rigid cavity model, and will be extended later to more general types of joints.

6.1.1 The rigid cavity model

This model represents a special joint clearance, understood as a rigid cavity attached to one of the connected bodies. The connection points of the other bodies are restricted to move inside this cavity, as shown in Figure 4. This model can accurately represent a real spherical joint where one of the bearings is out of the nominal spherical shape.

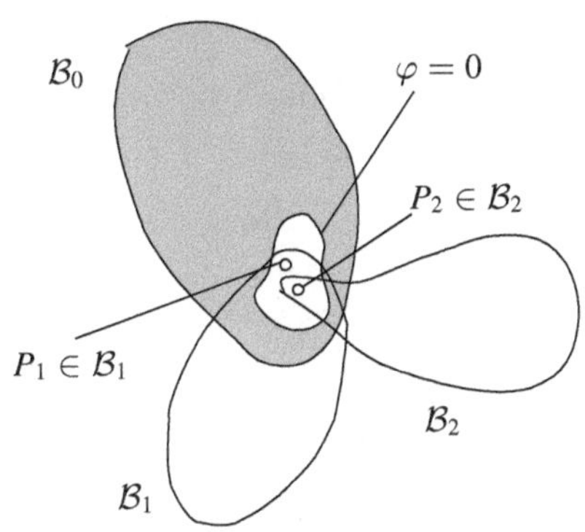

Fig. 4. Rigid cavity model

In Figure 4, the rigid body to which the rigid cavity is attached is denoted by $\mathcal{B}_0$. There are two other bodies $\mathcal{B}_1$ and $\mathcal{B}_2$ linked to $\mathcal{B}_0$ at points P_1 and P_2. These two points are constrained to move inside the cavity defined by the surface $\varphi : \mathbb{R}^3 \longrightarrow \mathbb{R}$, such that their coordinates verify

$$\varphi \leq 0\,. \tag{40}$$

The proposed methodology regularises the problem, designing an equality constraint (Φ) that replaces the inequality constraint given by (40). We require this function Φ to possess the following features:

i. It has to vanish at the clearance surface: verify $\Phi = 0$ when evaluated at $\varphi = 0$.
ii. The related contact force modulus has to be at least C^1, in order to avoid stability problems during persistent contact. It can be proved (20) that this requirement is satisfied if the surface is regular and the modulus of the constraint gradient vanishes over it.

iii. The contact force must be formally similar to the Hertz model (14), which has been employed by several authors based on experimental investigations (17; 31). In the case of a perfect spherical clearance (the bearings are spherical but have different radius), we want to exactly recover the classical Hertz formulation of the contact force. This is given by $\mathbf{f}_{\text{Hertz}} = kg^n N$, being g the normal gap and N the unit normal to the spherical cavity.

A constraint function Φ that fulfils all these requirements is given by:

$$\Phi = \begin{cases} 0 & \text{if} \quad \varphi < 0 \\ \sqrt{\frac{2k}{n+1}\varphi^{n+1}} & \text{if} \quad \varphi \geq 0 \end{cases} \tag{41}$$

with $n > 1$ and $k > 0$. The related contact force, based on (6) with a penalty parameter ($\alpha = 1$), is given by:

$$\mathbf{f}_{\Phi} = \Phi(\mathrm{D}\Phi) = \begin{cases} 0 & \text{if} \quad \varphi < 0 \\ k\,\varphi^n(\mathrm{D}\varphi) & \text{if} \quad \varphi \geq 0 \end{cases} \tag{42}$$

where parameter k plays the role of the contact stiffness.

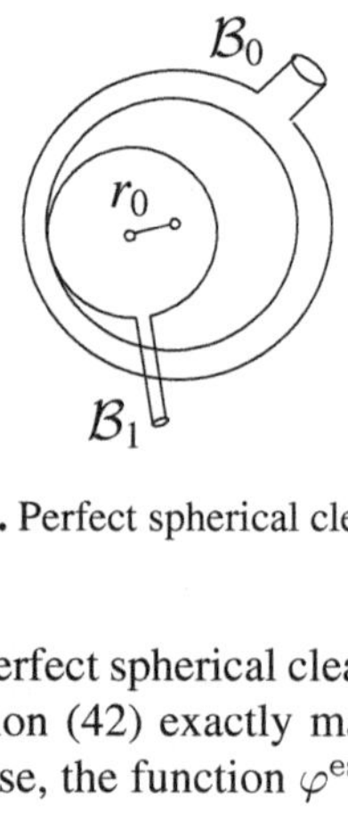

Fig. 5. Perfect spherical clearance

For the particular case of a perfect spherical clearance (Figure 5), it can be shown that the force given by expression (42) exactly matches the one obtained with the classical Hertz model. In this case, the function φ^{esph} that defines the surface of the cavity is given by:

$$\varphi^{\text{esph}} = r - r_0\,, \tag{43}$$

being $r = ||r||$ the distance from the centre of the spherical cavity and r_0 its radius. Introducing (43) in expression (42) the contact force match the Hertz model, given by:

$$\mathbf{f}_{\Phi}^{\text{esph}} = k\,(r - r_0)^n\,\frac{r}{r} = kg^n N \qquad \text{for} \quad g \geq 0\,, \tag{44}$$

being g the normal gap and N the unit normal to the spherical cavity.

Based on these considerations, it is possible to understand the contact force given by (42) as a generalised Hertz model for an arbitrary contact surface. It is important to note that this methodology does not use the gap as a primary variable, which can be computed in the post-processing stage if desired. Only the evaluation of the constraint function φ is required throughout the calculation.

6.1.2 A general clearance model

The previous ideas can be extended naturally to a more general situation where the function φ does not represent any physical surface in the euclidean $\mathbb{R}^3$ space.

In such a general case, the clearance is represented by a set of q inequality constraints φ_i, each one expressed in terms of the generalised coordinates that define the configuration of the connecting bodies. If we collect these coordinates within a vector $\mathbf{q} \in \mathbb{R}^p$, any constraint of the set is a function defined as:

$$\varphi_i : \mathbb{R}^p \ni \mathbf{q} \longrightarrow \varphi_i(\mathbf{q}) \in \mathbb{R} \quad , \qquad \text{with} \quad i = 1, .., q \, .$$

In this fashion, each constraint φ_i can be understood as a surface defined in the coordinate space $\mathbb{R}^p$, and it is possible to apply all the concepts presented in the rigid cavity model. The fact that this surface has not a direct geometrical interpretation in the geometrical three-dimensional space where the movement takes place has no effect from the computational point of view. Furthermore, this approach becomes specially attractive for complex joints where the actual gap determination can be computationally expensive, since it is completely avoided.

Based on these considerations and the results of section 6.1.1, it possible to define a set of new regularised constraints Φ_i in terms of the original functions φ_i as follows:

$$\Phi_i = \begin{cases} 0 & \text{if} \quad \varphi_i < 0 \\ \sqrt{\frac{2k}{n+1}\varphi_i^{n+1}} & \text{if} \quad \varphi_i \geq 0 \end{cases} \quad , \qquad \text{for} \quad i = 1, ..., q \tag{45}$$

The associated constraint forces possess the same properties described in section 6.1.1 for the rigid cavity model in terms of regularity, and have the benefits already discussed for intermittent contact situations.

Example 5. Clearance in a three dimensional revolute joint linking two rigid bodies. We consider now a revolute joint with a clearance (Figure 6), accounting for the following relative movements:

- Maximum relative displacement along the revolute axis (δ_z).
- Maximum relative displacement perpendicular to the revolute axis (δ_r).
- Maximum misalignment of the revolute axis (δ_θ).

A convenient approach to treat these clearances is to consider the first two movements as independent, and the third related to the others. Based on this consideration, it follows that a clearance located only in the longitudinal or only in the radial direction would not allow any misalignment of the revolute axis.

In this case, each body must be defined by 4 different points, and each one has a different vector defining the revolute axis, as shown in Figure 7. The position of the

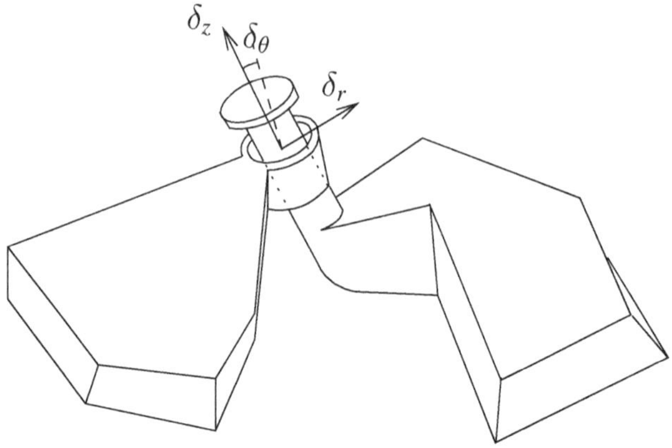

Fig. 6. Three-dimensional revolute joint with clearance

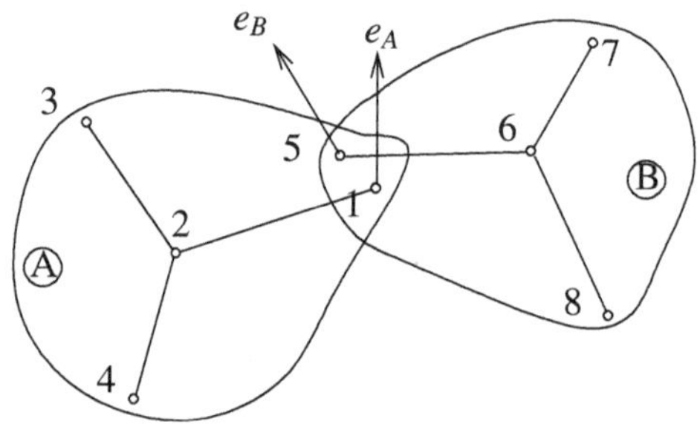

Fig. 7. Numerical model for a revolute joint with clearance

system composed by the two connected bodies is defined by vector $\mathbf{q} \in \mathbb{R}^{24}$, which contains the inertial cartesian coordinates of the eight points.

The possible movements at the joint clearance are the displacement vector $r = (x_5 - x_1)$ and the angle θ between the two revolute axes. Figure 8 shows these movements, where angle θ is visualised in the meridional plane of the inner bearing only for clarity. With this notation, the constraint functions related with this joint can be expressed as:

$$\varphi_1 = \cos\delta_\theta - (e_A \cdot e_B) \leq 0 \tag{46}$$

$$\varphi_2 = |r \cdot e_A| - \delta_z \leq 0 \tag{47}$$

$$\varphi_3 = ||r \wedge e_A|| - \delta_r \leq 0 \tag{48}$$

assuming that $|e_A| = |e_B| = 1$. Note that the clearances δ_r and δ_z can be considered constant and be calculated from the particular geometry of the joint if the bearings are perfectly cylindrical, but this is not possible for δ_θ in general; it depends on the geometry but also on the relative displacement (r, θ) of the joint. It means that relation (46) can be especially complex if all the possible movements are allowed.

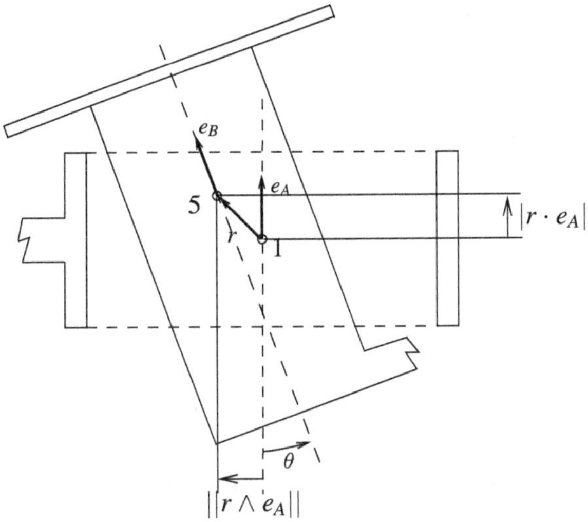

Fig. 8. Revolute joint movements.

Leaving aside the complexity introduced by the calculation of the clearances δ_θ, δ_z and δ_r in terms of the geometry of the joint and the coordinate vector $\mathbf{q}$ for an arbitrary position, most of the other terms can be stated in a simple and closed matrix form, and will be analysed with more detail below.

Angular clearance

The angular misalignment (θ) of the revolute axes is represented by the inequality constraint φ_1 given in (46).

We denote by $\mathbf{q}_A$ and $\mathbf{q}_B$ the vectors containing the cartesian inertial coordinates, in column format, of the four points belonging to each body A and B respectively. These vectors can be related to the global unknown vector $\mathbf{q}$ through constant projection matrices $\mathbf{P_A}$ and $\mathbf{P_B}$:

$$\mathbf{q}_A = \mathbf{P}_A\,\mathbf{q} \quad , \qquad \mathbf{q}_B = \mathbf{P}_B\,\mathbf{q} \quad , \tag{49}$$

with:

$$\mathbf{P}_A = \left(\mathbf{I}_{12}\middle|\mathbf{0}_{12}\right) \quad , \qquad \mathbf{P}_B = \left(\mathbf{0}_{12}\middle|\mathbf{I}_{12}\right) \quad , \tag{50}$$

being $\mathbf{0}_{12}$ and $\mathbf{I}_{12}$ the (12×12) null and identity matrices, respectively.

We can also obtain the cartesian inertial components of the revolute axes e_1 and e_2 from vectors $\mathbf{q}_A$ and $\mathbf{q}_B$ respectively. The revolute axis e_A can be expressed in the reference system $S_A = \{t_1, t_2, t_3\}$, rigidly attached to the first body, as:

$$e_A = X_1' t_1 + X_2' t_2 + X_3' t_3 \; .$$

Taking into account that $t_1 = (x_1 - x_2), t_2 = (x_3 - x_2)$ and $t_3 = (x_4 - x_2)$, the following matrix expression is obtained:

$$e_A = \mathbf{C}_A \mathbf{q}_A \quad , \qquad \text{with} \quad \mathbf{C}_A = \left(X_1' \mathbf{I}_3 \middle| X_0' \mathbf{I}_3 \middle| X_2' \mathbf{I}_3 \middle| X_3' \mathbf{I}_3 \right) \,, \tag{51}$$

being $X_0' = -\sum_{i=1}^{3} X_i'$.

The same procedure can be followed in order to obtain e_B from coordinate vector $\mathbf{q}_B$. In this case, the projection matrix $\mathbf{C}_B$ depends on the constant coordinates Y_i' of vector e_B relative to the reference system $S_B = \{n_1, n_2, n_3\}$ defined as $n_1 = (x_5 - x_6), n_2 = (x_7 - x_6)$ and $n_3 = (x_8 - x_6)$, obtaining:

$$e_B = \mathbf{C}_B \mathbf{q}_B \quad , \qquad \text{with} \quad \mathbf{C}_B = \left(Y_1' \mathbf{I}_3 \middle| Y_0' \mathbf{I}_3 \middle| Y_2' \mathbf{I}_3 \middle| Y_3' \mathbf{I}_3 \right) \,, \tag{52}$$

being $Y_0' = -\sum_{i=1}^{3} Y_i'$. Using relations (49), (51) and (52), the angular constraint φ_1 given in (46) can be expressed as:

$$\varphi_1 = \cos\delta_\theta - (e_A \cdot e_B) = \cos\delta_\theta - \mathbf{q}^{\mathrm{T}} \mathbf{\Gamma}_1 \mathbf{q} \leq 0 \,, \tag{53}$$

where the (24×24) constant matrix $\mathbf{\Gamma}_1$ is given by:

$$\mathbf{\Gamma}_1 = \mathbf{P}_A^{\mathrm{T}} \mathbf{C}_A^{\mathrm{T}} \mathbf{C}_B \mathbf{P}_B \tag{54}$$

Axial clearance

The constraint function is in this case (see Figure 8):

$$\varphi_2 = |r \cdot e_A| - \delta_z \leq 0 \,,$$

where $r = x_5 - x_1$ denotes the relative position of the points located at the joint. The cartesian inertial coordinates of x_1 and x_5 can be expressed in a compact matrix format through a constant projection matrix $\mathbf{P}$:

$$x_1 = \mathbf{P}\, \mathbf{q}_A \quad ; \qquad x_5 = \mathbf{P}\, \mathbf{q}_B \quad ; \qquad \mathbf{P} = \left(\mathbf{I}_3 \middle| \mathbf{0}_3 \middle| \mathbf{0}_3 \middle| \mathbf{0}_3 \right) \tag{55}$$

Using (55) and (49) is possible to obtain a matrix expression for vector r:

$$r = x_5 - x_1 = \mathbf{P}(\mathbf{q}_B - \mathbf{q}_A) = \mathbf{P}(\mathbf{P}_B - \mathbf{P}_A)\, \mathbf{q} \tag{56}$$

Finally, employing relation (51) for e_A and expression (56), it is possible to obtain the following compact matrix expression for the constraint function:

$$\varphi_2 = |\mathbf{q}^{\mathrm{T}} \mathbf{\Gamma}_2\, \mathbf{q}| - \delta_z \leq 0 \quad , \qquad \text{with} \qquad \mathbf{\Gamma}_2 = (\mathbf{P}_B^{\mathrm{T}} - \mathbf{P}_A^{\mathrm{T}}) \mathbf{P}^{\mathrm{T}} \mathbf{C}_A \mathbf{P}_A \tag{57}$$

being $\mathbf{\Gamma}_2$ a constant (24×24) matrix.

Radial clearance

The constraint function is in this case (see Figure 8):

$$\varphi_3 = ||r \wedge e_A|| - \delta_r \leq 0 \,, \tag{58}$$

where $r = x_5 - x_1$ denotes the relative position of the points located at the joint. There are alternative formulations that avoid the use of the vector cross product; e.g., the one given by:

$$\varphi_3^\star = ||r - (r \cdot e_A)e_A|| - \delta_z \leq 0 \,,$$

which may be reformulated as:

$$\varphi_3^\star = ||\,[1 - (e_A \otimes e_A)] \cdot r|| - \delta_z \leq 0 \,, \tag{59}$$

where 1 denotes the unit second-order tensor and $\otimes$ the standard tensor product of two vectors.

Despite the fact that expressions (58) and (59) can be efficiently treated within the general methodology presented in section 6.1, they can not be stated in such a compact matrix form as the other two types of movements (angular and axial); consequently, no more analytical details of their formulation will be presented.

6.2 Conservative formulation

Once a clearance constraint is formulated in terms of an equality relationship as in (41), it is possible to evaluate the contact force in such a way that the total energy is conserved, as presented in section 5.1. An intuitive interpretation for energy conservation is to ensure that the contact energy stored during penetration is exactly restored when the contact is released, such that the total energy remains constant. Contact force is evaluated with the algorithmic expression (30), calculating parameter β by means of finding the root of the scalar equation (31).

Some remarks must be made at this point:

Remark 11. The proof of the existence of β relies on the mean value theorem, which requires the regularised constraint function Φ to be at least C^1. If the original constraint function φ is C^1, the regularised function Φ inherits this property if the parameter n appearing in (41) is strictly greater than one.

Remark 12. A consequence from the previous remark is that the conservative formulation (30) may not be applicable if the standard penalty regularisation is employed ($n = 1$ in (41)). In this case, parameter β may not exist for an arbitrary function φ, due to the regularity requirement on function Φ.

Remark 13. As a consequence of previous remarks, the conservative formulation of the revolute joint clearance requires the numerical calculation of β from equation (31); the original constraint functions φ_i given by (53), (57) and (58) are at least quadratic and their first derivative does not vanish at the constraint surface.

Remark 14. If the original constraint function φ can be written in terms of an scalar r (typically, the distance between two points), the conservative formulation of the constraint force is always possible and given by (23):

$$\mathbf{f}_\Phi^c = -\alpha \frac{\Phi_{n+1}^2 - \Phi_n^2}{r_{n+1}^2 - r_n^2} \begin{Bmatrix} r_{n+\frac{1}{2}} \\ -r_{n+\frac{1}{2}} \end{Bmatrix}$$

where $r_{n+\frac{1}{2}} = \frac{1}{2}(r_{n+1} + r_n)$. This is the case of a perfect spherical clearance, as described in section 6.1.1.

Remark 15. When the constraint function Φ is not piecewise defined and smooth, the calculation of β from (31) is not so relevant, and $\beta = 1/2$ usually is a fair approximation that does not significantly spoils the conserving properties of the algorithm. However, the precise calculation of β is crucial with a piecewise function, specially at the interval containing the border of different branches of the function. This interval is where penetration and separation from the constraint surface occurs, typically associated to β values close to 1 or 0 respectively. In these situations, alternative methods to the Newton-Raphson scheme given by (32) could perform better, e.g. the bisection method.

7 Representative Numerical Application: Double Pendulum with Joint Clearances

The selected application is a double pendulum composed by two rigid bars of length $l = 1$ m and mass $m = 1$ kg. Both bars are initially horizontal and perpendicular to each other, as shown in Figure 9. The system is released from rest under the gravity $g = 10$ m/s^2, and the motion is integrated up to 5 s.

Two variations are presented from this basic mechanism. The first case incorporates spherical joints, and rigid-cavity-type clearances are introduced at both joints O and A. The second case incorporates revolute joints, with an angular-type clearance in one of them.

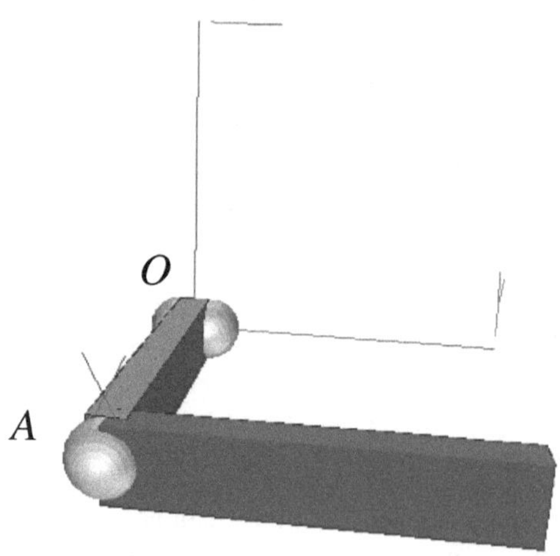

Fig. 9. Double pendulum with clearances. Initial configuration. (not to scale)

7.1 Rigid-cavity-type clearances

In this case the system has two smooth clearances represented by the rigid cavity model described in section 6.1.1, located at points O and A of Figure 9. The inertia of each bar around its longitudinal axis is supposed to be negligible, so each one is completely defined with its two end points.

The clearance located at point O has a "blob" shape, shown in Figure 10, defined by the function:

$$\varphi = e^{[1-(k_x x)^2-(k_y y+\delta)^2-(k_z z)^2]} + e^{[1-(k_x x)^2-(k_y y-\delta)^2-(k_z z)^2]} - 1\,,$$

with $k_x = k_y = k_z = 50$ and $\delta = 1.1$. The regularisation of this constraint is made with parameters $k = 10^7$ (contact stiffness) and $n = 4$.

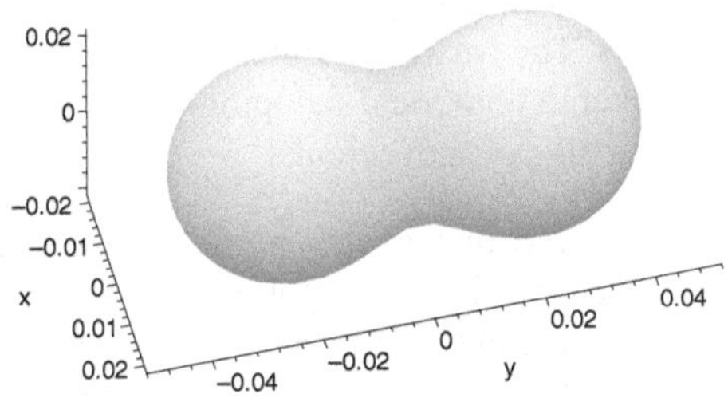

Fig. 10. Rigid cavity clearance at point O: "blob" shape.

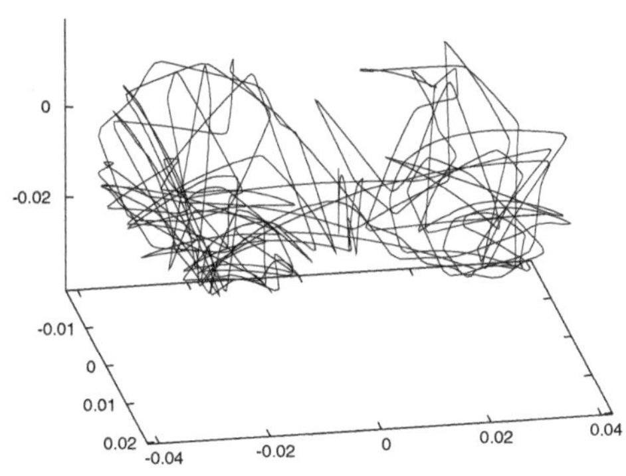

Fig. 11. Double pendulum: trajectory of point O

The clearance at point A, connecting both bars, is spherical with radius $r_0 = 0.04$ m, with parameters $k = 10^7$ and $n = 1.5$. A penalty parameter $\alpha = 10^{12}$ was employed for the enforcement of the constant-distance constraints of the rigid bars, and a constant time-step $\Delta t = 0.001$ s is adopted.

Figure 11 shows the trajectory of point O, which clearly resembles the blob-shaped volume where it is constrained to move. Figure 12 shows the effect of the clearances over the trajectory of the free end of the double pendulum (point B) during the first 3 s. Figure 13 shows the exact conservation of the total energy (kinetic + gravitational potential + constraint) during the integration.

Two remarks must be made:

Remark 16. Typically, the maximum allowable time step achieved with the conserving integration is significantly larger than the one achieved with any standard implicit integrator, unless considerable numerical damping is introduced. In the case of the trapezoidal rule, this maximum time step for successful integration of the total 5 s movement is $\Delta t = 7 \cdot 10^{-5}$ s. The backward Euler method allows up to $\Delta t = 0.003$, but introducing an unacceptable numerical damping that reduces the kinetic energy by half at the end of the first 10 s of movement.

Remark 17. It has been observed that in this particular application the maximum time-step is significantly influenced by the complex blob-shaped clearance at point O. If this is replaced by an spherical one, identical to the one located at joint A, the allowable time step goes up to $\Delta t = 0.007$ s with exact energy conservation.

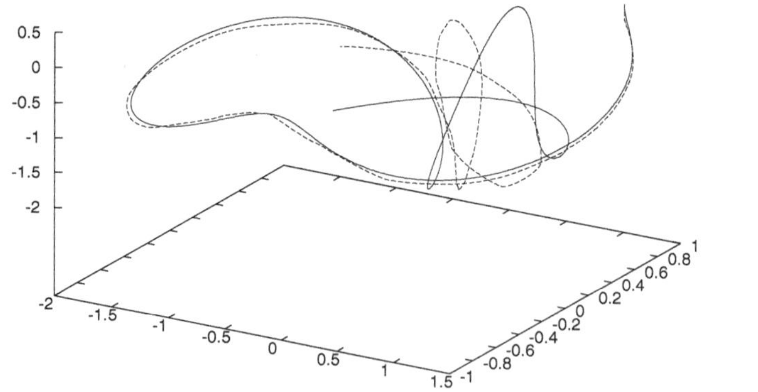

Fig. 12. Trajectory of the free end of the double pendulum (point B) during the first 3 seconds: effect of the clearances.

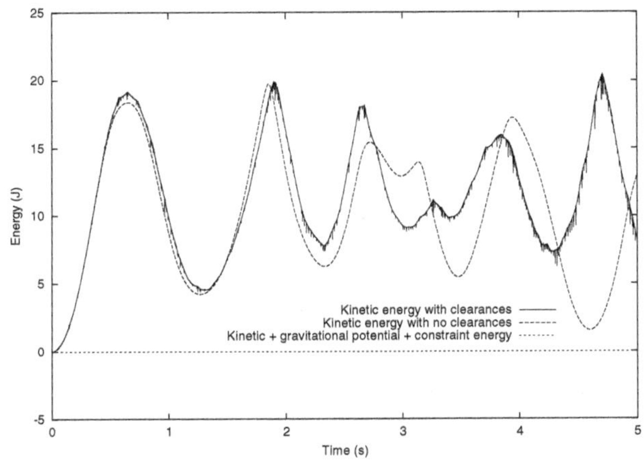

Fig. 13. Double pendulum: kinetic and total energy vs. time

7.2 Revolute joint clearance

In this second case the system incorporates two revolute joints at O and A, with horizontal and vertical axes respectively, as shown in Figure 14. Each bar is modelled as a prismatic homogeneous rigid body with dimensions $(1 \times 0.1 \times 0.02)$ m.

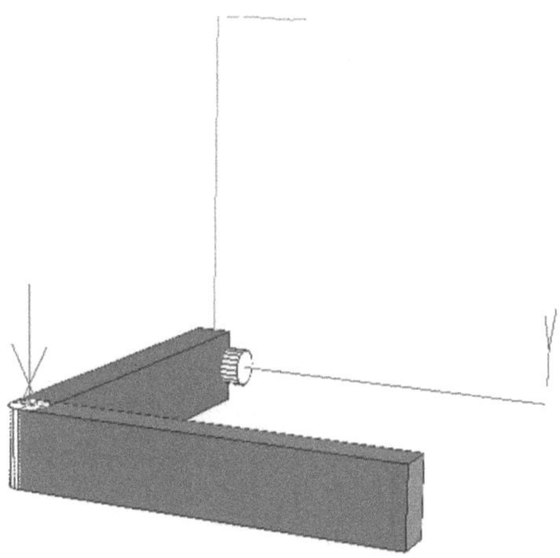

Fig. 14. Double pendulum with revolute joints

Joint O is a perfect revolute joint, and joint A incorporates an angular clearance $\delta_\theta = 0.2$ rad ($\simeq 11.5°$), with parameters $k = 10^7$ and $n = 2$. A penalty parameter $\alpha = 10^{12}$ was employed for the enforcement of other constraints (related to the rigid bodies and the perfect revolute joint).

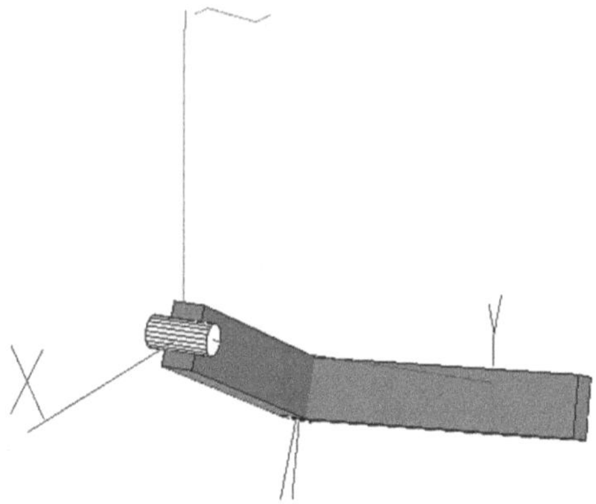

Fig. 15. Double pendulum with revolute joints: snapshot of the movement at $t = 4$ s

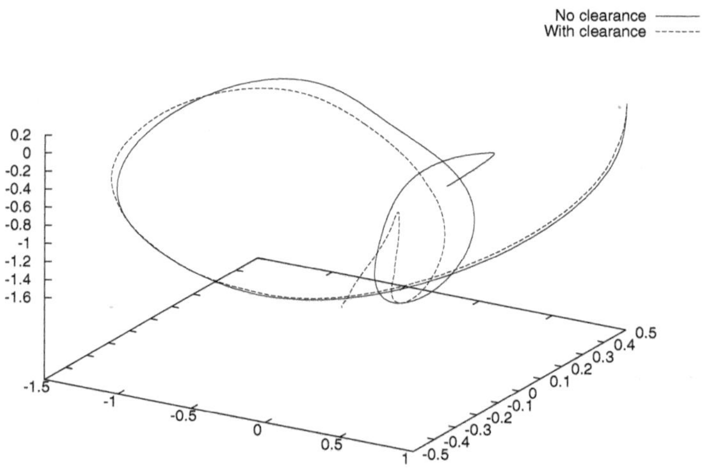

Fig. 16. Trajectory of the centre of mass of the end bar during the fist 3 seconds: effect of the revolute clearance

The motion is integrated with a time step $\Delta t = 0.004$ s. Figure 15 shows a snapshot of the movement at $t = 4$ s, where the angle between the two revolute joint axes is clearly visible. Figure 16 shows the effect of the clearance on the trajectory of the centre of mass of the end bar during the first 3 s, and Figure 17 the evolution of the energy over the total movement.

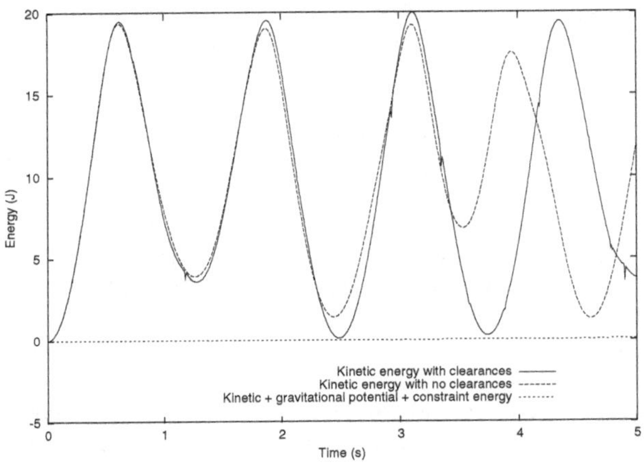

Fig. 17. Double pendulum with revolute joints: kinetic and total energy vs. time

8 Conclusions

The following general conclusions can be attained, based on the ideas presented in this chapter:

- The parametrisation based on inertial cartesian coordinates of selected points and the use of the penalty method for constraints, provide a simple formulation of the equations of motion of flexible multibody systems.
- The energy-momentum method overcomes the numerical ill-conditioning introduced by the large penalty factors and the high stiffness of the deformable bodies. It's possible to adopt rather large time steps while exactly conserving total energy and momentum.
- Holonomic constraints, representing perfect joints, can be consistently formulated with the energy-momentum method, which ensures energy preservation and overcome the numerical ill-conditioning and stiff character of the equations.
- A methodology for the analysis of real joints with clearances has been presented, which appears to be promising in terms of numerical efficiency and robustness. This approach takes advantage of the analytical definition of the clearance surfaces in order to define a model where the gap between the contacting surfaces is not a primary variable. The formulation is established in the context of an energy-conserving time integration which results specially well suited for applications where a high number of intermittent contacts are present.

References

1. Armero F, Petöcz E (1998) Formulation and analysis of conserving algorithms for frictionless dynamic contact/impact problems. Computer methods in applied mechanics and engineering 158:269–300

2. Armero F, Petöcz E (1999) A new dissipative time-stepping algorithm for frictional contact problems: formulation and analysis. Computer methods in applied mechanics and engineering 179:151–178
3. Armero F, Romero I (2001) On the formulation of high-frequency dissipative time-stepping algorithms for nonlinear dynamics. Part I: Low-order methods for two model problems and nonlinear elastodynamic. Computer Methods in Applied Mechanics and Engineering 190:2603–2649
4. Armero F, Romero I (2001) On the formulation of high-frequency dissipative time-stepping algorithms for nonlinear dynamics. Part II: High order methods. Computer Methods in Applied Mechanics and Engineering 190:6783–6824
5. Chen A J (1998) Energy-momentum conserving methods for three dimensional dynamic nonlinear multibody systems. PhD thesis, Stanford University, Department of mechanical engineering
6. Trainelli L, Bottasso C L, Borri M (2001) Integration of elastic multibody systems by invariant conserving/dissipating schemes. Part II. Computer methods in applied mechanics and engineering 190:3701–3733
7. Borri M, Bottasso C L, Trainelli L (2001) Integration of elastic multibody systems by invariant conserving/dissipating schemes. Part I. Computer methods in applied mechanics and engineering 190:3669–3699
8. Crisfield M A (1997) Non-linear finite element analysis of solids and structures, volume I and II. Wiley
9. Ferreira C (1997) Simulacion dinamica de mecanismos tridimensionales con elementos flexibles. PhD thesis, Centro Politecnico Superior de la Universidad de Zaragoza
10. García de Jalón J, Bayo E (1994) Kinematic and Dynamic Simulation of Multibody Systems. The Real Time Challenge. Springer-Verlag
11. Geradin M, Cardona A (2000) Flexible multibody dynamics. A finite element approach. Wiley
12. Goicolea J M, García Orden J C (2000) Dynamic analysis of rigid and deformable multibody systems with penalty methods and energy-momentum schemes. Computer Methods in Applied Mechanics and Engineering 188(4):789–804
13. Goicolea J M, García Orden J C (2002) Quadratic and higher-order constraints in energy-conserving formulations in flexible multibody systems. Multibody System Dynamics, 7:3–29
14. Goldsmith W (1960) Impact. The theory and Physical Behaviour of Coliding Solids. Dover
15. González O (1996) Design and analysis of conserving integrators for nonlinear hamiltonian systems with symmetry. PhD thesis, Stanford University Department of Mechanical Engineering
16. González O, Simó J C (1996) On the stability of symplectic and energy-momentum algorithms for non linear hamiltonian systems with symmetry. Computer methods in applied mechanics and engineering
17. Dobychin M N, Kragelsky I V, Kombalov V S (1982) Friction and wear. Calculation Methods. Pergamon Press
18. Laursen T A (2002) Computational contact and impact mechanics. Springer
19. García Orden J C (1999) Dinamica no lineal de sistemas multicuerpo flexibles mediante algoritmos conservativos. PhD thesis, ETSI Caminos, Canales y Puertos. Universidad Politcnica de Madrid
20. García Orden J C (2003) Analysis of joint clearances in multibody systems. Multibody System Dynamics. Submitted

21. García Orden J C, Goicolea J M (2000) Conserving properties in constrained dynamics of flexible multibody systems. Multibody System Dynamics 4:225–244
22. Pfeiffer F, Glocker C (Editors) (2000) Multibody dynamics with unilateral contacts. Internaltional Centre for Mechanical Sciences, SpringerWienNewYork
23. Puso M A (2002) An energy and momentum conserving method for rigid-flexible body dynamics. International Journal for Numerical and Analytical Methods in Engineering 53:1393–1414
24. Shabana A A (2001) Performance of non-linear finite element formulations in flexible multibody simulations. In J A C Ambrosio and M Kleiber (editors) Computational Aspects of Nonlinear Structural Systems with Large Rigid Body Motion, pp. 29–40. IOS Press
25. Shabana A A (1998) Dynamics of Multibody Systems. Wiley
26. Simó J C, Tarnow N (1992) The discrete energy-momentum method. Conserving algorithms for nonlinear elastodynamics. ZAMP
27. Simó J C, Vu-Quoc L (1986) A three-dimensional finite-strain rod model. Part II: computational aspects. Computer methods in applied mechanics and engineering 58:79–116
28. Simó J C, Vu-Quoc L (1988) On the dynamics in space of rods undergoing large motions - a geometrically exact approach. Computer methods in applied mechanics and engineering 66:125–161
29. Simó J C, Wong K K (1991) Unconditionally stable algorithms for rigid body dynamics that exactly preserve energy and momentum. International Journal of Numerical Methods in Engineering 31:19–52
30. Taylor R L (2001) Finite element analysis of rigid-flexible systems. In J A C Ambrosio and M Kleiber (editors), Computational Aspects of Nonlinear Structural Systems with Large Rigid Body Motion, pp. 63–84. IOS Press
31. Wriggers P Computational Contact Mechanics. Wiley

Modelling, Control and Validation of Flexible Robot Manipulators

Jorge M. Martins, José Sá da Costa and Miguel Ayala Botto

IST, DEM-GCAR, Technical University of Lisbon, Portugal
{martins,sadacosta,migbotto}@dem.ist.utl.pt

The dynamics equations describing the motion of a flexible manipulator are developed. It is assumed that the manipulator supports the gravitational force, external punctual forces and torques applied at specific points along each link's neutral axis, a punctual force and torque at the end effector, and loads resulting from the application of piezoelectric patches that are bonded at specific points along the elastic links. A discrete model of the Newton-Euler type capturing the fundamental dynamics required for flexible manipulator analysis is deduced for a generic link. A Eulerian formulation is used for the rigid body motion and a total Lagrangian formulation is used for the elastic deformation. To this end, Jourdain's Principle or the Principle of Virtual Powers is adopted, assuming a Rayleigh-Ritz expansion of the elastic variables. The elastic variables, which are the links curvature and shear deformation, are assumed to be infinitesimal. However, nonlinear displacements are considered due to the large length/width aspect ratio of the links. The dynamics model of the manipulator is obtained from the assembling of the individual links. Both the Articulated Body (AB) method, and the Composite Inertia (CI) method are obtained. A validation and control exercise is performed on a single flexible link. Frequency domain and time domain validation is performed in regard to the order of the cross section rotation matrix. Linear and quadratic assumptions are compared against each other, and against the experimental apparatus. Curvature feedback control is compared against classical joint (collocated) feedback, and it's improved performance is shown through the measurement of tip acceleration.

1 Introduction

The advantages of lightweight manipulators over the traditional heavy and rigid manipulators have for long been recognized: less power consumption, higher payload to weight ratio and faster motions are just a few. However, there is still a set back in the use of these better machines. This resides in the fact that lightweight taken to a certain level will give way to a natural loss of stiffness allowing the manipulator to

J.A.C. Ambrósio (ed.), Advances in Computational Multibody Systems, 239–268.

vibrate due to elastic deformation, highly deteriorating machine precision. A promising approach to compensate for this set back, is to incorporate into these machines more sophisticated control algorithms, with a more involving actuating and sensing network. A basis for this approach, is a manipulator model capable of reproducing the fundamental system dynamics in a given application, and amenable for real time computation.

The first problem one faces in the modelling of flexible manipulators is at the link level: how large are the elastic displacements? This problem has been topic of research for many years, and one answer is that the displacements should not be assumed only of first order. If one wishes a dynamics model capable of capturing centrifugal stiffening and a kinematic model capable of reproducing the link foreshortening, at least second order strain-displacement relations are needed. Within beam kinematics assumptions or rigid cross sections, this implies assuming at least a second order rotation matrix of a cross section due to elasticity. Parameterization of higher order rotation matrices on the other hand, require aditional care in the choice of which parameters to use. One possible choice, are the beam bending curvatures.

The second problem one faces is the global dynamics formulation and solution problem. Two approaches are available for dealing with the dynamics of rigid multibody systems: the Composite Rigid Body (CRB) method and the Articulated Body (AB) method (Featherstone, 1987). For systems of smaller dimension, the CRB method runs faster than the AB method, and as the size of the system increases the AB method becomes more efficient. In robotics applications, both approaches should be available, since the size of the system changes frequently (for example cooperating manipulators). The formulation and solution problem is designated as the formulation stiffness problem in (Ascher, 1997). These methods may be extended to flexible manipulator systems, the CRB method being now designated as the Composite Inertia (CI) method.

The control problem of flexible manipulator systems may be tackled by analyzing the advantages of curvature (bending strain) feedback, over the traditional joint feedback control. The high increase in vibration damping achieved at the cost of small joint error reveals high potential of the approach. Is is simple to implement, and totally signal based. Together with a higher level model based approach, it is the grounds for many existing control approaches for flexible manipulator systems.

Following the works of [1–3], and [4] we present our modelling and computational environment, which is aimed at analysis and control of flexible manipulator arms. To this end, in Sections 2 and 3 the dynamics equations of a rigid body and of a flexible beam are presented respectively. In Section 3, piezoelectric actuators are also modelled. In Section 4, the AB and CI solution methods are presented for a serial topology manipulator. Each link is assumed to be a flexible beam rigidly attached to a rigid body at each end. In Section 5, validation and strain feedback control experiments are presented for a single flexible link, and in Section 6 the final conclusions are drawn.

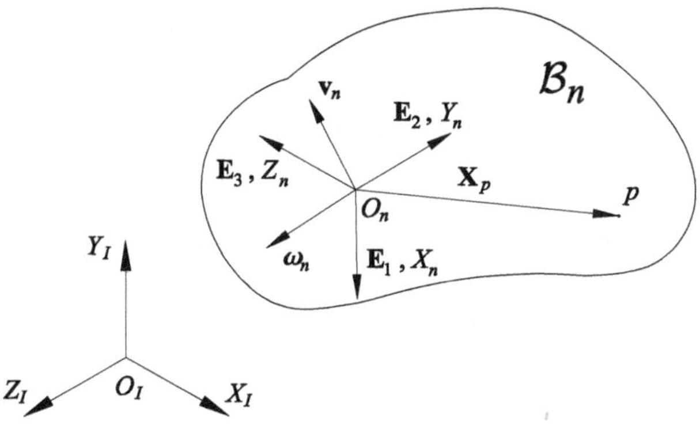

Fig. 1. Rigid body reference frames

2 Dynamics Equations of a Rigid Body

A *Eulerian* description is used to formulate the rigid body dynamics equations. The inertial reference frame is designated as $\{O_I, X_I Y_I Z_I\}$ and the body reference frame is designated as $\{O_n, X_n Y_n Z_n\}$ (Fig. 1). The latter is rigidly attached to a material point of the body. The orthogonal matrix expressing the orientation of $\{O_n, X_n Y_n Z_n\}$ relatively to $\{O_I, X_I Y_I Z_I\}$ is designated as $\mathbf{R}$, and is given by

$$\mathbf{R} = \begin{bmatrix} \mathbf{E}_1 \; \mathbf{E}_2 \; \mathbf{E}_3 \end{bmatrix} \tag{1}$$

where $\mathbf{E}_1$, $\mathbf{E}_2$ and $\mathbf{E}_3$ are the unit vectors along the axes of the body reference frame, expressed in the inertial reference frame. The position of a material point p of the body, relative to, and expressed in the body reference frame is defined as $\mathbf{X}_p$.

Due to the orthogonality of $\mathbf{R}$ the following is verified [5]

$$\mathbf{R}^T\mathbf{R} = \mathbf{I} \tag{2}$$

$$\mathbf{R}^T\dot{\mathbf{R}} = \widetilde{\boldsymbol{\omega}_n} = \begin{bmatrix} 0 & -\omega_{n3} & \omega_{n2} \\ \omega_{n3} & 0 & -\omega_{n1} \\ -\omega_{n2} & \omega_{n1} & 0 \end{bmatrix} \tag{3}$$

$$\mathbf{R}^T\ddot{\mathbf{R}} = \mathbf{\Lambda}_n = \frac{d}{dt}\left(\mathbf{R}^T\dot{\mathbf{R}}\right) - \dot{\mathbf{R}}^T\dot{\mathbf{R}} = \dot{\widetilde{\boldsymbol{\omega}_n}} + \widetilde{\boldsymbol{\omega}_n}\,\widetilde{\boldsymbol{\omega}_n} \tag{4}$$

$\boldsymbol{\omega}_n = vect(\widetilde{\boldsymbol{\omega}_n}) = \left[\omega_{n_1}\, \omega_{n_2}\, \omega_{n_3}\right]^T$ is the angular velocity vector of the body reference frame, relative to the inertial reference frame (absolute angular velocity), and expressed in the body reference frame. $\mathbf{\Lambda}_n$ is the absolute angular acceleration matrix of the body reference frame, and expressed in the body reference frame. The first part of $\mathbf{\Lambda}_n$ is skew symmetric and contains the transverse angular acceleration vector $\dot{\boldsymbol{\omega}}_n = vect(\dot{\widetilde{\boldsymbol{\omega}_n}})$, the second part is symmetric and contains the centripetal acceleration terms. The velocity vector, $\boldsymbol{\omega}_n$, may not be integrated with respect to time in order to obtain angular displacement since it is expressed in a non-inertial reference frame, i.e, it represents a quasi-coordinate [6, 7].

The absolute linear velocity and absolute linear acceleration of a material point p are given by

$$\mathbf{v}_p = \mathbf{v}_n + \widetilde{\boldsymbol{\omega}_n}\,\mathbf{X}_p \tag{5}$$

$$\mathbf{a}_p = \mathbf{a}_n + \dot{\widetilde{\boldsymbol{\omega}_n}}\,\mathbf{X}_p + \widetilde{\boldsymbol{\omega}_n}\,\widetilde{\boldsymbol{\omega}_n}\,\mathbf{X}_p \tag{6}$$

respectively. These are expressed in the body reference frame. $\mathbf{v}_n$ is also a quasi-coordinate and together with $\boldsymbol{\omega}_n$ forms the set of quasi-coordinates that describe the rigid body motion (*Eulerian* description). $\mathbf{a}_n = \dot{\mathbf{v}}_n + \widetilde{\boldsymbol{\omega}_n}\,\mathbf{v}_n$ is the absolute linear acceleration vector of the body reference frame.

The Newton-Euler equations of motion may now be deduced through Jourdain's Principle, also known as the Principle of Virtual Powers [8], which considering the above formulation is stated as

$$\int_{\mathcal{B}_n} \delta\mathbf{v}_p^T\,\mathbf{a}_p\,\rho_p\,d_{\mathcal{B}_n} = \int_{\mathcal{B}_n} \delta\mathbf{v}_p^T\,\mathbf{g}\,\rho_p\,d_{\mathcal{B}_n} + \int_{\Sigma\mathcal{B}_n} \delta\mathbf{v}_p^T\,\mathbf{f}_s\,d_{\Sigma\mathcal{B}_n} \tag{7}$$

The term on the left hand side represents the virtual power of the inertial force. The first term on the right hand side represents the virtual power of the gravitational force and the last term represents the virtual power of the external forces which are applied on the surface of the body. ρ_p represents the specific mass of the body at point p, $\mathbf{g}$ represents the gravitational acceleration vector and $\mathbf{f}_s$ represents the external force vector (force per unit of surface area) applied on the body surface $\Sigma\mathcal{B}_n$. All vectors in Eq. 7 are expressed in the body reference frame. The velocity variation is calculated by applying the variational operator, δ, to the expression of the absolute velocity vector (Eq. 5)

$$\delta\mathbf{v}_p = \delta\mathbf{v}_n + \delta\widetilde{\boldsymbol{\omega}_n}\,\mathbf{X}_p \tag{8}$$

Using Eq. 8 the external force term reduces to the effect of a punctual force $\mathbf{F}_s$ and punctual moment $\mathbf{M}_s$ applied at a specific point $\mathbf{X}_{ln}$

$$\begin{aligned}\int_{\Sigma\mathcal{B}_n} \delta\mathbf{v}_p^T\,\mathbf{f}_s\,d_{\Sigma\mathcal{B}_n} &= \delta\mathbf{v}_n^T \int_{\Sigma\mathcal{B}_n} \mathbf{f}_s\,d_{\Sigma\mathcal{B}_n} + \delta\boldsymbol{\omega}_n^T \int_{\Sigma\mathcal{B}_n} \widetilde{\mathbf{X}_p}\,\mathbf{f}_s\,d_{\Sigma\mathcal{B}_n} \\ &= \delta\mathbf{v}_n^T\,\mathbf{F}_s + \delta\boldsymbol{\omega}_n^T\left(\widetilde{\mathbf{X}_{ln}}\,\mathbf{F}_s + \mathbf{M}_s\right)\end{aligned} \tag{9}$$

Solving the other two terms, and collecting the coefficients of $\delta\mathbf{v}_n$ and of $\delta\boldsymbol{\omega}_n$ into a matrix form results in the Newton-Euler dynamics equations of a rigid body

$$\begin{bmatrix} m_b\mathbf{I} & m_b\widetilde{\mathbf{X}_{gn}}^T \\ m_b\widetilde{\mathbf{X}_{gn}} & \mathbf{J}_n \end{bmatrix} \begin{bmatrix} \mathbf{a}_n - \mathbf{g} \\ \dot{\boldsymbol{\omega}}_n \end{bmatrix} + \begin{bmatrix} m_b\,\widetilde{\boldsymbol{\omega}_n}\,\widetilde{\mathbf{X}_{gn}}^T\,\boldsymbol{\omega}_n \\ \widetilde{\boldsymbol{\omega}_n}\,\mathbf{J}_n\boldsymbol{\omega}_n \end{bmatrix} = \begin{bmatrix} \mathbf{F}_s \\ \widetilde{\mathbf{X}_{ln}}\,\mathbf{F}_s + \mathbf{M}_s \end{bmatrix} \tag{10}$$

The mass of the body, its center of mass and its second moment of inertia have been defined as

$$m_b = \int_{\mathcal{B}_n} \rho_p\,d_{\mathcal{B}_n} \tag{11}$$

$$\mathbf{X}_{gn} = \frac{1}{m_b}\int_{\mathcal{B}_n} \rho_p \mathbf{X}_p \, d_{\mathcal{B}_n} = \frac{1}{m_b}\int_{\mathcal{B}_n} \rho_p \left[X_1 \; X_2 \; X_3\right]^T d_{\mathcal{B}_n} \tag{12}$$

$$\mathbf{J}_n = \int_{\mathcal{B}_n} \rho_p \widetilde{\mathbf{X}_p} \widetilde{\mathbf{X}_p}^T d_{\mathcal{B}_n} = \int_{\mathcal{B}_n} \rho_p \begin{bmatrix} {X_2}^2 + {X_3}^2 & -X_1X_2 & -X_1X_3 \\ -X_2X_1 & {X_1}^2 + {X_3}^2 & -X_2X_3 \\ -X_3X_1 & -X_3X_2 & {X_1}^2 + {X_2}^2 \end{bmatrix} d_{\mathcal{B}_n} \tag{13}$$

respectively. The diagonal terms of the second moment of inertia tensor represent the principal moments of inertia of the rigid body, whereas the off-diagonal terms represent the products of inertia. If the body reference frame is chosen to be at the center of mass of the body, then $\mathbf{X}_{gn} = 0$ and Eq. 10 becomes decoupled. Furthermore, if the reference frame is oriented along the principal axes of inertia of the body then $\mathbf{J}_n$ becomes diagonal.

Assuming the case of a rigid body in a serial link manipulator (as in Figs. 4 and 5), the load term in Eq. 10 may be rewritten as

$$\begin{aligned} \begin{bmatrix} \mathbf{F}_s \\ \widetilde{\mathbf{X}_{ln}} \, \mathbf{F}_s + \mathbf{M}_s \end{bmatrix} &= \begin{bmatrix} \mathbf{F}_n \\ \mathbf{M}_n \end{bmatrix} - \begin{bmatrix} \mathbf{R}_{n+1/n} & 0 \\ \widetilde{\mathbf{r}_{n,n+1}} \mathbf{R}_{n+1/n} & \mathbf{R}_{n+1/n} \end{bmatrix} \begin{bmatrix} \mathbf{F}_{n+1} \\ \mathbf{M}_{n+1} \end{bmatrix} \\ &= \mathcal{F}_n - \mathbf{\Phi}_{rn+1,n}^T \mathcal{F}_{n+1} \end{aligned} \tag{14}$$

where $\mathcal{F}_n$ and $\mathcal{F}_{n+1}$ are the generalized force vectors applied on body $\mathcal{B}_n$ at O_n and O_{n+1} respectively. The former is expressed in $\{O_n, X_nY_nZ_n\}$ and the latter is expressed in $\{O_{n+1}, X_{n+1}Y_{n+1}Z_{n+1}\}$. $\mathbf{r}_{n,n+1}$ is the position vector from O_n to O_{n+1} expressed in $\{O_n, X_nY_nZ_n\}$, and $\mathbf{R}_{n+1/n}$ is the rotation matrix whose columns are the projections of the basis vector of reference frame $\{O_{n+1}, X_{n+1}Y_{n+1}Z_{n+1}\}$ on $\{O_n, X_nY_nZ_n\}$. In compact form one writes Eq. 10 as

$$\mathcal{M}_{rn}\mathbf{A}_n + \mathcal{N}_{rn}(\boldsymbol{\omega}_n) = \mathcal{F}_n - \mathbf{\Phi}_{rn+1,n}^T \mathcal{F}_{n+1} \tag{15}$$

3 Dynamics Equations of a Flexible Link

3.1 Kinematics

A *Eulerian* description is used for the rigid body motion and a *total Lagrangian* description is used for the deformation [9]. Similarly to the rigid body case, the inertial reference frame is designated as $\{O_I, X_IY_IZ_I\}$, and the body reference frame is designated as $\{O_n, X_nY_nZ_n\}$ (Fig. 2). The latter is rigidly attached to the first (base) cross section of the beam. It gives the orientation of the beam in space in its reference undeformed configuration designated as $\mathcal{B}_{n_o}$. The elastic motion of the beam is based on the following deformation assumptions:

1. plane beam cross sections before deformation, remain plane after deformation (warping is not considered),
2. bending and shear deformation are considered (Timoshenko beam theory),

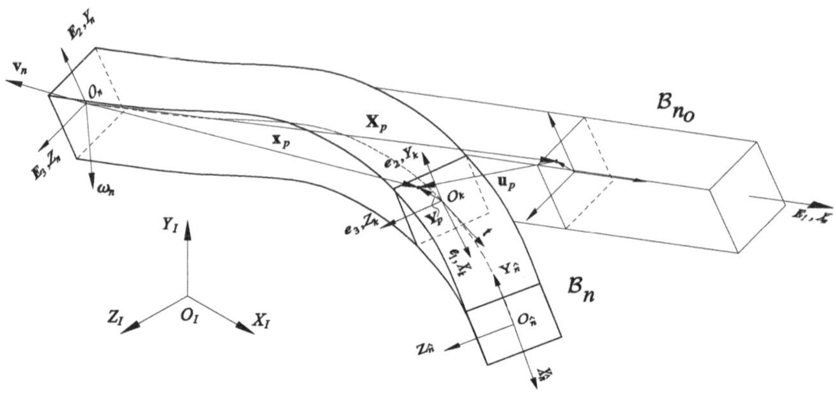

Fig. 2. Beam Kinematics

3. the beam neutral fiber does not suffer extension,
4. the beam neutral axis in the undeformed configuration is a straight line,
5. the beam cross sections are of constant specific mass, and are symmetrical relative to their principal axes,
6. and the shear strains and bending strains are considered to be small, that is, they are accounted for up to the first order, $O(1)$.

The kinematics of deformation is developed through the adoption of the cross section reference frame $\{O_k, X_k\, Y_k\, Z_k\}$. Its origin is placed at the point where the beam neutral axis intersects the beam cross section and its axes are directed along the axes of symmetry of the cross section. For the base cross section, reference frames $\{O_n, X_nY_nZ_n\}$ and $\{O_k, X_k\, Y_k\, Z_k\}$ coincide. For the last cross section (tip) the cross section reference frame is designated as $\{O_{\hat{n}}, X_{\hat{n}}Y_{\hat{n}}\, Z_{\hat{n}}\}$.

The orthogonal matrix expressing the orientation of $\{O_k, X_k\, Y_k\, Z_k\}$ relatively to, and expressed in the body reference frame is designated as $\mathbf{R}_{e_k}$. The position vector of a material point p of the beam belonging to a certain cross section, relatively to, and expressed in $\{O_k, X_k\, Y_k\, Z_k\}$ is defined as $\mathbf{Y}_p$. Due to the assumption of rigid cross sections the components of this vector remain constant during deformation. However, relatively to the moving reference frame, the position coordinates of point p change due to deformation. The reference position of point p is designated as $\mathbf{X}_p$ and the displaced position of point p is defined as $\mathbf{x}_p$. The displacement vector carrying point p from position $\mathbf{X}_p$ to position $\mathbf{x}_p$ is defined as $\mathbf{u}_p$. Vectors $\mathbf{X}_p$, $\mathbf{x}_p$ and $\mathbf{u}_p$ are all expressed in $\{O_n, X_nY_nZ_n\}$.

A material point in the beam neutral fibre is described by setting $\mathbf{Y}_p = [0\;0\;0]^T$. Accordingly, its reference position, displaced position and displacement vector become $\mathbf{X}_k = [X_1\;\;0\;\;0]^T$, $\mathbf{x}_k$ and $\mathbf{u}_k$ respectively. The vector tangent to the neutral fiber is given by

$$\mathbf{t} = \frac{d\mathbf{x}_k}{dX_1} \tag{16}$$

This vector, also of unit magnitude due to assumption 3, does not coincide with the vector perpendicular to the beam cross section due to shear deformation.

The basic deformation kinematic equation for a material point p of the flexible link is (*total Lagrangian* description).

$$\mathbf{x}_p = \mathbf{X}_k + \mathbf{u}_k + \mathbf{R}_{e_k}\,\mathbf{Y}_p \tag{17}$$

The terms that are due to the deformation of the link are the displacement vector of the material points of the neutral axis, $\mathbf{u}_k$, and the rotation matrix of a cross section $\mathbf{R}_{e_k}$.

3.2 Nonlinear Strain-Displacement Relations

The strain measures that are most commonly applied to the treatment of flexible beams are the Green-Lagrange strain measures [10] and the displacement gradient measure of deformation [5]. The former provide a general means to measure the deformation inside a 3-D continuum whereas the latter is a simplification specific to beam kinematics. In [5] it is shown that for the case of linear bending curvatures and linear shear angles these coincide. The displacement gradient measure of deformation may therefore be used. For the current deformation assumptions results

$$\mathbf{D}(\mathbf{X}_p) = \mathbf{\Gamma}_k + \widetilde{\mathbf{K}_k}\,\mathbf{Y}_p \tag{18}$$

where $\mathbf{\Gamma}_k$ is the vector of infinitesimal shear angles expressed in the cross section reference frame

$$\mathbf{\Gamma}_k = \begin{bmatrix} 0 \; \varphi_{12} \; \varphi_{13} \end{bmatrix}^T \tag{19}$$

and the skew symmetric matrix $\widetilde{\mathbf{K}_k}$ represents the curvature of the beam cross sections from which the curvature vector expressed in the cross section reference frame may be extracted, $\mathbf{K}_k = vect(\widetilde{\mathbf{K}_k}) = [K_1 \; K_2 \; K_3]^T$ with

$$\widetilde{\mathbf{K}_k} = \begin{bmatrix} 0 & -K_3 & K_2 \\ K_3 & 0 & -K_1 \\ -K_2 & K_1 & 0 \end{bmatrix} = \mathbf{R}_{e_k}^T \frac{d\mathbf{R}_{e_k}}{dX_1} \tag{20}$$

With the assumption of infinitesimal shear angles, the tangent vector to the neutral axis may be written in the cross section reference frame as

$$_k\mathbf{t} = \begin{bmatrix} 1 \; \varphi_{12} \; \varphi_{13} \end{bmatrix}^T \tag{21}$$

and in the moving reference frame as

$$\mathbf{t} = \frac{d\mathbf{x}_k}{dX_1} = \mathbf{R}_{e_k}\,{}_k\mathbf{t} \tag{22}$$

Noticing from Eq. 17 that

$$\frac{d\mathbf{x}_k}{dX_1} = \frac{d}{dX_1}(\mathbf{X}_k + \mathbf{u}_k) = \begin{bmatrix} 1 \\ 0 \\ 0 \end{bmatrix} + \frac{d\mathbf{u}_k}{dX_1} \tag{23}$$

then $\mathbf{u}_k$ can be calculated through integration

$$\mathbf{u}_k = \int_0^{X_1} \mathbf{R}_{e_k\,k}\mathbf{t}\, d\xi - \begin{bmatrix} X_1 \\ 0 \\ 0 \end{bmatrix} \tag{24}$$

Although the bending strains are infinitesimal, and the shear strains are infinitesimal, the strain displacement relations may be nonlinear if the rotation matrix $\mathbf{R}_{e_k}$ is expanded up to nonlinear terms in the kinematics Eq. 17. Eq. 20 yields that infinitesimal bending strains along the beam, $\mathbf{K}_k$, do not necessarily yield infinitesimal cross section rotations. Similarly to [11] and [8], expanding $\mathbf{R}_{e_k}$ and $\mathbf{K}_k$ in Taylor series in Eq. 20, and retaining only the first order term of the bending strains results

$$\underbrace{\frac{d\mathbf{R}_{e_{k1}}}{dX_1}}_{O(1)} + \underbrace{\frac{d\mathbf{R}_{e_{k2}}}{dX_1}}_{O(2)} + \cdots = (\underbrace{\mathbf{I}}_{O(0)} + \underbrace{\mathbf{R}_{e_{k1}}}_{O(1)} + \underbrace{\mathbf{R}_{e_{k2}}}_{O(2)} + \cdots)\underbrace{\widetilde{\mathbf{K}_{k1}}}_{O(1)} \tag{25}$$

where $\mathbf{I}$ is the identity matrix, $\mathbf{R}_{e_{ki}}$ is the term of order i of the rotation matrix $\mathbf{R}_{e_k}$ and $\mathbf{K}_{k1}$ is the first order term of the bending strains.

The rotation matrix can then be calculated up to order n, through integration of the same order terms in Eq. 25

$$\mathbf{R}_{e_k} = \mathbf{I} + \mathbf{R}_{e_{k1}} + \mathbf{R}_{e_{k2}} + \cdots + \mathbf{R}_{e_n} \tag{26}$$

where

$$\mathbf{R}_{e_i} = \int_0^{X_1} \mathbf{R}_{e_{i-1}}\widetilde{\mathbf{K}_{k1}}\, d\xi \tag{27}$$

In order to simplify the notation in the above parameterization of the rotation matrix, the following integral expressions are defined,

$$\upsilon_2 = \int_0^{X_1}\int_0^{\xi} K_3\, d\eta\, d\xi = \boldsymbol{\upsilon}_{2x}^T(X_1)\boldsymbol{\upsilon}_{2t}(t) \tag{28}$$

$$\upsilon_3 = \int_0^{X_1}\int_0^{\xi} -K_2\, d\eta\, d\xi = \boldsymbol{\upsilon}_{3x}^T(X_1)\boldsymbol{\upsilon}_{3t}(t) \tag{29}$$

$$\alpha = \int_0^{X_1} K_1\, d\xi = \boldsymbol{\alpha}_x^T(X_1)\boldsymbol{\alpha}_t(t) \tag{30}$$

$$\gamma_2 = \int_0^{X_1} \varphi_{12}\, d\xi = \boldsymbol{\gamma}_{2x}^T(X_1)\boldsymbol{\gamma}_{2t}(t) \tag{31}$$

$$\gamma_3 = \int_0^{X_1} \varphi_{13}\, d\xi = \boldsymbol{\gamma}_{3x}^T(X_1)\boldsymbol{\gamma}_{3t}(t) \tag{32}$$

where υ_2 and υ_3 represent pure bending deflections, α represents a pure torsion angle, and γ_2 and γ_3 represent pure shear deflections. The Rayleigh-Ritz approximation of these variables is written in the right hand side of Eqs.28-32. $\boldsymbol{\upsilon}_{2x}$, $\boldsymbol{\upsilon}_{3x}$, $\boldsymbol{\alpha}_x$, $\boldsymbol{\gamma}_{2x}$ and $\boldsymbol{\gamma}_{3x}$ are the vectors of shape functions for the elastic deflections, and $\boldsymbol{\upsilon}_{2t}$, $\boldsymbol{\upsilon}_{3t}$, $\boldsymbol{\alpha}_t$, $\boldsymbol{\gamma}_{2t}$ and $\boldsymbol{\gamma}_{3t}$ are the corresponding vectors of modal coordinates. For a chosen order of the rotation matrix in Equation 26 the displacement vector is calculated according to Eq. 24 considering the above discretization. The vector of elastic coordinates is defined as

$$\mathbf{q}_{e_n} = \left[\mathbf{q}_{K_n}^T \,\vdots\, \mathbf{q}_{\varphi_n}^T\right]^T = \left[\boldsymbol{v}_{2t}^T \;\; \boldsymbol{v}_{3t}^T \;\; \boldsymbol{\alpha}_t^T \,\vdots\, \boldsymbol{\gamma}_{2t}^T \;\; \boldsymbol{\gamma}_{3t}^T\right]^T \tag{33}$$

3.3 The Equations of Motion

Due to the orthogonality of $\mathbf{R}_{e_k}$, the following is verified [5]

$$\mathbf{R}_{e_k}\mathbf{R}_{e_k}^T = \mathbf{I} \tag{34}$$

$$\dot{\mathbf{R}}_{e_k}\mathbf{R}_{e_k}^T = \widetilde{\boldsymbol{\Omega}_k} \tag{35}$$

$$\ddot{\mathbf{R}}_{e_k}\mathbf{R}_{e_k}^T = \mathbf{A}_k = \frac{d}{dt}\left(\dot{\mathbf{R}}_{e_k}\mathbf{R}_{e_k}^T\right) - \dot{\mathbf{R}}_{e_k}\dot{\mathbf{R}}_{e_k}^T = \dot{\widetilde{\boldsymbol{\Omega}_k}} + \widetilde{\boldsymbol{\Omega}_k}\,\widetilde{\boldsymbol{\Omega}_k} \tag{36}$$

where $\boldsymbol{\Omega}_k = vect(\widetilde{\boldsymbol{\Omega}_k}) = [\Omega_{k_1}\; \Omega_{k_2}\; \Omega_{k_3}]^T$ is the angular velocity vector of the cross section reference frame, $\{O_k, X_k\, Y_k\, Z_k\}$, relative to and expressed in the body reference frame (relative angular velocity). This is the angular velocity due to deformation alone. $\mathbf{A}_k$ is the relative angular acceleration matrix of the cross section reference frame, also expressed in the body reference frame. The first part of $\mathbf{A}_k$ is skew symmetric and contains the angular acceleration vector $\dot{\boldsymbol{\Omega}}_k = vect(\dot{\widetilde{\boldsymbol{\Omega}_k}})$, the second part is symmetric and contains the centrifugal acceleration terms. Contrarily to the rigid body description, these vectors may be integrated in time to give the rotation of the beam cross sections due to deformation (*total Lagrangian* description).

After spatial discretization, the first and second time derivative of the displacement vector in Eq. 24 may be written as

$$\dot{\mathbf{u}}_k = \mathbf{J}_{T_k}\,\dot{\mathbf{q}}_{e_n} = \left[\mathbf{J}_{TK_k}\;\; \mathbf{J}_{T\varphi_k}\right]\begin{bmatrix}\dot{\mathbf{q}}_{K_n}\\ \dot{\mathbf{q}}_{\varphi_n}\end{bmatrix} \tag{37}$$

$$\ddot{\mathbf{u}}_k = \dot{\mathbf{J}}_{TK_k}\,\dot{\mathbf{q}}_{K_n} + \dot{\mathbf{J}}_{T\varphi_k}\,\dot{\mathbf{q}}_{\varphi_n} + \mathbf{J}_{TK_k}\,\ddot{\mathbf{q}}_{K_n} + \mathbf{J}_{T\varphi_k}\,\ddot{\mathbf{q}}_{\varphi_n} \tag{38}$$

where $\mathbf{J}_{T_k}$ is the translation Jacobian. Similarly, the angular velocity due to deformation of a cross section, written in the body reference frame, may be written as

$$\boldsymbol{\Omega}_k = vect(\widetilde{\boldsymbol{\Omega}_k}) = \mathbf{J}_{\mathbf{R}_{e_k}}\,\dot{\mathbf{q}}_{K_n} \tag{39}$$

and its first time derivative becomes

$$\dot{\boldsymbol{\Omega}}_k = \dot{\mathbf{J}}_{\mathbf{R}_{e_k}}\,\dot{\mathbf{q}}_{K_n} + \mathbf{J}_{\mathbf{R}_{e_k}}\,\ddot{\mathbf{q}}_{K_n} \tag{40}$$

where $\mathbf{J}_{\mathbf{R}_{e_k}}$ is the rotation Jacobian

Due to the kinematic assumptions for the beam deformation, the motion of a material point may be dealt with in a more practical form where linear and angular motion of a cross section are separated [9, 11]. The absolute linear and angular velocity vectors of the body reference frame are given by $\mathbf{v}_n$ and $\boldsymbol{\omega}_n$ as in the rigid body case. The absolute linear velocity vector of a material point, $\mathbf{v}_p$, expressed in the body reference frame becomes

$$\mathbf{v}_p = \mathbf{v}_k + \widetilde{\boldsymbol{\omega}_k}\,\mathbf{R}_{e_k}\mathbf{Y}_p \tag{41}$$

where $\mathbf{v}_k$ is the absolute linear velocity vector of the cross section and $\boldsymbol{\omega}_k$ is its absolute angular velocity vector,

$$\mathbf{v}_k = \mathbf{v}_n + \widetilde{\boldsymbol{\omega}_n}\,(\mathbf{X}_k + \mathbf{u}_k) + \dot{\mathbf{u}}_k \tag{42}$$

$$\boldsymbol{\omega}_k = \boldsymbol{\omega}_n + \boldsymbol{\Omega}_k \tag{43}$$

In a similar fashion, the absolute linear acceleration vector of a material point, $\mathbf{a}_p$, may be written as

$$\mathbf{a}_p = \mathbf{a}_k + \boldsymbol{\Lambda}_k\,\mathbf{R}_{e_k}\mathbf{Y}_p \tag{44}$$

where $\mathbf{a}_k$ is the absolute linear acceleration vector of the cross section and $\boldsymbol{\Lambda}_k$ is the absolute angular acceleration matrix,

$$\mathbf{a}_k = \mathbf{a}_n + \left(\dot{\widetilde{\boldsymbol{\omega}_n}} + \widetilde{\boldsymbol{\omega}_n}\,\widetilde{\boldsymbol{\omega}_n}\right)(\mathbf{X}_k + \mathbf{u}_k) + 2\widetilde{\boldsymbol{\omega}_n}\,\dot{\mathbf{u}}_k + \ddot{\mathbf{u}}_k \tag{45}$$

$$\boldsymbol{\Lambda}_k = \left(\dot{\widetilde{\boldsymbol{\omega}_n}} + \dot{\widetilde{\boldsymbol{\Omega}_k}} + \widetilde{\boldsymbol{\omega}_n\boldsymbol{\Omega}_k}\right) + \left(\widetilde{\boldsymbol{\omega}_n} + \widetilde{\boldsymbol{\Omega}_k}\right)^2 \tag{46}$$

with $\mathbf{a}_n = \dot{\mathbf{v}}_n + \widetilde{\boldsymbol{\omega}_n}\,\mathbf{v}_n$.

Jourdain's Principle may now be stated as follows:

$$\int_{\mathcal{B}_{n_o}} \delta\mathbf{v}_p^T\,\mathbf{a}_p\,\rho_p\,d\mathcal{B}_{n_o} + \int_{\mathcal{B}_{n_o}} \delta\dot{\mathbf{D}}^T\,\boldsymbol{\sigma}\,d\mathcal{B}_{n_o} = \int_{\mathcal{B}_{n_o}} \delta\mathbf{v}_p^T\,\mathbf{g}\,\rho_p\,d\mathcal{B}_{n_o} + \int_{\Sigma\mathcal{B}_{n_o}} \delta\mathbf{v}_p^T\,\mathbf{f}_s\,d\Sigma\mathcal{B}_{n_o} \tag{47}$$

The first term on the left hand side of this equation represents the virtual power of the inertial force and the second term represents the virtual power of the internal elastic force. The first term on the right hand side represents the virtual power of the gravity force and the last term represents the virtual power of the external forces which are applied on the boundary of the beam. In this equation, ρ_p represents the specific mass of the beam at point p in the reference configuration, vector $\boldsymbol{\sigma} = [\sigma_{11}\,\sigma_{12}\,\sigma_{13}]^T$ contains the Piola-Kirchoff stresses acting on a cross section, $\mathbf{g}$ represents the gravitational acceleration vector and $\mathbf{f}_s$ represents the external force vector (force per unit of undeformed surface area) applied on the beam. All vectors in Eq. 47 are expressed in the body reference frame with the exception of the internal elastic term which vectors are expressed in the cross section reference frame. It is assumed that the material law is linear and obeys Hook's law.

The variational operator, δ, when applied to the velocity expression yields

$$\begin{aligned}\delta\mathbf{v}_p &= \delta\mathbf{v}_k + \delta\widetilde{\boldsymbol{\omega}_k}\,\mathbf{R}_{e_k}\mathbf{Y}_p \\ &= \delta\mathbf{v}_n + \delta\widetilde{\boldsymbol{\omega}_n}\,(\mathbf{X}_k + \mathbf{u}_k) + \delta\dot{\mathbf{u}}_k + \left(\delta\widetilde{\boldsymbol{\omega}_n} + \delta\widetilde{\boldsymbol{\Omega}_k}\right)\mathbf{R}_{e_k}\mathbf{Y}_p\end{aligned} \tag{48}$$

with

$$\delta \dot{\mathbf{u}}_k = \mathbf{J}_{T_k}\, \delta \dot{\mathbf{q}}_{e_n} = \mathbf{J}_{TK_k}\, \delta \dot{\mathbf{q}}_{K_n} + \mathbf{J}_{T\varphi_k}\, \delta \dot{\mathbf{q}}_{\varphi_n} \tag{49}$$

and

$$\delta \mathbf{\Omega}_k = \mathbf{J}_{\mathbf{R}_{e_k}}\, \delta \dot{\mathbf{q}}_{K_n} \tag{50}$$

And when applied to the time derivative of Eq. 18 yields

$$\delta \dot{\mathbf{D}} = \delta \dot{\mathbf{\Gamma}}_k + \delta \dot{\widetilde{\mathbf{K}_k}} \mathbf{Y}_p \tag{51}$$

3.3.1 The External Force Term

The external force term may be separated into three contributions: the loads on the first cross section, the loads on the tip cross section and the loads on the edges of the in between cross sections.

$$\begin{aligned}&\int_{\Sigma\mathcal{B}_{n_o}} \delta \mathbf{v}_p^T\, \mathbf{f}_s\, d_{\Sigma\mathcal{B}_{n_o}} \\ &= \int_{A_n} \delta \mathbf{v}_p^T\, \mathbf{f}_n\, d_{A_n} + \int_{\overline{\Sigma\mathcal{B}_{n_o}}} \delta \mathbf{v}_p^T\, \mathbf{f}_{\overline{\Sigma\mathcal{B}_{n_o}}}\, d_{\overline{\Sigma\mathcal{B}_{n_o}}} + \int_{A_{\hat{n}}} \delta \mathbf{v}_p^T\, \mathbf{f}_{\hat{n}}\, d_{A_{\hat{n}}}\end{aligned} \tag{52}$$

where A_n represents the first cross section, $A_{\hat{n}}$ represents the last cross section and $\overline{\Sigma\mathcal{B}_{n_o}}$ represents the boundary of the beam excluding the first and last cross sections, i.e, the lateral faces of the beam excluding the edges of the first and last cross section. In a typical application, the first and last cross sections would be attached to the neighboring bodies, and the lateral faces of the beam may have forces applied on them either from external disturbances or for control purposes. The latter may be achieved for example through the use of control-moment-gyros or proof-mass actuators [12].

In a robotic manipulator, the last cross section of a flexible beam is typically connected to another body, either rigid or another flexible beam. Therefore, reference frames $\{O_{\hat{n}}, X_{\hat{n}} Y_{\hat{n}} Z_{\hat{n}}\}$ and $\{O_{n+1}, X_{n+1} Y_{n+1} Z_{n+1}\}$ are placed coincident. Using Eq. 48 in Eq. 52 under these assumptions yields

$$\begin{aligned}&\begin{bmatrix} \delta \mathbf{v}_n \\ \delta \boldsymbol{\omega}_n \\ \delta \dot{\mathbf{q}}_{K_n} \\ \delta \dot{\mathbf{q}}_{\varphi_n} \end{bmatrix}^T \left(\begin{bmatrix} \mathbf{I} & 0 \\ 0 & \mathbf{I} \\ \hline 0 & 0 \\ 0 & 0 \end{bmatrix} \begin{bmatrix} \mathbf{F}_n \\ \mathbf{M}_n \end{bmatrix} + \int_{\bar{0}}^{\bar{L}} \begin{bmatrix} \mathbf{I} & 0 \\ \widetilde{\mathbf{x}_k} & \mathbf{I} \\ \mathbf{J}_{TK_k}^T & \mathbf{J}_{\mathbf{R}_{e_k}}^T \\ \mathbf{J}_{T\varphi_k}^T & 0 \end{bmatrix} \begin{bmatrix} \mathbf{F}_{\overline{\Sigma\mathcal{B}_{n_o}}} \\ \mathbf{M}_{\overline{\Sigma\mathcal{B}_{n_o}}} \end{bmatrix} dX_1 \right. \\ &\qquad \left. - \begin{bmatrix} \mathbf{R}_{e_{\hat{n}}} & 0 \\ \widetilde{\mathbf{x}_{\hat{n}}}\, \mathbf{R}_{e_{\hat{n}}} & \mathbf{R}_{e_{\hat{n}}} \\ \hline \mathbf{J}_{TK_{\hat{n}}}^T \mathbf{R}_{e_{\hat{n}}} & \mathbf{J}_{\mathbf{R}_{e_{\hat{n}}}}^T \mathbf{R}_{e_{\hat{n}}} \\ \mathbf{J}_{T\varphi_{\hat{n}}}^T \mathbf{R}_{e_{\hat{n}}} & 0 \end{bmatrix} \begin{bmatrix} \mathbf{F}_{n+1} \\ \mathbf{M}_{n+1} \end{bmatrix} \right) \\ &= \begin{bmatrix} \delta \mathbf{v}_n \\ \delta \boldsymbol{\omega}_n \\ \delta \dot{\mathbf{q}}_{K_n} \\ \delta \dot{\mathbf{q}}_{\varphi_n} \end{bmatrix}^T \left(\begin{bmatrix} \mathbf{I} \\ \hline 0 \end{bmatrix} \mathcal{F}_n + \mathcal{F}_{\overline{\Sigma\mathcal{B}_{n_o}}} - \begin{bmatrix} \mathbf{\Phi}_{r\,n+1,n}^T \\ \hline \mathbf{\Phi}_{e\,n+1,n}^T \end{bmatrix} \mathcal{F}_{n+1} \right)\end{aligned} \tag{53}$$

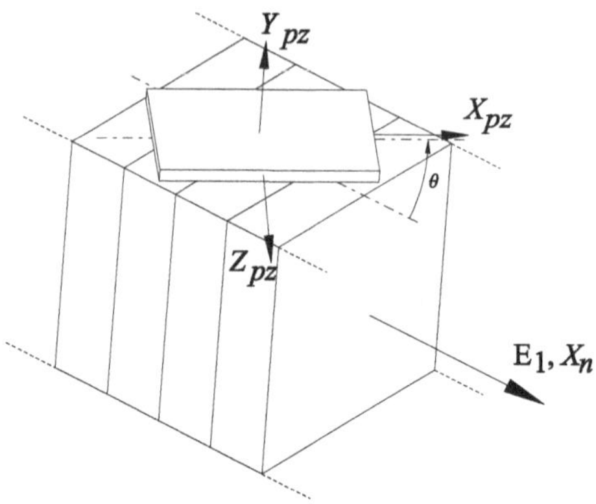

Fig. 3. A piezopatch bonded on a lateral face of a beam

3.3.2 The Piezoelectric Actuator

In this section, laminar shape actuators are considered. It is assumed that the piezoelectric patch is rigidly bonded to the lateral faces of the beam, and that it is sufficiently thin when compared to the beam in order for the strain distribution to be considered constant throughout the thickness of the patch [13] (Fig. 3). The deformation problem is therefore a plane stress problem. The patch design, consisting of the geometry and properties of the piezoelectic material and of the electrodes, that yields the plane stress deformation problem for the piezoelectric patch is left generic. Such can be achieved for example through the traditional design that takes advantage of the in plane actuation mode, or through more complex designs that exploit the d_{33} piezoelectric coupling term in a laminar patch [14].

The generation of longitudinal and shear stresses on the beam will depend mainly on the geometric distribution of the piezoelectric patches on the lateral faces of the beam. For an effective design, this distribution should favor the deformation and geometric assumptions introduced in Sec. 3.1. Specifically, patches distributed as to generate bending, torsion and shear, should induce zero strain on the longitudinal axis, due to the assumption of inextensibility of the longitudinal axis of the beam, and should also not alter the symmetry of the cross sections.

A generic expression for a laminar piezoactuator is of the form

$$\boldsymbol{\sigma}_{pz} = \mathbf{E}_{pz}\, \boldsymbol{\varepsilon}_{pz} - \mathbf{d}_{pz}\, \mathbf{V}_{pz} \tag{54}$$

where $\boldsymbol{\sigma}_{pz}$ is the vector of stresses in the piezomaterial, $\mathbf{E}_{pz}$ is a matrix containing Young's modulus and Poisson's ratio, $\boldsymbol{\varepsilon}_{pz}$ is the corresponding vector of strains, $\mathbf{d}_{pz}$ is a matrix containing Young's modulus, Poisson's ratio and the piezoelectric coupling terms, and finally, $\mathbf{V}_{pz}$ is the vector containing the applied voltage that creates the electric field. The electric field is equal to the applied voltage devided by the distance between the electrodes.

The stresses and strains in the piezoelectric material are transformed into stresses and strains of the beam through a transformation into a plane parallel to the beam cross section. These transformations are calculated according to the placing of the piezopatch relative to the beam (they depend on the orientation angle θ), and also on which face of the beam the patch is bonded. The transformation matrices are defined as $\mathbf{K}_\sigma$ and $\mathbf{K}_\varepsilon$ for the stress and strain respectively and are written as

$$\boldsymbol{\sigma} = \mathbf{K}_\sigma \, \boldsymbol{\sigma}_{pz} \tag{55}$$

and

$$\boldsymbol{\varepsilon}_{pz} = \mathbf{K}_\varepsilon \, \mathbf{D} \tag{56}$$

where $\boldsymbol{\sigma}$ is the vector of Piola-Kirschoff stresses, and $\mathbf{D}$ are the strains from Eq. 18. $\mathbf{K}_\sigma$ is a 3×2 matrix and $\mathbf{K}_\varepsilon$ is a 2×3 matrix. Applying these relations to Eq. 54 yields

$$\boldsymbol{\sigma} = \mathbf{K}_\sigma \, \mathbf{E}_{pz} \, \mathbf{K}_\varepsilon \, \mathbf{D} - \mathbf{K}_\sigma \, \mathbf{d}_{pz} \, \mathbf{V}_{pz} \tag{57}$$

The effect of the piezomaterial on a flexible link is accounted for by calculating the virtual power of the elastic forces inside the piezoelectric patch similarly to the calculation of the virtual power of the beam elastic forces; integration being performed over the volume of the piezoelectric material in the undeformed configuration,

$$\begin{aligned}\int_{\mathcal{B}_{pz}} \delta\dot{\mathbf{D}}^T \, \boldsymbol{\sigma} \, d_{\mathcal{B}_{pz}} = \int_{\mathcal{B}_{pz}} & \left(\delta\dot{\mathbf{\Gamma}}_k + \delta\dot{\widetilde{\mathbf{K}}}_k \, \mathbf{Y}_p\right)^T \mathbf{K}_\sigma \, \mathbf{E}_{pz} \, \mathbf{K}_\varepsilon \left(\mathbf{\Gamma}_k + \widetilde{\mathbf{K}}_k \mathbf{Y}_p\right) \\ & - \left(\delta\dot{\mathbf{\Gamma}}_k + \delta\dot{\widetilde{\mathbf{K}}}_k \, \mathbf{Y}_p\right)^T \mathbf{K}_\sigma \, \mathbf{d}_{pz} \, \mathbf{V}_{pz} \, d_{\mathcal{B}_{pz}}\end{aligned} \tag{58}$$

The first term of the right hand side of this equation represents the structural effect of the added material on the beam. It expresses an increase in the stiffness of the overall structure. The inertial effect of the piezopatch on the other hand, is included by making the adequate provisions in the inertial term, i.e, the piezomaterial specific mass and its inertia tensor must be considered. If acceptable, both the structural stiffness and inertia effect term may be neglected due to the small thickness of the piezoelectric material. The last term of the above equation represents the contribution of the electric field to the deformation of the structure. This is the manipulated variable that is used for the control of the elastic deformation. Its effect is that of actively changing the local strain in the beam surface by inducing stresses in the piezoelectric material.

The control term of Eq. 58 may be further simplified,

$$\begin{aligned}&\int_{\mathcal{B}_{pz}} - \left(\delta\dot{\mathbf{\Gamma}}_k + \delta\dot{\widetilde{\mathbf{K}}}_k \, \mathbf{Y}_p\right)^T \mathbf{K}_\sigma \, \mathbf{d}_{pz} \, \mathbf{V}_{pz} \, d_{\mathcal{B}_{pz}} \\ &= \int_a^b \delta\dot{\mathbf{\Gamma}}_k{}^T \underbrace{\int_{A_{pz}} -\mathbf{K}_\sigma \, \mathbf{d}_{pz} \, \mathbf{V}_{pz} \, d_{A_{pz}}}_{\mathbf{F}_{pz} = [0 \; F_{pz_2} \; F_{pz_3}]^T} + \delta\dot{\mathbf{K}}_k{}^T \underbrace{\int_{A_{pz}} \widetilde{\mathbf{Y}}_p{}^T \, \mathbf{K}_\sigma \, \mathbf{d}_{pz} \, \mathbf{V}_{pz} \, d_{A_{pz}}}_{\mathbf{M}_{pz} = [M_{pz_1} \; M_{pz_2} \; M_{pz_3}]^T} \, dX_1\end{aligned} \tag{59}$$

where a and b are the beginning and ending point along X_1 of the piezomaterial, and A_{pz} is the area of the piezoactuator in the plane of the beam cross sections. The first term of the above equation represents the virtual power of the induced shear and the second term represents the virtual power of the induced curvature.

The resulting discretized dynamics equation of the flexible beam is obtained by solving Eq. 47 and collecting into a matrix system the coefficients of $\delta\mathbf{v}_n$, $\delta\boldsymbol{\omega}_n$, $\delta\dot{\mathbf{q}}_{K_n}$ and $\delta\dot{\mathbf{q}}_{\varphi n}$

$$\begin{bmatrix} \mathcal{M}_{en,rr} & \mathcal{M}_{en,re} \\ \mathcal{M}_{en,re}^T & \mathcal{M}_{en,ee} \end{bmatrix} \begin{bmatrix} \mathbf{A}_n \\ \ddot{\mathbf{q}}_{e_n} \end{bmatrix} + \mathcal{N}_{en}\left(\mathbf{v}_n, \boldsymbol{\omega}_n, \dot{\mathbf{q}}_{e_n}\right) + \mathcal{K}_{en}\left(\mathbf{q}_{e_n}\right)$$
$$= \begin{bmatrix} \mathbf{I} \\ 0 \end{bmatrix} \mathcal{F}_n + \mathcal{F}_{\Sigma\overline{\mathcal{B}_{n_o}}} + \mathcal{F}_{pz} - \begin{bmatrix} \boldsymbol{\Phi}_{rn+1,n}^T \\ \boldsymbol{\Phi}_{en+1,n}^T \end{bmatrix} \mathcal{F}_{n+1} \tag{60}$$

where $\mathcal{M}_{en}$ is the mass matrix of the flexible beam, $\mathcal{N}_{en}$ is the generalized nonlinear inertial force vector, $\mathcal{K}_{en}$ is the linear generalized elastic force vector, $\mathcal{F}_{\Sigma\overline{\mathcal{B}_{n_o}}}$ is the generalized force vector of external forces applied on the sides of the beam, and $\mathcal{F}_{pz}$ is the generalized force vector due to the piezoelectric material. The piezoelectric loading vector is given by

$$\mathcal{F}_{pz} = \int_a^b \begin{bmatrix} 0 \\ 0 \\ \boldsymbol{\upsilon}_{2x}'' M_{pz_3} \\ -\boldsymbol{\upsilon}_{3x}'' M_{pz_2} \\ \boldsymbol{\alpha}_x' M_{pz_1} \\ \boldsymbol{\gamma}_{2x}' F_{pz_2} \\ \boldsymbol{\gamma}_{3x}' F_{pz_3} \end{bmatrix} dX_1 \tag{61}$$

where $(.)' = \frac{\partial}{\partial X_1}(.)$, and the linear elastic force vector is given by

$$\mathcal{K}_{en} = \int_0^L \begin{bmatrix} 0 \\ 0 \\ EI_3 \boldsymbol{\upsilon}_{2x}'' \boldsymbol{\upsilon}_{2x}''^T \boldsymbol{\upsilon}_{2t} \\ EI_2 \boldsymbol{\upsilon}_{3x}'' \boldsymbol{\upsilon}_{3x}''^T \boldsymbol{\upsilon}_{3t} \\ GJ \boldsymbol{\alpha}_x' \boldsymbol{\alpha}_x'^T \boldsymbol{\alpha}_t \\ GA \boldsymbol{\gamma}_{2x}' \boldsymbol{\gamma}_{2x}'^T \boldsymbol{\gamma}_{2t} \\ GA \boldsymbol{\gamma}_{3x}' \boldsymbol{\gamma}_{3x}'^T \boldsymbol{\gamma}_{3t} \end{bmatrix} dX_1 \tag{62}$$

To optimize the control authority of the piezoactuator on the structure, $\mathcal{F}_{pz}$ should be analyzed. To this end, the control moments M_{pz_1} M_{pz_2} and M_{pz_3}, and the control forces F_{pz_2} and F_{pz_3} due to the piezoactuator between point a and point b may also be separated into a product of a spatial function vector multiplied by a time function vector (Rayleigh-Ritz separation). The spatial term is a function of the geometry (width and thickness) of the piezopatch between point a and point b and the time varying term is a function of the applied voltage. Typically the shape function would have the same form as the associated curvature function. Similarly, the problem of optimum placement of small patches may be tackled by setting the integral between point a and b to between the base of the beam and its tip. The effectiveness of the

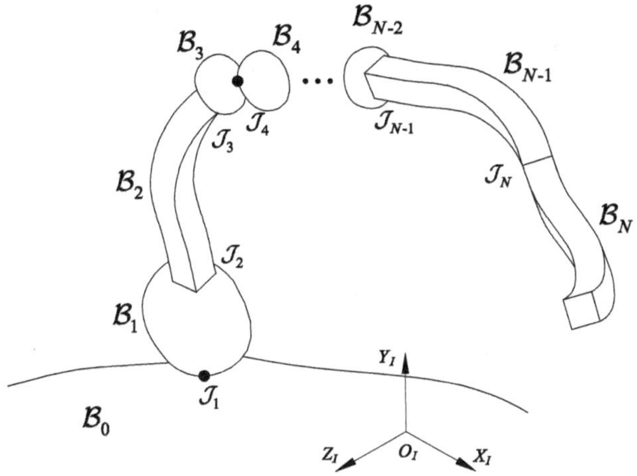

Fig. 4. Rigid-flexible multi-link serial manipulator

piezocontrol on the beam will ultimately be a function of the difference between the elastic force vector $\mathcal{K}_{en}$ and the piezoactuator control vector $\mathcal{F}_{pz}$.

In order to obtain a dynamics equation for the elastic beam with the same number of terms as the dynamics equation of a rigid body, Eq. 60 may be rewritten in the form

$$\begin{bmatrix} \mathcal{M}_{en,rr} & \mathcal{M}_{en,re} \\ \mathcal{M}_{en,re}^T & \mathcal{M}_{en,ee} \end{bmatrix} \begin{bmatrix} \mathbf{A}_n \\ \ddot{\mathbf{q}}_{e_n} \end{bmatrix} + \begin{bmatrix} \overline{\mathcal{N}}_{en,r} \\ \overline{\mathcal{N}}_{en,e} \end{bmatrix} \left(\mathbf{v}_n, \omega_n, \dot{\mathbf{q}}_{e_n}, \mathcal{K}_{en}, \mathcal{F}_{\Sigma \overline{\mathcal{B}_{n_o}}}, \mathcal{F}_{pz} \right)$$
$$= \begin{bmatrix} \mathbf{I} \\ 0 \end{bmatrix} \mathcal{F}_n - \begin{bmatrix} \mathbf{\Phi}_{r n+1,n}^T \\ \mathbf{\Phi}_{e n+1,n}^T \end{bmatrix} \mathcal{F}_{n+1} \tag{63}$$

This allows a simpler analysis for the multibody case.

4 Dynamics of a Multi-Link Manipulator

A flexible manipulator consists of rigid bodies and flexible beams connected by joints in a serial chain topology (Fig. 4). The body at one of the extremities of the chain is designated as the base body $\mathcal{B}_1$, and the body at the other end of the chain is designated as the tip body $\mathcal{B}_N$. The in-between bodies of the chain are connected to only two joints. Body zero, $\mathcal{B}_0$, is the designation reserved for the inertial body.

The joints are numbered in a similar fashion as the bodies. For the nth body $\mathcal{B}_n$, the inboard joint is designated as $\mathcal{J}_n$, and the outboard joint is designated as $\mathcal{J}_{n+1}$. The designation of inboard (outboard) refers to the topological position of the body or joint in the chain which is closer (further) from the inertial body. The first joint, $\mathcal{J}_1$, is the joint connecting the chain to the inertial body, $\mathcal{B}_0$.

The joints may have from zero to six degrees of freedom. If the multibody system is floating in space for example, then $\mathcal{J}_1$ is set to allow the six degrees of freedom. Similarly, if the multibody system consists of a mobile robot to which a flexible

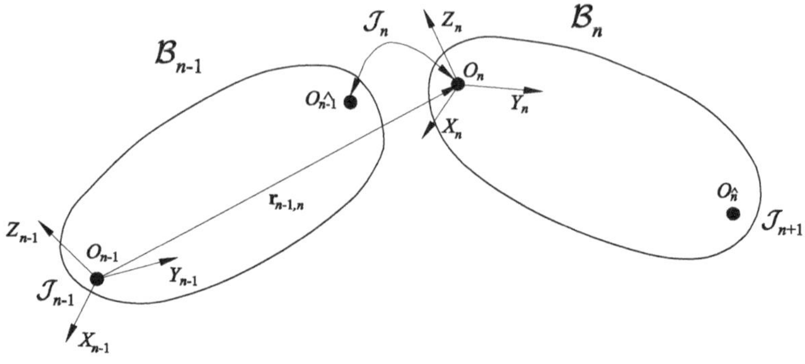

Fig. 5. Joint kinematics

manipulator is attached, then $\mathcal{J}_1$ must possess translation degrees of freedom. For the typical case of a multibody system with a fixed base as in industrial manipulators, then $\mathcal{J}_1$ only possesses rotational degrees of freedom.

A joint with zero degrees of freedom serves the purpose of rigidly connecting flexible beams with flexible beams, as in $\mathcal{J}_N$ of Fig. 4 for example, and flexible beams with rigid bodies, as in $\mathcal{J}_2$, $\mathcal{J}_3$ and $\mathcal{J}_{N-1}$ for example. By rigidly connecting flexible beams, flexible bodies of greater complexity may be modelled; higher displacements may be achieved and the complexity of the shape functions reduced, by assuming a flexible body composed of several flexible beams connected base to tip. The dimension of the model however, is increased.

4.1 Joint kinematics

Joint $\mathcal{J}_n$ connecting bodies $\mathcal{B}_{n-1}$ and $\mathcal{B}_n$ is represented in Fig. 5. We use a joint representation similar to that of [15], but with the increasing body numbering from base to tip as in [16]. The reference position of $\mathcal{J}_n$ on $\mathcal{B}_{n-1}$ is designated as the inboard point of $\mathcal{J}_n$, $O_{\widehat{n-1}}$. Similarly, the reference position of $\mathcal{J}_n$ on $\mathcal{B}_n$ is designated as the outboard point of $\mathcal{J}_n$, O_n, the origin of body reference frame $\{O_n, X_nY_nZ_n\}$. The interbody position vector, which describes the position of O_n relative to O_{n-1} expressed in $\{O_{n-1}, X_{n-1}Y_{n-1}Z_{n-1}\}$, is expressed as $\mathbf{r}_{n-1,n}$.

The rotation of $\mathcal{B}_n$ relative to $\mathcal{B}_{n-1}$ is defined though the orthogonal rotation matrix $\mathbf{R}_{n/n-1}$. The columns of $\mathbf{R}_{n/n-1}$ are the projections of the basis vectors of reference frame $\{O_n, X_nY_nZ_n\}$ on $\{O_{n-1}, X_{n-1}Y_{n-1}Z_{n-1}\}$. Similarly to Eq. 3, the angular velocity of $\mathcal{B}_n$ relative to $\mathcal{B}_{n-1}$ expressed in $\{O_n, X_nY_nZ_n\}$ is obtained from

$$\widetilde{\boldsymbol{\omega}_{n/n-1}} = \mathbf{R}^T_{n/n-1}\dot{\mathbf{R}}_{n/n-1} \tag{64}$$

The absolute angular velocity of body $\mathcal{B}_n$ expressed in $\{O_n, X_nY_nZ_n\}$ may then be written as

$$\boldsymbol{\omega}_n = \mathbf{R}^T_{n/n-1}\boldsymbol{\omega}_{n-1} + \boldsymbol{\omega}_{n/n-1} \tag{65}$$

The relative angular velocity between the two rigid bodies may be expressed in terms of the time derivative of Euler angles [17] (the case of typical robotic manipulators), therefore, $\omega_{n/n-1}$ may be written as a product of a rotational joint matrix, $\mathbf{H}_{n\omega}(\mathbf{q}_{n\omega})$ (rotational Jacobian matrix of joint n), multiplied by the vector of angular velocity parameters, $\dot{\mathbf{q}}_{n\omega}$

$$\omega_{n/n-1} = \mathbf{H}_{n\omega}(\mathbf{q}_{n\omega})\,\dot{\mathbf{q}}_{n\omega} \tag{66}$$

The absolute linear velocity of $\mathcal{B}_n$, expressed in $\{O_n, X_nY_nZ_n\}$ is given by

$$\begin{aligned}\mathbf{v}_n &= \mathbf{R}^T_{n/n-1}\left(\mathbf{v}_{n-1} + \widetilde{\omega_{n-1}}\mathbf{r}_{n-1,n} + \dot{\mathbf{r}}_{n-1,n}\right)\\ &= \mathbf{R}^T_{n/n-1}\left(\mathbf{v}_{n-1} + \widetilde{\omega_{n-1}}\mathbf{r}_{n-1,n}\right) + \mathbf{H}_{nv}(\mathbf{q}_{n\omega})\,\dot{\mathbf{q}}_{nv}\end{aligned} \tag{67}$$

where $\mathbf{H}_{nv}(\mathbf{q}_{n\omega})$ is the translation joint matrix (translation Jacobian matrix of joint n), and $\dot{\mathbf{q}}_{nv}$ is the vector of linear velocity parameters. $\mathbf{H}_{nv}(\mathbf{q}_{n\omega})$ is dependent on $\mathbf{q}_{n\omega}$ if joint translation occurs in $\mathcal{B}_{n-1}$. If translation occurs in $\mathcal{B}_n$ it is a constant matrix. Writing Eq. 65 together with Eq. 67 yields [2]

$$\begin{aligned}&\begin{bmatrix}\mathbf{v}_n\\ \omega_n\end{bmatrix} = \begin{bmatrix}\mathbf{R}^T_{n/n-1} & \mathbf{R}^T_{n/n-1}\widetilde{\mathbf{r}_{n-1,n}}^T\\ 0 & \mathbf{R}^T_{n/n-1}\end{bmatrix}\begin{bmatrix}\mathbf{v}_{n-1}\\ \omega_{n-1}\end{bmatrix} + \begin{bmatrix}\mathbf{H}_{nv} & 0\\ 0 & \mathbf{H}_{n\omega}\end{bmatrix}\begin{bmatrix}\dot{\mathbf{q}}_{nv}\\ \dot{\mathbf{q}}_{n\omega}\end{bmatrix}\\ &\Leftrightarrow \mathbf{V}_n = \mathbf{\Phi}_{rn,n-1}\mathbf{V}_{n-1} + \mathbf{H}_n\dot{\mathbf{q}}_n\end{aligned} \tag{68}$$

The absolute linear acceleration vector of $\mathcal{B}_n$, expressed in $\{O_n, X_nY_nZ_n\}$ may be obtained as in Eq.45, and using the definition of the translation joint matrix $\mathbf{H}_{nv}$. On the other hand, the absolute angular acceleration vector may be obtained by taking the time derivative of Eq.65 and using the definition of the rotational joint matrix $\mathbf{H}_{n\omega}$. The result is

$$\begin{aligned}&\begin{bmatrix}\mathbf{a}_n\\ \dot{\omega}_n\end{bmatrix} = \begin{bmatrix}\mathbf{R}^T_{n/n-1} & \mathbf{R}^T_{n/n-1}\widetilde{\mathbf{r}_{n-1,n}}^T\\ 0 & \mathbf{R}^T_{n/n-1}\end{bmatrix}\begin{bmatrix}\mathbf{a}_{n-1}\\ \dot{\omega}_{n-1}\end{bmatrix} + \begin{bmatrix}\mathbf{H}_{nv} & 0\\ 0 & \mathbf{H}_{n\omega}\end{bmatrix}\begin{bmatrix}\ddot{\mathbf{q}}_{nv}\\ \ddot{\mathbf{q}}_{n\omega}\end{bmatrix}\\ &\quad + \begin{bmatrix}\mathbf{R}^T_{n/n-1}\widetilde{\omega_{n-1}}\widetilde{\omega_{n-1}}\mathbf{r}_{n-1,n} + \cdots\\ \mathbf{R}^T_{n/n-1}\widetilde{\omega_{n-1}}\mathbf{H}_{n\omega}\dot{\mathbf{q}}_{n\omega} + \cdots\end{bmatrix.\\ &\qquad \left.\cdots\left(2\left(\widetilde{\mathbf{R}^T_{n/n-1}\omega_{n-1}}\right) + \widetilde{\omega_{n/n-1}}\right)\mathbf{H}_{nv}\dot{\mathbf{q}}_{nv} + \dot{\mathbf{H}}_{nv}\dot{\mathbf{q}}_{nv}\right]\\ &\qquad \cdots\dot{\mathbf{H}}_{n\omega}\dot{\mathbf{q}}_{n\omega}\\ &\Leftrightarrow \mathbf{A}_n = \mathbf{\Phi}_{rn,n-1}\mathbf{A}_{n-1} + \mathbf{H}_n\ddot{\mathbf{q}}_n + \mathbf{N}_n\end{aligned} \tag{69}$$

The control force and the control torque at the joints may be obtained by performing a virtual power balance across joint n. Since a joint has no mass, for the linear relative motion results

$$\delta\left(\mathbf{H}_{nv}\dot{\mathbf{q}}_{nv}\right)^T\mathbf{F}_n = \delta\dot{\mathbf{q}}^T_{nv}\mathbf{H}^T_{nv}\mathbf{F}_n \tag{70}$$

and for the angular relative motion

$$\delta\left(\mathbf{H}_{n\omega}\dot{\mathbf{q}}_{n\omega}\right)^T\mathbf{M}_n = \delta\dot{\mathbf{q}}_{n\omega}^T\mathbf{H}_{n\omega}^T\mathbf{M}_n \tag{71}$$

The externally applied generalized force vector (torques and forces) is then written as

$$\begin{bmatrix}\mathcal{T}_{nv}\\ \mathcal{T}_{nw}\end{bmatrix} = \begin{bmatrix}\mathbf{H}_{nv}^T & 0\\ 0 & \mathbf{H}_{n\omega}^T\end{bmatrix}\begin{bmatrix}\mathbf{F}_n\\ \mathbf{M}_n\end{bmatrix}$$
$$\Leftrightarrow \mathcal{T}_n = \mathbf{H}_n^T\mathcal{F}_n \tag{72}$$

In the case of a rigid connection where $\mathcal{B}_{n-1}$ is a flexible beam and $\mathcal{B}_n$ is either a flexible beam or a rigid body, then reference frames $\{O_{\widehat{n-1}}, X_{\widehat{n-1}}Y_{\widehat{n-1}}Z_{\widehat{n-1}}\}$ and $\{O_n, X_nY_nZ_n\}$ coincide, and Eqs. 68 and 69 are obtained by evaluating $\mathbf{v}_k$, $\boldsymbol{\omega}_k$, $\mathbf{a}_k$ and $\dot{\boldsymbol{\omega}}_k$ at $O_{\widehat{n-1}}$:

$$\begin{bmatrix}\mathbf{v}_n\\ \boldsymbol{\omega}_n\end{bmatrix} = \left[\begin{array}{cc:cc}\mathbf{R}_{n/n-1}^T & \mathbf{R}_{n/n-1}^T\widetilde{\mathbf{r}_{n-1,n}}^T & \mathbf{R}_{n/n-1}^T\mathbf{J}_{TK_{\widehat{n-1}}} & \mathbf{R}_{n/n-1}^T\mathbf{J}_{T\varphi_{\widehat{n-1}}}\\ 0 & \mathbf{R}_{n/n-1}^T & \mathbf{R}_{n/n-1}^T\mathbf{J}_{\mathbf{R}_{e_{\widehat{n-1}}}} & 0\end{array}\right]\left[\begin{array}{c}\mathbf{v}_{n-1}\\ \boldsymbol{\omega}_{n-1}\\ \hdashline \dot{\mathbf{q}}_{K_{n-1}}\\ \dot{\mathbf{q}}_{\varphi_{n-1}}\end{array}\right]$$
$$\Leftrightarrow \mathbf{V}_n = \left[\boldsymbol{\Phi}_{rn,n-1} \,\vdots\, \boldsymbol{\Phi}_{en,n-1}\right]\left[\begin{array}{c}\mathbf{V}_{n-1}\\ \hdashline \dot{\mathbf{q}}_{e_{n-1}}\end{array}\right] \tag{73}$$

and

$$\mathbf{A}_n = \left[\boldsymbol{\Phi}_{rn,n-1} \,\vdots\, \boldsymbol{\Phi}_{en,n-1}\right]\left[\begin{array}{c}\mathbf{A}_{n-1}\\ \hdashline \ddot{\mathbf{q}}_{e_{n-1}}\end{array}\right] + \mathbf{N}_{en} \tag{74}$$

where

$$\mathbf{N}_{en} = \begin{bmatrix}\mathbf{R}_{n/n-1}^T\left(\widetilde{\boldsymbol{\omega}_{n-1}}\widetilde{\boldsymbol{\omega}_{n-1}}\mathbf{r}_{n-1,n} + 2\widetilde{\boldsymbol{\omega}_{n-1}}\mathbf{J}_{T_{\widehat{n-1}}}\dot{\mathbf{q}}_{e_{n-1}} + \dot{\mathbf{J}}_{T_{\widehat{n-1}}}\dot{\mathbf{q}}_{e_{n-1}}\right)\\ \mathbf{R}_{n/n-1}^T\left(\widetilde{\boldsymbol{\omega}_{n-1}}\mathbf{J}_{\mathbf{R}_{e_{\widehat{n-1}}}}\dot{\mathbf{q}}_{K_{n-1}} + \dot{\mathbf{J}}_{\mathbf{R}_{e_{\widehat{n-1}}}}\dot{\mathbf{q}}_{K_{n-1}}\right)\end{bmatrix} \tag{75}$$

and $\mathbf{R}_{n/n-1} = \mathbf{R}_{e_{\widehat{n-1}}}$, $\mathbf{r}_{n-1,n} = \mathbf{x}_{\widehat{n-1}}$.

4.2 Dynamics of a Rigid-Flexible-Rigid RFR body

A flexible beam attached to two rigid bodies, one at each end, may be seen as a building block for a flexible manipulator. The rigid bodies are the supporting structures for actuators and sensors at the joints, and are typically of significant mass. However, if that is not the case, their mass may be easily neglected, and what remains are their geometric properties. To this end, let us consider

Rigid Body: $\mathcal{B}_n$ (from Eq. 15, and setting $\dot{\mathbf{q}}_n = 0$ and $\ddot{\mathbf{q}}_n = 0$ in Eqs. 68 and 69)
Flexible Beam: $\mathcal{B}_{n+1}$ (from Eqs. 63, 73 and 74)
Rigid Body: $\mathcal{B}_{n+2}$ (from Eq. 15)

The generalized coordinates of the RFR body are the absolute linear and angular velocity vectors of $\mathcal{B}_n$, $\mathbf{v}_n$ and $\boldsymbol{\omega}_n$ respectively, and the elastic bending displacements, torsion angle, and shear angles of body $\mathcal{B}_{n+1}$, $\mathbf{q}_{K_{n+1}}$ and $\mathbf{q}_{\varphi_{n+1}}$. The dynamics model of the RFR body may then be obtained by reapplying the Principle of Virtual Powers. The generalized virtual velocities are $\delta\mathbf{V}_n$ and $\delta\dot{\mathbf{q}}_{e_{n+1}}$, and therefore

$$\delta\mathbf{V}_{n+1} = \boldsymbol{\Phi}_{rn+1,n}\delta\mathbf{V}_n \tag{76}$$

and

$$\delta\mathbf{V}_{n+2} = \left[\boldsymbol{\Phi}_{rn+2,n+1}\boldsymbol{\Phi}_{rn+1,n} \,\vdots\, \boldsymbol{\Phi}_{en+2,n+1}\right]\begin{bmatrix}\delta\mathbf{V}_n \\ \hdashline \delta\dot{\mathbf{q}}_{e_{n+1}}\end{bmatrix} \tag{77}$$

The acceleration of the flexible beam is given in relationship to the acceleration of $\mathcal{B}_n$ by Eq. 69, and the acceleration of $\mathcal{B}_{n+2}$ is obtained from Eqs. 69, and 74 as

$$\mathbf{A}_{n+2} = \left[\boldsymbol{\Phi}_{rn+2,n+1}\boldsymbol{\Phi}_{rn+1,n} \,\vdots\, \boldsymbol{\Phi}_{en+2,n+1}\right]\begin{bmatrix}\mathbf{A}_n \\ \hdashline \ddot{\mathbf{q}}_{e_{n+1}}\end{bmatrix} + \boldsymbol{\Phi}_{rn+2,n+1}\mathbf{N}_{n+1} + \mathbf{N}_{en+2} \tag{78}$$

Adding the contribution of the three bodies according to the Principle of Virtual Powers, and then adding the coefficients of $\delta\mathbf{V}_n^T$ and $\delta\dot{\mathbf{q}}_{e_{n+1}}^T$ and writing in matrix form results in the following system of dynamics equations

$$\mathcal{M}_{rfr_n}\begin{bmatrix}\mathbf{A}_n \\ \ddot{\mathbf{q}}_{e_{n+1}}\end{bmatrix} + \mathcal{N}_{rfr_n} = \begin{bmatrix}\mathbf{I} \\ 0\end{bmatrix}\mathcal{F}_n - \begin{bmatrix}\left(\boldsymbol{\Phi}_{rn+3,n+2}\boldsymbol{\Phi}_{rn+2,n+1}\boldsymbol{\Phi}_{rn+1,n}\right)^T \\ \left(\boldsymbol{\Phi}_{rn+3,n+2}\boldsymbol{\Phi}_{en+2,n+1}\right)^T\end{bmatrix}\mathcal{F}_{n+3} \tag{79}$$

where

$$\begin{aligned}\mathcal{M}_{rfr_n} = {} & \begin{bmatrix}\mathcal{M}_{rn} & 0 \\ 0 & 0\end{bmatrix} + \begin{bmatrix}\boldsymbol{\Phi}_{rn+1,n}^T\mathcal{M}_{en+1,rr}\boldsymbol{\Phi}_{rn+1,n} & \boldsymbol{\Phi}_{rn+1,n}^T\mathcal{M}_{en+1,re} \\ \mathcal{M}_{en+1,re}^T\boldsymbol{\Phi}_{rn+1,n} & \mathcal{M}_{en+1,ee}\end{bmatrix} \\ & + \left[\begin{array}{c:c}\left(\boldsymbol{\Phi}_{rn+2,n+1}\boldsymbol{\Phi}_{rn+1,n}\right)^T\mathcal{M}_{rn+2}\boldsymbol{\Phi}_{rn+2,n+1}\boldsymbol{\Phi}_{rn+1,n} & \left(\boldsymbol{\Phi}_{rn+2,n+1}\boldsymbol{\Phi}_{rn+1,n}\right)^T\mathcal{M}_{rn+2}\boldsymbol{\Phi}_{en+2,n+1} \\ \boldsymbol{\Phi}_{en+2,n+1}^T\mathcal{M}_{rn+2}\boldsymbol{\Phi}_{rn+2,n+1}\boldsymbol{\Phi}_{rn+1,n} & \boldsymbol{\Phi}_{en+2,n+1}^T\mathcal{M}_{rn+2}\boldsymbol{\Phi}_{en+2,n+1}\end{array}\right]\end{aligned} \tag{80}$$

and

$$\begin{aligned}\mathcal{N}_{rfr_n} = {} & \begin{bmatrix}\mathcal{N}_{rn} \\ 0\end{bmatrix} + \begin{bmatrix}\boldsymbol{\Phi}_{rn+1,n}^T\left(\mathcal{M}_{en+1,rr}\mathbf{N}_{n+1} + \overline{\mathcal{N}}_{en+1,r}\right) \\ \mathcal{M}_{en+1,re}^T\mathbf{N}_{n+1} + \overline{\mathcal{N}}_{en+1,e}\end{bmatrix} \\ & + \begin{bmatrix}\left(\boldsymbol{\Phi}_{rn+2,n+1}\boldsymbol{\Phi}_{rn+1,n}\right)^T\left(\mathcal{M}_{rn+2}\left(\boldsymbol{\Phi}_{rn+2,n+1}\mathbf{N}_{n+1} + \mathbf{N}_{en+2}\right) + \mathcal{N}_{rn+2}\right) \\ \boldsymbol{\Phi}_{en+2,n+1}^T\left(\mathcal{M}_{rn+2}\left(\boldsymbol{\Phi}_{rn+2,n+1}\mathbf{N}_{n+1} + \mathbf{N}_{en+2}\right) + \mathcal{N}_{rn+2}\right)\end{bmatrix}\end{aligned} \tag{81}$$

4.3 Dynamics of a serial multi Rigid-Flexible-Rigid body system

Let us now consider an *articulated chain* of RFR bodies representing a flexible manipulator. From Eq. 68, and $\mathbf{V}_{n+2}$ as given in Eq. 77, the velocity of a fourth body connected to $\mathcal{B}_{n+2}$ through an articulated joint may by written as

$$\mathbf{V}_{n+3} = \left[\mathbf{\Phi}_{rn+3,n+2}\mathbf{\Phi}_{rn+2,n+1}\mathbf{\Phi}_{rn+1,n} \,\vdots\, \mathbf{\Phi}_{rn+3,n+2}\mathbf{\Phi}_{en+2,n+1} \right] \begin{bmatrix} \mathbf{V}_n \\ \dot{\mathbf{q}}_{e_{n+1}} \end{bmatrix} + \mathbf{H}_{n+3}\dot{\mathbf{q}}_{n+3} \tag{82}$$

and from Eqs. 69 and 78

$$\mathbf{A}_{n+3} = \left[\mathbf{\Phi}_{rn+3,n+2}\mathbf{\Phi}_{rn+2,n+1}\mathbf{\Phi}_{rn+1,n} \,\vdots\, \mathbf{\Phi}_{rn+3,n+2}\mathbf{\Phi}_{en+2,n+1} \right] \begin{bmatrix} \mathbf{A}_n \\ \ddot{\mathbf{q}}_{e_{n+1}} \end{bmatrix} + \mathbf{H}_{n+3}\ddot{\mathbf{q}}_{n+3} + \mathbf{N}_{n+3} + \mathbf{\Phi}_{rn+3,n+2}\left(\mathbf{\Phi}_{rn+2,n+1}\mathbf{N}_{n+1} + \mathbf{N}_{en+2}\right) \tag{83}$$

Renumbering Eqs.79, 82 and 83 in terms of the nth RFR body, and rewriting the generalized velocity, acceleration and force vectors in order to include the elastic contributions results in the dynamics equation,

$$\begin{aligned} &\mathcal{M}_{rfr_n} \begin{bmatrix} \mathbf{A}_n \\ \ddot{\mathbf{q}}_{e_{n+1}} \end{bmatrix} + \mathcal{N}_{rfr_n} \\ &\quad = \begin{bmatrix} \mathcal{F}_n \\ 0 \end{bmatrix} - \begin{bmatrix} \left(\mathbf{\Phi}_{rn+3,n+2}\mathbf{\Phi}_{rn+2,n+1}\mathbf{\Phi}_{rn+1,n}\right)^T & 0 \\ \left(\mathbf{\Phi}_{rn+3,n+2}\mathbf{\Phi}_{en+2,n+1}\right)^T & 0 \end{bmatrix} \begin{bmatrix} \mathcal{F}_{n+3} \\ 0 \end{bmatrix} \\ &\Leftrightarrow \mathcal{M}_{rfr_n}\mathbf{A}_{rfr_n} + \mathcal{N}_{rfr_n} = \mathcal{F}_{rfr_n} - \mathbf{\Phi}_{rfr_{n+1,n}}^T \mathcal{F}_{rfr_{n+1}} \end{aligned} \tag{84}$$

the velocity equation,

$$\begin{aligned} \begin{bmatrix} \mathbf{V}_n \\ \dot{\mathbf{q}}_{e_{n+1}} \end{bmatrix} &= \begin{bmatrix} \mathbf{\Phi}_{rn,n-1}\mathbf{\Phi}_{rn-1,n-2}\mathbf{\Phi}_{rn-2,n-3} & \mathbf{\Phi}_{rn,n-1}\mathbf{\Phi}_{en-1,n-2} \\ 0 & 0 \end{bmatrix} \begin{bmatrix} \mathbf{V}_{n-3} \\ \dot{\mathbf{q}}_{e_{n-2}} \end{bmatrix} \\ &\quad + \begin{bmatrix} \mathbf{H}_n & 0 \\ 0 & \mathbf{I} \end{bmatrix} \begin{bmatrix} \dot{\mathbf{q}}_n \\ \dot{\mathbf{q}}_{e_{n+1}} \end{bmatrix} \\ \Leftrightarrow \mathbf{V}_{rfr_n} &= \mathbf{\Phi}_{rfr_{n,n-1}}\mathbf{V}_{rfr_{n-1}} + \mathbf{H}_{rfr_n}\dot{\mathbf{q}}_{rfr_n} \end{aligned} \tag{85}$$

and the acceleration equation

$$\begin{aligned} \begin{bmatrix} \mathbf{A}_n \\ \ddot{\mathbf{q}}_{e_{n+1}} \end{bmatrix} &= \begin{bmatrix} \mathbf{\Phi}_{rn,n-1}\mathbf{\Phi}_{rn-1,n-2}\mathbf{\Phi}_{rn-2,n-3} & \mathbf{\Phi}_{rn,n-1}\mathbf{\Phi}_{en-1,n-2} \\ 0 & 0 \end{bmatrix} \begin{bmatrix} \mathbf{A}_{n-3} \\ \ddot{\mathbf{q}}_{e_{n-2}} \end{bmatrix} \\ &\quad + \begin{bmatrix} \mathbf{H}_n & 0 \\ 0 & \mathbf{I} \end{bmatrix} \begin{bmatrix} \ddot{\mathbf{q}}_n \\ \ddot{\mathbf{q}}_{e_{n+1}} \end{bmatrix} + \begin{bmatrix} \mathbf{N}_n + \mathbf{\Phi}_{rn,n-1}\left(\mathbf{\Phi}_{rn-1,n-2}\mathbf{N}_{n-2} + \mathbf{N}_{en-1}\right) \\ 0 \end{bmatrix} \\ \Leftrightarrow \mathbf{A}_{rfr_n} &= \mathbf{\Phi}_{rfr_{n,n-1}}\mathbf{A}_{rfr_{n-1}} + \mathbf{H}_{rfr_n}\ddot{\mathbf{q}}_{rfr_n} + \mathbf{N}_{rfr_n} \end{aligned} \tag{86}$$

Furthermore, the externally applied generalized force vector in Eq. 72 may be rewritten as

$$
\begin{bmatrix} \mathcal{T}_n \\ 0 \end{bmatrix} = \begin{bmatrix} \mathbf{H}_n^T & 0 \\ 0 & \mathbf{I} \end{bmatrix} \begin{bmatrix} \mathcal{F}_n \\ 0 \end{bmatrix}
$$
$$
\Leftrightarrow \mathcal{T}_{rfr_n} = \mathbf{H}_{rfr_n}{}^T \mathcal{F}_{rfr_n} \tag{87}
$$

Equations 84, 86 and 87 are in a form compatible with the problem formulation and solution methods for rigid manipulators presented in [18] and [19]. The solution method consists in writing a large algebraic system as in [20] or [21], which is solved through elimination methods leading to either the global dynamics Composite Inertia (CI) method (the Composite Rigid Body (CRB) method for rigid multibody systems), or the Articulated Body (AB) method. The former has complexity $O(N^3)$ due to the need of inverting the global system mass matrix, whereas the latter has complexity $O(N)$. To illustrate, following [18], the algebraic system for an unconstrained ($\mathcal{F}_{rfr_3} = 0$) 2 RFR body manipulator is written as

$$
\left[\begin{array}{ccc|ccc}
\mathcal{M}_{rfr_2} & 0 & \mathbf{I} & & & \\
0 & 0 & \mathbf{H}_{rfr_2}{}^T & & & \\
\mathbf{I} & \mathbf{H}_{rfr_2} & 0 & -\mathbf{\Phi}_{rfr_{2,1}} & & \\
\hline
 & & -\mathbf{\Phi}_{rfr_{2,1}}{}^T & \mathcal{M}_{rfr_1} & 0 & \mathbf{I} \\
 & & & 0 & 0 & \mathbf{H}_{rfr_1}{}^T \\
 & & & \mathbf{I} & \mathbf{H}_{rfr_1} & 0
\end{array}\right]
\left[\begin{array}{c}
-\mathbf{A}_{rfr_2} \\ \ddot{\mathbf{q}}_{rfr_2} \\ \mathcal{F}_{rfr_2} \\ \hline -\mathbf{A}_{rfr_1} \\ \ddot{\mathbf{q}}_{rfr_1} \\ \mathcal{F}_{rfr_1}
\end{array}\right]
=
\left[\begin{array}{c}
\mathcal{N}_{rfr_2} \\ \mathcal{T}_{rfr_2} \\ -\mathbf{N}_{rfr_2} \\ \hline \mathcal{N}_{rfr_1} \\ \mathcal{T}_{rfr_1} \\ -\mathbf{N}_{rfr_1}
\end{array}\right] \tag{88}
$$

The global dynamics CI method consists in rearranging the block rows and columns of the above system into the form

$$
\left[\begin{array}{cc|cc|cc}
\mathcal{M}_{rfr_2} & & & & \mathbf{I} & \\
 & \mathcal{M}_{rfr_1} & & & -\mathbf{\Phi}_{rfr_{2,1}}{}^T & \mathbf{I} \\
\hline
 & & & & \mathbf{H}_{rfr_2}{}^T & \\
 & & & & & \mathbf{H}_{rfr_1}{}^T \\
\hline
\mathbf{I} & -\mathbf{\Phi}_{rfr_{2,1}} & \mathbf{H}_{rfr_2} & & & \\
 & \mathbf{I} & & \mathbf{H}_{rfr_1} & &
\end{array}\right]
\left[\begin{array}{c}
-\mathbf{A}_{rfr_2} \\ -\mathbf{A}_{rfr_1} \\ \hline \ddot{\mathbf{q}}_{rfr_2} \\ \ddot{\mathbf{q}}_{rfr_1} \\ \hline \mathcal{F}_{rfr_2} \\ \mathcal{F}_{rfr_1}
\end{array}\right]
=
\left[\begin{array}{c}
\mathcal{N}_{rfr_2} \\ \mathcal{N}_{rfr_1} \\ \hline \mathcal{T}_{rfr_2} \\ \mathcal{T}_{rfr_1} \\ \hline -\mathbf{N}_{rfr_2} \\ -\mathbf{N}_{rfr_1}
\end{array}\right]
$$
$$
\Leftrightarrow \begin{bmatrix} \mathcal{M} & 0 & \mathbf{\Phi}^{-T} \\ 0 & 0 & \mathbf{H}^T \\ \mathbf{\Phi}^{-1} & \mathbf{H} & 0 \end{bmatrix} \begin{bmatrix} -\mathbf{A} \\ \ddot{\mathbf{q}} \\ \mathcal{F} \end{bmatrix} = \begin{bmatrix} \mathcal{N} \\ \mathcal{T} \\ -\mathbf{N} \end{bmatrix} \tag{89}
$$

and solving through block-row elimination in order to arrive at

$$
\left(\mathbf{H}^T \mathbf{\Phi}^T \mathcal{M} \mathbf{\Phi} \mathbf{H}\right) \ddot{\mathbf{q}} + \mathbf{H}^T \mathbf{\Phi}^T \left(\mathcal{M} \mathbf{\Phi} \mathbf{N} + \mathcal{N}\right) = \mathcal{T} \tag{90}
$$

which is the joint space dynamics equation of the flexible manipulator. This is the solution one would obtain through a reapplication of the Principle of Virtual Powers to the chain of RFR bodies. From Eq. 85 one may write

$$\begin{bmatrix} \mathbf{V}_{rfr2} \\ \mathbf{V}_{rfr1} \end{bmatrix} = \begin{bmatrix} \mathbf{I} & \mathbf{\Phi}_{rfr2,1} \\ 0 & \mathbf{I} \end{bmatrix} \begin{bmatrix} \mathbf{H}_{rfr2} & 0 \\ 0 & \mathbf{H}_{rfr1} \end{bmatrix} \begin{bmatrix} \dot{\mathbf{q}}_{rfr_2} \\ \dot{\mathbf{q}}_{rfr_1} \end{bmatrix}$$
$$\Leftrightarrow \mathbf{V} = \mathbf{\Phi}\mathbf{H}\dot{\mathbf{q}} \tag{91}$$

thus identifying the product $\mathbf{\Phi}\mathbf{H}$ as a Jacobian matrix.

The AB method on the other hand, solves the algebraic system in Eq. 88, taking advantage of its block diagonal structure. Each block corresponds to a RFR body, which is coupled to the next body in the chain through matrix $\mathbf{\Phi}_{rfr}{}_{n+1,n}^{T}$. The solution procedure consists in eliminating these matrices in order to decouple the diagonal blocks. Therefore, for $n = N, N-1, ..., 1$, eliminate the middle row of each block using the first row and then the last row of the block. Then, using the resulting n^{th} block block rows, eliminate matrix $\mathbf{\Phi}_{rfr}{}_{n+1,n}^{T}$, coupling block $n-1$ with block n. The resulting system is

$$\left[\begin{array}{ccc|ccc} \mathcal{M}_{rfr2} & 0 & \mathbf{I} & & & \\ 0 & \mathcal{D}_{rfr2} & 0 & -\mathbf{H}_{rfr2}^{T}\mathcal{M}_{rfr2}\mathbf{\Phi}_{rfr2,1} & & \\ \mathbf{I} & \mathbf{H}_{rfr2} & 0 & -\mathbf{\Phi}_{rfr2,1} & & \\ \hline & & & \breve{\mathcal{M}}_{rfr_1} & 0 & \mathbf{I} \\ & & & 0 & \mathcal{D}_{rfr1} & 0 \\ & & & \mathbf{I} & \mathbf{H}_{rfr1} & 0 \end{array}\right] \left[\begin{array}{c} -\mathbf{A}_{rfr2} \\ \ddot{\mathbf{q}}_{rfr_2} \\ \mathcal{F}_{rfr2} \\ \hline -\mathbf{A}_{rfr1} \\ \ddot{\mathbf{q}}_{rfr_1} \\ \mathcal{F}_{rfr1} \end{array}\right] = \left[\begin{array}{c} \mathcal{N}_{rfr2} \\ \breve{\mathcal{T}}_{rfr_2} \\ -\mathbf{N}_{rfr2} \\ \hline \breve{\mathcal{N}}_{rfr_1} \\ \breve{\mathcal{T}}_{rfr_1} \\ -\mathbf{N}_{rfr1} \end{array}\right] \tag{92}$$

where

$$\breve{\mathcal{T}}_{rfr_n} = \mathcal{T}_{rfrn} - \mathbf{H}_{rfr}{}_{n}^{T}\left(\breve{\mathcal{N}}_{rfr_n} + \breve{\mathcal{M}}_{rfr_n}\mathbf{N}_{rfrn}\right) \tag{93}$$

$$\mathcal{D}_{rfrn} = \mathbf{H}_{rfr}{}_{n}^{T}\breve{\mathcal{M}}_{rfr_n}\mathbf{H}_{rfrn} \tag{94}$$

$$\begin{aligned} \breve{\mathcal{N}}_{rfr_n} = {} & \mathcal{N}_{rfrn} + \mathbf{\Phi}_{rfr}{}_{n+1,n}^{T}\left(\breve{\mathcal{N}}_{rfr_{n+1}} + \breve{\mathcal{M}}_{rfr_{n+1}}\mathbf{N}_{rfrn+1}\right) \\ & + \mathbf{\Phi}_{rfr}{}_{n+1,n}^{T}\breve{\mathcal{M}}_{rfr_{n+1}}\mathbf{H}_{rfrn+1}\mathcal{D}_{rfr}{}_{n+1}^{-1}\breve{\mathcal{T}}_{rfr_{n+1}} \end{aligned} \tag{95}$$

and

$$\begin{aligned} \breve{\mathcal{M}}_{rfr_n} = {} & \mathcal{M}_{rfrn} + \mathbf{\Phi}_{rfr}{}_{n+1,n}^{T}\breve{\mathcal{M}}_{rfr_{n+1}}\mathbf{\Phi}_{rfrn+1,n} \\ & - \mathbf{\Phi}_{rfr}{}_{n+1,n}^{T}\breve{\mathcal{M}}_{rfr_{n+1}}\mathbf{H}_{rfrn+1}\mathcal{D}_{rfr}{}_{n+1}^{-1}\mathbf{H}_{rfr}{}_{n+1}^{T}\breve{\mathcal{M}}_{rfr_{n+1}}\mathbf{\Phi}_{rfrn+1,n} \end{aligned} \tag{96}$$

with $\breve{\mathcal{M}}_{rfr_N} = \mathcal{M}_{rfrN}$. Eq. 92 represents a linear system for $\ddot{\mathbf{q}}_{rfr_n}$ and $\mathbf{A}_{rfrn}$. For $n = 1, ..., N$, one now may calculate $\ddot{\mathbf{q}}_{rfr_n}$. Eq. 96 is the composite rigid body inertia introduced in [22] in the context of rigid multibody dynamics.

5 Validation and Control of a Single Flexible Link

The IST planar flexible manipulator, Fig.6, was designed for the purpose of position and force control algorithm testing [23, 24]. It consists of a modular structure where the joints and links can be easily exchanged. To transform it into a single link flexible manipulator, the first link is a very stiff steal beam and the respective joint is blocked. Only the second joint of the manipulator is allowed to rotate and the respective link is made of a very flexible spring-steal beam. The actuation mechanism is a Harmonic Drive RH-14-6002 servo system, current driven by a 12A8 servo amplifier from Advanced Motion Controls. The sensors used in the measurements are a 2000 pulses per revolution shaft encoder for the motor rotor position, a high rate acquisition camera system for the tip displacement, a strain gauge full bridge for the beam curvature at its base, and an accelerometer for the tip acceleration. The shaft encoder is integrated in the Harmonic Drive servo system, and the visual system acquiring 1000 frames per second is a Kodak motion corder analyzer SR-series, model PS220. The hub velocity and hub acceleration are obtained though differentiation of the hub position. The relevant characteristics are as follows: beam length $L = 0.5$m, beam

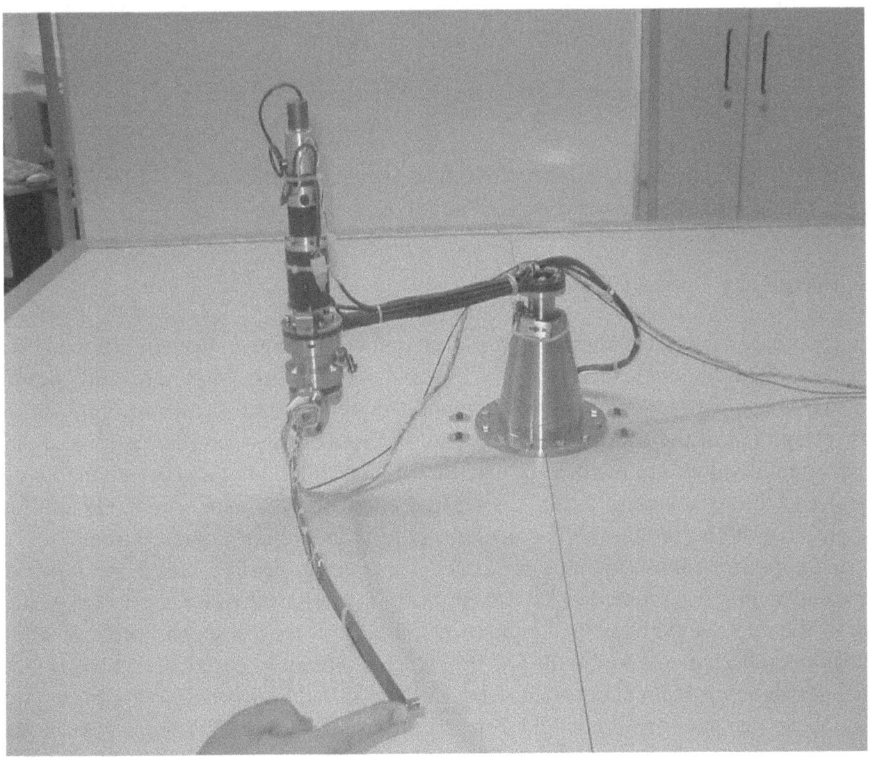

Fig. 6. The IST planar flexible manipulator

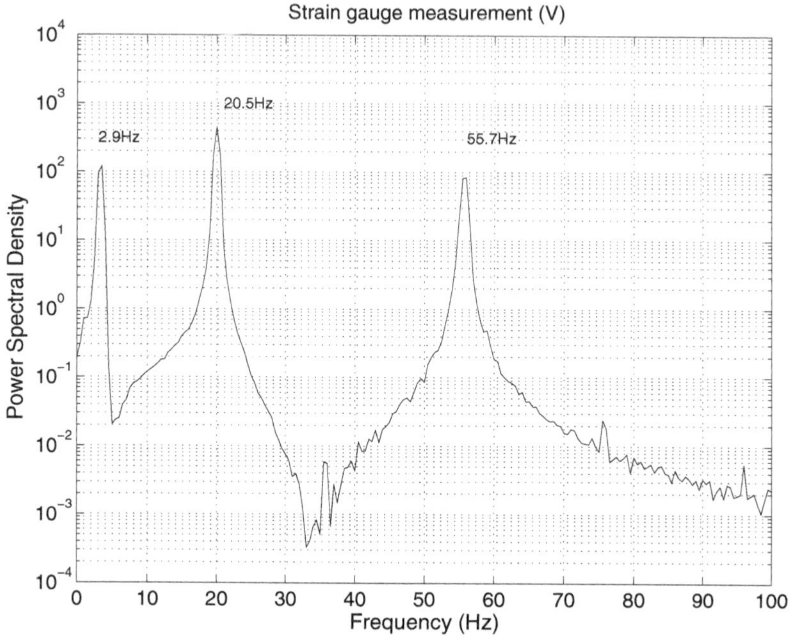

Fig. 7. Power spectral density of strain gauge measurements

width $w = 0.001$m, beam height $h = 0.02$m, beam specific mass $\rho = 7850\text{Kg/m}^3$, beam Young modulus $E = 209 \times 10^9$Pa, hub radius $r = 0.075$m.

5.1 Frequency Domain Validation

The first three theoretical clamped-free (constrained) natural frequencies of the beam are $\varpi_1 = 20.95$rad/s (3.33Hz), $\varpi_2 = 131.28$rad/s (20.89Hz) and $\varpi_3 = 367.6$rad/s (58.5Hz). The experimental values (damped frequencies) are shown in Fig.7 for an impact applied at the tip of the beam. The differences between the theoretical and experimental values are justified by friction (material damping and aerodynamic effects) and the existence of strain gauges and cables on the experimental beam that are not taken into account in the calculation of the theoretical values. Crucial to the correct capture of the frequency behavior of a rotating flexible link is the order of the rotation matrix assumed in Eq. 26. In order to capture the increase in frequency due to the centrifugal force (*centrifugal stiffening*), at least a second order rotation matrix must be considered, Fig.8. The frequencies plotted were calculated by solving the eigenvalue problem at constant velocities from 0rad/s to 10rad/s. Comparing the frequencies at 0rad/s and at 10rad/s shows an increase of 4.7% for the first mode, 1.9% for the second mode, and 0.8% for the third mode.

If a rotation matrix of order one is assumed, then a decrease in frequency is obtained with the increase of hub velocity. This is of course contrary to the physical

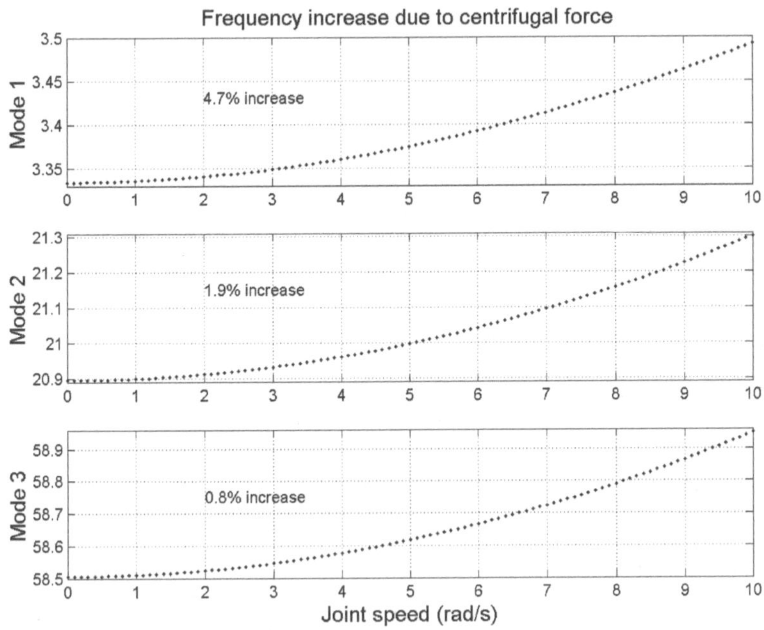

Fig. 8. Effect of the centrifugal force on the natural frequencies

reality, and is known as *centrifugal softening* [25]. In order to capture the (sometimes reasonable) zero increase of frequency with the increase of hub velocity this term in the dynamics equations must be neglected.

5.2 Time Domain Validation

Displacement driven manoeuvres are used for time domain validation [26]. This allows us not to have to model joint friction. The system input for the displacement driven models is the hub acceleration. Therefore, the approach adopted is to initially implement a Proportional Derivative (PD) control loop around the hub position, and apply a reference motion of a certain form to the manipulator. The real hub position and velocity is then recorded and the acceleration estimated. The acceleration response is then used as the input signal for the displacement driven models. The applied maximum rotation speed is 7.6 rad/s as shown in Fig.9. Fig.10 shows the measured end-point displacement of the manipulator and the simulated values. The experimental values were measured with the camera system, and the simulated values were calculated assuming a quadratic rotation matrix in Eq. 26, and assuming a linear rotation matrix, but neglecting the centrifugal softening term (the simplified model). The quadratic model shows closer performance to the real manipulator, and the increase in frequency is visible in the decrease in oscillation period when compared to the simplified model.

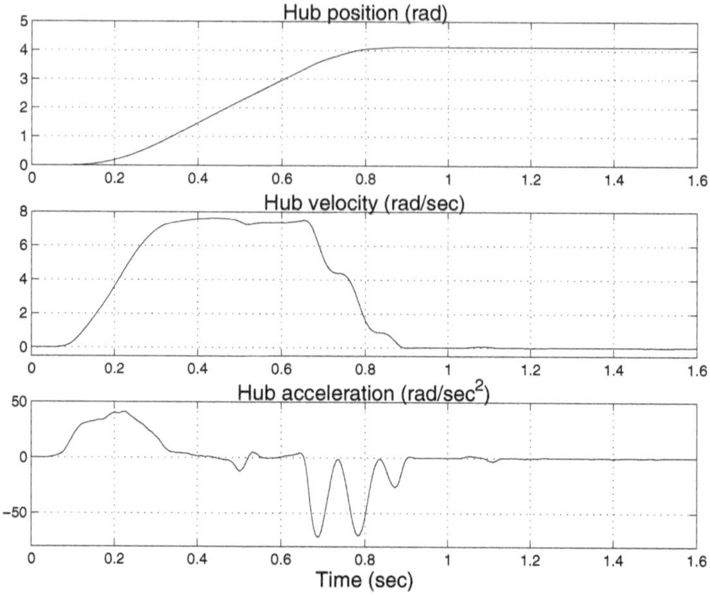

Fig. 9. Joint manoeuvre

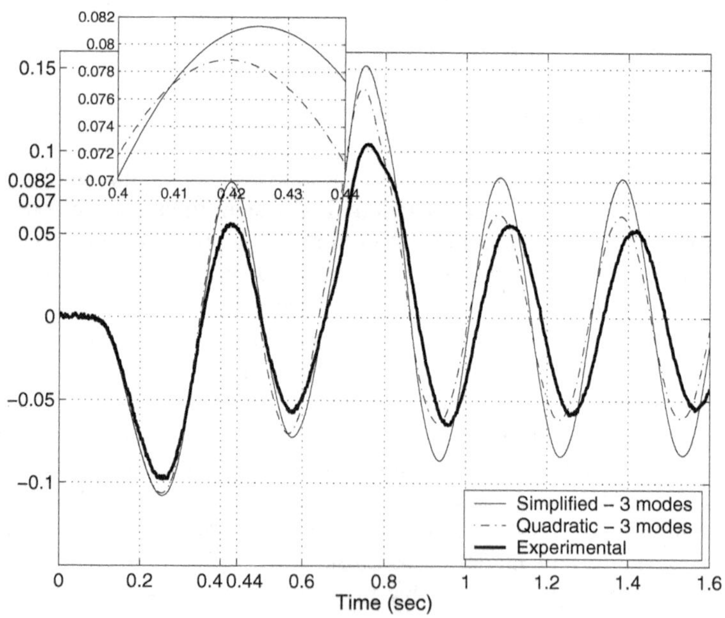

Fig. 10. Tip displacement [m]

5.3 Bending strain feedback control

The strain feedback approach adopted in this investigation follows the main route found in works such as [11, 27] and [28]. The underlying idea is that vibration damping cannot be achieved with collocated control (joint feedback) because of the Coulomb friction present at the joint. Basically, an observability problem occurs when the torque at the clamping point of the beam on the hub is smaller than the Coulomb friction torque at the joint. An approach to overcome this problem is to place a strain gauge close to the hub of the manipulator.

In this work, direct strain feedback is used for control of the vibration of the flexible link [29]. The control structure comprises two feedback loops: (1) The hub angle and hub velocity as inputs to a collocated PD control for rigid body motion control. (2) A direct strain feedback signal from a strain gauge for vibration control. These two loops are then combined to give a torque input to drive the system. A block diagram of the control scheme is shown in Fig.11 where q_3 represents the joint angle and K_3 is the beam curvature measured with the strain gauge. Thus, the control law is given by

$$u(t) = \mathsf{K_p}\left(q_{3ref} - q_3\right) - \mathsf{K_v}\,\dot{q}_3 - \mathsf{K_s}\,K_3 \tag{97}$$

The PD controller parameters $\mathsf{K_p}$ and $\mathsf{K_v}$ were deduced as 0.22 and 0.001 respectively using the root locus analysis. The corresponding hub angle and end-point acceleration responses of the manipulator using the PD control are shown in Fig. 12. It is noted that an acceptable hub angle response was achieved. The manipulator reached the demanded angle with a rise and settling times and overshoot of 0.098sec, 0.143 sec and 2.2% respectively. However, a significant amount of vibration occurred during the motion of the link as demonstrated in the end-point acceleration response. Moreover, the oscillation does not settle within 4 seconds with magnitude of acceleration of ± 50 m/sec^2. When the strain feedback loop is turned on, the tip vibration is significantly improved with a small penalty on the hub angle Fig. 13.

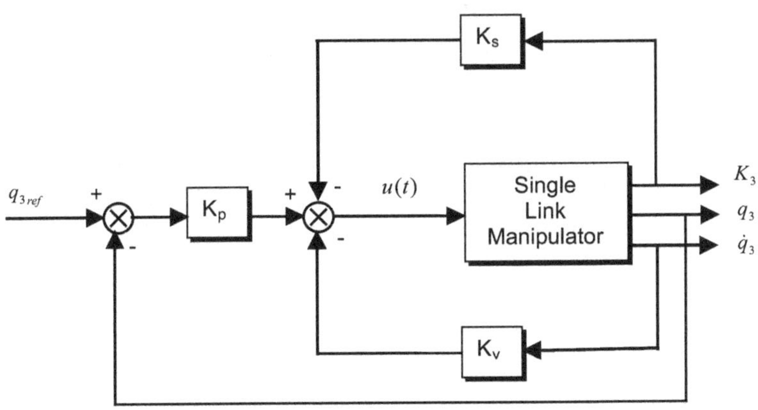

Fig. 11. Block diagram of strain feedback control

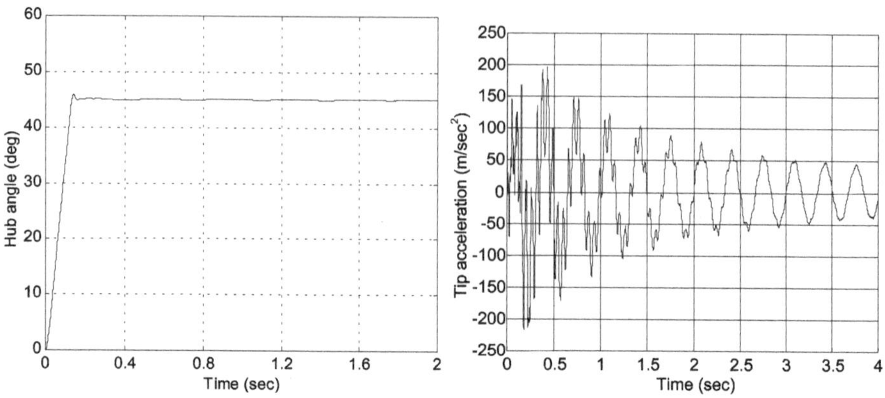

Fig. 12. Joint angle and tip acceleration without strain feedback

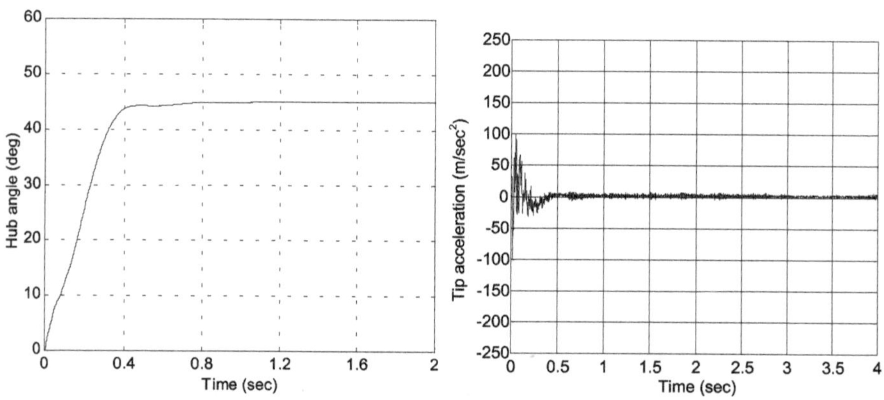

Fig. 13. Joint angle and tip acceleration with strain feedback

6 Conclusions

A systematic modelling approach leading to a general modelling environment, which may be used either in analysis or in the development of real time control schemes of flexible manipulator arms has been presented. A flexible beam attached to two rigid bodies, one at each end, has been assumed as a building block for a flexible manipulator. The rigid bodies are the supporting structures for actuators and sensors at the joints, and are typically of significant mass. However, if that is not the case, their mass may be easily neglected, and what remains are their geometric properties. The cross section rotations due to elasticity of the flexible beam may be of higher order as needed and shear deformation according to the Timoshenko beam theory is included. The formulation method for the dynamics of a flexible manipulator may be either the global dynamics composite inertia (CI) method or the recursive articulated body (AB) method. A validation and control exercise was performed on a single flexible link. Frequency domain and time domain validation were performed in regard to the order of the cross section rotation matrix. Linear and quadratic assumptions were compared against each other, and against the experimental apparatus. Curvature feedback control was compared against classical joint (collocated) feedback, and it's improved performance is shown through the measurement of the tip acceleration.

Acknowledgements

This work was financially supported by POCTI/EME/41690/2001, Ministério da Ciência e do Ensino Superior, Fundação para a Ciência e a Tecnologia, Portugal, and POCTI Unidade 46 Linha 3, subsidized by FEDER.

References

1. D'Eleuterio G M T (1992) Dynamics of an Elastic Multibody Chain: Part C–Recursive Dynamics. Dynamics and Stability of Systems, 7(2):61–89
2. Jain A, Rodriguez G (1992) Recursive Flexible Multibody System Dynamics Using Spatial Operators. AIAA Journal of Guidance, Control, and Dynamics, 15(6):1453–1466
3. Bremer H, Pfeiffer F (1995) Experiments with Flexible Manipulators. Control Engineering Practice, 3(9):1331–1338
4. Piedboeuf J-C (1998) Recursive Modeling of Serial Flexible Manipulators. The Journal of the Austronautical Sciences, 46(1):1–24
5. Géradin M, Cardona A (2001) Flexible Multibody Dynamics: A Finite Element Approach. John Wiley & Sons Ltd, Chichester, England
6. Meirovitch L (1991) Hybrid State Equations of Motion for Flexible Bodies in Terms of Quasi-Coordinates. AIAA Journal of Guidance Control and Dynamics, 14(5):1008–1013
7. Stemple T J (1998) Dynamics and Control of Flexible Multibody Structures. PhD thesis, Virginia State University, Virginia, USA
8. Schwertassek R, Wallrapp O (1999) Dynamic Flexible Multibody Systems. Vieweg, Braunschweig, Germany *(In German)*
9. Boyer F, Coiffet P (1996) Generalization of Newton-Euler Model for Flexible Manipulators. Journal of Robotic Systems, 13(1):11–24
10. Sharf I (1996) Geometrically Non-Linear Beam Element for Dynamics Simulation of Multibody Systems. International Journal for Numerical Methods in Engineering, 39:763–786
11. Bremer H, Pfeiffer p (1992) Elastic Multibody Systems. Teubner, Stuttgart, Germany *(In German)*
12. Preumont A (1997) Vibration Control of Active Structures: An Introduction. Kluwer Academic Publishers, Dordrecht, The Netherlands
13. Leleu S, Abou-Kandil H, Bonnassieux Y (2001) Piezoelectric Actuators and Sensors Location for Active Control of Flexible Structures. IEEE Transactions on Instrumentation and Measurement, 50(6):1577–1582
14. Giurgiutiu V (2000) Active-Materials Induced-Strain Actuation for Aeroelastic Vibration Control. The Shock and Vibration Digest, 32(5):355–368
15. Rodriguez G, Jain A, Kreutz-Delgado K (1992) Spatial Operator Algebra for Multibody System Dynamics. The Journal of the Austronautical Sciences, 40:27–50

16. Hardt M W (1999) Multibody Dynamical Algorithms, Numerical Optimal Control, with Detailed Studies in the Control of Jet Engine Compressors and Biped Walking. PhD Thesis, Electrical engineering (intelligent systems, robotics, and control), University of California, San Diego
17. Sincarsin G B , D'Eleuterio G M T, Hughes P C (1993) Dynamics of an Elastic Multibody Chain: Part D–Modelling of Joints. Dynamics and Stability of Systems, 8(2):127–146
18. Ascher U, Pai D K, Cloutier B (1997) Forward Dynamics, Elimination Methods, and Formulation Stiffness in Robot Simulation. International Journal of Robotics Research, 16(2):749–758
19. Pai D K, Ascher U M, Kry P G (2000) Forward Dynamics Algorithms for Multibody Chains and Contact. In: IEEE International Conference on Robotics and Automation pp. 857–863
20. Lubich C, Nowak U, Pohle U, Engstler Ch (1992) MEXX - Numerical Software for the Integration of Constrained Mechanical Multibody Systems. Technical Report SC 92-12, Konrad-Zuse-Zentrum für Informationstechnik, Berlin
21. von Schwerin R (1999) MultiBody System SIMulation - Numerical Methods, Algorithms, and Software. Lecture Notes in Computational Science and Engineering. Springer, Berlin
22. Featherstone R (1987) Robot Dynamics Algorithms. Kluwer Academic Publishers, Massachusetts, USA
23. Martins J, Ventura R, Sá da Costa J (1998) Design of a 2DOF Planar Manipulator for Force/Position Control Algorithms Test. In 3rd Portuguese Conference on Automatic Control - Special Session on Robotics and Automation, pp. 850–852, Coimbra, Portugal
24. Martins J M, Mohamed Z, Tokhi M O, Sá da Costa J, Botto M A (2003) Approaches for dynamic modelling of flexible manipulator systems. IEE Proc.-Control Theory Appl., 150(4):401–411
25. Kane T R, Ryan R R, Banerjee A K (1987) Dynamics of a Cantilever Beam Attached to a Moving Base. AIAA Journal of Guidance, Control and Dynamics, 10(2):139–151
26. Martins J, Ayala Botto M, Sá da Costa J (2002) Modeling of flexible beams for robotic manipulators. Multibody System Dynamics, 7(1):79–100
27. Luo Z-H (1994) Direct Strain Feedback Control of Flexible Robot Arms: New Theoretical and Experimental Results. IEEE Transactions on Automatic Control, 38(11):1610–1622
28. Luo Z-H, Feng D-H (1999) Nonlinear Torque Control of a Single-Link Flexible Robot. Journal of Robotic Systems, 16(1):25–35
29. Zaharuddin M, Martins J M, Tokhi M, Sá da Costa J, Botto M A (2005) Vibration Control Schemes of a Very Flexible Manipulator System. Control Engineering Practice, 13:267-277

An Elastic Simulation Model of a Metal Pushing V-Belt CVT

Markus Bullinger, Kilian Funk and Friedrich Pfeiffer

Institute of Applied Mechanics, Technical University of Munich, Germany
{mbu,pfeiffer}@amm.mw.tu-muenchen.de

This contribution presents the modelling of a metal pushing V-belt CVT to be used in numerical simulations. The system is subdivided into the pulleys, the belt and different types of contacts. The modelling is described in detail.

Since the deformation of the CVT is of major importance for the mechanical behaviour, the elasticity of the colliding bodies has to be taken into account. The deflection of the pulleys is split up into three parts. First the shaft is bent by radial forces. Here a spatial model of an elastic beam is used. Second the pulley sheaves tilt due to elasticity and clearance of the shaft-to-collar connection. This is modelled by a force element. And third, as a consequence of the asymmetrical loading, the elastic sheaves deform. This is calculated by different approaches, in which the best results are achieved by the use of CASTIGLIANO's Strain Energy Theorem.

The motion of the belt is specified in EULER-coordinates for the axial plane by separate longitudinal and transversal approaches. Therefore continuous RITZ-approximations are used in combination with hierarchical shape functions. For the kinematics of the moving elements a transformation to LAGRANGE-description is used.

The system contains a large number of contacts. The contact between the pulley sheaves and the belt elements is modelled spatial. The contact between two adjacent elements is modelled one dimensional. Both types of contacts consider unilateral constraints. In the contact between two ring layers the radial expansion has to be modelled accurately.

The simulation model is implemented in the object-orientated programming language C^{++}. This allows to calculate the kinematics of the belt and the local distribution of all contact forces. It permits to compute the necessary pulley thrust, the maximum transmittable torque and the efficiency for given loading cases. Some results of the numerical simulation are presented at the end of this contribution. Although this work deals with a metal pushing V-belt, the models can be modified for other CVTs.

J.A.C. Ambrósio (ed.), Advances in Computational Multibody Systems, 269–293.

1 Introduction

Nowadays automatic transmission is dominating the Japanese and American automotive market. Besides conventional automatic shift transmissions, continuously variable transmissions (CVT) have gained importance. Due to a step-less speed ratio they have the potential to be an ideal link between the engine and the power-train. Figure 1 shows a belt-CVT being used in a power train of a motorcar. The interface

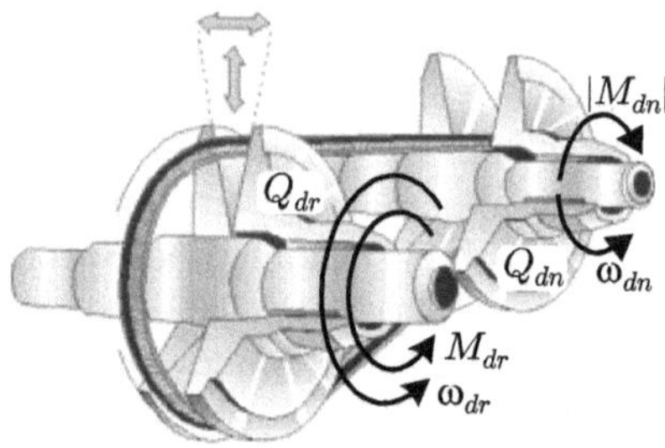

Fig. 1. Functionality of a CVT-system

of the described mechanical system is defined by the rotation speed ω, the external torque M and the pulley thrust Q. The indices dr, dn denote the **dr**iving and the **d**rive**n** pulley. The belt runs inside two V-pulleys and transmits the power by friction. One sheave of each pulley can be displaced axially by a hydraulic cylinder, the belt is forced to a given input and output radius. Thus a continuously variable adjustment of the transmission ratio is possible. This report deals with a metal pushing V-belt shown in Fig. 2. It consists of 300 to 600 flat steel elements that are held by two packs of thin steel rings. The flanks of the elements are periodically in contact to the

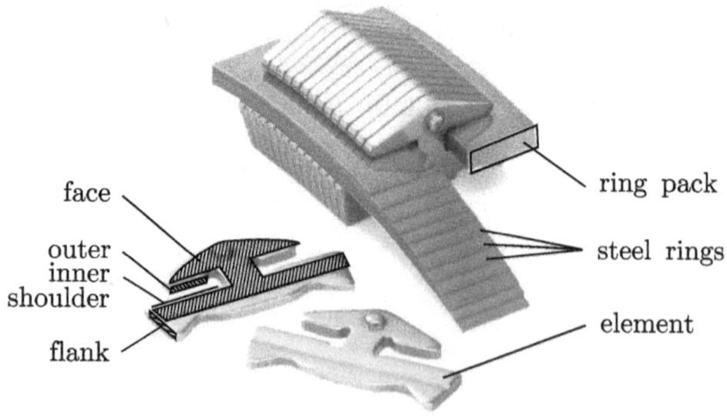

Fig. 2. Metal pushing V-belt

driving and driven pulley. The torque of the pulleys is effected by the tangential friction forces in the flanks. Intra belt the element transmits the tangential flank friction into two parts. One is transmitted by compression of the elements, the other by the tension of the rings. The history of the CVT push belt started in 1971 with the foundation of Van Doorne's Transmissie in the Netherlands. The customer requirements concerning transmission capacity, durability and efficiency have increased. Therefore it is important to analyse and to predict the behaviour of power transmissions in an early state of design. In this process numerical simulation models provide strong support.

2 Power Transmission

This section explains the fundamental mode of operation of the system and presents some important effects, which have to be taken into consideration modelling a push-belt. Figure 3 shows the basic mechanism of the power transmission of a push-belt

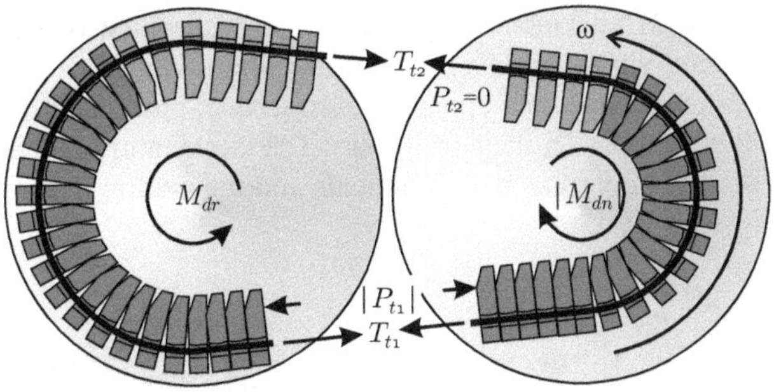

Fig. 3. Power transmission of a metal pushing V-belt CVT

CVT. The transmitted torque M results from the difference of the span forces, in detail the compression $P < 0$ and the ring tension $T > 0$. Using the contact radius r it can be estimated by

$$\frac{M_{dr}}{r_{dr}} \approx P_{t2} - P_{t1} + T_{t2} - T_{t1} \approx \frac{M_{dn}}{r_{dn}} \, . \tag{1}$$

In case $M_{dr} > 0$ the main part of power is transmitted by a compression force P_{t1} in the span leaving the driving pulley whereas the elements in the opposite span are separated. For $M_{dr} < 0$ it is contrary. The maximum transmittable torque M^* depends on the compression Q, the friction coefficient μ and the half wedge angle δ_0 of the pulley groove. It can be estimated by

$$\frac{M^*_{dr}}{r_{dr}} \approx \frac{M^*_{dn}}{r_{dn}} \approx \frac{2\mu}{\cos\delta_0} \min\{Q_{dr}, Q_{dn}\} \, . \tag{2}$$

The approximations assume that the friction forces are tangential with respect to the pulley. In reality the thrust Q does not only transmit power by friction, but it also deforms the pulley sheaves and the elements due to elasticity. This affects a radial velocity due to the penetration of the belt into the pulley groove. The velocities of the pulley $\mathbf{v}_P$ and the belt $\mathbf{v}_B$ as well as the relative velocity $d\mathbf{v}$ and the friction force f are shown in Fig. 4.

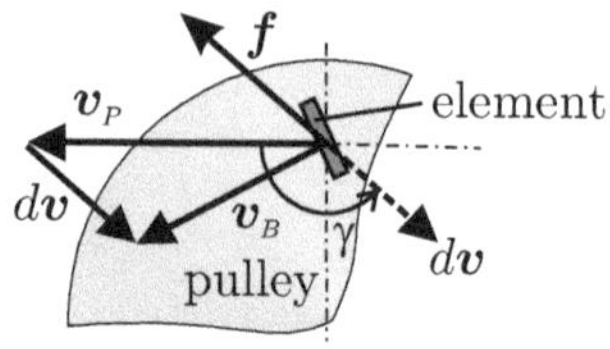

Fig. 4. Velocities of the pulley sheaves and the element

At a speed ratio $i \neq 1$ and a smaller torque M with respect to the maximum transmittable torque M^* a kinematic phenomenon occurs that is typical for push-belts. As experiments show, the inversion of the compression $P_{t1} \leftrightarrow P_{t2}$ from one span to the opposite one does not appear at zero torque, but it happens at a specific torque $M^0 \neq 0$ depending on the speed ratio. The explanation is found in a radial offset δr between the pitch line of the elements and the neutral lines of the rings. It affects a relative velocity

$$v_{rel} = \omega_{dr} \left(\frac{1}{i} (r_{dn} + \delta r) - (r_{dr} + \delta r) \right) \tag{3}$$

that is too big to be compensated by the elasticity of the rings. The phenomenon is illustrated in Fig. 5 for underdrive ($r_{dr} < r_{dn}$) and overdrive ($r_{dr} > r_{dn}$) with

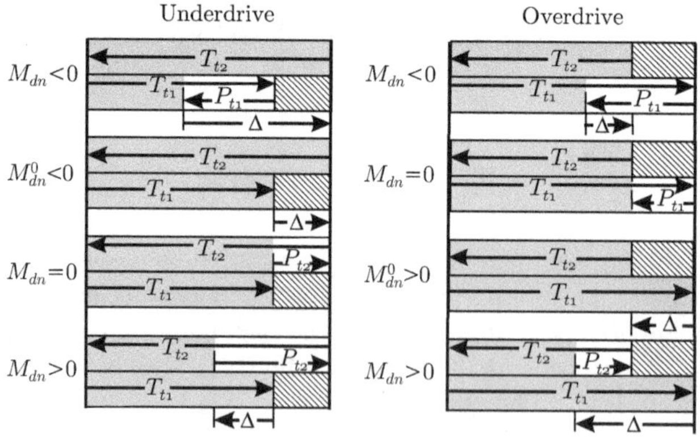

Fig. 5. Inversion of compression

four load cases each. The resulting transmission force is specified by $\Delta = \frac{M}{r}$. The two grey bars present the total tangential force of the spans $t1$ and $t2$ including ring tension T and compression P. In the case of underdrive the rings slip in the driving pulley opposite to the rotation direction. Due to this effect the tension T_{t2} rises always above the tension T_{t1}. To transmit zero torque it needs a compression force P_{t2} to equalise the difference of ring tension. In the overdrive condition the opposite effect occurs with a compression force P_{t1}. To simulate an accurate behaviour of the power transmission, the following model considers the elastic deformation of the pulleys and the elements as well as the unilateral constraints, such that separation of the elements can be modelled.

3 Simulation Model

Before the CVT-system can be characterised by algebraic equations, an abstraction of the real system to a physical model is necessary. This has to capture all important aspects in detail, but at the same time it has to be simple enough to guarantee reasonable simulation times. First of all the system is delimited with respect to the power train and the external excitations have to be defined. The presented model contains the driving and the driven pulleys, the belt and a controller. By the choice of a defined interface it is possible to include this model in a simulation program for complete power trains [7]. The rotation of the pulleys is excited by external kinematical or kinetical boundary conditions E_{dr}, E_{dn}. A target speed ratio i_0 and a torque ratio r [3], the latter being defined as the transmitted torque M divided by the maximum transmittable torque M^*, are assigned to the controller. The structure of the simulation model is shown in Fig. 6. To achieve a well structured model and implementation, the CVT-system is divided into sub-systems. These communicate with

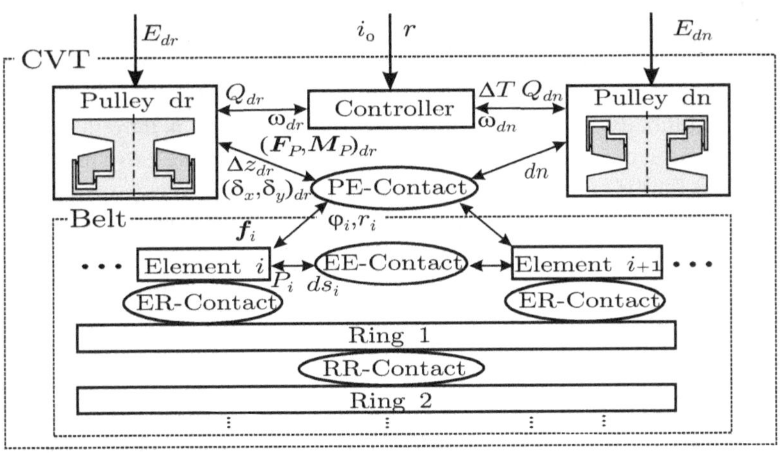

Fig. 6. Structure of the simulation model

each other via defined interfaces and are realized as classes in an object-orientated programming language (C^{++}). By this the sub-systems can be modified and tested easily and it is possible to reuse them for other simulation programs (e.g. simulation of chain-drives). The sub-systems are classified in force elements and bodies. *Bodies* contain a set of rigid and/or elastic degrees of freedom. The equations of motion result in the principle of virtual power, using the well known projection method by JOURDAIN [6].

$$\sum_i \int_{m_i} {}_B\mathbf{J}_i^T \left({}_B\mathbf{a}_i - {}_B\mathbf{f} \right) \mathrm{d}m_i = 0\,, \quad {}_B\mathbf{J}_i = \frac{\partial_B \mathbf{v}_i}{\partial \dot{\mathbf{q}}} \tag{4}$$

Force elements calculate the interaction between bodies, e.g. the relative kinematics and forces in a contact. In the following sections the sub-systems are described in detail.

4 Model of the Pulleys

A CVT-system always contains a driving and a driven pulley, which transform the input- and output-torque into a tangential friction force. In belt-CVTs a set of V-pulleys is used, where one sheave of each pulley is axially movable by a hydraulic cylinder forcing the belt to a prescribed radius (s. Fig. 1). A control algorithm affects the pulley thrust Q_{dr}, Q_{dn} to adjust the speed ratio. The embedding of the pulleys into the CVT-system is shown in Fig. 6. The boundary conditions are given by the external excitation E the control pressure to affect the pulley thrust Q and the belt contact forces $\mathbf{f}_i$. The external excitation E can either be a torque M or an angular velocity ω as a constraint for the rotation. The pulley sheaves, the shaft and the shaft-to-collar connection are deformed according to their elasticity. This causes a radial penetration of the belt into the pulley groove, which is of major importance for the mechanical behaviour of the CVT system. In contrast to rubber V-belts the deformation of the sheaves and the steel elements are of the same magnitude, so both must be taken into consideration while modelling a metal V-belt system. This is included in the present multibody system. First an elastic RITZ-approach is introduced and discussed. Because some eigenfrequencies of the elastic bodies are extremely high and not of interest for the investigated problems, an alternative approach is introduced to achieve reasonable times for the numerical simulation. This splits up the degrees of freedom into dynamical and quasi-static ones.

4.1 Dynamical Ritz-approach

To evaluate Eq. (4) the absolute acceleration of the mass elements is needed. Modelling an elastic multibody system, it is a common procedure [1] to describe the position $\mathbf{r}$ of a mass point of the deformed configuration by the undeformed configuration $\mathbf{r}_0$ and a displacement vector $\mathbf{r}_{el}$. In order to limit the degrees of freedom of the infinitesimal mass points a RITZ-approximation is used.

$$\mathbf{r}(\mathbf{r}_0, t) = \mathbf{r}_0(t) + \mathbf{r}_{el}(\mathbf{r}_0, t), \quad \mathbf{r}_{el}(\mathbf{r}_0, t) = \mathbf{W}(\mathbf{r}_0)\mathbf{q}_{el}(t) \tag{5}$$

The matrix $\mathbf{W}$ contains the shape functions, the eigenforms of the pulley sheaves respectively that are evaluated by FEM-calculations. The RITZ-approximation (5) leads to an ordinary differential equation in standard form.

$$\mathbf{M}\ddot{\mathbf{q}} - \mathbf{h}(\mathbf{q}, \dot{\mathbf{q}}, t) = \mathbf{0} \tag{6}$$

The matrix $\mathbf{M}$ contains the mass integrals of shape functions. Due to small elastic deformations, these integrals are independent of time and need to be evaluated only once.

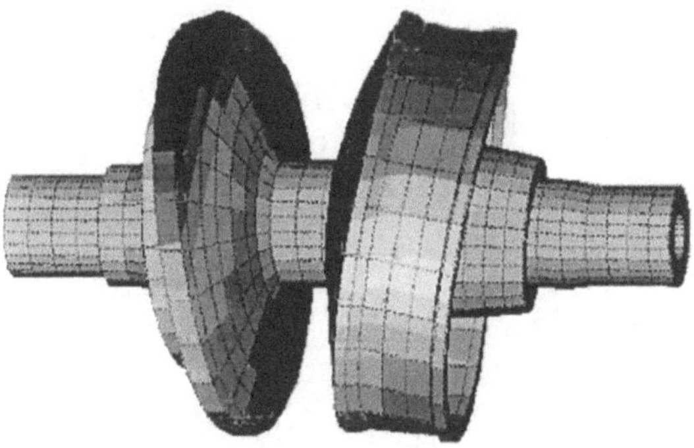

Fig. 7. FE-Model of the pulley

The RITZ-approximation is a powerful and often used method. To model the dynamics of CVT pulleys it was first used by SRNIK [9]. In the implementation only a confined number of shape functions can be evaluated. This means the computation of the pulley surface implies a position error. Because the contact area between the pulley sheaves and the belt spreads across the total arc of contact this error has a big influence on the distribution of the contact forces. Many eigenforms have to be superposed to compute an accurate force distribution. In [9] the eigenfrequencies belonging to the higher eigenforms lead to nearly unacceptable computation times for the numerical simulation. In order to eliminate these high frequencies and still obtain an adequate deformations, a quasi-static approach for the elastic deformation has been developed. Alternatively it would be possible to use hierarchical shape functions in the same manner as in Sec. 5.1.

4.2 Mixed Dynamical- and Quasi-Static-Approximation

Due to the high stiffness, a CVT-system involves a diversified frequency spectrum. In order to obtain an appropriate mathematical model, it is important to define the limit of frequencies that can be resolved. Frequencies that exceed the limit are eliminated, by calculating their degrees of freedom with a quasi-static formulation. At the beginning of this section the dynamical degrees of freedom are introduced. Then the quasi-static degrees of freedom are described, while the deformation of the elastic bodies are discussed in detail.

4.2.1 Rigid-Body-Model

The elastic deformation of the pulley system is small, and therefore the global dynamics can be approached by a model of two rigid-bodies shown in Fig. 8. Its state is defined by four degrees of freedom. The bearings of the shaft are assumed to be inelastic. The rotation is specified by the angle φ and the angular velocity ω. The sec-

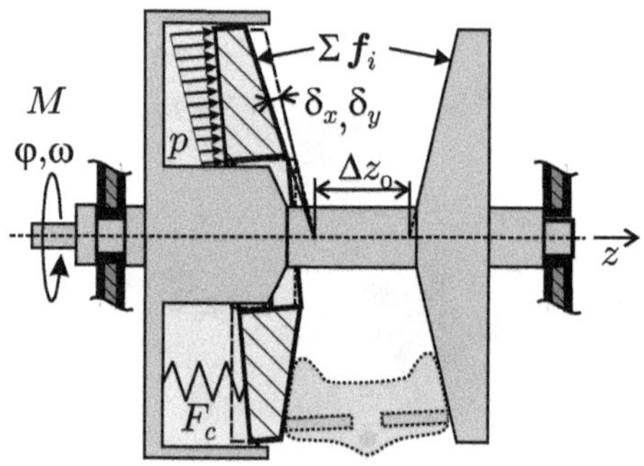

Fig. 8. Rigid body model of the pulleys

ond sheave is bedded by a shaft-to-collar connection. Its axial position is quantified by the distance Δz_0 between the fixed and the movable sheave. Due to the elasticity and clearance of the shaft-to-collar connection, the movable sheave can tilt. Its orientation is quantified by the angles δ_x and δ_y with respect to an orthogonal axes system that is perpendicular to the rotation axis. The rotation ω is either prescribed kinematically, or it results from the principle of angular momentum.

$$J_z \dot{\omega} = M - \sum_i (\mathbf{r}_i \times \mathbf{f}_i)_z \,, \qquad \varphi = \int \dot{\omega} \, \mathrm{d}t \tag{7}$$

Here the moment of inertia J_z contains all masses of the pulley. In the steady state the external torque M and the axial components of the torques resulting from the contact forces $\mathbf{f}_i$ and the lever arms $\mathbf{r}_i$ are in equilibrium. The motion of the adjustable pulley sheave is strongly damped by the hydraulic oil. When a second order differential equation is used to model the dynamics of the sheave the resulting eigenvalues are of different order of magnitude. To avoid numerical problems the order is reduced by one so that we obtain a PT_1 behaviour. The distance Δz_0 is calculated by Eq. (8) with a time delay T_1.

$$T_1 \dot{z} = \sum_i f_{i,ax} - F_c - Ap - F_\omega \, , \qquad z = \int \dot{z} \, \mathrm{d}t \tag{8}$$

The time constant T_1 in Eq. (8) is scaled by a numerically calculated stiffness $\frac{\partial F_c}{\partial z}$ to transform the equation to a standard form. In the steady state case the axial components of the contact forces $f_{i,ax}$ are balanced by the spring prestress F_c and the piston force. It consists of two parts. The first part Ap_0 is independent of the rotational velocity. It is proportional to the surface A of the piston and the applied pressure p_0. The second part F_ω considers the centrifugal forces that influences the local piston pressure. The tilting angles δ_x and δ_y are calculated by Eq. (9) using a rotational stiffness c_δ.

$$\begin{pmatrix} T_1 & 0 \\ 0 & T_1 \end{pmatrix} \begin{pmatrix} \dot{\delta}_x \\ \dot{\delta}_y \end{pmatrix} = \begin{pmatrix} \frac{1}{c_\delta} \sum_i (\mathbf{r}_i \times \mathbf{f}_i)_x - \delta_x \\ \frac{1}{c_\delta} \sum_i (\mathbf{r}_i \times \mathbf{f}_i)_y - \delta_y \end{pmatrix}, \qquad \begin{pmatrix} \delta_x \\ \delta_y \end{pmatrix} = \int \begin{pmatrix} \dot{\delta}_x \\ \dot{\delta}_y \end{pmatrix} dt \tag{9}$$

This rigid-body model considers only the elasticity of the shaft-to-collar connection, while the elasticity of the shaft and the sheaves are not included yet. This is done by superposing the tilting angles δ_x, δ_y and the elastic deformation of the shaft and the sheaves. The deflection models are presented in the next sections.

4.2.2 Deflection of the Shaft

The bending of the shaft is approximated by the deflection curve of a spatial beam. The boundary conditions consider the deflection and clearance of the bearing points. The contact forces of the belt are summed up to a resultant couple ($\mathbf{F}_P$, $\mathbf{M}_P$). The deflection can be calculated analytically by subdividing the shaft into segments of constant cross sections. Figure 9 shows the mechanical model of the shaft. Assuming rigid sheaves, the wedge angle is changed by the local bending δ_B. In case of the movable pulley the tilting angles δ_x, δ_y and the clearance δ_S is added. The local wedge angle depending on the rotational position is calculated by both parts of x- and y-axis.

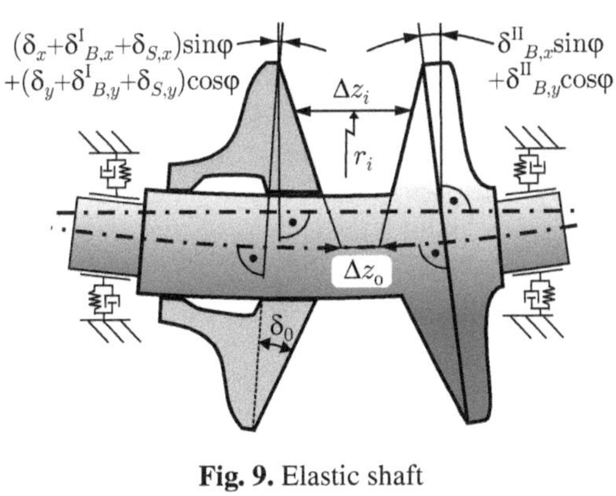

Fig. 9. Elastic shaft

4.2.3 Deflection of the Pulley Sheaves

To reduce acceleration losses, the moment of inertia of the power train should be as small as possible. The CVT-pulleys contribute to the inertia significantly and therefore the pulley sheaves should be as thin as possible. Here the reduction of weight is limited by the increasing deformation. It is important to model the deformations as accurate as possible. In the following sections two approaches are presented.

A simple approach with which the deformation of the pulley sheaves can be considered is to assume that the cone surface is tilting as a whole. In this case an equivalent torsional spring is used. The stiffness can be treated as a spring in series with the stiffness of the shaft-to-collar connection. Figure 10 shows the model of

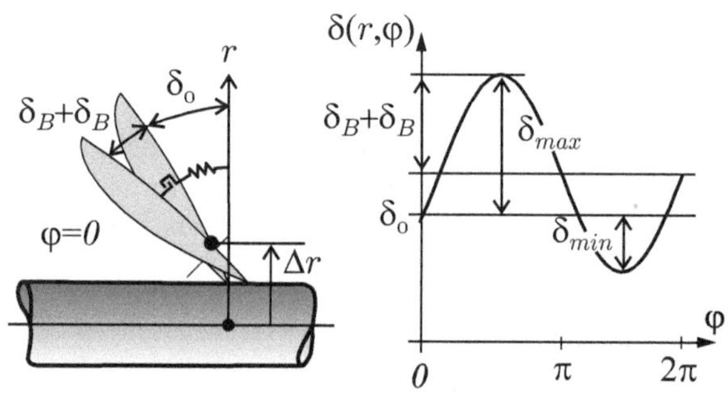

Fig. 10. Sinus approach

this approach. The results can be improved if the axis of tilting is shifted from the centre point ($r = 0$) towards the contact (Δr). The approach leads to good results in modelling CVT-chains [9] [8] but the distribution of the contact force is not accurate when this approach is used for metal V-belts since the axial stiffness of the belt is much higher in this case.

Thus a better approximation for the deformation has to be formulated in which the high eigenfrequencies are eliminated. To this end CASTIGLIANO's Strain Energy Theorem is used, which is a useful tool for determining displacements of a linear elastic system [10]. The change of the elastic potential is equal to the work performed at the surface of the body. The partial derivative of the strain energy V by an external force f_i, acting on the surface point i, leads to its displacement u_i along the direction of f_i.

$$V = \frac{1}{2} \sum_{i=1}^{n} \sum_{j=1}^{n} w_{ij} f_i f_j \quad \rightarrow \quad \frac{\partial V}{\partial f_i} = \sum_{j=1}^{n} w_{ij} f_j = u_i \tag{10}$$

The flexibility coefficients w_{ij} characterise the interaction of the surface points by elasticity. Here they give the displacement of the i-th point of contact due to a generalised force f_j acting at a contact point j. The flexibility coefficients w_{ij} can easily

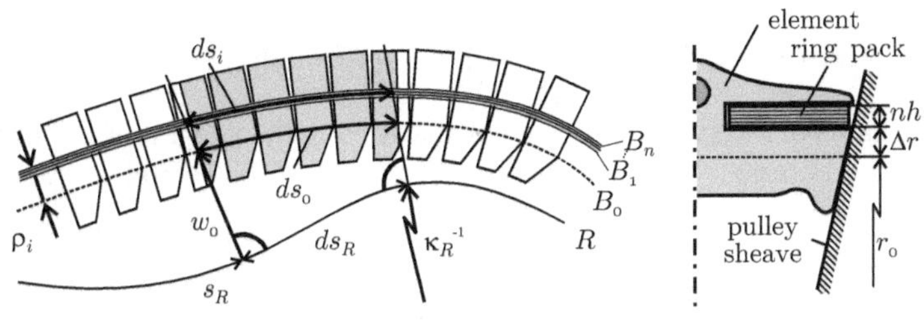

Fig. 11. Reference systems of the belt

be calculated using a linear finite element analysis. The calculation of the contact forces f_j is explained in Sec. 6.1.

5 Model of the Belt

Before modelling the belt, one has to decide whether the behaviour of the metal V-belt should be described by a discrete or a continuous approach. A discrete model of the elements can be realised easily, but it results in extreme high frequencies ($\approx 500kHz$) due to the small mass and the high stiffness. To achieve reasonable simulation times, only frequencies up to a certain limit, that is of technical interest, are included and the frequencies above the limit have to be eliminated. For this objective the motion of the belt is specified by separate longitudinal and transversal approaches in combination with hierarchical shape functions. The position of the belt is specified in a reference system R by a path coordinate s_R and the transversal displaced $w(s_R, t)$. Due to the radial expansion of the elements and the rings a radial reference mark is needed to which the transversal displacement w and the tangential velocity v refers to. The pitch line B_0 is used for the elements and the neutral line B_i for the rings. Figure 11 shows the systems of the belt. Because the transversal position of elements and rings are coupled, the systems B_0 and B_i are parallel and the spacial derivatives w', w'' and time derivative $\dot{w}$ are the same for all layers. The motion of the belt is calculated in EULER-coordinates. To describe the kinematics of the belt-fixed elements, a transformation to the LAGRANGE-description is used. Equation (11) is valid for any field-variable x, e.g. the transversal displaced w.

$$\dot{x} = \frac{\partial x}{\partial t} + v_i \frac{\partial x}{\partial s} \tag{11}$$

Furthermore transformations between the reference system and the belt systems have to be considered. An infinitesimal length ∂s_R of the reference system R differs in the dedicated length ∂s_i in a belt system B_i. The transformation depends on the curvature κ_R, the transversal position w_i and its first derivative in space w'. The planar transformation between the reference system R and a belt system B_i as well as the transformation between two single rings $i = 1..n$ with the thickness h are given by

$$\frac{\mathrm{d}s_i}{\mathrm{d}s_R} = \sqrt{(1+\kappa_R w_i)^2 + w'^2}, \qquad \frac{\mathrm{d}s_{i+1}}{\mathrm{d}s_i} = 1 + \kappa_R h \quad i = 1 \dots n-1\,. \quad (12)$$

The path lengths of the systems B_i are calculated with the integral of Eq. (12) along the reference coordinate s_R, whereas the length l_0 of the reference system R is known.

$$l_0 := \int ds_R, \quad l_{B,0} = \int_0^{l_0} \frac{\mathrm{d}s_0}{\mathrm{d}s_R} \mathrm{d}s_R, \quad l_{B,i+1} = l_{B,i} + 2\pi h, \; i = 1 \dots n-1 \quad (13)$$

5.1 Transversal Dynamics

The transversal position of the belt depends on the local deflection of the pulley sheaves and results in the equilibrium of contact forces acting on the elements and rings. A continuous approach is used. The LAGRANGE equations of motion are derived from the kinetic energy T and the potential energy V of a stretched, moving belt with the tangential velocity v. Here μ is the mass per length, EI the bending stiffness and F the tension force. In this section the reference coordinate s_R is shortened by s.

$$T = \frac{\mu}{2} \int \left((\dot{w} + v w')^2 + (1 + w\kappa_R)^2 v^2 \right) \mathrm{d}s \quad (14)$$

$$V = \frac{1}{2} \int \left(F \left(w'^2 + 2w\kappa_R \right) + EI \left(\kappa_R - w\kappa_R^2 - w'' \right)^2 \right) \mathrm{d}s \quad (15)$$

The displacement w and its derivatives depend on both, time and space. In order to solve the LAGRANGE's equations, a RITZ approximation is introduced to separate both dependencies.

$$w(s,t) = \bar{\mathbf{w}}^T(s)\mathbf{q}(t) \quad (16)$$

This leads to the standard form of equations of motion (ODE, 2nd order), that are integrated numerically by transforming to a first order system.

$$\mathbf{M}\ddot{\mathbf{q}} + \mathbf{D}\dot{\mathbf{q}} + \mathbf{C}\mathbf{q} + \mathbf{b} = \mathbf{Q}_k + \mathbf{Q}_m \quad (17)$$

The mass matrix $\mathbf{M}$, the damping matrix $\mathbf{D}$, the matrix of stiffness $\mathbf{C}$ and the vector $\mathbf{b}$ contain integrals of the shape functions and have to be calculated only once, if the reference system is invariant in time.

$$\mathbf{M} = \bar{m} \int \bar{\mathbf{w}}\bar{\mathbf{w}}^T \mathrm{d}s \quad (18)$$

$$\mathbf{D} = \bar{m}v \int \bar{\mathbf{w}}\bar{\mathbf{w}}'^T - \bar{\mathbf{w}}'\bar{\mathbf{w}}^T \mathrm{d}s \quad (19)$$

$$\mathbf{C} = EI \int \bar{\mathbf{w}}''\bar{\mathbf{w}}''^T + \frac{3}{r^4}\bar{\mathbf{w}}\bar{\mathbf{w}}^T + \frac{1}{r^2}\bar{\mathbf{w}}'\bar{\mathbf{w}}'^T + \frac{3}{r^2}\bar{\mathbf{w}}''\bar{\mathbf{w}}^T + \frac{3}{r^2}\bar{\mathbf{w}}\bar{\mathbf{w}}''^T \mathrm{d}s$$
$$+F \int \bar{\mathbf{w}}'\bar{\mathbf{w}}'^T \mathrm{d}s - \bar{m}v^2 \int \bar{\mathbf{w}}'\bar{\mathbf{w}}'^T + \frac{1}{r^2}\bar{\mathbf{w}}\bar{\mathbf{w}}^T \mathrm{d}s \quad (20)$$

$$\mathbf{b} = -EI \int \frac{1}{r^3}\bar{\mathbf{w}} + \frac{1}{r}\bar{\mathbf{w}}'' \mathrm{d}s + F \int \frac{1}{r}\bar{\mathbf{w}} \mathrm{d}s - \bar{m}v^2 \int \frac{1}{r_i}\bar{\mathbf{w}} \mathrm{d}s \quad (21)$$

The generalised force vector $\mathbf{Q}_k$ contains the radial contact forces of the pulley. The vector $\mathbf{Q}_m$ comprises the torque that results from the compression force between two interconnecting faces of elements. This is modelled with discrete force elements with the distance δ.

$$\mathbf{Q}_k = \sum \bar{\mathbf{w}}(s_i) F_i \tag{22}$$

$$\mathbf{Q}_m = \sum \left(\mathbf{w}'(s_i + \frac{\delta}{2}) - \mathbf{w}'(s_i - \frac{\delta}{2}) \right) M_i \tag{23}$$

The bending torque in the belt-plane is given by Eq. (24), while $k^{\pm}$ is the outer and inner border of the contact face based on the pitch line of the belt.

$$M_i = \begin{cases} Pk^- & \text{for} \quad EI\kappa_i > -Pk^- \\ -EI\kappa_i & \text{for } Pk^- \leq -EI\kappa_i \leq -Pk^+ \\ -Pk^+ & \text{for} \quad -EI\kappa_i > -Pk^+ \end{cases} \tag{24}$$

The current position of the belt is calculated by the RITZ-approximation in Eq. (16) with a confined number of eigenforms. So the set of possible positions is restricted by the linear combination of shape functions $\bar{\mathbf{w}}$. The position error has a big influence on the distribution of the contact forces. A small modelling error in position leads to a large error in the contact forces. For a realistic distribution of the contact forces, smooth local shape functions are necessary to describe the deformation of the contact area. Then an ordinary RITZ-approach results in high eigenfrequencies that lead to the same numerical problems as a discrete element by element model.

To avoid the high eigenfrequencies the method of hierarchical bases is used, which represents finite functions with differently sized supports. So the whole belt is discretised by a rough grid (Fig. 12a). B-Splines are used as shape functions. The

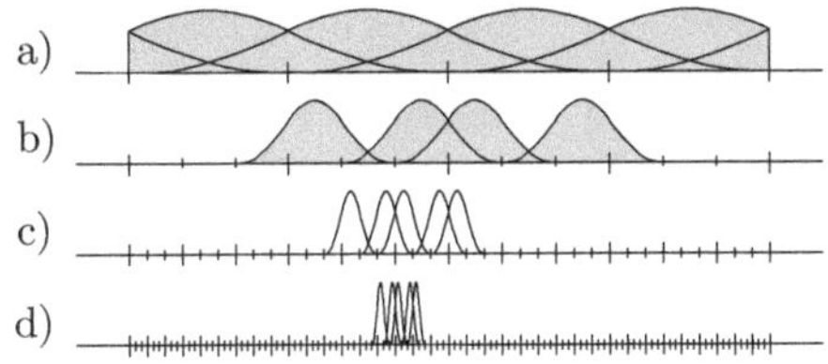

Fig. 12. Hierarchical bases

contact areas are refined with additional splines (Fig. 12b) which do not change the existing shape functions. In the borderline of the contact areas there are high gradients of curvature w''', so more refinements are needed (Fig. 12c,d). Because every base corresponds to a certain eigenfrequency, it is possible to trim the frequency spectrum that is considered in the simulation results. If the eigenfrequency of a shape function should be eliminated, the weighting is not calculated dynamically but quasi-staticly ($PT1$). By this the state vector $\mathbf{z} = (\mathbf{q}, \dot{\mathbf{q}})$ is composed of the position $\mathbf{q} = (\mathbf{q}_{dyn}, \mathbf{q}_{PT1})$ and the velocity $\dot{\mathbf{q}} = \dot{\mathbf{q}}_{dyn}$.

5.2 Longitudinal Dynamics

The longitudinal dynamics of the belt is fundamental for the power transmission. Therefore only frequencies up to a certain limit ($\approx 1kHz$) have to be taken into consideration. In almost the same manner as in Sec. 5.1 a dynamical and a quasi-static approach are superposed. In this section the elements (index 0) as well as the rings (index $1..n$) are labelled by *layers*, to shorten the writing.

The tangential velocities v_i of the elements and the rings are calculated by the Continuity Equation. The equation is based on the line density $\rho_i = (1+\varepsilon_i)^{-1}$ that is a function of the local strain $\varepsilon_i(t, s_R)$.

$$\frac{\partial}{\partial t}(1+\varepsilon_i)^{-1} + \frac{\partial}{\partial s_R}\left(\frac{v}{1+\varepsilon_i}\right) = 0 \tag{25}$$

The local strain ε_i of the belt layers is calculated from the sum of the mean strain $\bar{\varepsilon}_i$ and a local deviation $\Delta\varepsilon_i$. The mean strain $\bar{\varepsilon}_i$ is defined by the proportion of the present length of the layer and the length of the unstrained material. The present length is equal to the length $l_{B,i}$ of the belt system in Eq. (13). The mean strain of the elements $\bar{\varepsilon}_0$ and the rings $\bar{\varepsilon}_i$ is given by

$$\bar{\varepsilon}_0 = \frac{l_{B0}}{nb} - 1, \qquad \bar{\varepsilon}_i = \frac{l_{Bi}}{k_i(l_1 + 2\pi(i-1)h)} - 1\,, \quad i = 1..n\,. \tag{26}$$

Here nb is the sum of the thickness b of all elements and l_1 is the ideal length of the inner ring. In industrial manufacturing, the length of the rings contains a tolerance k_i. The local deviations $\Delta\varepsilon_i$ of the strain with respect to the mean values $\bar{\varepsilon}_i$ is caused by the local friction between the rings. It is approximated by a RITZ-approach in Eq. (27) using linear shape functions, to get clear partition between separation and compression of the elements. At this the closing condition has to be considered.

$$\varepsilon_i(s_R, t) = \bar{\varepsilon}_i + \bar{\mathbf{w}}^T(s_R)\,\mathbf{q}_i(t)\,, \qquad \mathbf{q}_i^T(t)\int_0^{l_0} \bar{\mathbf{w}}(s_R)\,\mathrm{d}s_R = 0 \tag{27}$$

Although the closing condition can be treated as a constraint for the numerical integration (DAE), in this case it is possible to consider the closing condition in the choice of the generalised coordinates $\mathbf{q}_i$ in advance. This reduces the dimension of $\mathbf{q}_i$ by one and the constraints need not be checked separately (ODE). To eliminate vibrations of single elements, the field of the deviations $\Delta\varepsilon_i(\mathbf{q}_i)$ is calculated neglecting local acceleration. This leads to an ordinary differential equation of order one.

$$\mathbf{T}\dot{\mathbf{q}}_i = \mathbf{H}_i\mathbf{q}_i + \mathbf{h}_i \qquad \mathbf{q} \in I\!R^{n-1} \tag{28}$$

The matrix $\mathbf{T}$ of characteristic time looses its band structure when the closing condition is considered. Since it is constant, it needs to be inverted only once. In addition, it is equal for all layers. The matrix $\mathbf{H}_i$ represents the linearised interaction in a single layer. In case of elements it includes the unilateral constraints, specified in Sec. 6.2.

The vector $\mathbf{h}_i$ contains the generalised friction forces. Here the friction forces acting on the surface of the layer have to be generalised to the neutral line of the layer. Details for this transformation are given in Sec. 6.3.

The local velocity v_i is defined by Eqs. (25) and (27). For a constant loading the local strain ε depends primarily on the path coordinate s_R, while the influence of the time coordinate t is low. So we assume the left term of Eq. (25) to be zero.

$$v_i = C\,(1+\varepsilon_i) \quad \rightarrow \quad \bar{v}_i = \frac{C}{l_0}\int_0^{l_0} (1+\varepsilon_i)\,\mathrm{d}s_R = C\,(1+\bar{\varepsilon}_i) \tag{29}$$

The constant C is computed by eliminating the closing condition. Then the local velocity v_i is given by

$$v_i(t, s_R) = \bar{v}_i(t)\frac{1+\varepsilon_i(t, s_R)}{1+\bar{\varepsilon}_i(t)} \ . \tag{30}$$

The mean velocity $\bar{v}_i$ is computed in the equation of the linear momentum p in tangential direction.

$$\dot{p}_i = l_{B,i}\mu_i\dot{\bar{v}}_i = \int_{s=0}^{l_{B,i}} \left(J_i^{i+1} f_{t,i+1} - J_i^i f_{t,i}\right)\mathrm{d}s \tag{31}$$

Here μ_i is the mass per length of the layer i and $f_{t,i}$ is the friction force between the layer $i-1$ and the layer i for an infinitesimal belt length ds. The friction force $f_{t,j}$ acts on the surface j of the layer and has to be generalised to the neutral line B_i by J_i^j. The details for the transformations are given in Sec. 6.3.

In addition to the local velocity v_i the compression force P and the tension forces T_i can simply be calculated from the local strain ε_i. The elements can only transmit pressure forces P but no tension. The opposite is the case for the rings.

$$P(s_R) = \begin{cases} EA\varepsilon_0 & \text{for } \varepsilon_0 < 0 \\ 0 & \text{for } \varepsilon_0 \geq 0 \end{cases}, \qquad T(s_R) = \begin{cases} 0 & \text{for } \varepsilon_i \leq 0 \\ EA\varepsilon_i & \text{for } \varepsilon_j > 0 \end{cases} \tag{32}$$

6 Modelling of the Contacts

The behaviour of the power transmission is primarily defined by the contact conditions between the sheaves, the elements and the rings. Because of the high number of contacts it is necessary to model only the important functionality of the contacts in detail. Here state transitions and the chosen dimension of the transmitted forces are important. If the state of a contact changes between *closed* and *open*, the unilateral constraints have to be taken into consideration. A tangential relative velocity along with a normal force generates a friction force in the contact plane. Depending on the change of the relative velocity, one or two components of the friction force have to be taken into account. Figure 13 shows the contact forces acting on an element. The modelling is described in the following sections.

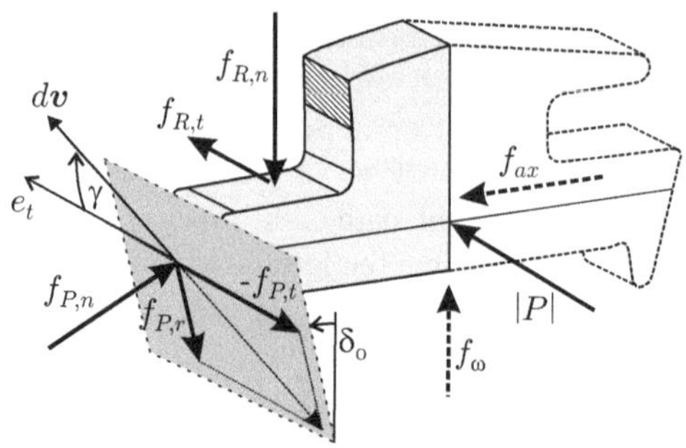

Fig. 13. Forces acting on an element

6.1 Contact between the Pulley Sheaves and the Belt

The contact between the pulley sheaves and the belt is modelled by an unilateral normal force $f_{P,n}$ and and a two-dimensional friction force in the plane of the flanks. In tangential direction it is labelled by $f_{P,t}$, in radial direction by $f_{P,r}$. The calculation of the unilateral normal force $f_{P,n}$ and the friction forces is presented in this section.

Using the models described in Sec. 4.2, the tilting of the rigid sheaves consists of the local bending δ_B of the shaft and the elastic deformation δ and the clearance δ_S of the shaft-to-collar connection. The position of a contact point i is defined by the coordinates $({}_P\varphi_i, {}_Pr_i)$ in a polar pulley system P. The distance, penetration $\tilde{g}_i$ of rigid sheaves and a rigid element with the axial width z_e, is calculated by

$$\begin{aligned} \tilde{g}_i = \Delta z_0 + 2r_i \tan\delta_0 - r_i\left(\delta_x + \delta_{B,x} + \delta_{S,x}\right)\sin\varphi_i \\ + r_i\left(\delta_y + \delta_{B,y} + \delta_{S,y}\right)\cos\varphi_i - z_e \,. \end{aligned} \tag{33}$$

Next the geometrical value $\tilde{g}_i$ has to be separated into three parts. The first part is the elastic deformation of the pulley sheaves, the second part is the axial strain resulting from the compression of the elements and the last part is a positive distance g_i of the deformed bodies. Either the element i is in contact with the pulley sheave so that the distance g_i is zero and an axial force $f_{ax,i} > 0$ is transmitted or the deformed bodies are separated $g_i > 0$, so the contact force $f_{ax,i} = 0$ disappears. Algebraically this is formulated by the complementarity

$$f_{ax,i} \geq 0, \quad g_i \geq 0, \quad f_{ax,i}\, g_i = 0 \,. \tag{34}$$

As shown in Sec. 4.2 every contact force $f_{ax,i}$ leads to a global deformation of the sheave that is described by the flexibility coefficients w_{ij}. A negative value of $\tilde{g}$ is neither necessary nor sufficient to decide if a contact is open or closed. This is illustrated in Fig. 14 using a cantilever beam. The contact on the right is open ($g_i > 0$) in spite of the fact that $\tilde{g}_i$ is negative. The state of the contacts has to be identified by solving all forces and deformations simultaneously. This leads to a linear system of equations containing the flexibility coefficients w_{ij} and the axial stiffness c_e of the elements.

$$\tilde{g}_i = \sum_j w_{ij} f_{ax,j} + c_e^{-1} f_{ax,i} + g_i \tag{35}$$

The linear system of equations (35) has to be solved considering the unilateral constraints in Eq. (34). This results in a linear complementarity problem [6].

The contact force f_{ax} calculated above acts in axial direction. The axial force f_{ax} has to be in equilibrium with all other forces in the cross section of the belt. As can be seen in Fig. 13 the axial force f_{ax}, the normal ring force $f_{R,n}$ and the dynamic force f_ω are equal to the normal and radial pulley forces $f_{P,n}$, $f_{P,r}$. Here the direction of the friction force $f_{P,r}$ influences the quantity of the axial and normal force. Fig. 15 shows the polygon of forces for both directions of the friction force. On the upper side the element slides into the pulley grove ($\dot{r} < 0$) and the friction force $f_{P,r}$ attributes to the axial force f_{ax}. The other polygon shows an element with increasing contact radius ($\dot{r} > 0$). Here the friction force $f_{P,r}$ works contrary to the axial force f_{ax}.

The contact forces $f_{P,n}$, $f_{P,t}$, $f_{P,r}$ are evaluated in a suitable system C that lies in the flank of the element. The friction forces $f_{P,t}$, $f_{P,r}$ are computed by COULOMB's friction law. Fig. 16 shows the contact forces in the system C. The direction of the relative velocity $d\mathbf{v}$ is specified by the sliding angle γ with respect to the tangential direction $\mathbf{e}_t$. It is calculated by

$$\gamma = \begin{cases} \arcsin \frac{dv_r}{dv} & \text{for } dv_t > 0 \\ \pi - \arcsin \frac{dv_r}{dv} & \text{for } dv_t < 0 \end{cases} . \tag{36}$$

The absolute value of the friction force is transformed in a tangential and radial component defined by the sliding angle γ. Here we assume no stick-slip effects, because a disappearance of both, the tangential and the radial relative velocity is infrequent.

$$f_{P,t} = -f_{P,n}\,\mu \cos\gamma = -f_{P,n}\,\mu \frac{dv_t}{dv} \tag{37}$$

$$f_{P,r} = -f_{P,n}\,\mu \sin\gamma = -f_{P,n}\,\mu \frac{dv_r}{dv} \tag{38}$$

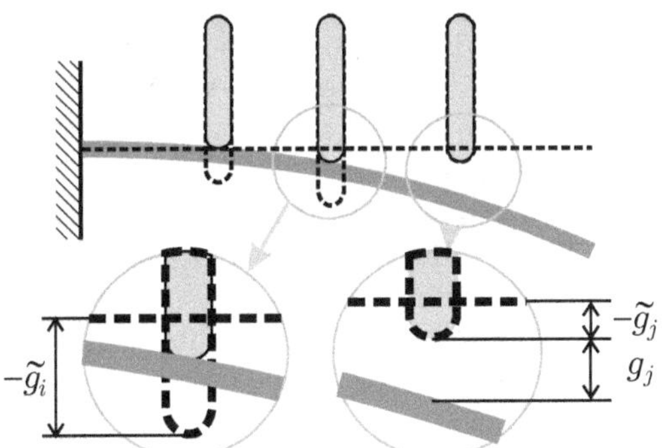

Fig. 14. Unilateral constraints

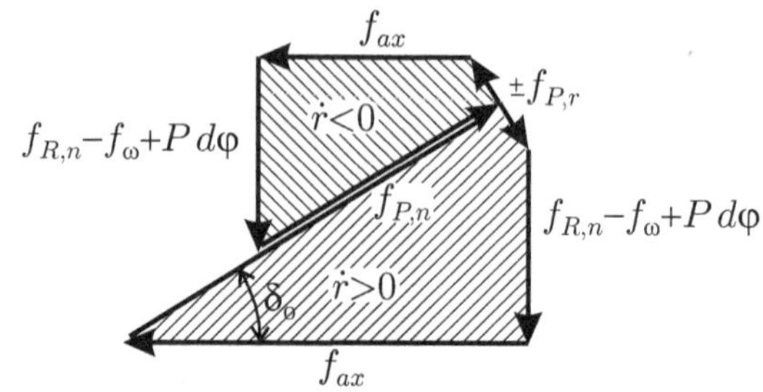

Fig. 15. Polygon of forces in the cross section of the belt

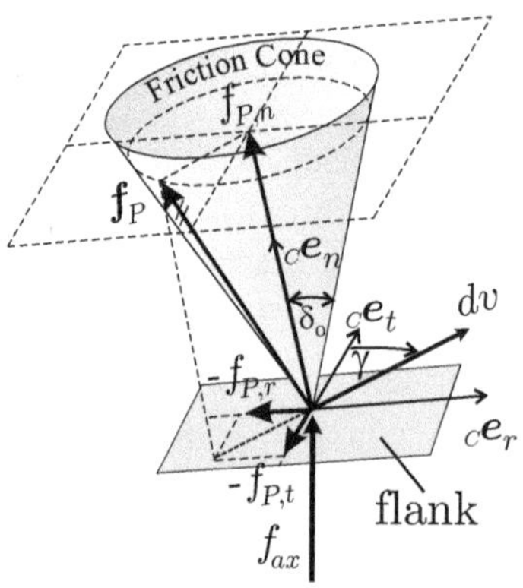

Fig. 16. Pulley contact forces in the system C

The contact force $f_{P,n}$ in normal direction is deduced from Fig. 15 using the friction force defined in Eq. (38).

$$f_{P,n} = \frac{f_{ax}}{\cos\delta_0 - \mu\sin\delta_0\sin\gamma} \tag{39}$$

Next the relative velocity $d\mathbf{v}$ is determined. The direction of the relative velocity, the sliding angle γ respectively is necessary to assign the direction of the friction force. Besides the contact system C, four additional coordinate systems are used for the transformations. They are shown in Fig. 17. The reference system R and the belt system B_0 have already been introduced in Sec. 5. For the pulleys an inertial system I and a polar system P is used in addition. The relative velocity $d\mathbf{v}$ is calculated by the difference between the local velocities of the pulley sheaves $\mathbf{v}_P$ and the element $\mathbf{v}_B$. It has to be transformed into the contact plane C of the element's flank.

$$\begin{aligned} {}_C d\mathbf{v} &= {}_C\mathbf{v}_B - {}_C\,\mathbf{v}_P \\ &= \mathbf{A}_{CB}\mathbf{A}_{BR}\left(\begin{pmatrix} v_0\,(1+w\kappa) \\ -(\dot{w}+v_0 w') \\ v_{ax,1}\end{pmatrix} - \mathbf{A}_{RP}\begin{pmatrix}\omega\,(r_i+w) \\ 0 \\ v_{ax,2}\end{pmatrix}\right)\end{aligned} \tag{40}$$

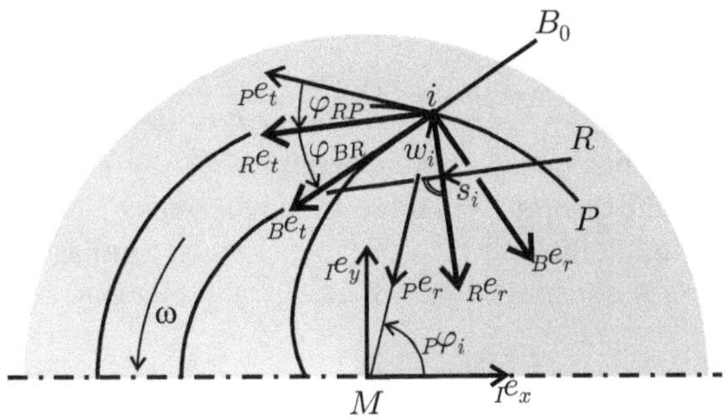

Fig. 17. Coordinate systems

Since the model of the belt is two-dimensional, the axial component in Eq. (40) is not defined. On the other hand, it needs the relative velocity $d\mathbf{v}$ only in the case of a closed contact. In this case the relative velocity in normal direction $_C dv_n$ is zero. This constraint allows to calculate the unknown value v_{ax}. Note that there is only one unknown value in Eq. (40), because the transformation $\mathbf{A}_{RP}$ is a rotation around the axial axis.

In order to compute the longitudinal dynamics, the contact force $_C\mathbf{f}_P$, defined by Eqs. (37), (38), (39) has to be transformed into the B-system by the transformation $\mathbf{A}_{BC}$. For the transversal dynamics a further transformation $\mathbf{A}_{RB}$ to the reference system R is needed. And finally the excitation of the pulley model in Sec. 4.2 is calculated in the initial system I. Therefore an additional transformation $\mathbf{A}_{IP}\mathbf{A}_{PR}$ is used.

6.2 Contact between the Elements

The contact between two elements is modelled in one dimension, since there is no relative velocity in radial and axial direction. This results from the construction of the elements shown in Fig. 18. The normal force P is unilateral and allows a separation of the elements. Therefore a correlation of stress, force respectively and strain is

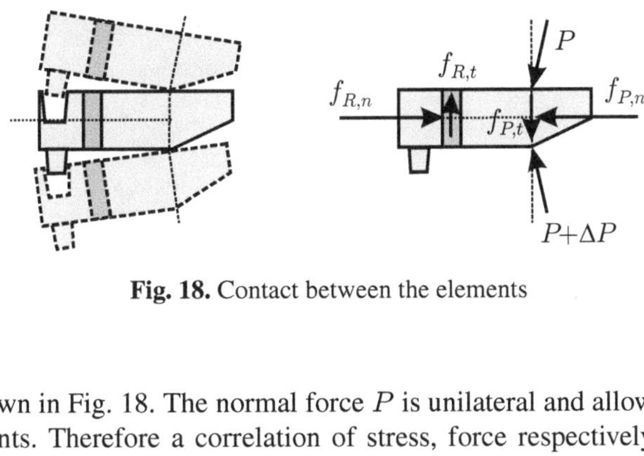

Fig. 18. Contact between the elements

used.

$$\varepsilon_0(s_R) = \varepsilon^+(s_R) - \frac{P(s_R)}{EA} \tag{41}$$

The local strain of the elements ε_0, calculated by Eq. (27) is split up into an elastic strain, resulting from the compression force P and the gaps ε^+ between the elements. At this EA is the longitudinal stiffness of the elements. Similar to Eq. (34) either a contact force $P < 0$ is transmitted or if it is zero the elements are separated and a positive distance ε^+ occurs.

$$-P(s_R) \geq 0,\ \varepsilon(s_R)^+ \geq 0,\ P(s)\,\varepsilon^+(s_R) = 0 \tag{42}$$

The information, if the contacts are *open* or *closed* is stored in a permutation matrix, that is multiplied by the matrix $\mathbf{H}_i$ in Eq. (28) to switch on/off the interaction of adjacent elements.

6.3 Contact of Rings

The contact between the elements and the innermost or outermost ring are calculated in the same way as the contact between two adjacent rings. In this section the elements (index 0) as well as the rings (index $1..n$) are named *layers*, to shorten the writing. We assume that there is no separation of the contacts. To get realistic relative motions between the layers their radial expansion has to be modelled accurately. Fig. 19 shows the assembly of the rings in a finite section of the element. The rela-

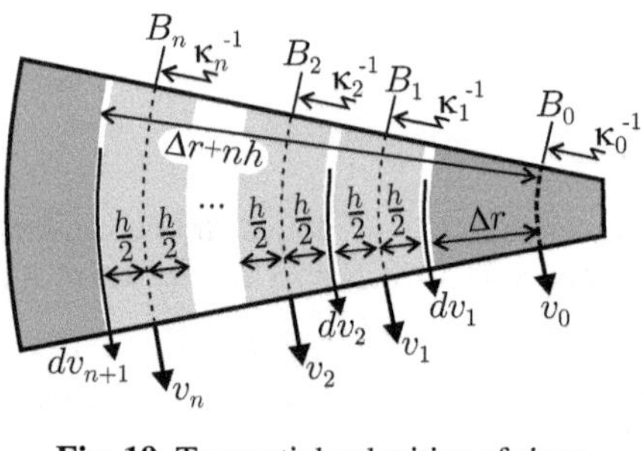

Fig. 19. Tangential velocities of rings

tive velocity dv is calculated by the difference of the local velocities of the layers' contact surfaces. Here it is important to distinguish the neutral line (index i) and the surface (index k) of a layer. For this reason a transformation

$$J_i^k = \frac{\partial v_k}{\partial v_i} \tag{43}$$

is necessary. The relative velocity dv_k is defined by the difference of velocities of the inner surface of layer $i = k$ and the outer surface of layer $i = k - 1$. In case of very high transmission velocities a contact between the outer ring n and the element $i = 0$ may occur at the outlet of the pulley. The relative velocity becomes

$$dv_{n+1} = J_0^{n+1} v_0 - J_n^{n+1} v_n \tag{44}$$

$$dv_k = J_k^k \, v_k \quad - J_{k-1}^k v_{k-1} \, , \quad k = 1..n \tag{45}$$

The transformations J_i^k depend on the radial displacement of the neutral line and the surface in respect of the curvature κ of the belt systems B_i. In detail they are defined by

$$J_0^1 = (1 + \Delta r \kappa_0), \; J_0^{n+1} = (1 + (\Delta r + nh)\kappa_0) \tag{46}$$

$$J_k^k = (1 - \frac{1}{2} h \kappa_k), \; J_k^{k+1} = (1 + \frac{1}{2} h \kappa_k) \, , \; k = 1..n \, . \tag{47}$$

The tangential force is calculated by continuous STRIBECK-friction laws depending on the relative velocity and the normal force. In Sec. 5.1 we assume the transversal dynamics of the elements and rings to be coupled. Due to this constraint the normal force is eliminated. Since the normal force is needed to quantify the friction forces it is calculated analytically. For a convex curved belt section $\mathrm{d}\varphi$ the normal force $f_{R,n}$ decrease from inner to outer ring due to the ring tension T. It is approximated by

$$_B f_{R,n}^{k+1} = _B f_{R,n}^k + \left(\bar{m}_k v_k^2 - T_k\right) \mathrm{d}\varphi \, , \quad \mathrm{d}\varphi \approx b\kappa_0 \, . \tag{48}$$

Here $\bar{m}_k$ is the mass per length of the layer k and the angle $\mathrm{d}\varphi$ is estimated by the radian, the thickness of the element respectively, and the curvature κ_0 of the pitch line B_0.

7 Simulation Results

This section shows results of numerical simulations using the model described above. They are compared with published measurements if available. There are very few publications presenting local measurements on metal V-belt CVT systems and there is none that also specifies the exact geometry of the system. Thus the comparison of measurements and results of simulations are effected only qualitatively.

Fist measurements of the Doshisha University of Kyoto in collaboration with Honda *R&D* are used for validation. The measurements have been effected by micro strain gauges that have been assembled in the elements. Even if the local measurements presented in a series of papers [2][3][4] are only qualitative, they provide a good insight in the belt mechanics.

Figure 20 compares the simulation results and the measurements for three different torque ratios r, which gives the transmitted torque M with respect to the maximum transferable torque M^* (s. Eq. 2). It is affected by varying the transmitted torque M, while the thrust Q_{dn} of the driven pulley is kept constant. The plots are divided into four parts. The first specifies the areas of the driving pulley, the second the span $t1$, the third the driven pulley and the last the span $t2$. Three forces are shown, on the left the compression force P, in the middle the tangential friction force $f_{P,t}$ between the pulley sheaves and the element and on the right the tangential friction force $f_{R,t}$ between the element and the innermost ring (s. Fig. 13). The comparison of all plots give good consistence.

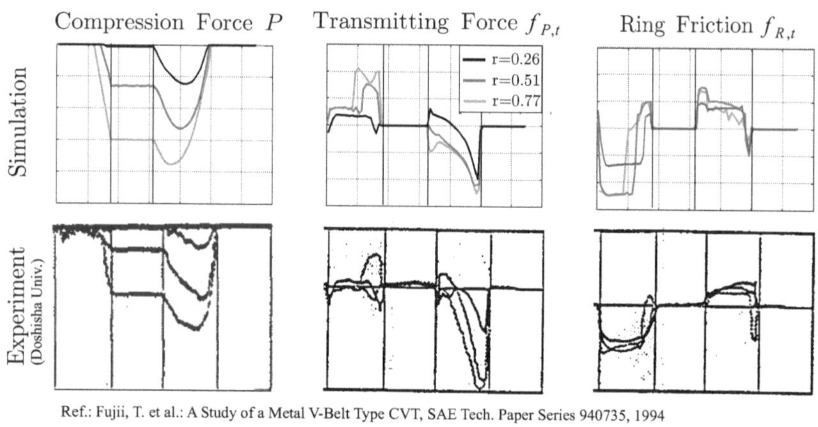

Fig. 20. Comparison with experiment ($i = 1.0$, torque ratio $r = 0.26, 0.51, 0.77$)

The elements enter the driving pulley being separated and uncompressed $P = 0$. So the tangential forces $f_{P,t}$ and $f_{R,t}$ have to be in equilibrium to each other. At a load depending position in the driving pulley the compression force P appears and glows rapidly up to the exit of the pulley. This increases the tangential forces $f_{P,t}$ and influences the ring friction $f_{R,t}$ by a jump in the velocity of the elements. In the span $t1$ the compression force stays constant due to the ring friction is small and there is no contact with the pulleys ($f_{P,t} = 0$). In the driven pulley the compression area continuous up to the exit. The torque ratios r used in Fig. 20 are rather low. So the tangential slip is small. In this case the compression force P increases after entering the driven pulley before it decreases to zero at the exit. This is affected by the penetration of the belt into the pulley groove due to the elastic deformation. The local contact radius decreases passing through the driven pulley. This can be seen by the transversal position in Fig. 21. The velocity of the pulley in the contact becomes slower with decreasing radius and the tangential relative velocity increases. The sliding angle (s. Eq. 36) decreases starting with an angle above $\frac{\pi}{2}$, which means the pulley drives the belt to an angle below $\frac{\pi}{2}$, so the belt drives the pulley. Due to this the elements are distorted around the area where the tangential velocity changes sign ($\frac{\pi}{2}$). Fig. 20 shows results for a speed ratio $i = 1$. Here the inversion of the compression force P proceeds around the change of sign of the transmitted torque. At no-load the compression force disappear in both spans. For a speed ratio $i \neq 1$ we observe a kinematic phenomenon that is described in Sec. 2. If no torque is transmitted in case of overdrive (s. Fig. 22, left) a compression force P_{t1} occurs

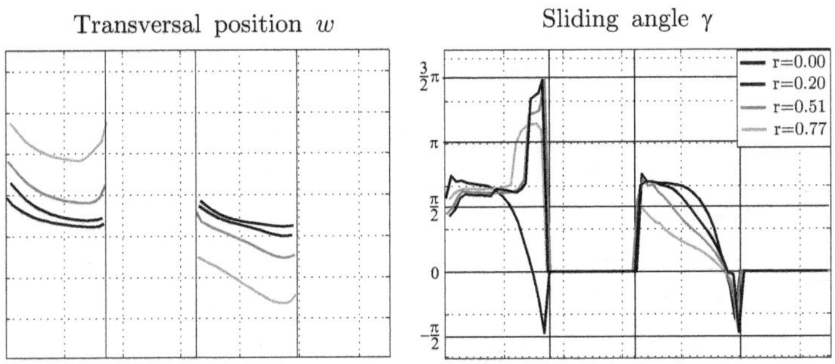

Fig. 21. Transversal position and sliding angle ($i = 1.0$, $r = 0, 0.26, 0.51, 0.77$)

in the span $t1$ to equalise the difference of ring tension. In underdrive conditions

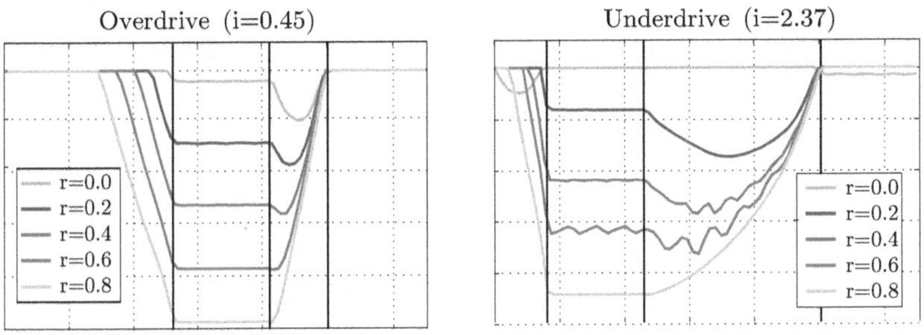

Fig. 22. Compression force P varying the torque ratio r

(s. Fig. 22, right) the difference of ring tension is contrary, so it needs a compression force P_{t2} in the span $t2$. For both speed ratios the increase of the compression force P after entering the driven pulley disappears for high torque ratios r due to a rise of the tangential slip, so the sliding angle is below $\frac{\pi}{2}$.

Finally the normal contact force $f_{P,n}$ is presented. Measurements [5] have been done by IDE/TANAKA using an ultrasonic technique. Figure 23 compares the simulation results and the measurements for different torque ratios r. If the torque ratio r is negative, this indicates a reversal of the power transmission e.g. dynamic braking. The plots show the contact force $f_{P,n}$ in the pulley that drives the belt if $r > 0$. Due to the exact geometry is unknown, it is possible to compare the measurements to the simulation results only qualitatively. If no torque is transmitted and the rotation speed is zero (not shown in Fig. 23) the distribution of the normal force between the inlet and the outlet is symmetrically. Because the normal force acts on only one half of the pulley sheave there are peaks at the inlet and the outlet due to neighbour support. For a rotating pulley the friction force $f_{P,r}$ decreases the normal force $f_{P,n}$ while sliding into the pulley groove and increases the normal force $f_{P,n}$ sliding to a

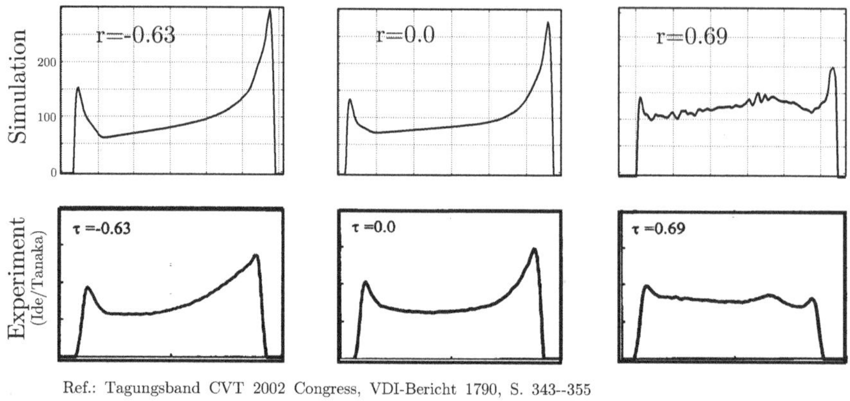

Fig. 23. Normal force $f_{P,n}$ in the driving pulley ($i = 2.16$, $r = -0.63, 0.0, 0.63$)

bigger radius (s. Fig. 15). This decreases the peak at the inlet and increases the peak at the outlet. Further the force distribution is influenced by the tangential span force. The normal force $f_{P,n}$ is proportional to the sum of ring tension T and compression force P (s. Fig. 15 and Eq. 48).

$$f_{P,n} \approx C_1 \left(\sum T - P \right) \mathrm{d}\varphi + C_2 \tag{49}$$

In case of $r > 0$ the normal force $f_{P,n}$ increases at the inlet and decreases at the outlet. In the case of $r < 0$ it is contrary. Since the tangential force in the case of $r < 0$ and the rotation have the same influence on the normal force the left and middle plot in Fig. 23 are similar. In the case of $r > 0$ the influences erasure so the right plot in Fig. 23 appears in a different way.

8 Conclusion

To describe the behaviour of the power transmission of a metal pushing V-belt CVT it is important to model the deflection of the pulleys and the belt. Furthermore all relevant contact properties e.g. unilateral constraints have to be considered to obtain accurate results. To avoid numerical problems the frequency spectrum is separated by a limit-frequence ($\approx 1kHz$). The frequencies that are of technical relevance are calculated dynamicly. Frequencies above the limit have to be eliminated by a quasi-static approach. By this we get accurate results together with reasonable simulation times.

The presented multibody model allows to calculate the distribution of all contact forces. In addition it permits to evaluate necessary pulley thrust, the maximum transmittable torque and the efficiency. The comparisons between the simulation results and the measurements that can be found in the public literature give good consistence.

References

1. Bremer H (1988) Dynamik und Regelung mechanischer Systeme. Teubner Studienbuecher, Stuttgart
2. Fujii T, Kurokawa T, Kanehara S (1993) A Study of Metal Pushing V-Belt Type CVT - Part 1: Relation between Transmitted Torque and Pulley Thrust. Int. Congress and Exposition Detroit, SAE Technical Paper Series, Nr. 930666, SAE, pp. 1–11
3. Fujii T, Takemasa K, Kanehara S (1993) A Study of Metal Pushing V-Belt Type CVT - Part 2: Compression Force between Metal Blocks and Ring Tension. Int. Congress and Exposition Detroit, SAE Technical Paper Series, Nr. 930667, SAE, pp. 13–22
4. Fujii T, Kitagawa T, Kanehara S (1993) A Study of Metal Pushing V-Belt Type CVT - Part 3: What Forces Act on Metal Blocks. Int. Congress and Exposition Detroit, SAE Technical Paper Series, Nr. 940735, SAE, pp. 95–105
5. Ide T, Tanaka H (2002) Contact Force Distribution Between Pulley Sheave and Metal Pushing V-Belt. Proceedings of CVT 2002 Congress, VDI-Bericht 1790 pp. 343–355, VDI-Verlag, Duesseldorf
6. Pfeiffer F, Glocker C (1996) Multibody Dynamics with Unilateral Contacts. Wiley & Sons, New York
7. Post J (2003) Objektorientierte Softwareentwicklung zur Simulation von Antriebsstraengen. VDI-Fortschrittberichte, Reihe 11, Nr. 317, VDI-Verlag, Duesseldorf
8. Sedlmayr M, Pfeiffer F (2002) Spatial Dynamics of CVT Chains. Proceedings of APM 2002, Petersburg
9. Srnik J (1999) Dynamik von CVT-Keilkettengetrieben. Fortschritt-Berichte VDI, Reihe 12, Nr. 372, VDI-Verlag, Duesseldorf
10. Timoshenko S, Young H (1968) Elements of Strength of Materials. D. Van Nostrand Company

Multicriteria Optimization of Train Structures for Crashworthiness

João P. Dias and Manuel S. Pereira

Instituto de Engenharia Mecânica (IDMEC), Instituto Superior Técnico, Av. Rovisco Pais, 1049-001 Lisboa, Portugal.
{jdias,mpereira}@dem.ist.utl.pt

In the first stages of the design of vehicle structures for crashworthiness, multicriteria optimization tools can be very useful in order to evaluate how conflicting requirements compete. However, crashworthiness problems are by nature time consuming, if the standard non-linear finite elements simulation tools are used. Simplified models based on multibody dynamics can be developed and used for crashworthiness simulations as proposed in this work. With these simplified models, the computational time can be drastically reduced and evolutionary or genetic algorithms, that require large numbers of simulations, can be used, as proposed.

The vehicle or structures are described as a set of rigid or flexible bodies connected by non-linear springs. Different models are presented hereunder dealing with different needs or stages of the projects of railway vehicles, namely, 1D models that are mass-spring models, without kinematic constraints developed for the simulation of train sets and 2D rigid-flexible bodies developed for the study of structures and energy absorption devices with planar motion. These models include a Hertz contact-impact model (2D) for the simulation of the contact between the bodies and barriers and Coulomb friction for the wheel-rail contact. These two simulation tools are linked with evolutionary optimization algorithms to the multicriteria design of train structures. The design methodology developed is general and allows evaluating different conflicting design functions together, such as accelerations or measures of the acceleration, deformations, energies, masses or velocities. Also and in particular for the simulation of collisions of train sets different scenarios can be considered, which corresponds to the evaluation of different models in a single iteration. Three examples are presented in order to illustrate the design methodology developed herein. A single criteria optimization problem of the setup of an experimental train crash test; a multicriteria optimization problem of the energy absorption devices of a train in a multi-scenario collision with 1D models; and the multicriteria optimization of an energy absorption structure with a 2D flexible model.

1 Introduction

Vehicle crashworthiness is a complex problem and requires, in general, large computer resources. The use of finite elements programs such as LS-DYNA or PAM-CRASH leads to several drawbacks: the time necessary to develop the models is large and, in general, these programs do not include optimization procedures. In the first stages of the design project, when the characteristics of the structures and energy absorption devices are not known in detail, simplified models that allow simulation and optimization can be used with success. Mass-spring models have been used for the crash simulation of automotive vehicles [1,2,3], and for the crash simulation of train sets [4]. Models based on 1D, 2D and 3D multibody dynamics formulations, with rigid and flexible bodies, have been used successfully in the crash simulation of train structures. These models have been validated with experimental results [5,6,7,8,9], and train designers use some of them industrially. Multibody dynamics models have shown to be suitable to simulate the behavior of the structures, energy absorption devices, and vehicles in crashworthiness.

The increasing demands in passive safety of trains are putting new challenges in the design methodologies. The designers nowadays, require not only simulation tools with high levels of complexity, based on nonlinear finite element models, but also simplified simulation tools with optimization procedures that can be used efficiently in the earlier design stages. These simplified models based in multibody dynamics formulations, have been integrated in design methodologies with deterministic algorithms to evaluate the optimal characteristics of energy absorption devices and structures in crashworthiness in single design function problems [6,7,9]. However, in the first stages of the design process, the designer wants to know not only a solution, but also the spread of possible solutions and understand the behavior of conflicting functions. More powerful multicriteria design tools can nowadays, replace single objective functions well accepted in the past.

Issues related with sensitivities are critical in the selection of the optimization algorithms. In general, deterministic algorithms, that use sensitivities information, are more efficient than zero-order algorithms i.e. optimization algorithms that do not use sensitivity information. The development of analytical sensitivities for multibody systems such as sensitivities obtained by the adjoint method [10] or by direct differentiation [11] represents a difficult task, that in the last years has became easier with the development of automatic differentiation tools such as ADIFOR [12]. It is well known that numerical sensitivities may exhibit inaccuracies, especially in non-linear problems. Recent progress in the area of genetic and evolutionary algorithms [13] made possible their application, in many fields, namely in structural dynamics and in multibody dynamics problems [14,15]. Crashworthiness problems are non-linear in nature and the system response can exhibit discontinuities or noise. The deterministic algorithms that have proved to be efficient in structural optimization can however be stacked in local minima that can arise in crashworthiness problems. Evolutionary algorithms can overcome this local minima problem but require efficient simulation tools in order that the Pareto

Fronts can be obtained in a reasonable time. When evolutionary algorithms are compared with deterministic algorithms that use analytical or numerical sensitivities the major advantages and drawback of each type of algorithm is as follows: The number of simulations required by the evolutionary algorithms is much larger than with deterministic algorithms; Evolutionary algorithms are better for functions that have local minima in the system response. Deterministic algorithms can be stacked in the local minima, because each new solution is evaluated, in general, from the last one. Evolutionary algorithms work with a set of solutions and if the population and the number of generations are large enough a global solution can be found; the convergence of the deterministic algorithms depends on the initial solution provided. With evolutionary algorithms, no initial guess for the design function is required. In conclusion, evolutionary algorithms in single objective optimization are recommended only, for problems not computationally expensive and if the objective functions have local minima or noise. For multicriteria optimization and when the different conflicting functions are not combined in one single design function, evolutionary algorithms are very attractive, however they require a large number of simulations, which leads these algorithms prohibitive for problem with large simulation times.

In the past, evolutionary or genetic algorithms were considered not suitable for crashworthiness optimization, but nowadays with accurate simplified models, with computers that are more powerful and with evolutionary algorithms more efficient, it is possible now to perform crashworthiness optimization in a few hours in a modern personal computer. The great advantage of the use of simplified models is that the simulation times are quite small and then these models can be used together with evolutionary algorithms in the design process. The Pareto fronts are very important in the first stages of the design process when the trade-offs are to be decided. From the Pareto optimal front, and using higher order information, the designers can pick the best compromising solution between the different conflicting objectives. In crashworthiness problems, maximal accelerations or a measure of the accelerations and deformations are conflicting functions that can be used in multicriteria optimization. Accident fatalities are mainly caused, by large accelerations, normally observed when structures are too stiff, or loss of survival space when excessive structural deformation is observed [7]. Therefore, a trade-off between acceleration levels and deformations is necessary in order to protect efficiently the vehicle passengers. Other conflicting design functions can be used in train crashworthiness such as, the minimization of the mass or the maximization of the energy absorbed. In these cases, the design variables are associated with the relevant structural characteristics of the energy absorption devices and structures.

In this work, the evolutionary optimization algorithm NSGA II [16] is selected to be integrated in the design methodology, dealing with multicriteria constrained optimization problems.

2 Dynamic Analysis of Multibody Systems

The simplified simulation models used in this work are based on the dynamic analysis of multibody systems. The dynamic equations for a 2D rigid-flexible system are [7]

$$\begin{bmatrix} \mathbf{M}_{rr} & \mathbf{M}_{rf} & \mathbf{\Phi}_{\mathbf{q}_r}^T \\ \mathbf{M}_{fr} & \mathbf{M}_{ff} & \mathbf{\Phi}_{\mathbf{q}_f}^T \\ \mathbf{\Phi}_{\mathbf{q}_r} & \mathbf{\Phi}_{\mathbf{q}_f} & 0 \end{bmatrix} \begin{Bmatrix} \ddot{\mathbf{q}}_r \\ \ddot{\mathbf{q}}_f \\ \boldsymbol{\lambda} \end{Bmatrix} = \begin{Bmatrix} \mathbf{f}_r + \mathbf{g}_r \\ \mathbf{f}_f + \mathbf{g}_f - \mathbf{K}_{ff}\mathbf{q}_f - \mathbf{C}_{ff}\dot{\mathbf{q}}_f \\ \boldsymbol{\gamma} \end{Bmatrix} \tag{1}$$

where $\mathbf{M}_{rr}$ contains the rigid body inertia $\mathbf{M}_{ff}$ and $\mathbf{K}_{ff}$ are the standard finite element matrices, $\mathbf{M}_{rf}$ is the inertia coupling between rigid ($\mathbf{q}_r$) and flexible coordinates ($\mathbf{q}_f$), is the Lagrange multiplier vector, $\mathbf{f}_r$ and $\mathbf{f}_f$ are the generalized external forces, $\mathbf{g}_r$ and $\mathbf{g}_f$ are the quadratic forces including gyroscopic, Coriolis forces and other terms associated with kinetic energy. The terms $\mathbf{\Phi}_{qr}{}^T$, $\mathbf{\Phi}_{qf}{}^T$ are the jacobians of the kinematic constraints. These formulations are developed using a moving frame approach to describe the kinematics of the deformable bodies. The large rigid motion is described using Cartesian coordinates and the flexibility is introduced using the finite element method. To reduce the number of elastic degrees of freedom, the component mode synthesis is used. The mean axis condition method is applied to the reference conditions to reduce the dynamic coupling between rigid and flexible coordinates. The system of equations (1) is transformed using the following coordinate transformation from the modal coordinates to the physical elastic coordinates

$$\mathbf{q}_f = \mathbf{U}\,\mathbf{p}_f \tag{2}$$

where $\mathbf{U}$ is the modal transformation matrix whose columns are the *nm* low-frequency eigenvectors and $\mathbf{p}_f$ is the vector of modal coordinates. The equations of motion (1) are now written in terms of the modal coordinates, as

$$\begin{bmatrix} \bar{\mathbf{M}}_{rr} & \bar{\mathbf{M}}_{rf} & \bar{\mathbf{\Phi}}_{\mathbf{q}_r}^T \\ \bar{\mathbf{M}}_{fr} & \bar{\mathbf{M}}_{ff} & \bar{\mathbf{\Phi}}_{\mathbf{p}_f}^T \\ \bar{\mathbf{\Phi}}_{\mathbf{q}_r} & \bar{\mathbf{\Phi}}_{\mathbf{p}_f} & \mathbf{0} \end{bmatrix} \begin{Bmatrix} \ddot{\mathbf{q}}_r \\ \ddot{\mathbf{p}}_f \\ \boldsymbol{\lambda} \end{Bmatrix} = \begin{Bmatrix} \bar{\mathbf{f}}_r + \bar{\mathbf{g}}_r \\ \bar{\mathbf{f}}_f + \bar{\mathbf{g}}_f - \bar{\mathbf{K}}_{ff}\mathbf{p}_f - \bar{\mathbf{C}}_{ff}\dot{\mathbf{p}}_f \\ \bar{\boldsymbol{\gamma}} \end{Bmatrix} \tag{3}$$

The transformations from nodal to modal coordinates can be found in [17]. Equation (3) with the initial conditions

$$\left\{\mathbf{q}_r(t^0)=\mathbf{q}_r^0\ ,\mathbf{p}_f(t^0)=\mathbf{p}_f^0\ ,\dot{\mathbf{q}}_r(t^0)=\dot{\mathbf{q}}_r^0\ ,\ \dot{\mathbf{p}}_f(t^0)=\dot{\mathbf{p}}_f^0\right\} \tag{4}$$

can be integrated with respect to time, in order to obtain all the state variables of the system, i. e. positions, velocities, accelerations and Lagrange multipliers.

For the simulation of train collisions, 1D rigid body models can be used. In this case, Equation (1) is simplified as

$$\mathbf{M}\,\ddot{\mathbf{q}} = \mathbf{f} \tag{5}$$

where **M** is the mass matrix containing the body lumped masses, $\ddot{\mathbf{q}}$ the acceleration vector and **f** the forces vector. The differential equations (5) or the algebraic-differential equations (3) can be solved for the determination of the accelerations (1D) or the accelerations and Lagrange multipliers vector (2D). Then the accelerations are integrated with respect to time in order to obtain velocities and positions. A variable order, variable step size algorithm has been used.

2.1 Plastic Deformations

The adequate modeling of the plastic deformations at the energy absorption devices located at the extremities of the cars and crashworthy structures is crucial to simulate accurately crashworthiness with multibody models. Two types of plastic deformations are modeled, axial and bending. The axial plastic deformations are modeled as non-linear actuators coupled with translational joints or in 1D without any kinematic joints. The energy absorption devices such as, couplers, honeycomb structures and the end-structure are represented by non-linear actuators. These actuators take into account the energy absorption and loading-unloading effects. A typical force-displacement curve and the location of the energy absorption devices are shown in Figure 1.

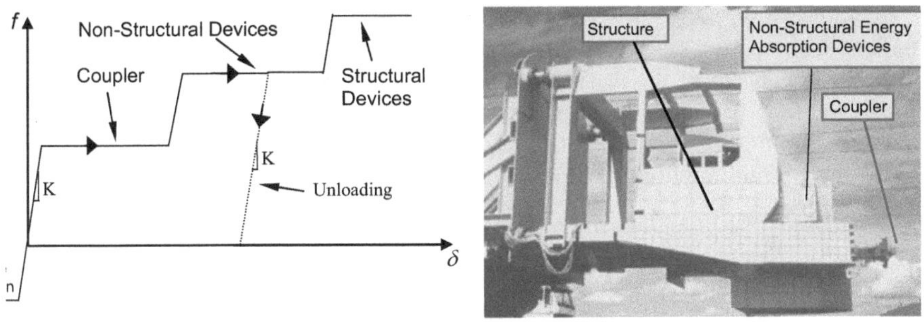

Fig. 1. Typical force-displacement curve and location of the energy absorption devices in a train extremity

A detailed force-displacement curve is indicated in Figure 2. Basically each one of the energy absorption devices is represent by two straight line segments representing the elastic and plastic behaviour, where d_i is the crushing displacement, f_i is the plastic force level and k_i is the elastic stiffness.

During the impact, four situations can arise: Elastic loading, plastic loading, elastic unloading and plastic unloading. Let us define d as the current crushing displacement and d(t-Δt) as the previous crushing displacement. The crushing displacement is calculated from the position of the connected lumped masses. By convention, positive displacements correspond to crushing, $d > 0$ is compression loading and $d < 0$ is traction. In addition, positive forces correspond to compressive forces. For traction phases, only an elastic stiffness equal to the compression elastic stiffness and a traction force (f_T) is defined. Unloading may start at a point

(f_u,d_u). For the compression phase, the number of force levels to be considered depends on the complexity of the end extremity of the train. In general, three plastic levels are used to represent the global crashworthiness behavior, giving a good approach for the deformations and energies absorbed. However, in these models, the number of plastic levels can be increased in order to improve the accuracy of the accelerations, but with only a few stages, a reasonable idea of the sustained acceleration levels is achieved. During the simulation, three conditions are verified:

1) EPC: Elastic-Plastic Condition
2) CTC: Compression-Traction Condition
3) LUC: Loading-Unloading Condition

These 3 conditions are stated as

$$\mathrm{CTC}=\begin{cases}1 & \text{if } \mathrm{d}>0 \text{ (Compression)}\\ 0 & \text{if } \mathrm{d}<0 \text{ (Traction)}\end{cases}$$

$$\mathrm{EPC}=\begin{cases}1 & \text{if } \mathrm{d}_{(i-1)i}<\mathrm{d}<\mathrm{d}_i \text{ (Elastic)}\\ 0 & \text{if } \mathrm{d}_{(i-1)}<\mathrm{d}<\mathrm{d}_{(i-1)i} \text{ (Plastic)}\end{cases}$$

$$\mathrm{LUC}=\begin{cases}1 & \text{if } \mathrm{d}>\mathrm{d}(t-\Delta t) \text{ (Loading in compression phase)}\\ 0 & \text{if } \mathrm{d}<\mathrm{d}(t-\Delta t) \text{ (Unloading in compression phase)}\end{cases}$$

Structural unloading follows a path parallel to the elastic slope. When a new loading occurs after an unloading, the plastic deformation restarts from the last plastic point (f_u,d_u) of the original curve. In each time step, the displacements, the forces and the current segment of the curve are stored. So in the next time step the new location and the determination of the current situation is verified and according to the current situation the forces are calculated. The forces for each situation are given by

$$f=\begin{cases}\mathrm{f}_T \text{ (if EPC=1 and LUC=1 and CTC=0)}\\ \mathrm{k}_1 d \text{ (if EPC=0 and } \mathrm{d}_T<d<\mathrm{d}_1\text{)}\\ \mathrm{f}_i \text{ (if EPC=1 and LUC=1 and CTC=1 and } d>\mathrm{d}_1\text{)}\\ \mathrm{f}_i+\mathrm{k}_{(i+1)}(d-\mathrm{d}_{i(i+1)}) \text{ (if EPC=0 and CTC=1 and } d>\mathrm{d}_1\text{)}\\ \mathrm{f}_i\text{-}\mathrm{k}_i((\mathrm{du}-d)) \text{ (if EPC=1 and LUC=0 and CTC=1 and } d>\mathrm{d}_1\text{)}\end{cases} \tag{6}$$

where k_i is the stiffness of the elastic slope and f_i is the plastic level for the i element of the curve that represents the last stage of the curve reached. Because a predictor-corrector integration algorithm is used, a computational procedure to avoid misunderstands of the unloading phenomena during the corrector phase, as been implemented.

The force displacement curves are obtained from, experimental tests, finite elements models or known characteristics of the standard devices. The detailed curve obtained from the non-linear finite element analysis, is represented by a stepped curve representing in the least squares sense the original curve.

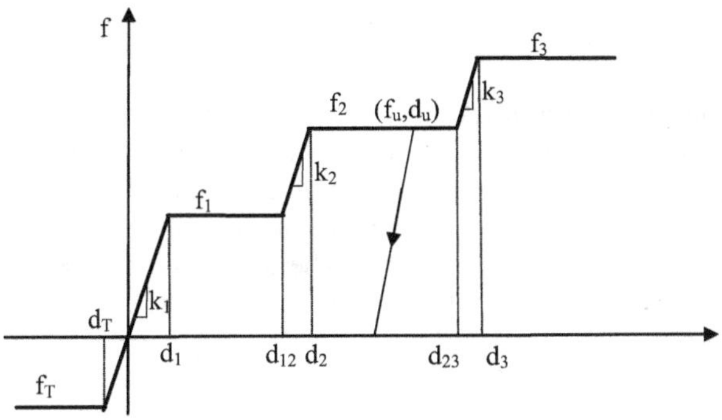

Fig. 2. Detailed force-displacement curve for the energy absorption elements

For the bending plastic deformations, the plastic hinge concept is used with revolute joints and generalized spring elements, to represent the constitutive characteristics of localized bending plastic deformation of beams as shown in Figure 3.

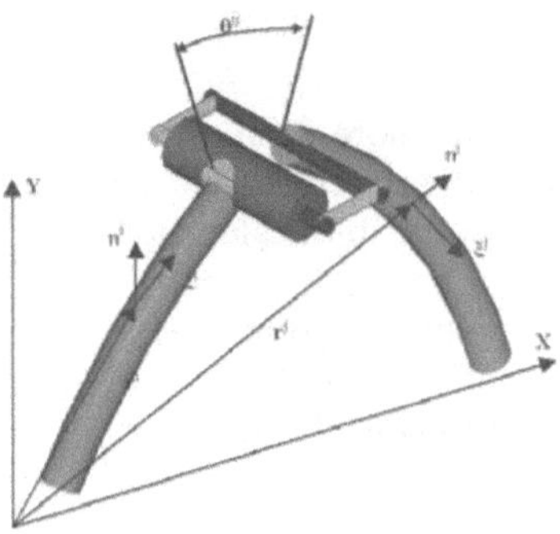

Fig. 3. Plastic Hinge concept

The revolute joint must be simultaneously perpendicular to the neutral axis of the beam and to the plastic hinge bending plane. From Figure 3 the following relationship can be written

$$\theta^{ij} = \theta^{i} + \varepsilon_{k^i} - \theta^{j} - \varepsilon_{k^j} \tag{7}$$

which shows the dependency of the plastic hinge angle on the rigid body orientation positions θ^i and θ^j and on the elastic rotations of body i and body j, ε_{ki} and ε_{kj} at the attachment node k. The angle values are directly obtained as relative coordinates from the integration process and correspond to the relative degree of freedom θ^{ij} of the revolute joint under consideration. The bending moments are calculated in a similar form as in Equation (7).

2.2 1D Models

For the collision simulation of train sets, each car is modeled as a series of lumped masses, connected by actuators that represent the different energy absorption devices. The typical force-displacement curve of a car end extremity is shown in Figure 1. Using these simulation models, a collision between two trains or a train and a truck can be performed in a few seconds. These models include braking forces resulting from the application of Coulomb's friction law for the wheel-rail contact as

$$\begin{cases} f_T = -\mu(s) f_N & \text{for } v_T = 0 \\ f_T = -\dfrac{v_T}{|v_t|}\mu(s) f_N & \text{for } v_T \neq 0 \end{cases} \tag{8}$$

where v_t is the tangential velocity, $\mu\,(s)$ is the coefficient of friction and f_N and f_T are the normal and tangential forces, respectively.

2.3 2D Models

With 2D tools, the models are more detailed and have been developed and used for simulation and design of the structures and energy absorption devices. In addition to the overall behavior of the train calculated by 1D rigid models, 2D flexible models can provide accelerations that are more accurate. 2D models can be used for the design of vehicle structures or other energy absorption structures when the deformation occurs in a plane. In the 2D models a proportional damping matrix based on the Rayleigh model is considered as

$$\mathbf{C}_{\text{ff}} = \alpha\, \mathbf{K}_{\text{ff}} + \beta\, \mathbf{M}_{\text{ff}} \tag{9}$$

where the constants α and β are calculated from the vibration frequencies of the structure, and β is limited by 0.1. The structural damping is important in order to have accurate acceleration results [5]. Also for the contact between structural members and barriers a Hertz contact model that takes into account the energy dissipation is included, as

$$f = K\delta^n \left[1 + \frac{3(1-e^2)}{4} \frac{\dot{\delta}}{\dot{\delta}^{(-)}} \right] \tag{10}$$

where δ is the penetration, $\dot{\delta}^{(-)}$ is the initial velocity of penetration, e is the coefficient of restitution, n is a constant typically 2/3 and K is a constant stiffness coefficient that depends on the material and geometrical characteristics of the contact surfaces, that in the case of flat surfaces is given by

$$K = \frac{2\sqrt{ab}}{m\left[\frac{1-\nu^{i^2}}{E^i} + \frac{1-\nu^{j^2}}{E^j}\right]} \tag{11}$$

where ν^i and E^i are the Poisson's ratio and Young Modulus associated to each surface and a and b are the dimensions of the smallest contact surface.

3 Multicriteria Optimization Problem

A general multi-objective or multicriteria optimization problem has a number *ndo* of competing design functions f(**b**) that are to be minimized or maximized and a number *ndc* of inequality design constraints g(**b**) and *nedc* equality constraints h(**b**) that any feasible solution must satisfy. The design functions and design constraints are a function of the vector of design variables **b**. Each design variable has a lower bound $\mathbf{b}^{L}$ and an upper bound $\mathbf{b}^{U}$. In the particular situation of single criteria optimization, the parameter *ndo* is set to 1. The general objective optimization problem can be stated as

$$\text{Minimize/Maximize} \quad f_i(\mathbf{b}) \quad i=1, ndo \tag{12}$$

$$\text{Subject to} \quad g_j(\mathbf{b}) \leq 0 \quad j=1,ndc \tag{13}$$

$$h_k(\mathbf{b})=0 \quad k=1,nedc \tag{14}$$

$$b_l^{L} \leq b_l \leq b_l^{U} \quad l=1,nb \tag{15}$$

With the evolutionary algorithm NSGA-II [16] all the design constraints should be "greater than" and all the design functions should be minimized. To fulfill this condition the duality principle is used so that

$$\text{maximize } f(\mathbf{b}) = \text{minimize } -f(\mathbf{b}) \tag{16}$$

The main idea in this work is to solve multicriteria problems with several case models. A model can be a different multibody system or the same multibody system simulated with different initial conditions, such as, for example different collision scenarios. The multicriteria multi-model flowchart is indicated in Figure 4.

The design methodology for evolutionary multicriteria optimization of multibody systems has been developed in a general form allowing any design function and design constraint related with the relevant characteristics of the system. In the particular case of the train crashworthiness problem the possible design functions are the maximal accelerations or a measure of the accelerations, deformations, energies absorbed, masses and impact velocities. The possible design parameters are the characteristics of the force-displacement curves or in general any geometric or material property of the energy absorption devices or structures. The multicriteria design methodology for train crashworthiness is applied in two different simulation codes based on multibody dynamics formulations. The code Crash1d for the simulation of train set collisions and the code Moms2D for the simulation of any structure or energy absorption device for crashworthiness when the deformation occurs in a plane have been developed.

The design methodology as presented in Figure 4 starts from an initial population that is selected randomly. This initial population (the size is specified by the user) corresponds to a random selection of the values for the design variables. However this initial population can contain predefined cases which already proven to be reasonable solutions, but the genetic algorithm used herein is an elitist algorithm where the best solutions always are passed to the next generation and this can lead to a Pareto Front with a reduced spread. Once the population has been specified, then the objective functions and design constraints are evaluated. For railway vehicles crashworthiness it is sometimes necessary to consider different collision scenarios that will be explained later. Therefore, in this case the evaluation of a member of the population (which corresponds to a specific set of design variables) can correspond to several simulations with different models. These different collision scenarios are independent and can be simulated using parallel processing. In the next step, the fitness is calculated. The fitness of a solution is a measure of its quality, or in general case and for non-constrained problems, the value of the objective function. However, for constrained optimization problems, as in the present case, the genetic algorithm uses a generalized fitness measure involving the value if the design function and a penalty term associated to the constraint violation. Therefore, when two solutions are compared in the selection process, a feasible solution is always preferred to an unfeasible one, leading to the rejection of solutions involving design constraint violations.

After the fitness is calculated, a termination condition is checked. In the case of single or weighted design functions, this condition is associated with a convergence tolerance. However, in the general case were different design functions are evaluated independently, the process will be halted when a predefined number of generations is obtained. Once the termination condition is not verified, the typical process of evolution of the genetic algorithm starts and the following steps are performed:

1) Selection. The population selection is based on the principle of “survival of the fittest”. The tournament method is used in the present algorithm. Each solution participates in two tournaments. The fittest solution will collect two victories and the worst two defeats. According the number of victories each solution will have 0, 1 or 2 and will serve as progenitors for the next generations.
2) Crossover. The crossover is the process of creating offspring from parents and is illustrated in Figure 5. Two pairs of solutions are selected based on a crossover probability specified and then two new offsprings are created by exchanging chromosomes from the parents.
3) Mutation. Is the process of exchanging with a specified probability select bits in the chromosome to create a new individual in the population.

Then the optimization process continues until the stopping condition is fulfilled.

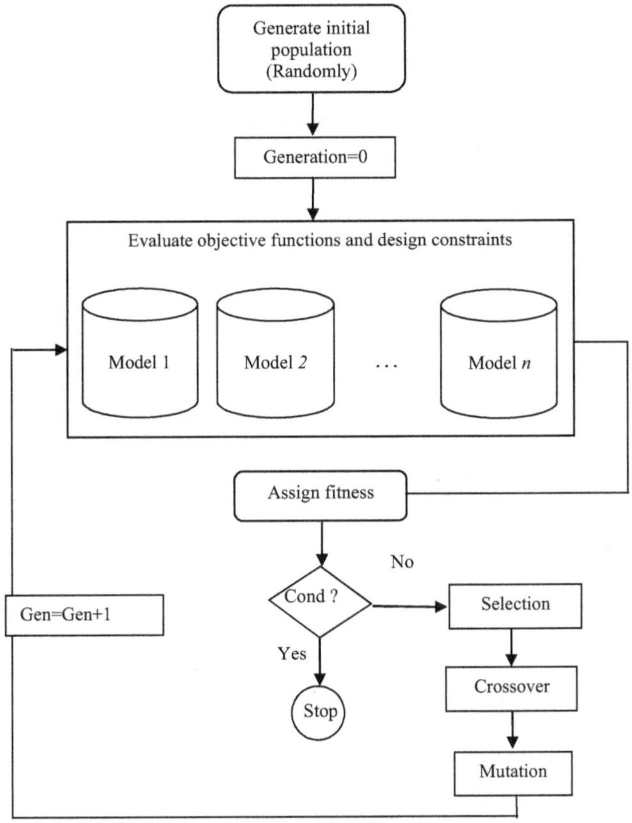

Fig. 4. Flowchart of the design methodology for multicriteria optimization of multibody systems

$$\left.\begin{array}{ll} 010|00\ 01010 & (8,10) \\ 011|10\ 00110 & (14,6) \end{array}\right\} \Rightarrow crossover \Rightarrow \left\{\begin{array}{ll} 010|10\ 00110 & (10,6) \\ 011|00\ 01010 & (12,10) \end{array}\right.$$

$$\{010|00\ 01010\} \Rightarrow mutation \Rightarrow \{010|10\ 01010\}$$

Fig. 5. Crossover and mutation operators for the genetic algorithm

The solution of Multi Objective Optimization Problem (12-15) can be classified in three categories [18]: 1) priori methods; 2) posteriori methods; 3) interactive methods. Traditionally in multicriteria optimization all, the design functions are combined in a unique weighted design function, which is called the priori method. However, these weights are very difficult to specify in the first stages of the design process, especially when conflicting design functions are considered. In the design methodology presented herein, the weights can be specified and two or more design functions are combined in a single objective function. Nevertheless, the main purpose here is to show the advantages of the evaluation of the Pareto Front, or ideally the Optimal Pareto Front. From the Pareto front, the designer obtains a set of solutions and using high order information then a specific solution can be selected. This is called the posteriori method. The determination of the Pareto Front is one of the greatest advantages of the genetic or evolutionary algorithms. Due to the population concept, in each iteration several solutions are evaluated simultaneously and if the genetic algorithm' parameters are chosen adequately, a Pareto Front with a good spread of solutions is obtained.

4 Design Examples

To illustrate the multicriteria evolutionary optimization methodology developed for design for crashworthiness, three examples are presented. These examples illustrate the applicability of the current methodologies, from single criteria problems that can be used for the specification of the conditions of experimental crash-tests or even to the determination of the optimal characteristics of energy absorption devices and structures, to multicriteria problems simulated with 1D rigid models or 2D flexible models. The characteristics of the low-energy absorption devices to be tested are presented in Figure 6.

4.1 Single-Objective Optimization of a Train Crash Test Setup

In this example, the solution obtained with the presented methodology is compared with the solution obtained with deterministic algorithms [9]. This is a single design function problem, and the use of both types of algorithms is discussed. This example deals with a full-scale crash test carried out in framework of the European Project Safetrain [19]. In this experimental test (Figure 7), a wagon B collides with two wagons (A and C) (Figure 8) connected by a Low-Energy device (LE), which needed to be tested and validated. In the front of wagon C, a honeycomb structure (Figure 7), simulating a High-Energy (HE), is located. The characteristics of the low-energy absorption devices to be tested are presented in Figure 6.

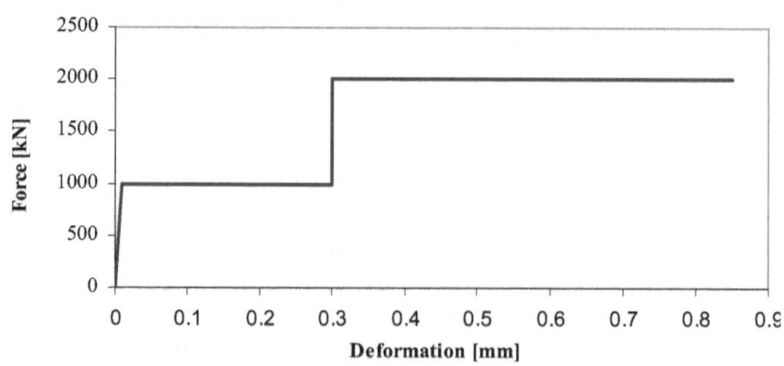

Fig. 6. Characteristics of the Low-Energy absorption devices

Fig. 7. Experimental test: Physical set-up and honeycomb structure

The optimization problem consists in evaluate the definition of the experimental test (masses, velocity of the impact wagon and characteristics of the High Energy absorption device), in order to absorb 1.4 MJ in the LE extremity. Without design tools the determination of the experimental test conditions, is very difficult, due to the strong interaction of the different physical parameters. In Table 1, the optimization problem is presented.

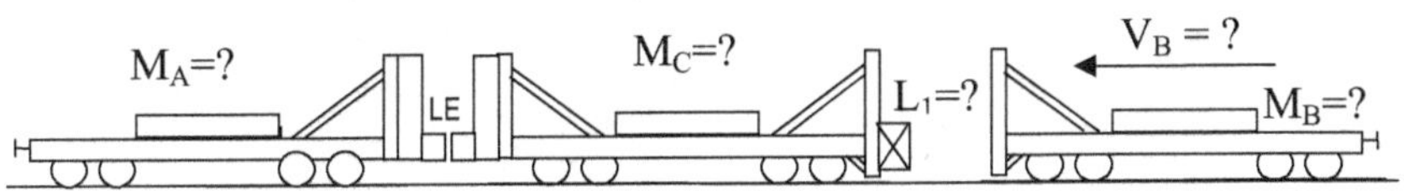

Fig. 8. Experimental test: Collision scenario and conditions

The specified design constraints and the limits for the design variables are related with structural and manufacturing constraints of the experimental set-up. The optimization results obtained with the presented methodology are compared with the results obtained with the quadratic programming deterministic algorithm [9]. The results for the Genetic algorithm (GA) have been obtained with a population of 18, 35 generations, crossover index of 60, mutation index of 20, crossover probability of 0.9 and mutation probability of 0.2.

One of the problems related with genetic algorithms use, is the limits imposed in the design variables. Zero value limits for the design variables are quite acceptable in deterministic algorithms, but can lead to errors in the simulation code or even very large computational times. To avoid this, the limits of the design variables have been changed and are presented between parentheses in Table 2. The results obtained are quite similar. The results presented for the masses for the deterministic algorithm are truncated due to practical reasons related with the execution of the test.

Table 1. Optimization problem for the determination of the experimental test set-up

Design Function	$\psi_0 = (LE-1.4)^2$
Design Constraints	$\psi_1 = HE<3MJ$
Design Variables	$b = [M_B, V_B, M_C, M_A, L_1]^T$

With: M_B: Mass of the impact wagon (Car B), V_B: Velocity of the impact wagon, M_A and M_C: Mass of the wagons (Car A and C), L_1: Force level for the Honeycomb.

Table 2. Optimization results for the determination of the experimental test set

	M_B (Ton.)	M_C (Ton.)	M_A (Ton.)	V_B (km/h)	L_1 (KN)	ψ_1 HE (MJ)	ψ_0 LE (MJ)
Lower Limits	30	30	30	0 (10)	0 (1000)	-	-
Upper Limits	70	50	60	72	3000	-	-
Optimal DA [9]	70	30	59.5	53.7	3000	3	1.41
Optimal GA	69.6	30.1	60.0	53.4	2999.4	2.98	1.41

4.2 Multicriteria Optimization of the Energy Absorption Devices

This example describes a typical multicriteria problem that arises in the design of the energy absorption devices located in the front of the trains (High Energy devices) and between cars (Low Energy devices) as indicated in Figure 9.

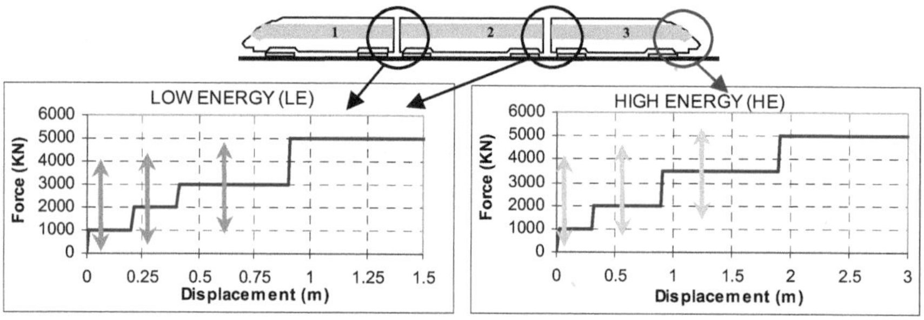

Fig. 9. Characteristics and data of the front-front collision between two 3-car trains

These devices are optimized for a 3-car train with masses: 50 (front car), 45 (middle car) and 40 ton (rear car), in a head-on train collision for a relative collision speed of 55 km/h. The accelerations in two other collision scenarios, a collision with a 30 ton truck at 90 km/h and a collision against a line end at 30 km/h are also included as design constraints. The three different collision scenarios considered are schematically represented in Figure 10.

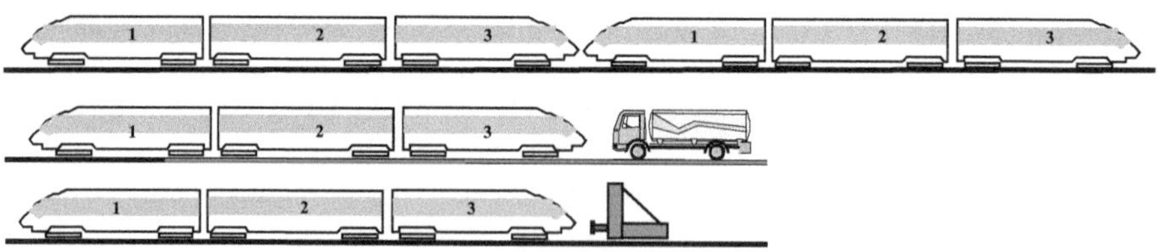

Fig. 10. Collision scenarios for multicriteria optimization

The conflicting design functions selected are the overall deformation in the train and the sum of the sustained accelerations along the train. The design variables are the plastic levels of the energy absorption devices located in the front of the cars (indicated by arrows in Figure 9). Each one of the energy absorption devices (HE and LE) is represented by 3 design variables corresponding to the different force levels. The design constraints are related with the maximal accelerations in each one of the cars and with the difference between consecutive plastic levels. The optimization problem is summarized in Table 3. This problem has 2 design functions, 16 design constraints, and 6 design variables. The Pareto front for a population size of 70 (meaning 70 independent simulations each one corresponding to a different set of design variables) and 30 generations (meaning 30 iterations) is shown in Figure 11. The Pareto front has been obtained in about 8 hours in a PC 1.4 GHz.

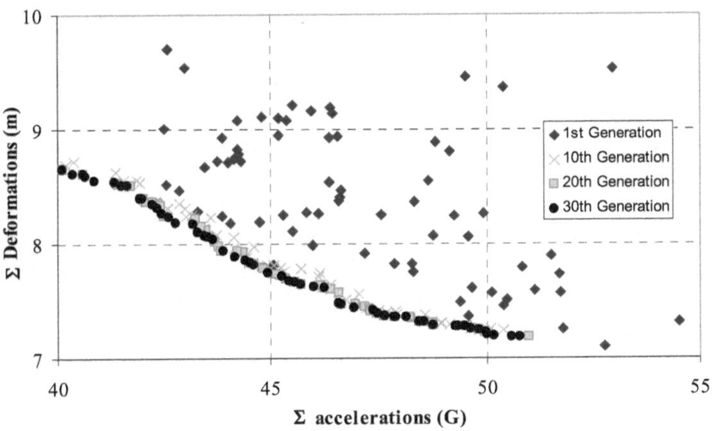

Fig. 11. Pareto front for the 3-car train set in a set of collision scenarios

The results in Figure 11 show that, when the sum of the accelerations in the cars decrease the sum of the deformations increase. Each point in the curve corresponds to specific characteristics of the energy absorption devices. The accelerations levels are one of the most important parameters to be considered in the project, because they are directly related with vehicle occupant injuries. On the other hand lower deformation in cars means more space for the passenger's compartment. Once the designer selects a relative weight for each design function, the correspondent force-deformation curve is selected. A weighted design can then be optimized as a single criteria design problem or from the optimal values for the plastic levels, the structures and energy absorption devices can then be designed using detailed models, such as finite element models. In the case of 1D models and for this type of problems it was found that is better to have a larger population instead of a large number of generations. It can be seen in Figure 11, that with 20 generations, the designer of the structures can have a reasonable idea of the Pareto front.

Table 3. Multiobjective optimization problem of the Energy Absorption Devices in a train

Design Functions	$\min \sum_{i=1}^{n_{cars}} acc_i \;;\; \min \sum_{i=1}^{n_{cars}} def_i$
Design Constraints	$acc_i < 5g \;\; i=1, ..., n_{cars}$; $L_{xe_i} > L_{xe_{i-1}} + 500KN$
Design Variables	$[\{L_{Le1}, L_{Le2}, L_{Le3}\};\; \{L_{He1}, L_{He2}, L_{He3}\}]$

With: n_{cars} - Total number of cars; acc_i - Maximal acceleration level in car *i*;
Def_i - Maximal allowed deformation in interface *i*;
$L_{LE1}, L_{LE2}, L_{LE3}$ - Low-Energy plastic limits; $L_{HE1}, L_{HE2}, L_{HE3}$ - High-energy plastic limits.

4.3 Multi-Objective Optimization of a S-Shape Beam for Crashworthiness

In this example, an energy absorption structure, usually known as S-Shape beam, is present. The simulation setup and configuration of the S-shape beam are presented in Figure 12. The mass of the impact car is 1500 kg and the velocity of impact 50 km/h. Due to the symmetry only half structure has been considered in the simulation. The characteristics of the structure are presented in In this study and for the multi-optimization optimization several conflicting design functions, as follows, can be considered and analyzed.

Table 4. Characteristics of the S-Shape Beam

Material	A36 Steel
Section of the beams	Hollow Square
Total height of the structure	0.879 m
Length of bodies type A	0.2635 m
Length of bodies type B	0.404 m
Angle between bodies type A and B	30°

- Minimization of the mass;
- Minimization of the VCSI (Vehicle Crash Severity Index);
- Maximization of the energy absorbed;
- Minimization of the deformation.

. The S-Shape beam has been divided in three flexible bodies (2 bodies' type A and one body type B as indicated in Figure 12). Each node of these flexible bodies is represented with 5 nodes and 4 beam elements. The first four vibration modes have been considered. The model has 4 plastic hinges, located in the connections between the bodies of the S-Shape beam and two additional plastic hinges, one connecting the beam and a plate that impact the rigid wall and the other connecting the impact car. This model has been already studied with other characteristics and the results compared with experimental results [5]. The simulation takes about one minute in a PC 2.0 GHz.

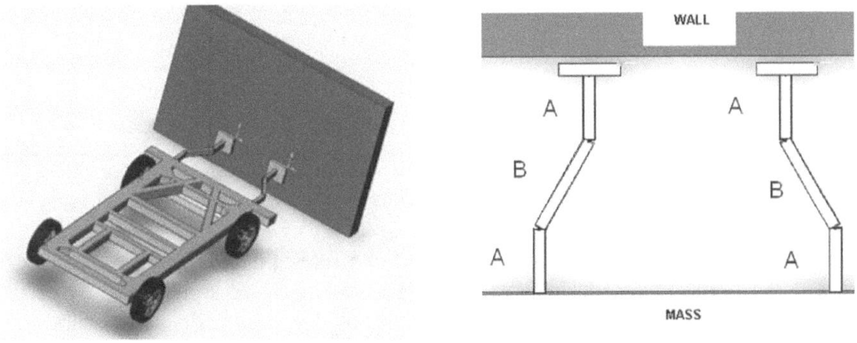

Fig. 12. Simulation setup for the S-Shape beam

In this study and for the multi-optimization optimization several conflicting design functions, as follows, can be considered and analyzed.

Table 4. Characteristics of the S-Shape Beam

Material	A36 Steel
Section of the beams	Hollow Square
Total height of the structure	0.879 m
Length of bodies type A	0.2635 m
Length of bodies type B	0.404 m
Angle between bodies type A and B	30°

- Minimization of the mass;
- Minimization of the VCSI (Vehicle Crash Severity Index);
- Maximization of the energy absorbed;
- Minimization of the deformation.

These four functions can be combined two by two forming a set of 6 multicriteria optimization problems. This combination in some cases seems not to be conflicting, but the maximal deformation of the structure is limited, the overlapping of the bodies is not allowed, and each geometric part of the beam structure is considered as a design variable, a Pareto Front can be obtained for each case. As a first example, the minimization of the mass and the minimization of the VCSI are considered as design functions. Three design constraints related with the deformation of the structure have been specified. These design constraints that correspond to the overlapping and penetration of the bodies can be specified in the simulation with contact models, but the contact-impact between bodies drastically increases the simulation time and high acceleration peaks occurs. These design constraints are illustrated in Figure 13.

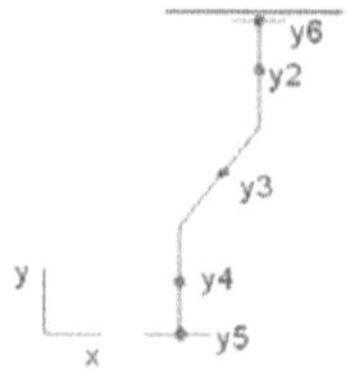

$$y_6 \geq y_2$$
$$y_4 \geq y_5$$
$$y_6 \geq y_5 + 0.3\,\text{m}$$

Fig. 13. Geometric design constraints for the S-Shape Beam

This design problem, which is summarized in Table 5, has 2 design functions, 3 design constraints, and 3 design variables. The beams have square hollow cross sections and the width (H) for each beam is considered as a design variable. A constant ratio H/t=6 is considered. The plasticity model corresponds to an elasto-plastic behavior where the plastic moment and the plastic angle are given by

$$M_P = \frac{19}{54}\sigma_y H^3 \tag{17}$$

$$\theta_P = \frac{M_P \mathrm{L}}{3\mathrm{EI}} \tag{18}$$

where σ_y is the yielding stress (300 MPa), "L" is the length of the corresponding beam, "E" the Young Modulus (Steel=219 GPa) and "I" the moment of inertia (calculated from "H"). Also because the beams' width is changed during the optimization process, all the simulation parameters dependent on the geometry are automatically updated such as areas and mass moments of inertia, structural damping parameters.

Table 5. Multiobjective optimization problem of the S-Shape Beam

Design Functions	$\min\left\{\mathrm{VCSI}=1/\mathrm{T}\int_0^{\mathrm{T}}(\ddot{y}_5)^2 dt\right\}$; $\min\{\text{mass}\}$
Design Constraints	$y_6 \geq y_2$; $y_4 \geq y_5$; $y_6 \geq y_5 + 0.3$
Design Variables	$\{\mathrm{H}_2, \mathrm{H}_3, \mathrm{H}_4\}$

With: y_i - Displacement in longitudinal direction of body *i*, $\ddot{y}_i$ - Acceleration of body *i*, θ_i - Rotation of body *i*.

The evolution of the Pareto front is indicated in Figure 14 for a population size of 20. The crossover probability is 0.9 and the mutation probability is 0.5. Each simulation takes about one minute in a PC 2.0 GHz. The solutions obtained are presented in Figure 15.

Table 6 contains the feasible designs of the S-shaped beam, in terms of the cross section widths. As can be observed the designs obtained are somehow unrealistic. This is because for keeping the deformation under the limits specified, the structure must be rigid at the extremities but reducing the mass in the central structural member the structure becomes sufficiently compliant in order to have lower accelerations levels. Note that the characteristics of the plastic hinges when two contiguous members have different dimensions, corresponds to the weakest member.

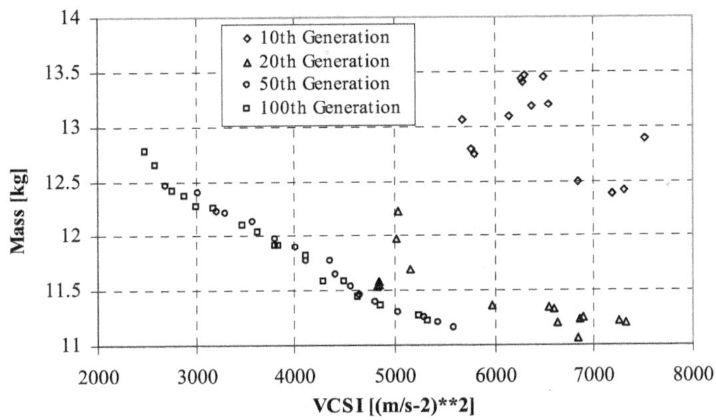

Fig. 14. Pareto front for the S-Shape Beam structure for mass and VCSI

The type of configurations presented in Table 6, can be eliminated by adding design constraints related with the relative dimensions between the structural members or choosing other design functions. This is illustrated in the second combination of design functions. In this case, the deformation and the accelerations are selected. These two widely used functions in crashworthiness have already been explained in the example of the design of energy absorption devices of trains. The results for these two design functions are presented in Figure 13. These results have been obtained with the same parameters for the genetic algorithm that in previous example.

From Figure 15 it can be observed that when the deformation tends to zero the acceleration levels tends to infinite. In this case and for this small population, the solution tends to be concentrated along a vertical line, but the more interesting solutions at an applicability point of view are located along the horizontal line. In addition, in this case the convergence is very fast, so it is advisable to use a larger population and a lower number of generations. With these two design functions, a wide variety of solutions is obtained as presented in Table 7.

The results in Figs. 14 and 15 illustrate the trade-offs that are necessary in the first stages of the design project between the design parameters. For the type of problems presented, and based on previous tests to find the more efficient parameters of NSGA II, it is advantageous to raise the population size and the number of generations, but in a case where the simulation time cannot be much higher it is more convenient to raise only the population size. The increase of the mutation probability reveals a stronger, and still positive, influence in this type of simulations. In these two cases, no limits for the design functions have been specified. However, this can be done in order to reduce the design space and forcing the population solutions to be concentrated in a narrower zone.

Table 6. Feasible designs of the S- Shape Beam for the Mass and VCSI problem

H2 (m)	H3 (m)	H4 (m)
0.0661	0.0100	0.0788
0.0661	0.0100	0.0779
0.0661	0.0100	0.0816
0.0661	0.0100	0.0797
0.0669	0.0100	0.0724
0.0667	0.0100	0.0764
0.0661	0.0100	0.0790
0.0670	0.0101	0.0740
0.0661	0.0100	0.0775
0.0670	0.0101	0.0740
0.0661	0.0100	0.0809
0.0673	0.0101	0.0731
0.0661	0.0100	0.0794
0.0669	0.0100	0.0729
0.0669	0.0100	0.0721
0.0638	0.0100	0.0689
0.0638	0.0100	0.0689
0.0667	0.0100	0.0764
0.0661	0.0100	0.0764
0.0664	0.0100	0.0806

Common configuration

5 Conclusions

In this work, a design framework for the multicriteria optimization of multibody systems, with application to vehicle crashworthiness, has been proposed. With the aim of multibody dynamics, simplified models for crashworthiness simulation are proposed and linked with the evolutionary algorithm NSGA-II. With these tools, the Pareto front for conflicting objectives is obtained in a few hours on a PC. The Pareto front is very important in the first stages of the design process when the trade-offs between conflicting objectives are to be decided. From the Pareto front, the design can select the best compromising solution or then, solve a single criteria constrained optimization problem with weighted functions to know the optimal characteristics of the structure or energy absorption devices. These aspects have been illustrated in the examples presented. The parameters provided to the genetic algorithm play an important role in the efficiency of the optimization process. In the examples presented has been demonstrated that a reasonable idea of the Pareto front can be obtained with a population size of 20-50 and for 20-30 generations. The effect in the results of the population size is more important than the effect of the number of iterations.

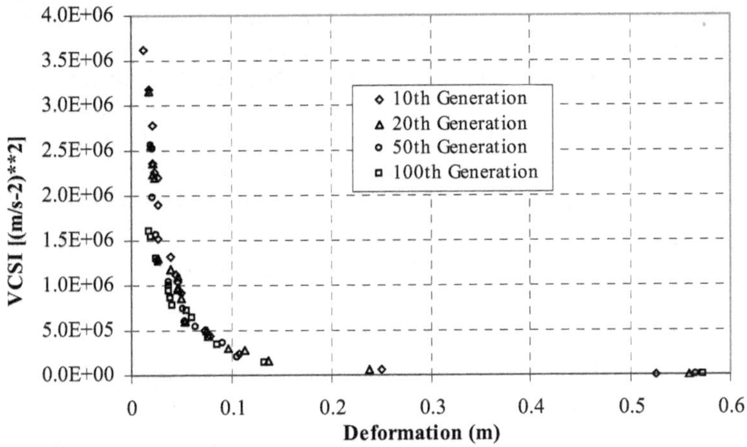

Fig. 15. Pareto front for the S-Shape Beam structure for Deformation and VCSI

Table 7. Feasible designs of the S-Beam Shape for the Deformation and VCSI problem

H2 (m)	H3 (m)	H4 (m)
0.1402	0.1321	0.0980
0.1638	0.0253	0.1891
0.1623	0.0230	0.1944
0.1345	0.1232	0.1199
0.0893	0.1326	0.1187
0.0640	0.1317	0.0985
0.1586	0.0218	0.1944
0.1623	0.0230	0.1944
0.1651	0.0214	0.1923
0.0902	0.1294	0.1188
0.1586	0.0218	0.1944
0.1343	0.0257	0.0672
0.0925	0.1326	0.1188
0.0925	0.1326	0.1187
0.0661	0.1361	0.1979
0.0596	0.1365	0.1887
0.1409	0.1307	0.0973
0.0647	0.1231	0.1195
0.0756	0.1318	0.1041
0.1358	0.1231	0.1199

The methodology has been implemented two general-purpose multibody codes and any response parameter can be used as the design function or a design constraint. The design methodologies presented have been applied for crashworthiness problems; however, they are quite general and can be applied for other field related with multibody dynamics such as, mechanisms and vehicle dynamics. The main limitation of the methodologies presented is that is the complexity of the models increases in the case of 2D flexible models, it will not be possible to obtain the Pareto front in a reasonable time. However, the simulations required in each generation are independent and the efficiency can be improved with parallel processing.

Acknowledgments

The support by the Fundação para a Ciência e Tecnologia (Portuguese Science Foundation) under project POCTI/EME/34340/00 is gratefully acknowledged. In addition, the help of the under-graduate students Ricardo Corrêa and Filipe Antunes is greatly appreciated.

References

1. Lust RV (1992) Structural optimization with crashworthiness constraints. Structural Optimization 4:85-89
2. Kim CH, Mijar AR, Arora, JS (2001) Development of simplified models for design and optimization of automotive structures for crashworthiness. Struct Multidisc Optim 22-4:307-321
3. Hamza K, Saitou K (2003) Design for Structural Crashworthiness using Equivalent Mechanism Approximations. Proceedings of the 2003 ASME Design Engineering Technical Conferences, Chicago, Illinois, September 2-6, DETC2003/DAC-48751.
4. Oyan C (1998) Dynamic simulation of Taipei EMU train. Vehicle System Dynamics 30:143-167
5. Ambrósio J, Pereira M, Dias J (1996) Distributed and discrete nonlinear deformations on multibody dynamics. Nonlinear Dynamics 10:359-379
6. Pereira M, Ambrósio J, Dias J (1997) Crashworthiness analysis and design using rigid-flexible multibody dynamics with application to train vehicles. International Journal for Numerical Methods in Engineering 40:655-687
7. Dias J, Pereira M (2002) Optimal Design of Train Structures for Crashworthiness using a Multi-load Approach. I. J. of Crashworthiness 7:331-343
8. Milho J, Ambrosio J, Pereira M (2003) Validated multibody model for train crash analysis. I. J. of Crashworthiness 8(4):339-352
9. Pereira M, Dias J (2003) Analysis and Design for Train Crashworthiness Using Multibody Models. Vehicle System Dynamics (to appear)
10. Bestle D, Eberhard P (1996) Multi-Criteria Multi-Model Design Optimization. In: Bestle D, Schiehlen W (eds) IUTAM Symposium on Optimization of Mechanical Systems. Kluwer Academic Publishers, Dordrecht, pp 863-876

11. Dias J, Pereira M (1997) Sensitivity Analysis of Rigid Flexible Multibody Systems. Multibody System Dynamics, 1:303-322
12. Bischof C, Carle A, Hovland P, Khademi P, Mauer A (1998) ADIFOR 2.0 User's Guide (revision D)". Technical Report CRPC-95516-S, Center for Research on Parallel Computation, Rice University
13. Zitzler E, Deb K, Thiele L, Coello Coello CA, Corne D (eds) (2001) Evolutionary multi-criterion optimization EMO 2001. Lecture Notes in Computer Science (LNCS) vol 1993, Springer-Verlag, Berlin
14. Eberhard P, Dignath F, Kübler L (2003) Parallel evolutionary optimization of multibody systems with application to railway dynamics. Multibody System Dynamics 9-2:143-164
15. Sélim D, Olivier V, Calogéro C (1999) Application of evolutionary strategies to optimal design of multibody systems. In: Ambrósio J (ed) EUROMECH Colloquim 404, Lisbon, Portugal, pp 645-661
16. Deb K (2001) Multi-Objective Optimization Using Evolutionary Algorithms. John Wiley & Sons
17. Shabana AA (1989) Dynamics of Multibody Systems. John Wiley & Sons, New York
18. Narayanan S, Azaram S (1999) On improving multiobjective genetic algorithms for design optimization. Structural Optimization 18:146-155
19. http://europa.eu.int/comm/research/growth/gcc/projects/safe-train.html

Multi-Criteria Optimization of a Hexapod Machine

Lars Kübler[1], Christoph Henninger[2], and Peter Eberhard[2]

[1] Institute of Applied Mechanics, University of Erlangen, Egerlandstr. 5, 91058 Erlangen, Germany
kuebler@ltm.uni-erlangen.de

[2] Institute B of Mechanics, University of Stuttgart, Pfaffenwaldring 9, 70550 Stuttgart, Germany
{eberhard,henninger}@mechb.uni-stuttgart.de

Alternative designs of a hexapod machine are proposed and investigated with the aims to reduce flexibility and to eliminate singular kinematic configurations that appear in the workspace for the current design of the machine. The hexapod is modeled as a rigid multibody system. Actuator amplitudes in the struts associated with desired tool trajectories are computed by inverse kinematics. Hence, dynamic forces and torques are not considered and, as there is no closed-loop control realized so far in the model, the actual rotational and translational position of the tool deviates from the desired position due to machining loads. These deviations serve as objective functions during a multi-criteria optimization in order to determine the best design regarding stiffness/flexibility of the machine. Further, a general approach for evaluating flexibility behavior of the machine in the complete workspace is introduced and the results from the optimization are verified. Besides flexibility, a crucial point for machining tools is the size of the feasible workspace. Therefore, the influence of the design modification on the workspace is also taken into account.

1 Introduction

Machines with parallel kinematics feature low inertia forces due to low masses of the structure combined with possible high accuracy and stiffness. Such machines are currently under investigation in various fields of engineering like robotics, measurement systems and manufacturing technologies, see e. g. [19, 6, 22]. The need of shorter process times with even increasing demands in accuracy leads to a great research interest in the area of machine tools with parallel kinematics, e.g. for high speed cutting purposes, cp. [15]. In this paper the hexapod robot HEXACT is investigated. HEXACT is a research machine tool with parallel kinematics, developed by Prof. U. Heisel and his coworkers at the Institute for Machine Tools at the University of Stuttgart, Germany, see Fig. 1.

Hexapod machines feature the advantages of parallel kinematics and are characterized by a simple, cost-effective design [16]. The tool carrier has six degrees of

J.A.C. Ambrósio (ed.), Advances in Computational Multibody Systems, 319–343.

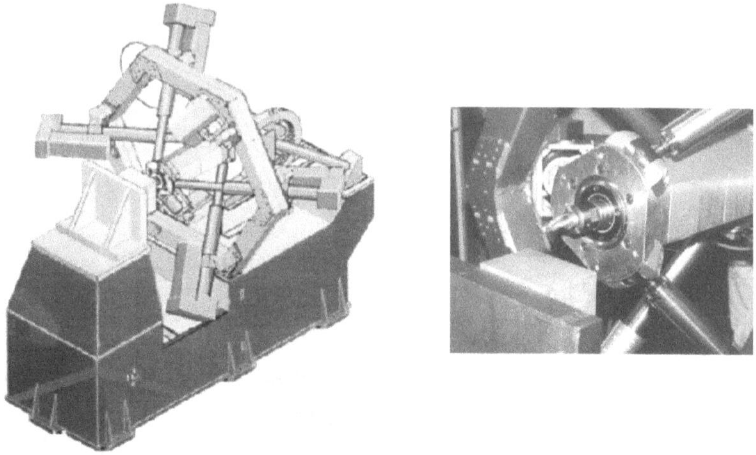

Fig. 1. Hexapod machine HEXACT (see http://www.ifw.uni-stuttgart.de)

freedom in space and is positioned by actuators integrated in the six telescope struts. Furthermore, the specific construction of this machine allows for compensation of joint clearance by pre-stressing of the symmetrically arranged struts with an initial tension. However, pre-stressing leads to loss of one degree of freedom, as the rotation about the tool axis is not controllable anymore. Nevertheless 5-axis processing of cubic parts is possible in the pre-stressed case with high static and dynamic accuracy.

A general drawback of parallel kinematic robots is the appearance of kinematic singular configurations that have to be avoided during operation and, hence, reduce the usable workspace. For the current design of the investigated hexapod machine an especially unpleasant kinematic singularity is located along the tool axis in the central position of the machine, cp. [18], where the rotational stiffness about the tool axis decreases to zero. One way to overcome this problem and to reduce the manifold of singularities is to use a higher number of actuators than the degree of freedom of the end-effector, so-called redundant actuation, see e. g. [7, 21, 26]. However, this approach leads to additional difficulties. For example special techniques for the control of the over-constrained systems are required. In this article, redundant actuation is not considered. The idea followed here is to remove unpleasent singularities in the workspace by changing the design of the machine.

Modeling and optimization of the hexapod machine follow the integrated design approach for dynamical systems proposed by [3], as illustrated in Fig. 2.

A brief description of the modeling of the hexapod is given in the first part of this contribution. The dynamic behavior of the system is described by nonlinear differential equations of motion obtained from the multibody system approach. In order to define tool trajectories, the actuator amplitudes in the struts have to be derived by inverse kinematics. Both topics are described in [8], where the hexapod robot was modeled and optimizations were carried out. We utilize a re-engineered, extended

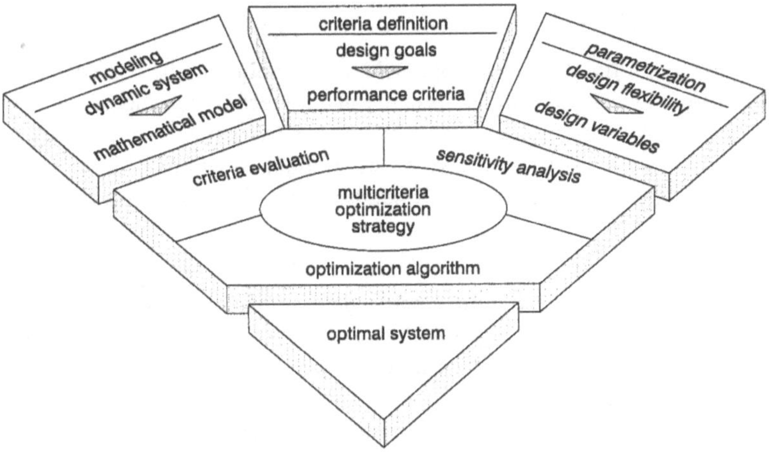

Fig. 2. Integrated modeling and design approach [3]

model in order to do further optimizations and to evaluate alternative designs, proposed here, with less or even no singular configurations in the interesting workspace.

The description then continues with the design process optimization, see Fig. 2. After definition of performance criteria and design variables a multi-criteria optimization approach is applied in order to find an 'optimal' system, see [1, 10, 11]. The design goals considered here are to reduce flexibility and improve accuracy under consideration of a sufficiently large workspace of the machine utilizing our alternative designs. This implies the decrease of the manifold of singular configurations of the machine.

Further, a general approach to evaluate the stiffness/flexibility behavior of the machine is introduced. Here, the tangential stiffness or flexibility matrices, respectively, are determind for the entire workspace of the machine. The results found with this method are compared with those from the previous optimization.

2 Modeling of the Hexapod Machine

A crucial point within the modeling of the hexapod machine is the description of the telescope struts. A main question is whether elastic deformations of the relatively slender struts have to be considered or not. In [8] flexible struts were investigated with the conclusion that in the low frequency range the strut elasticity has a negligible influence on the vibration behavior of the machine. In contrast, the elasticity of the joints turned out to be important to get sound results corresponding closely to real measured data. Hence, a rigid multibody system (MBS) model with elastic joints for the connection of the struts is utilized, Sect. 2.2. In order to eliminate singular configurations and reduce flexibility, alternative design variants are proposed in Sect. 2.3.

2.1 Multibody Systems

In the common multibody system approach, deformations are neglected and the bodies are considered as rigid. The bodies are interconnected with ideal joints and coupling elements like springs, dampers and elements for active control. The dynamics of multibody systems is described by the differential equations of motion, see [23]. Using f generalized coordinates $\mathbf{y}$ they can be written in general form as differential equations

$$\mathsf{M}(t,\mathbf{y}) \cdot \ddot{\mathbf{y}} + \mathbf{k}(t,\mathbf{y},\dot{\mathbf{y}}) = \mathbf{q}(t,\mathbf{y},\dot{\mathbf{y}}), \tag{1}$$

where $\mathsf{M} \in \mathbb{R}^{f\times f}$ is the symmetric, positive definite mass matrix, $\mathbf{k} \in \mathbb{R}^f$ the vector of generalized centrifugal and Coriolis forces and $\mathbf{q} \in \mathbb{R}^f$ the vector of generalized applied forces. Equation (1) can be derived from Newton's and Euler's laws and d'Alembert's principle.

Starting from initial conditions for position and velocity $\mathbf{y^0}$, $\dot{\mathbf{y}}^\mathbf{0}$ the equations of motion have to be solved by numerical time integration, where various algorithms exist, e.g. Runge-Kutta methods or the Shampine-Gordon method that is also applied to the presented example.

2.2 MBS Model of the Hexapod

The rigid MBS model of the hexapod used here is similar to the one in [8], except that it was re-engineered for computational reasons and in order to extend the model to further analysis and optimization tasks. The nonlinear equations of motion (1) are created symbolically with NEWEUL [17]. In Fig. 3 a section of the spatial rigid multibody model of the hexapod machine in central position is illustrated, compare also the real machine in Fig. 1. A front view of the model in the xy-plane is shown and for lucidity only three of the six telescope struts are drawn.

The model has $f = 36$ degrees of freedom. The generalized coordinates

$$\mathbf{y} = \begin{bmatrix}\mathbf{y}_{\text{tc}}\ \mathbf{y}_{\text{p1}}\ \cdots\ \mathbf{y}_{\text{p6}}\end{bmatrix} \in \mathbb{R}^f \tag{2}$$

describe the motion of the tool carrier $\mathbf{y}_{\text{tc}} = [x_{\text{tc}}\ y_{\text{tc}}\ z_{\text{tc}}\ \alpha_{\text{tc}}\ \beta_{\text{tc}}\ \gamma_{\text{tc}}]$ and the six strut plungers $\mathbf{y}_{\text{p}i} = [x_{\text{p}i}\ y_{\text{p}i}\ z_{\text{p}i}\ \eta_{\text{p}i}\ \vartheta_{\text{p}i}]$. Since the telescope struts cannot rotate about the strut axis only five coordinates are required for each plunger. The strut quils are connected to the plungers via position actuators, which realize the positioning travel of the telescope struts.

All parameters of the model like geometry, masses, spring and damper coefficients, etc. are taken from [8] or from the Institute of Machine Tools (IfW) at the University of Stuttgart. Details on the measured elasticities of the bearings are given in [9].

The actuator amplitudes are treated as time dependent variables, or rather rheonomic constraints. For desired tool trajectories they are computed by inverse kinematics. As the dynamic forces and torques are not considered and there is no closed-loop control realized so far, the actual position of the tool may deviate from the desired position. Hence, it is desirable to improve the stiffness of the machine. For this purpose alternative designs are proposed in the following section.

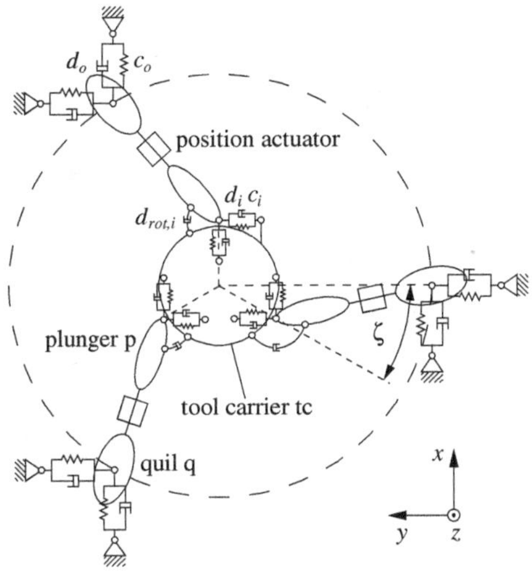

Fig. 3. MBS model of the hexapod machine (front struts only)

2.3 Alternative Designs

In the current design of the hexapod machine the joints are placed equidistantly on the perimeter of the tool carrier. Therefore, the struts are placed radial at the tool carrier in its central position. This leads to a kinematic singularity where the flexibility around the tool axis approaches infinity and is very high at the nearby surrounding. For machining processes this is very undesirable as high flexibility leads to loss of accuracy and low surface quality of the produced parts.

In order to improve the flexibility behavior of the hexapod and to eliminate singular configurations two options for alternative designs are proposed and a third variant, described in [9], is discussed. All variants, including the current design, are illustrated in Fig. 4.

The angle between the struts and the tool can be varied in different ways. As a first possibility one may rotate the joint positions at the tool carrier by an angle ζ, see Fig. 4 (design I). Thus, the singular point in the center vanishes and a lower flexibility is expected. However, the mentioned pre-stressing is no more possible. A second option is to rotate the front and back joints in counterwise directions, respectively, Fig. 4 (design II). It is the advantage of this design that a pre-stressing of the struts for eliminating joint clearance is still possible. However, this modification leads to a kinematic coupling of the tool rotation around the z-axis (tool axis) and its translation in z-direction. A third possibility, proposed in [8], is to change the positions of the two upper and lower joints each counterwise while the left and right joints remain in their original positions, as illustrated in Fig. 4 (design III).

All alternatives involve a reduction of the workspace in the xy-plane as the pivoting angle of the inner joints is restricted. This has to be considered when modifying the design of the hexapod.

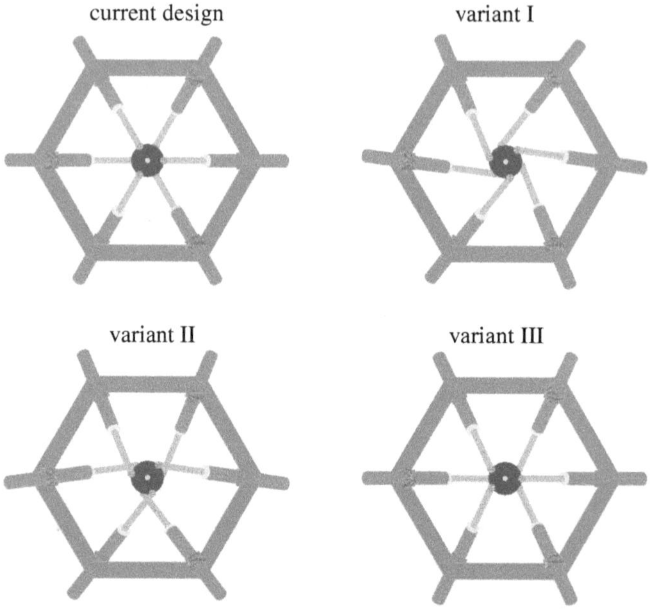

Fig. 4. Alternative designs: current design with radial struts; variant I with equally rotated joint positions; variant II with counter-rotated joint positions; variant III with counterwise rotated joint positions for four struts

3 Optimization of MBS

It is the purpose of optimizing multibody systems to find a mechanical design that leads to improved dynamic behavior of the technical system, e.g., increased damping of undesired vibrations. In order to evaluate the dynamic behavior, the simulation of the motion by numerical integration of the equations of motion is required. During the optimization process these time integrations are repeatedly executed using different sets of design variables $\mathbf{p}$. These design variables are the parameters of the system that can be changed within given ranges for optimizing the dynamic behavior, for example, parameters of the structure, geometric quantities, stiffnesses or active controls of the technical system.

More information concerning the procedure and available solution methods is given in the textbooks [1, 13] or the article [4].

3.1 Optimization Criteria

Use of numerical optimization methods requires that technical restrictions and design goals are formulated as mathematical functions. From the motion of the system, which is determined by the time trajectories of the generalized coordinates, the optimization criteria can be evaluated. For the evaluation of vibrations, energies, etc. these criteria are usually of the form

$$\xi_i = G_i(t_{\text{end}}, \mathbf{y}_{\text{end}}, \dot{\mathbf{y}}_{\text{end}}, \mathbf{p}) + \int_{t_0}^{t_{\text{end}}} F_i(t, \mathbf{y}, \dot{\mathbf{y}}, \ddot{\mathbf{y}}, \mathbf{p}) dt \tag{3}$$

which is closely related to optimal control. In (3) the dynamic behavior during the complete time interval $[t_0, t_{\text{end}}]$ is taken into account by F_i, while G_i only considers the final state and, hence, accounts for cases where special values for the final state or a minimum time or energy must be achieved. The final time t_{end} may be fixed or given implicitly by the final state

$$t_{\text{end}} : \; H_{\text{end}}(t_{\text{end}}, \mathbf{y}_{\text{end}}, \dot{\mathbf{y}}_{\text{end}}, \mathbf{p}) = 0. \tag{4}$$

Simple explicit functions $\xi_i = \xi_i(\mathbf{p})$ or user-defined functions are also utilized. For example, the bounds of the design variable space can be defined as

$$p_{i\text{l}} \leq p_i \leq p_{i\text{u}} \tag{5}$$

or implicitly

$$\begin{aligned} \xi_1 &= p_i - p_{i\text{l}} \geq 0, \\ \xi_2 &= p_{i\text{u}} - p_i \geq 0. \end{aligned} \tag{6}$$

All criteria ξ_i may be used either as equality constraints g_i, inequality constraints h_i, or optimization criteria ψ_i.

3.2 Multi-Criteria Optimization

The optimization of technical systems usually cannot be performed by the optimization of only one criterion. Instead, several conflicting criteria must often be considered for optimization. The theory of multi-criteria optimization, [10], helps to avoid such discrepancies. Several criteria can be minimized simultaneously, and optimal compromise solutions can be derived if conflicts arise. In general, these solutions are not unique but a whole set of solutions exists which are not comparable to each other, so-called Edgeworth-Pareto-optimal (EP-optimal) solutions. All of these solutions have to be considered as optimal since no strict order exists in vector optimization. In general, the multi-criteria optimization task can be written as follows

$$\text{find} \quad \underset{\mathbf{p} \in \mathbb{P}}{\text{opt}}\, \boldsymbol{\psi}(\mathbf{p}) \quad \text{where} \quad \mathbb{P} := \left\{ \mathbf{p} \in \mathbb{R}^h \,|\, \mathbf{g}(\mathbf{p}) = \mathbf{0}, \mathbf{h}(\mathbf{p}) \geq \mathbf{0} \right\} \tag{7}$$

with several optimization criteria ψ_i summarized in the vector $\boldsymbol{\psi}$ and analogous the vectors $\mathbf{g}$ and $\mathbf{h}$ of equality and inequality constraints. Goal of the optimization of the vector criterion $\boldsymbol{\psi}(\mathbf{p})$ is to minimize all criteria ψ_i. A simultaneous minimization of all criteria is rarely possible as generally no point is feasible in the design space where all criteria have their minimal values simultaneously. Such a point in the criteria space is called *utopian solution*. Instead, multi-criteria approaches try to find a compromise that favorably lies on the set of EP-optimal solutions. Which solution to choose from the set of EP-optimal solutions then depends on the preferences of the designer.

Common to most of the strategies is the reduction of the vector optimization problem to nonlinear programming problems. This reduction is based on two fundamental principles: scalarization and hierarchization.

In the case of scalarization the optimization criteria are combined to a scalar utility function $u(\boldsymbol{\psi})$ which is then optimized instead of the original vector criterion. Often the weighted criteria method is applied which can be problematic for dynamic problems, as the weighting coefficients are difficult to define physically and the dependency of the result on this choice is highly nonlinear.

Hierarchical methods have in common, that the designer has to assign a level of importance to each optimization criterion ψ_i. Optimization starts with a scalar optimization of the first, most important, criterion. Then the next important criterion can be minimized, where the results reached in the previous levels are considered as inequality constraints and so on. An often useful modification introduces worsening factors that allow a defined worsening of results reached in previous levels in order to give the optimization of the criterion in the actual level some margin.

3.3 Optimization Algorithms

During the multi-criteria optimization every scalar optimization task reads

$$\text{find} \quad \min_{\mathbf{p}\in\mathbb{P}} \psi(\mathbf{p}) \quad \text{where} \quad \mathbb{P} := \left\{\mathbf{p} \in \mathbb{R}^h \,|\, \mathbf{g}(\mathbf{p}) = 0, \mathbf{h}(\mathbf{p}) \geq 0\right\} \; . \tag{8}$$

In order to solve this nonlinear optimization problem with equality and inequality constraints deterministic or stochastic optimization algorithms can be applied. Deterministic algorithms usually use gradient information and have the advantage of locating a minimum within a few iteration steps. On the other hand, they usually only find local minima and require, at least in theory, a smooth optimization criterion. Stochastic methods have the advantage that they may find a global minimum without posing severe restrictions on the differentiability or convexity of the optimization criterion, however, they often need many criteria evaluations. Therefore, they are not well suited for systems which require a high computational effort for evaluation. This drawback can be reduced by using parallel criteria evaluation in combination with a stochastic evolution strategy, see e.g. [12].

For the use of gradient-based optimization algorithms, like sequential quadratic programming (SQP) algorithms that are implemented in NEWOPT/ AIMS [2], gradients of the optimization criteria with respect to the design variables $\mathbf{p}$ are required. Since the state variables $\mathbf{y}_0$, $\dot{\mathbf{y}}_0$ can not generally be given as explicit functions of $\mathbf{p}$, direct differentiation methods cannot be applied. In NEWOPT/AIMS gradients can be derived by purely numerical methods like finite differences or more sophisticated methods like the adjoint variables method [10], the direct differentiation approach [1] and automatic differentiation using e.g. ADIFOR [5].

4 Optimization of the Hexapod Machine

In this section the spatial MBS model of the hexapod machine described in Sect. 2 is optimized with respect to flexibility and workspace. The optimization is carried out under consideration of the alternative designs as proposed in Sect. 2.3. In a first step, the flexibility of the system is investigated without considering geometric restrictions in order to obtain the unconstrained theoretical minimum. This is advisable as a premature restriction of the design space often leads to loss of innovative solutions of the problem. Afterwards, the geometric boundary conditions are taken into account and the influence of the new configuration on the workspace is discussed, where a possible compromise is proposed. All optimizations are carried out using an SQP algorithm with gradients computed by finite differences or automatic differentiation.

4.1 Optimization Criteria and Design Variables

It is the goal of the optimization to increase the stiffness, i.e. to reduce the flexibility of the tool for machining purposes. Therefore, we define an exemplary tool trajectory with relative large, but still realistic machining forces and moments. When running the trajectory, the actual translational and rotational position is computed. The flexibility can be evaluated by comparing the actual trajectory with the target values. This can be formulated applying an integral optimization criterion, see (3). Regarding the rotational flexibility about the tool axis, the average deflection of the tool rotation can be used

$$\psi_1 = \frac{1}{t_{\text{tot}}} \int_{t_0}^{t_{\text{end}}} \left| \gamma_{\text{tc}} - \gamma_{\text{tc,target}} \right| dt. \tag{9}$$

The translational flexibility can be evaluated analogously

$$\psi_2 = \frac{1}{t_{\text{tot}}} \int_{t_0}^{t_{\text{end}}} \sqrt{\left(x_{\text{tc}} - x_{\text{tc,target}}\right)^2 + \left(y_{\text{tc}} - y_{\text{tc,target}}\right)^2 + \left(z_{\text{tc}} - z_{\text{tc,target}}\right)^2}. \tag{10}$$

As design variable p the joint rotation angle ζ is chosen. Due to geometric restrictions the admissible range of values for the design variable is bounded by $0 \le p \le \pi/2$.

4.2 Optimizing Design Variant I

During the first optimization run, only criterion ψ_1 evaluating the rotational flexibility is taken into account. In order to achieve an optimal angle ζ^* in the central position, the struts have to be fixed tangentially at the tool, here $\zeta^* = 80.2°$. The question is whether this is also the optimum for a typical machining trajectory. In order to answer this question, the following trajectory is defined, see Fig. 5.

- 0 – 1 sec: adjust joint rotation angle ζ,

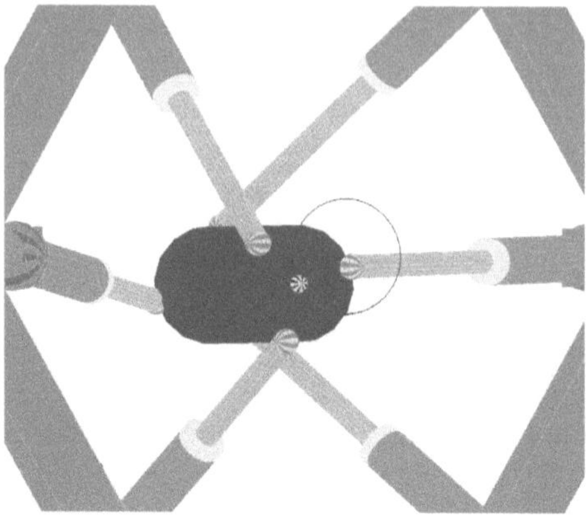

Fig. 5. Exemplary machining trajectory for evaluating flexibility behavior

- $1-2$ sec: point to point motion to [-0.1 0 0] m ,
- $2-7$ sec: circle with radius $r = 0.1$ m and midpoint at [0 0 0] m with an applied torque of $l_z = 1000$ Nm around the tool axis.

The optimum is found for approximately the same value for ζ as stated for the zero position, $\zeta^* = 82.5°$. Figure 6 shows the deviation of the tool rotation γ_{tc} both for the original design ($\zeta = 0°$) and the optimized design ($\zeta^* = 82.5°$). For the "stationary" state ($3 < t < 6$ sec), the average rotational flexibility around the tool axis calculates as $c_{\gamma\gamma} = \Delta\gamma_{tc}/l_z = 1.77 \cdot 10^{-2}$ °/Nm. The average rotational flexibility decreases to $c_{\gamma\gamma} = 9.00 \cdot 10^{-4}$ °/Nm which is an improvement by a factor of approximately 20. This factor can be seen in Fig. 6 from the ratio $\Delta\gamma_{tc}(\zeta = 0°)$ to $\Delta\gamma_{tc}(\zeta = 82.5°)$.

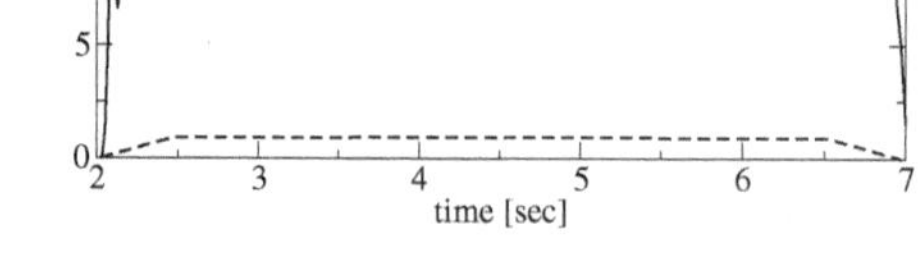

Fig. 6. Deviation of tool rotation γ_{tc} due to an applied torque for the original design ($\zeta = 0°$) and the optimized design ($\zeta = 82.5°$)

At real machining processes, also force loads are applied to the tool, of course. Therefore, the influence of the joint position angle ζ on the translational flexibility, evaluated by ψ_2, must also be considered. For the original design, the resulting translational flexibility $c_{\text{trans}} = |\Delta\mathbf{x}|/F$ has a relative high value of $3.39 \cdot 10^{-6}$ m/N. For the optimization, the above described trajectory is chosen again with an additional force loading of $F = [1\ 1\ 1]$ kN. Now a vector optimization problem with criteria ψ_1 and ψ_2 has to be solved. For this purpose a hierarchical optimization is performed. In the first level, criterion ψ_1 is optimized. The solution is identical with the result already derived for the first optimization ($\zeta = 82.5°$). The resulting translational flexibility, which is not considered in ψ_1, also improves by a factor of about 7 to a value of $4.96 \cdot 10^{-7}$ m/N, see Fig. 7.

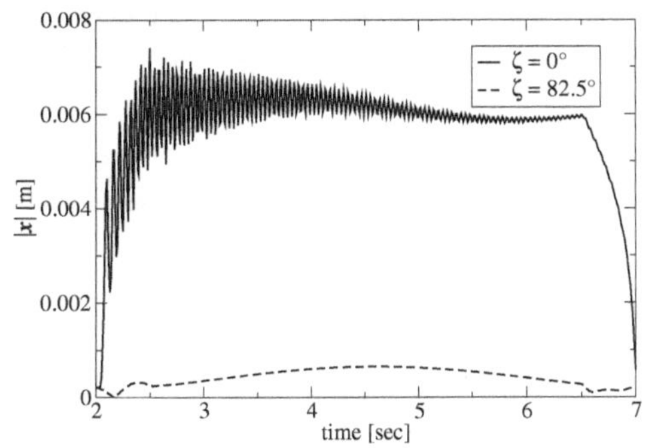

Fig. 7. Deviation of the absolute tool position **x** due to applied torque and force for the original design ($\zeta = 0°$) and the optimized design at level 1 ($\zeta = 82.5°$)

In the second level of the hierarchical optimization, the criterion ψ_2 evaluating the translational flexibility is optimized. Now the result for criterion ψ_1 is considered as an inequality constraint. In order to give the optimization of ψ_2 some margin in the design space, the optimal value for ψ_1 may increase, i.e. may be worsened by 50 %. The final optimum is stated for $\zeta = 68.2°$ where the translational flexibility c_{trans} decreases by only 4.3 % to $4.76 \cdot 10^{-7}$ m/N. Figure 8 (left) shows the decrease of deviation of the translational position for the final optimum ($\zeta = 68.2°$) compared to the result of the first optimization level ($\zeta = 82.5°$).

The rotational flexibility increases to $c_{\gamma\gamma} = 9.68 \cdot 10^{-4}$ °/Nm, see Fig. 8 (right). As this is still far away from the maximum allowed worsening, i.e. the constraints given by the worsening factor are not active, the optimum of ψ_2 corresponds to the unconstrained case.

4.3 Optimizing Design Variant II

Contrary to the expected behavior, a counter-wise rotation of the strut joints at the tool by an angle ζ does not improve the rotational stiffness of the tool, but even worsens it, see Fig. 9. For the desired trajectory, the average rotational flexibility increases to a value of $3.3 \cdot 10^{-2}$ °/Nm for $\zeta = 82.5°$ which is an increase by a

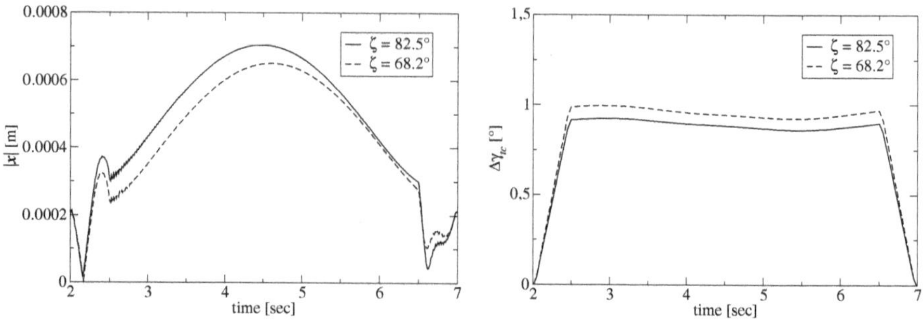

Fig. 8. Deviation of resulting tool position $\mathbf{x}$ (left) and tool rotation γ_{tc} (right) due to applied torque and force load for the optimal design of variant I at level 1 ($\zeta = 82.5°$) and the final design ($\zeta = 68.2°$)

factor of about 2. In fact, from an optimization of ψ_1 the optimal solution $\zeta = 0°$ is found.

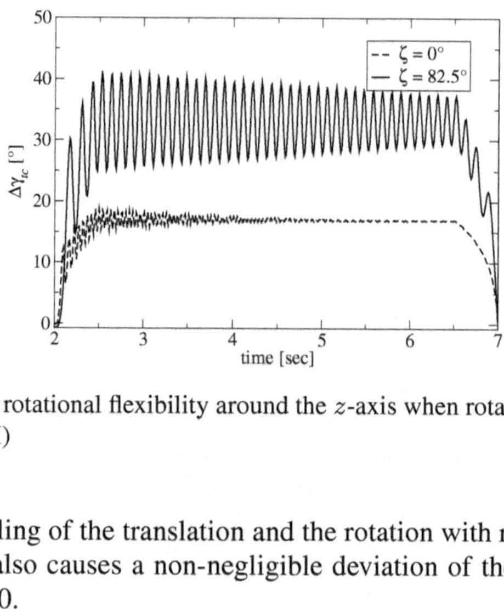

Fig. 9. Increase of the rotational flexibility around the z-axis when rotating the struts counterwise (design variant II)

Due to the coupling of the translation and the rotation with respect to the z-axis, the applied torque also causes a non-negligible deviation of the tool position in z-direction, see Fig. 10.

This behavior is not acceptable when considering machining processes with the hexapod. Hence, the alternative design II is discarded.

4.4 Results from an Optimization of Design Variant III

Design variant III has already been optimized for a spiral trajectory with a combined force and torque load [8]. As additional design variable also the pre-stressing force was taken into account. The optimum for the joint position angle was found at $\zeta = 34.1°$. However, for the different loadcase and trajectory observed here, this design does not show a better behavior regarding rotational flexibility than the original

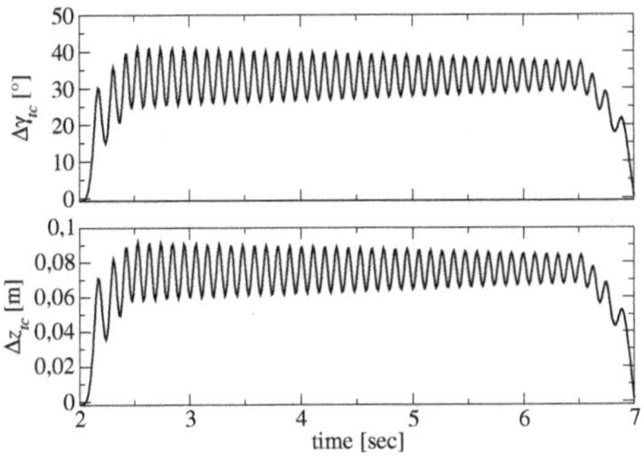

Fig. 10. Deviation of the tool's z-position due to the coupling of the tool's z-rotation and z-translation

design, cp. Fig. 11. The average rotational flexibility here is $c_{\gamma\gamma} = 1.81 \cdot 10^{-2}$ °/Nm. In addition, a very strong coupling between the z- and y-rotation can be observed, as illustrated in Fig. 12.

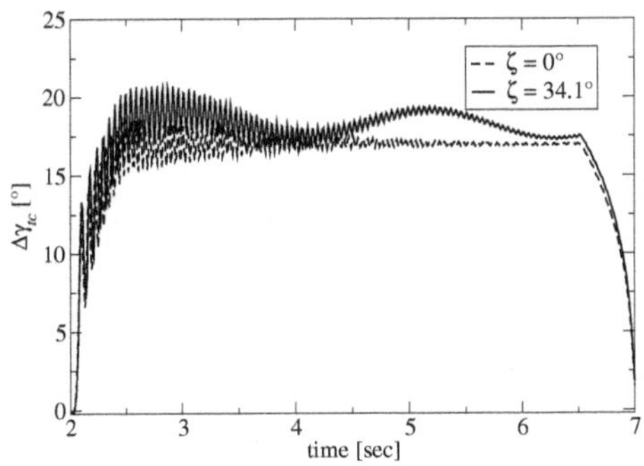

Fig. 11. Rotational flexibility around the z-axis for design variant III

4.5 Workspace

The workspace is restricted in two different ways. On the one hand, it is limited by the minimum and maximum stroke of the telescope struts. Additionally, the angle limitation of the inner joints also reduces the workspace, exemplary illustrated in Fig. 13 for design variant I. When the joint positions are rotated by an angle ζ, the workspace decreases.

A reasonable quantity for evaluating the size of the workspace is given by the radius r of the inscribed circle. When no rotation is applied to the joint positions, the workspace is bounded by the strut lengths ($r = 200$ mm). Above a rotation of the joints by $\zeta = 13.2°$, the restriction due to the angle limitation turns active. The radius

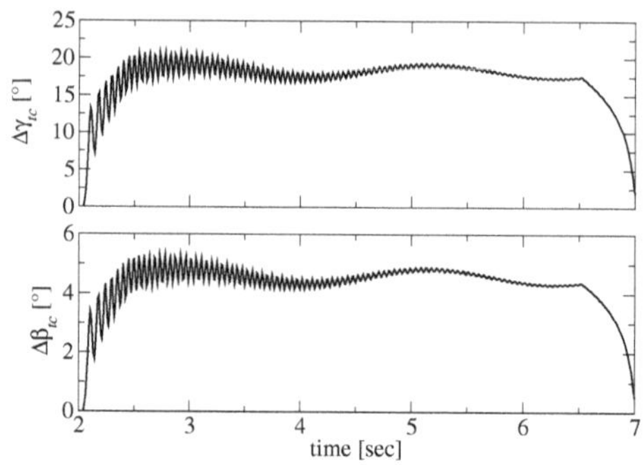

Fig. 12. Deviation of the tool's y-rotation due to the coupling of the y- and z-rotations

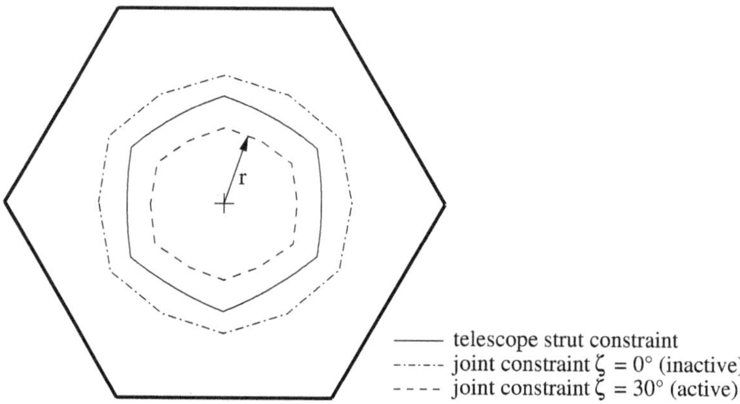

Fig. 13. Limitation of workspace in xy-plane due to the strut stroke limit of 400 mm and the admissible pivoting angle of the joints of $\pm 40°$ (design variant I)

of the workspace now decreases almost linearly and becomes zero at $\zeta = 33.8°$, see Fig. 14.

As a conclusion, the initial tool angle can be set to $\zeta = 13.2°$ without any loss of workspace. For this configuration both rotational and translational flexibilities already decrease by a factor of approximately 2.5 to values of $7.45 \cdot 10^{-3}$ °/Nm and $1.52 \cdot 10^{-6}$ m/N, respectively. Any further increase must be balanced with the process specific requirement of workspace. However, the reduction of the workspace due to the angle limitations at the inner joints can be circumvented by changes in the design of the machine. At the moment the alternative design I is realized by simply rotating the tool carrier by the angle ζ to its initial position so that the pivoting angle range is $\pm 40°$ still with respect to the radial direction. This implies that not the whole range of angles can be used for the motions. A better solution would be to rotate the joints by the initial angle ζ so that the limitation of the pivoting angle accounts to the initial strut orientation. Then, the design with the maximum stiffness can be realized without any loss of workspace.

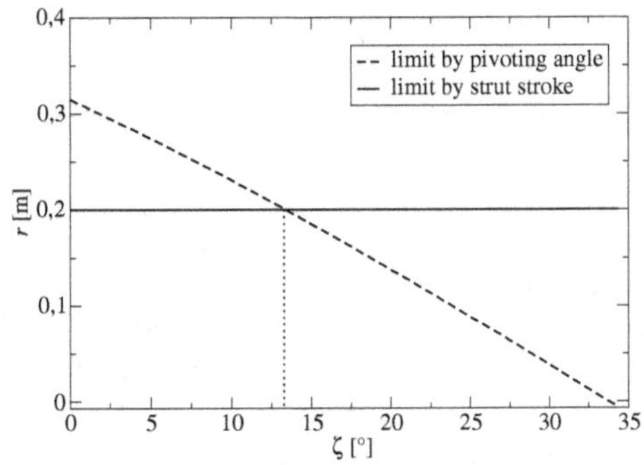

Fig. 14. Limitation of workspace radius depending on the joint rotation angle ζ (design variant I)

5 Investigation of Flexibility – General Approach

So far the flexibility behavior of the hexapod robot was investigated by observing the deviation of the tool position for a certain trajectory within a numerical simulation. Therefore, the results are only valid for this specific trajectory. Now, an approach is introduced that allows investigation of the flexibility behavior all over the workspace.

In robotics, besides flexibility a very important criterion for the evaluation of different design variants is the location of singular points in the workspace. At these points, at least one direction exists where a small variation of the tool carrier position is associated with an arbitrary change in the actuator amplitudes. Or vice versa, a small variation of one or several actuator amplitudes would cause a very intensive reaction of the tool carrier. These positions are disadvanteagous when located in the workspace as there does not exist a definite relationship between the tool carrier's position and the actuator amplitudes.

For the stiffness behaviour this means that there exists one combination of rotation and translation for which the stiffness vanishes or the flexibility approaches infinity, respectively. Mathematically, these positions can be received by regarding the eigenvalues of the tool carrier's Jacobian Matrix which is given by the derivatives of the actuator amplitudes $\mathbf{\Theta} = [\Theta_1, \Theta_2, \ldots, \Theta_6]$ with respect to the tool carrier's position $\mathbf{y}_{\mathrm{tc}}$,

$$\mathsf{J}_{\mathrm{tc}} = \frac{\partial \mathbf{\Theta}}{\partial \mathbf{y}_{\mathrm{tc}}}. \tag{11}$$

Positions where at least one eigenvalue approaches zero are kinematic singular positions. The associated eigenvector describes the motion of the tool carrier for which stiffness disappears. As there also exist very small eigenvalues in the neighborhood of singular positions, not only the exact singular position but also the neighborhood around it is critical. For parallel kinematics much effort is given on the investigation of singular constellations and their neighborhood [24, 25, 20, 14].

For the determination of kinematic singularities, the evaluation of the Jacobian is useful. However, when investigating the neighborhood of kinematic singularities, the point of interest is the flexibilitity behavior of the machine. The direct usage of the Jacobian matrix for investigations is not recommendable here as the results are not significant as neither resulting flexibilities nor the corresponding axes are available. Therefore, it is necessary to evaluate the flexibility behavior directly.

The stiffness and flexibility of the tool carrier depend on its translational and rotational position in the workspace. The relation between applied forces and torques on the one hand side and the evasive displacements of the tool carrier on the other one is highly nonlinear. At each point in the 6-dimensional workspace this relation can be linearized to obtain the tangential stiffness matrix K_t and the tangential flexibility matrix C_t. The tangential stiffness matrix K_t describes the forces caused by a given rotational and translational displacement where the tangential flexibility matrix C_t (often also called compliance matrix) gives the relationship between a given force and torque load and the resulting tool displacement. Both quantities are derived in the following section.

5.1 Matrices of Tangential Stiffnesses and Compliances

Each deviation $\Delta\mathbf{y}_\mathrm{tc}$ of the tool carrier from its original position $\mathbf{y}_{\mathrm{tc}0}$ causes a change in the strut lengths s_i,

$$\Delta s_i = f(\mathbf{y}_{\mathrm{tc}0}, \Delta\mathbf{y}_\mathrm{tc}) \ . \tag{12}$$

This relation can be derived by geometric consideration. The absolute values of the strut forces due to the resulting elasticity k of the joints are then given by

$$f_{\mathrm{s}i} = k\Delta s_i \ . \tag{13}$$

As also the strut vectors $\mathbf{s}_i$ are known by the tool carrier's position, the computation of the force vectors is straightforward,

$$\mathbf{f}_{\mathrm{s}i} = f_{\mathrm{s}i} \cdot \frac{\mathbf{s}_i}{s_i} \ . \tag{14}$$

The resulting forces $\mathbf{f}_\mathrm{tc}$ and torques $\mathbf{l}_\mathrm{tc}$ at the tool carrier follow from the equilibra of forces and moments,

$$\begin{aligned} \mathbf{f}_\mathrm{tc} &= \sum_{i=1}^{6} \mathbf{f}_{\mathrm{s}i} \ , \\ \mathbf{l}_\mathrm{tc} &= \sum_{i=1}^{6} \mathbf{a}_i \times \mathbf{f}_{\mathrm{s}i} \ , \end{aligned} \tag{15}$$

with $\mathbf{a}_i$ denoting the moment arm of $\mathbf{f}_{\mathrm{s}i}$ w.r.t. the tool carrier's coordinate system. The tangential stiffness matrix calculates as

$$\mathsf{K}_\mathrm{t} = \begin{bmatrix} \dfrac{\partial \mathbf{f}_\mathrm{tc}}{\partial \Delta\mathbf{y}_\mathrm{tc}} \\ \dfrac{\partial \mathbf{l}_\mathrm{tc}}{\partial \Delta\mathbf{y}_\mathrm{tc}} \end{bmatrix} \tag{16}$$

and the corresponding tangential flexibility matrix as

$$\mathsf{C_t} = \mathsf{K_t}^{-1} \,. \tag{17}$$

In the general case both matrices are fully occupied. This means that a load in one direction causes rotational and translational evasive motions in every direction.

5.2 Principal Stiffnesses and Flexibilities

A direct evaluation of the compliance matrix $\mathsf{C_t}$ is often not possible as the stiffnes matrix $\mathsf{K_t}$ ist badly conditioned in the neighborhood of kinematic singular positions and, of course, singular at singular positions. Therefore it is useful to transform the stiffness matrix into its system of principal axes. Here the eigenvalues are the resulting stiffnesses for the motions defined by the corresponding eigenvectors. The corresponding principal flexibilities are then the reciprocals of the principal stiffness constants as the transformed stiffness matrix is diagonal. In order to achieve more insight into the global flexibility behavior of the machine, it is necessary to evaluate either the principal stiffneses or the compliances at many points all over the workspace. Here exemplary points on regular grids lying in planes $x =$ const., $y =$ const. and $z =$ const. are regarded.

As criterion for the stiffness, it is both possible to regard the lowest normalized principal stiffnesses $\min(k_i^*)$ or the mean value of the normalized principal compliances $\mathrm{mean}(c_i^*)$ at each point, see also [27]. The normalization to unitless coefficients is necessary as the tangential stiffness matrix consists of four blocks having different units. For the normalization, characteristic reference values for forces ($F_0 = 1\,\mathrm{N}$), torques ($l_0 = 1\,\mathrm{Nm}$), lengths ($s_0 = 10^{-6}\,\mathrm{m}$) and angles ($\varphi_0 = 10^{-6}\,\mathrm{rad}$) are used. However, the numerical evaluation of compliances at kinematic singular points is problematic because they approach infinity. Therefore, it is often recommendable to regard the principal stiffnesses even if the compliances are the point of interest.

In the following section, the flexibilities and principal stiffnesses of the discussed design variants are investigated.

5.3 Evaluation of the Alternative Design Variants

In this section the flexibility/ stiffness behavior of the alternative design variants is evaluated with the described approach. In order to be able to explain the results from Sect. 4 variants I and II are discussed for the theoretically optimal joint position angle $\zeta = 82.5°$ and variant III for $\zeta = 34.1°$. In the following, several pictures are given, some showing the flexibility-constants c_{ij} along the vertical axes, some the minimum principal stiffnesses $\min(k_i^*)$. It is desirable to have very low flexibilities, i.e. low values of c_{ij} or high stiffnesses, i.e. high values of $\min(k_i^*)$.

5.3.1 Current Design

At the original design the kinematic singularity at the center position is obvious, compare Sects. 2.3 and 4.2. The compliance matrix at this point also confirms this observation

$$\mathsf{C}_\mathsf{t} = \begin{bmatrix} 0.9739 & 0 & 0 & 0 & 0 & 0 \\ 0 & 0.9739 & 0 & 0 & 0 & 0 \\ 0 & 0 & \infty & 0 & 0 & 0 \\ 0 & 0 & 0 & 0.1772 & 0 & 0 \\ 0 & 0 & 0 & 0 & 0.1772 & 0 \\ 0 & 0 & 0 & 0 & 0 & 0.1330 \end{bmatrix} \cdot 10^{-6}. \qquad (18)$$

At all other points on the described planes, always one principal stiffness vanishes. This means that every point in the workspace is kinematic singular with one principal compliance approaching infinity. The axis of maximum flexibility (which is the eigenvector corresponding to the lowest eigenvalue of K_t and therefore the highest one of C_t) depends on the position in workspace. In the central position the respective eigenvector is given by $[0\ 0\ 1\ 0\ 0\ 0]$ corresponding to the kinematic singularity observed in Sect. 4.2. The fact that at all other points also a singular direction is existent accounts for the bad behaviour of this design variant observed for the circular sample trajectory.

5.3.2 Design Variant I

During the optimization and evaluation of the designs in Sect. 4, the alternative design variant I achieved a strongly superior flexibility behavior in comparison to all other designs discussed here. A look at the tangential stiffnesses and flexibilities for $\zeta = 82.5°$ explains the advantage of this design. Here, no kinematic singularities appear in the workspace at all. This further allows the general conclusion that the favorable behavior can be achieved not only for the sample trajectory but also for arbitrary motions in the overall workspace. However, due to space limitations it is impossible to show the reader enough plots to sufficiently mirror the encouraging stiffness behavior for the entire workspace. Therefore, only some exemplary planes in the workspace are discussed. In Fig. 15 (left) the minimum principal stiffnesses $\min(k_i^*)$ in xy- and yz-planes with $z = 0\,\mathrm{m}$ and $x = 0\,\mathrm{m}$, respectively, are illustrated (xz-planes show similar results to yz-planes). These are critical planes during the following investigation of the other variants.

Even the minimum principal stiffnesses are encouragingly large, which will be seen in particular when compared to the other variants, discussed in the following sections. In order to indicate the global stiffness behavior in the complete workspace in Fig. 15 (right) exemplary also the minimum principal stiffnesses $\min(k_i)$ in xy- and yz-planes are plotted for $x = 0.1\,\mathrm{m}$ and $z = 0.1\,\mathrm{m}$, respectively, where again substantially large values are obtained.

Further, the flexibility behavior of the machine in the xy-plane at $z = 0\,\mathrm{m}$ for an applied torque around the z-axis l_z is illustrated in Fig. 16, where the tool carrier's displacement is shown for each coordinate of $\mathbf{y}_\mathrm{tc}$. This configuration corresponds to the sample trajectory and the loading case applied in Sect. 4.2. It can be seen that low flexibilities appear in all directions which is consistent with the results found in Sect. 4.2.

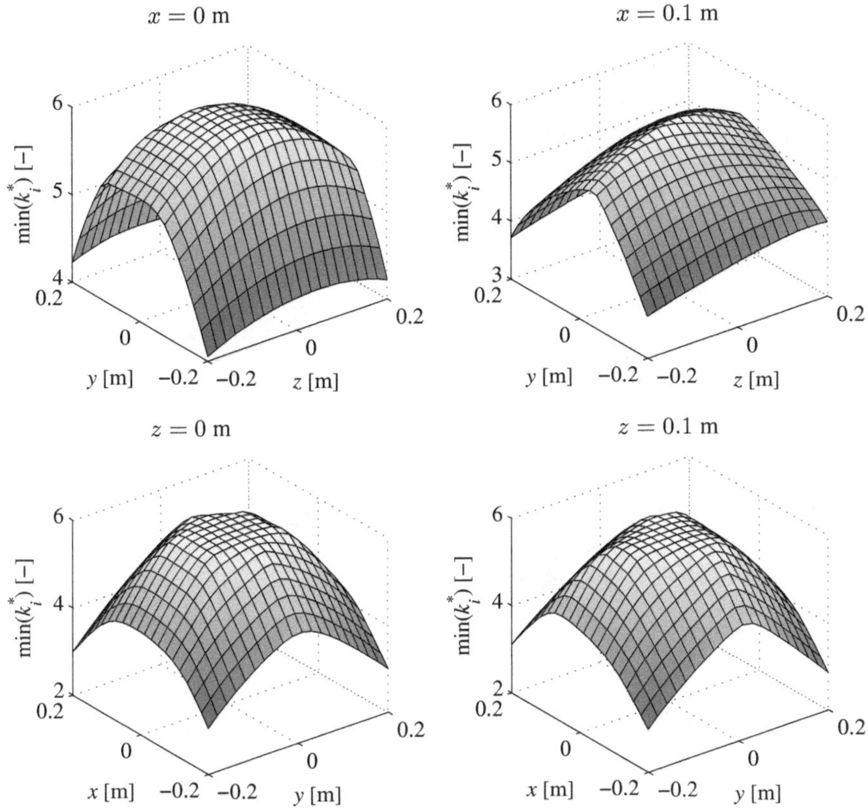

Fig. 15. Minimum principal stiffnesses $\min(k_i^*)$ in yz-planes at $x = 0$ and 0.1 m (top) and xy-planes at $z = 0$ and 0.1 m (bottom) for design variant I with $\zeta = 82.5°$

5.3.3 Design Variant II

At design variant II, the workspace contains a plane of kinematic singular positions located at $z = 0$, which divides the workspace into two usable domains, see Fig. 17 (lower left). Within these domains, the prinicipal stiffnesses are relatively low, as can be seen by comparison of Fig. 17 with the respective plots of variant I in Fig. 15.

The sample trajectory used during the investigation of design II in Sect. 4.3 is completely located in the singular plane, which explains the very bad behavior of this design. However, the flexibilities are also disappointingly high in the usable domains ($z \neq 0$) and the observed coupling of the translation and the rotation with respect to the z-axis remains. This is illustrated in Fig. 18 by the corresponding flexibility constants $c_{\gamma\gamma}$ and $c_{\gamma z}$ at $z = 0.1$ m, which are still much higher than the analogous flexibilities for design I at $z = 0$ m, see Fig. 16.

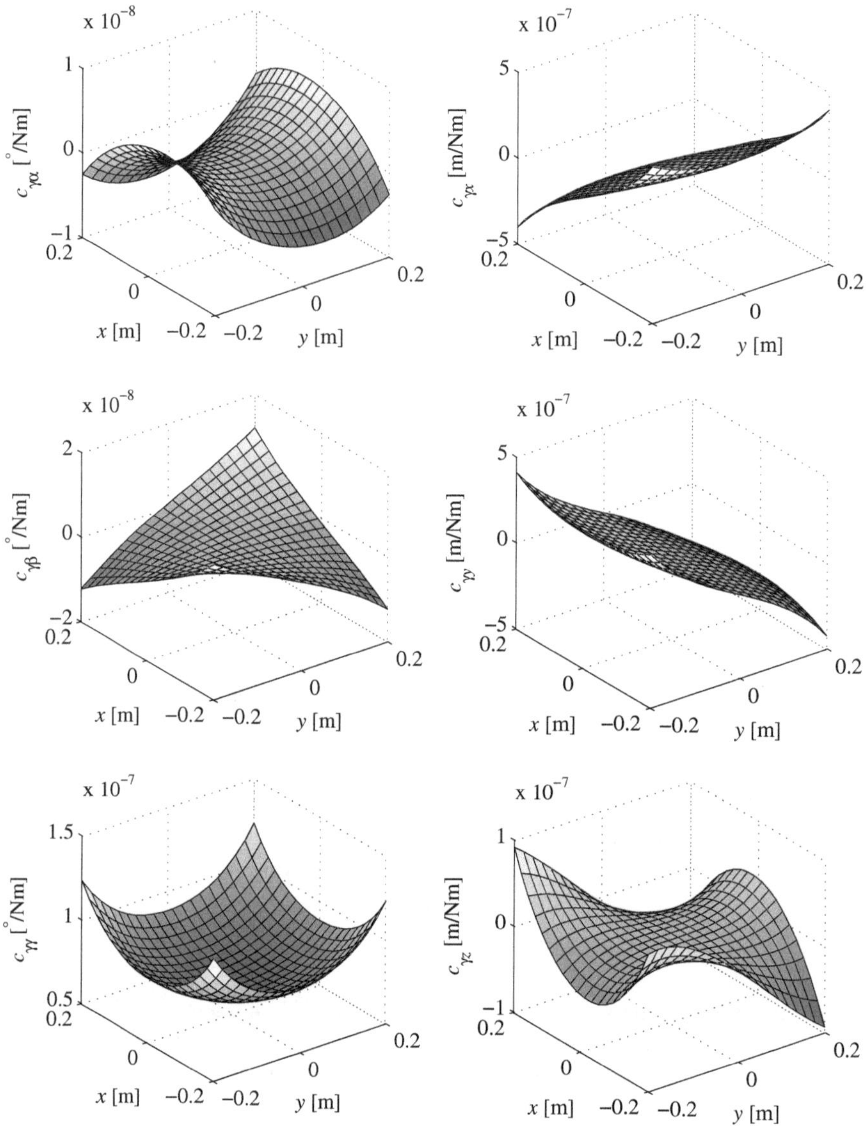

Fig. 16. Flexibility constants c_{γ_i} describing the tool carrier's displacements for an applied torque around the z-axis in the xy-plane at $z = 0\,\mathrm{m}$ for design variant I

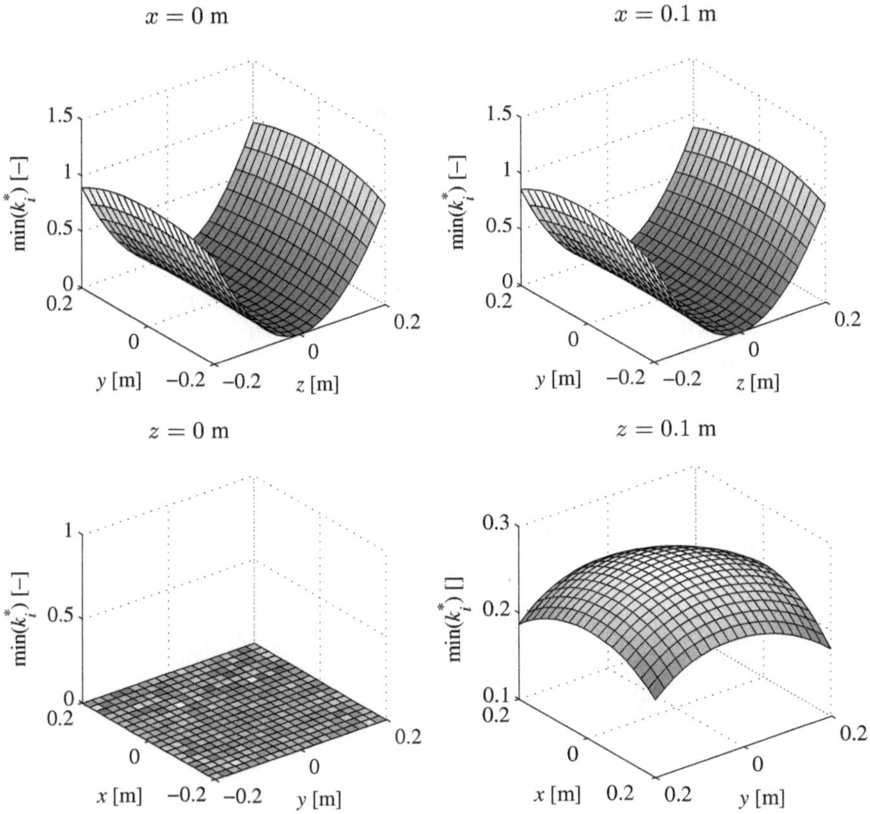

Fig. 17. Minimum principal stiffnesses $\min(k_i^*)$ in yz-planes at $x = 0$ and $0.1\,\mathrm{m}$ (top) and xy-planes at $z = 0$ and $0.1\,\mathrm{m}$ (bottom) for design variant II with $\zeta = 82.5°$

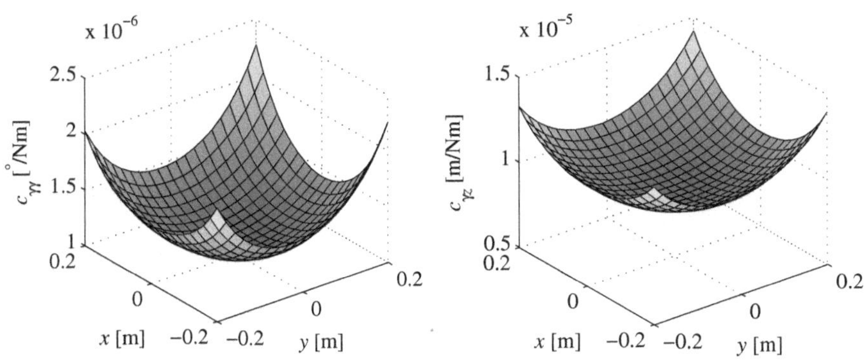

Fig. 18. Flexibility constants $c_{\gamma\gamma}$ and $c_{\gamma z}$ in xy-plane at $z = 0.1\,\mathrm{m}$ for design variant II

5.3.4 Design Variant III

For this design variant the workspace contains three planes of kinematic singular positions located at $x = 0$, $y = 0$ and one twisted across the workspace. These planes divide the workspace into eight domains. In Fig. 19 (top), the minimum principal stiffnesses in yz-planes at $x = 0$ and 0.1 m are shown. Further, in Fig. 19 (bottom) the minimum principal stiffnesses in xy-planes at $z = 0$ and 0.1 m are given which are also divided by the singular planes (lines with zero stiffness). The separated, relative small domains without singular configurations are surrounded by three singular planes each which leads to disappointingly low minimum prinicipal stiffnesses which can in particular be emphasized by comparison of the plots in Fig. 19 with those of variant I, Fig. 15.

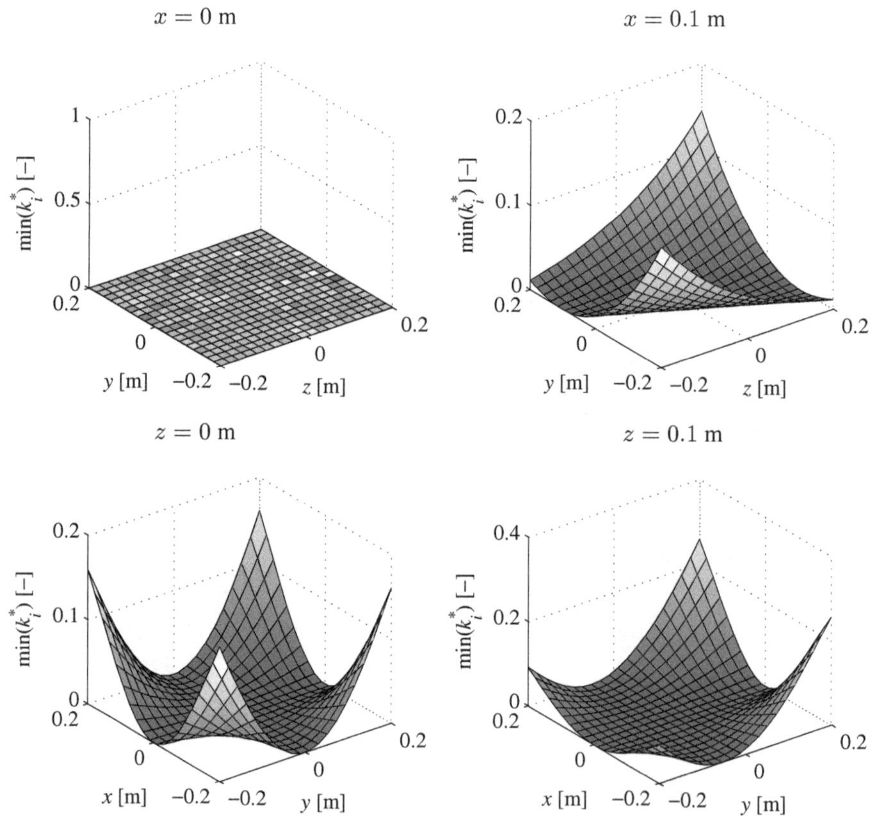

Fig. 19. Minimum principal stiffnesses $\min(k_i^*)$ in yz-planes at $x = 0$ and 0.1 m (top) and xy-planes at $z = 0$ and 0.1 m (bottom) for design variant III with $\zeta = 34.1°$

In Sect. 4.4 large deflections of the z- and y-rotation appeared for the applied torque around the z-axis. This behavior can also be seen when regarding the flexibilities $c_{\gamma\gamma}$ and $c_{\gamma\beta}$ in Fig. 20. High values of these flexibilities clearly show this strong coupling.

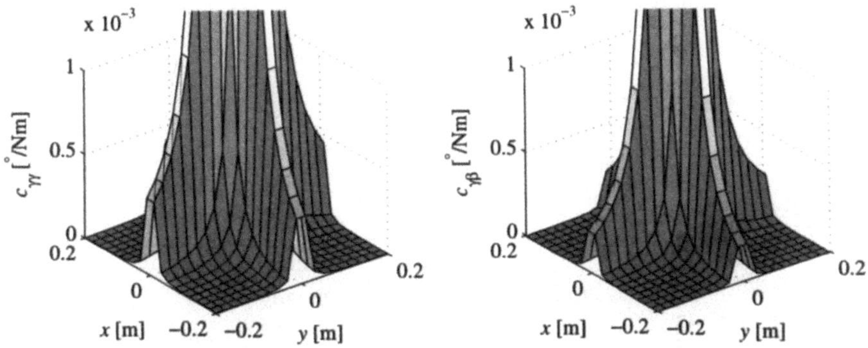

Fig. 20. Flexibility constants $c_{\gamma\gamma}$ and $c_{\gamma\beta}$ in xy-plane at $z = 0\,\mathrm{m}$ for design variant III

6 Conclusions and Outlook

In order to increase the stiffness of the hexapod machine HEXACT and to eliminate singular configurations three variants of alternative designs were discussed. Variant I with equally rotated joints leads to a substantial improvement in both translational and rotational flexibility in comparison to the current design. A global investigation of the tangential flexibilities in the entire workspace shows the absence of any singular configuration which confirms the results received during optimization. A drawback of this design is that pre-stressing, used for compensation of bearing clearance, is no more possible. The disadvantage of a workspace reduction can be eliminated by an alternative mounting of the joints at the tool carrier. Therefore, variant I is an attractive solution improving the flexibility behavior without reducing workspace.

In contrary to the expected behavior the flexibility cannot be improved by design variant II with counter-rotated joint positions, which is due to a singular plane in the workspace that also effects the flexibility behavior in the adjacent domains. Further, coupling of the translation along and the rotation around the tool axis causes a non-negligible deviation of the position in z-direction when torques are applied. Both results disqualify design variant II. Analogous results are found for variant III, now with three singular planes dividing the workspace into eight domains and a coupling of the z- and y-rotation.

The introduced model of the hexapod only considers linear elasticities of the joints. In future work, the joint clearance is introduced into the model, e.g. by using a nonlinear spring law. Then, the proposed optimum design can be compared with the current design with pre-stressing. Besides stiffness, further aspects like energy consumption or behavior under thermal loading are currently investigated. The selected exemplary tool trajectory and loading will be replaced by a typical machining benchmark. Further, as a more general optimization criterion the minima of the tangential principal stiffnesses in the entire workspace can be used. This allows for results independent from specific tool trajectories.

Acknowledgements

The authors want to acknowledge the help of Prof. U. Heisel and his coworkers from the IfW, University of Stuttgart, during modelling and parametrization of the machine HEXACT. Part of this work was financed by the German Research Council (DFG) under contract EB-195/3-1. This support within the framework of the SPP-1156 'Adaptronics for Machining Tools' is highly appreciated.

References

1. Bestle D (1994) Analyse und Optimierung von Mehrkörpersystemen. Berlin: Springer
2. Bestle D, Eberhard P (1994) NEWOPT/AIMS 2.2. Ein Programmsystem zur Analyse und Optimierung von mechanischen Systemen. Manual AN–35. Institute B of Mechanics, University of Stuttgart
3. Bestle D, Eberhard P (1997) Dynamic System Design via Multicriteria Optimization. In: Multiple Criteria Decision Making. Proc. of the Int. Conf. on MCDM, Hagen, 1995, pp. 467–478. Fandel G and Gal T (Eds.), Berlin: Springer
4. Bestle D, Eberhard P (1992) Analyzing and Optimizing Multibody Systems. Mechanics of Structures and Machines, 20:67–92
5. Bischof C, Carle A, Khademi P, Mauer A (1996) ADIFOR 2.0: Automatic Differentiation of Fortran 77 programs. IEEE Computational Science & Engineering, 3:18–32
6. Boër C, Molinari–Tosatti L, Smith K (Editors) (1999) Parallel Kinematic Machines: Theoretical Aspects and Industrial Requirements (Advanced Manufacturing). London: Springer
7. Dasgupta B, Mruthyunjaya T (1998) Force Redundancy in Parallel Manipulators: Theoretical and Practical Issues. Mechanisms and Machine Theory, 33(6):727–742
8. Dignath F (2002) Zur Optimierung mechatronischer Systeme mit nichtdifferenzierbaren Kriterien. PhD–Thesis (submitted 2002). Institute B of Mechanics, University of Stuttgart
9. Dignath F, Hempelmann D (2002) Grundlagenuntersuchungen zum thermischen Einfluss – Bericht 2002. Report ZB–131. Institute B of Mechanics, University of Stuttgart
10. Eberhard P (1996) Zur Mehrkriterienoptimierung von Mehrkörpersystemen. VDI–Fortschritt-Berichte, Series 11, No. 227, Düsseldorf: VDI-Verlag
11. Eberhard P, Bestle D (1994) Integrated Modeling, Simulation and Optimization of Multibody Systems. In: Integrated Systems Engineering. Johannsen G (Ed.), pp. 35–40, Oxford: Pergamon
12. Eberhard P, Dignath F, Kübler L (2003) Parallel Evolutionary Optimization of Multibody Systems with Application to Railway Dynamics. Multibody System Dynamics, 9:143–164
13. Fletcher R (1987) Practical Methods of Optimization. Chichester: John Wiley & Sons

14. Gosselin C, Angeles J (1990) Singularity Analysis of Closed–Loop Kinematic Chains. IEEE Transactions on Robotics and Automation, 6(3):281–290
15. Heisel U (1998) Precision Requirements of Hexapod-Machines and Investigation Results. In: Proceedings of the First European–American Forum on Parallel Kinematic Machines. Mailand
16. Heisel U, Maier V, Lunz E (1998) Auslegung von Maschinenkonstruktionen mit Gelenkstab-Kinematik-Grundaufbau, Tools, Komponentenauswahl, Methoden und Erfahrungen. wt – Werkstatttechnik 88, 4, pp. 75–78
17. Kreuzer E, Leister G (1991) Programmsystem NEWEUL'90. Manual AN–24. Institute B of Mechanics, University of Stuttgart
18. Liu J (2001) Kinematics and Dynamics of Spatial Hexapod Motions with Thermal Disturbances. Report IB–38. Institute B of Mechanics, University of Stuttgart
19. Merlet J-P (2000) Parallel Robots. Dordrecht: Kluwer
20. Merlet J-P (1989) Singular Configurations of Parallel Manipulators and Grassmann Geometry. Journal of Robotics Research, 8(5):45–56
21. O'Brien J, Wen J (1999) Redundant Actuation for Improving Kinematic Manipulability. In: Proceedings of the 1999 IEEE International Conference on Robotics and Automation. Detroit, Michigan, Vol. 2, pp. 1520–1525
22. Riebe S, Ulbrich H (2003) Modelling and Online Computation of the Dynamics of a Parallel Kinematic with Six Degrees-of-Freedom. Archive of Applied Mechanics, 72:817–829
23. Schiehlen W (1986) Technische Dynamik. Stuttgart: B.G. Teubner
24. Takeda Y, Funabashi H (1996) Kinematic and Static Characteristics of In–Parallel Actuated Manipulators at Singular Points and in Their Neighborhood. JSME International Journal, Series C, 39(1):85–93
25. Takeda Y, Funabashi H (1999) Kinematic Synthesis of In–Parallel Actuated Mechanisms Based on the Global Isotropy Index. Journal of Robotics and Mechatronics, 11(5):404–410
26. Valášek M, Šika Z, Bauma V, Vampola T (2001) Design Methodology for Redundant Parallel Robots. In: Proc. of AED 2001, 2nd Int. Conf. on Advanced Engineering Design. Glasgow, pp. 243–248
27. Zhengyi X, Fengfeng X, Mechefske C (2002) Kinetostatic Analysis and Optimization of a Tripod Attachment. In: Proc. of the Workshop on Fundamental Issues and Future Research Directions for Parallel Mechanisms and Manipulators. October 3–4, Quebec City, Canada, Gosselin C and Ebert-Uphoff I (Eds.), pp. 294–303

Multibody dynamics in advanced education

Ettore Pennestrì and Leonardo Vita

Università di Roma Tor Vergata Dipartimento di Ingegneria Meccanica via del Politecnico, 1, 00133 Roma - Italy
pennestri@mec.uniroma2.it

The use of multibody dynamics software in industrial firms is steadily increasing. Engineering curricula often do not include mandatory courses where multibody dynamics is taught. This paper discuss some relevant teaching issues in this specific cultural area. Noteworthy examples of the great variety of kinematic and dynamic formulations, available for teaching a basic course in multibody dynamics, are illustrated. The experience of the first author, when introducing a basic multibody dynamics course at an Italian university, is also reported.

1 Introduction

In 1999 the Education Ministers of 29 European countries signed in Bologna the document known as *Bologna Agreement*. The document was preceded by the *Sorbonne Agreement* signed in May 1998 and outlining cooperation between European countries in the field of higher education. The main goal of the Bologna Agreement is establishing, within 10 years, a standard European education system which would facilitate the implementation of clear and mutually recognized qualification standards. Thus, the purpose is to bring transparency to all national qualifications and to enable qualifications from one country to be recognised by another.

The proposed standard system of higher education would comprise 2 cycles:

- undergraduate bachelor degree (minimum 3 years), eligible also for professional employment;
- further studies, and complete higher education cycle with master and/or doctor degree.

The change is imposing a radical revision of the traditional program of studies in European countries. In many engineering faculties there is strong skepticism concerning the establishment of a bachelor's degree awarded after at least three years of study. In fact, it should be acknowledged that engineering teachers are complaining about the influence that this reform of studies is having on the level of technical skills of the European engineering graduates. We shall not discuss further on this topic, but

J.A.C. Ambrósio (ed.), Advances in Computational Multibody Systems, 345–370.

the mention was necessary in order to introduce the cultural environnement in which multibody dynamics education is likely to be developed in Europe.

The organisation of education and research activities in the United States of America, when compared to Europe, is very different. Since most prominent universities are "graduate schools", the professors can concentrate their teaching on topics closely related to their research [1]. All European universities offer instead both undergraduate and graduate curricula. This involves

- courses with more general contents and rather decoupled from state-of-the-art research topics;
- heavy teaching load and administrative burden for the professors.

For these reasons the number of advanced multibody dynamics courses offered in the USA is higher than in Europe.

Software packages based on multibody dynamics techniques are becoming common design tools. The availability of software where the cycle of design and mechanical system simulation is made all within the same environment not only speeds up the design overall process, but is making a radical change both in the way engineers are approaching problems and in their required skills.

The situation calls for a revision of the current engineering curricula. However, there are several issues and constraints involved in such revision. In this paper we will focus and discuss on the most common ones, but without any claim of giving a definite answer.

Although the education system is adapting itself to the new design methodologies and tools it seems that there is not a general agreement on the most effective response. Regarding university courses dedicated to multibody dynamics common arguments of debate are:

- which should be the level of familiarity of the student with the fundamentals of analytical mechanics and machine dynamics before being exposed to commercial software?
- at university level should be given more emphasis on the theory behind the code or it is better to concentrate on the development of the modeling skills of the student?

Shortly, it seems that there is not a general consensus on the *best* approach for teaching multibody dynamics. However, it is generalized the need for the introduction of courses where the derivation of the equations of motion is presented in a computer oriented manner together with time integration algorithms.

In a recent past, multibody dynamics was more a research topic rather than an established subject for mandatory university courses. Although many distinguished researchers in the field published textbooks describing the details of their methods, few universities offered courses in modern multibody dynamics. The current situation has similarities with the pre-finite elements era. In the sixties few university curricula included a course in finite elements. Nowdays the situation is radically changed. There is not any university offering a degree in mechanical or civil engineering without a mandatory course in finite elements.

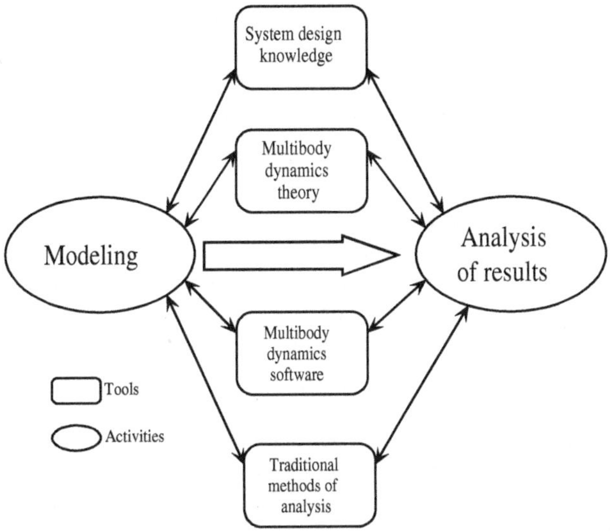

Fig. 1. Phases of system dynamics simulation

The content of textbooks traditionally used by undergraduate mechanical engineering students for learning machine dynamics (e.g. [2, 3]) focus on Newton-Euler approach and on simple planar inverse dynamics models. Forward dynamics is often limited to the classical one or two degrees-of-freedom linear models used in vibrations analysis. In some cases, advanced models of kinematic and dynamic systems, with several degrees-of-freedom, are based on clever formulations that take advantage of the specific problem to obtain a simplified form of equations of kinematics and dynamics.

When present, the computer coding of the problem is based on *ad hoc* models whose equations are deduced by hand.

In some instances the curricula reform which followed the Bologna agreement shrinked the number of credits allocated for mechanics courses. These time constraints stimulated a noteworthy attempt, due to Valasek [4], to introduce a multibody dynamics technique with a minimum or no prerequisite of analytical mechanics.

Thus, the teachers are forced to limit their lectures to the basics of machinery dynamics instead of extending them by introducing multibody dynamics techniques.

From colloquia with collegues it seems that the above situation is not uncommon at many European universities. This may slow down the introduction of multibody dynamics in engineering curricula.

As depicted in Fig. 1, a modern approach to modeling and analysis of results obtained from simulation involves knowledge and skills in different fields. In general the multibody dynamics approach to system simulation requires that the user is not just an analyst, but also a designer. This suggest that the student should have a basic knowledge of machinery dynamics before starting the study of multibody dynamics. In fact, he/she will have:

- a better appreciation of the power of systematic kinematic and dynamic formulations;
- the capability of a critical interpretation of results.

The widespread diffusion of multibody dynamics software calls for attention. Multibody dynamics modeling is applied on a greater extent within industrial companies or consulting firms. Thus, a significant portion of current European mechanical engineering graduates are potential users of such a software without having a minimal knowledge of its theoretical bases.

Moreover, many current users, although familiar with the graphic interface, lack the basic understanding of the theory behind the code. Unfortunately, software packages without a detailed theoretical manual are not rare and the courses organized by the companies, due also to time constraints, usually focus on the working features of the code and not on its theory.

A correct interpretation of the results obtained from a software requires a deep understanding of the theoretical bases used for its developement and of the numerical methods used. This is a widely accepted opinion in engineering. Errors and inaccuracies in data input, dynamic formulation limits, improper modeling, failure of numerical methods are common pitfalls for the user of multibody dynamics software. A formal education on multibody dynamics theory and software may reduce the probability of errors. However, considered the current status of mechanical engineering curricula, many users of multibody dynamics software may not be fully aware of the limits of their models. The potential danger of the above described situation requires some action from the multibody dynamics research community.

This action can take different forms according to the target of the instruction:

- students with no background in multibody dynamics and little or no experience of CAE software;
- engineers already using CAE software;

For this last cathegory of individuals, the continuing education programs offered by the universities or private companies should foresee lectures on theory and practice of multibody dynamics. Noteworthy attempts in this direction can be recorded.

Serious problems may arise by the use of simulation software, such as multibody dynamics codes, by CAD practitioners. Although extensive training is required to be expert in using a computer to create solid models, this does not makes the person an engineer [45]. Whoever is setting up the model not only must understand what kind of assumptions he/she is making, but should also be aware of the main pitfalls that can arise during simulation.

The difficulties of multibody dynamics modeling and simulation are very well expressed by Haug [14]

> "The insidious presence of nonlinearity in virtually all aspects of the kinematics and dynamics of machines leads to intricacies in mathematical and numerical analysis that are not easily solved as in the case in linear structural mechanics and associated finite element methods. Pathological forms of behavior, such as lock-up and branching of kinematic solutions, can best be

understoood and overcome by an engineer who develops both a firm mathematical foundation and a clear physical understanding of the behavior of mechanisms and machines."

Freeman [5] highlighted on some problems encountered by knowledgeable multibody analysts, with little background in vehicle dynamics, when modeling vehicles. The problems can be grouped as:

- Inappropriate application of models;
- Poor modeling assumptions;
- Incomplete model formulation or subsystem analysis;
- Methodology hides the completeness or lack thereof in the results.

Moreover, Freeman pointed out that

- it would be appropriate a direct interface of the numerical-multibody program with the CAD modeling program, making it easy the transition from model building to kinematic and dynamic simulation (much progress has been made in this direction);
- to simulate properly vehicle behavior the user must have knowledge of how different parameters and/or components affect system performance;
- only by understanding the relationship between the system components and system-level parameters one can unleash the power and flexibility of multibody dynamics software;
- ignorance of traditional analytical methods may lead easily to erroneous system behavior;
- these technical skills can only be achieved by education and experience.

The above conclusions are shared by most of the technical personnel that use multibody dynamics software within engineering firms.

Beside a broad theoretical and practical knowledge on mechanical system simulation to an engineer are also required many skills such as the ability:

- to interact with other team workers;
- to examine things critically and/or minutely, to separate the broad picture into its individual components;
- to write and speak clearly, to summarize and document information in a manner that other people can understand.

A modern education system must give the student the opportunity to develop and practice also the above skills. This require time, human and financial resources not always fully available.

The mechanical engineering community is currently benefitting from self-study systems and use of world wide web as a teaching tools [6]. Little is known on the effectiveness of such systems for learning multibody dynamics.

2 Multibody dynamics and mechanical systems modeling

At university level, different strategies can be adopted for teaching multibody dynamics and mechanical systems modeling.

One of the approaches, successfully experienced by Fisette and Samin [7], is based on a main course in classical mechanics which covers rigid body motion theory and 2D "academic" problems. In this course the students learn how to obtain analytical equations of motion. After that, with the active participation of a CAD course, the students are required to develop a multibody simulation of a realistic application.

The main steps of the project to be developed by a group of 6-8 undergraduate students are:

- Understanding the system to analyse
- Project planning
- Choice and practice of computer tools
- Formulation of relevant modeling hypotheses
- Symbolic multibody model deduction
- Data acquisition and/or computation
- Understanding and computation of environment forces
- Numerical program and first simulations
- System analysis and parameterization
- Project report, presentation and evaluation

According to Fisette and Samin [7] "this multi-disciplinary project really improves the student skills in the field of multibody system modeling". A somewhat different experience has been reported by Fanghella et al. [8] for teaching multibody dynamics

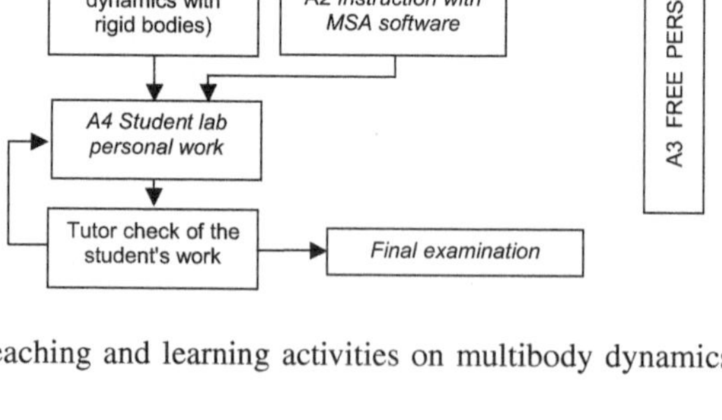

Fig. 2. Scheme of teaching and learning activities on multibody dynamics at University of Genoa [8]

to postgraduate students (i.e. attending the last year of a 3+2 engineering curriculum). In this case the lectures on multibody dynamics theory (32 hours) and hands on practice with commercial software (15 hours) are in parallel. At the end of these activities the student is required to solve an engineering analysis problem with numerical data defined. Fig. 2 depicts the flow of the described activities.

Industrial managers usually complain that most graduating engineers are very well qualified as far as a theoretical understanding is concerned, whereas few appear prepared for planning and implementing engineering projects which involve performance prediction for complex systems. In other terms, engineering graduates do not seem to have sufficient exposure to realistic modeling. Thus, it is important that the courses on multibody dynamics theory are followed or coupled with classes where are addressed realistic problems of mechanical system modeling. The structures of the courses can range from individual study to group projects, sometimes supervised jointly by faculty and industrial representatives.

The continuing education and training of engineers in system modeling by means of multibody dynamics software should be also of concern to industries. In this field it is auspicable a collaborative partnership between universities and industries.

3 Multibody dynamics in industry

Experienced engineers are challenged by technology moves beyond the levels that were current at the time of their formal education. Similarly, young engineers cannot achieve, through traditional mentoring and informal training, the levels of competence expected by a world-wide industrial competition.

In any case, gaining new skills and maintaining technical currency is a major concern for all professional engineers.

The introduction of powerful and versatile CAD programs signed a dramatic technological advancement. These tools allowed a significant reduction of the time required to move from initial design to full production. In this overall process dynamic simulation plays a fundamental role. The capability to predict the effects of design variables changes on system performance is necessary to optimize the mechanical system during the early phases of design. This allows also the reduction of costs of experimental analyses on physical prototypes. However, the mentioned capability rest on several factors such as:

- completeness and consistency of data;
- correcteness, sensitivity and reliability of the model;
- robustness of the numerical integration engine;
- expert interpretation of the results.

The discussed needs are all within the area of interest of multibody dynamics.

4 Preparation of the multibody dynamics course

An instructor willing to organize a multibody dynamics course will face different choices and options. The following items summarizes some of the main identified options.

- **Prerequisites**
  Multibody dynamics is an advanced topics. Students taking a course in this discipline should have a background in calculus, physics and machinery dynamics. Computer programming skills are also required. Knowledge of numerical analysis concepts would be preferable.
- **Kinematic formulation and dynamic principles**
  In multibody dynamics there is a wide choice of generalized coordinates (e.g. Euler angles, Cardan angles, Euler parameters, Denavit-Hartenberg, dual numbers,...) and methodologies for the systematic description of kinematic constraints (e.g. method of constraints, loop-closure equations,...)in mechanical systems.
  Moreover, the equation of dynamics may be deduced from different approaches (Newton-Euler, Lagrange, Gibbs-Appell, Jourdain, Gauss,...). Which kinematic formulation and dynamic principle is the most effective for teaching? Although this is a key question a definite answer cannot be given, and the final choice depends on the instructor's personal judgement and preferences. However, the teaching of dynamic formulations adopted in commercial software is recommended, at least in introductory courses. The student will have a better understanding of the theoretical bases and limitations of the software.
  In the bibliography a list of textbooks dedicated to multibody dynamics is included [10, 11, 12, 13, 14, 15, 16, 17, 18, 19, 20, 21, 22, 23, 24, 25, 26, 27, 28, 29, 30]. Likely the list is far to be exhaustive. However, it is a good starting point. A perspective on rigid multibody dynamics theory is presented in [16, 46].
- **Course length**
  How many hours are required to teach basic multibody dynamics course for graduate students?
  According to the authors' experience, 45 hours distributed in 15 lectures (2 lectures each week) should be enough. For experienced engineers, three days intensive course seems a reasonable length. In this case the instructor concentrates mostly on theory and numerical methods and less on programming.
- **Preparation of material to be distributed**
  Homework assignements and computer programming tasks should be clearly stated in handouts. Reference to the equations in textbooks will help the students. The reference to papers taken from technical literature may be useful when dealing with advanced topics.
- **Computer programming**
  The students must have normal skills of computer programming in a higher level language of their choice.

They should be required to develop software with the following modules: data input and output, automatic build up of equations and numerical solution. Through the computer programming of a multibody dynamics methodology the students usually reach a deep understanding of the subject. For this reason computer programming is strongly recommended in basic multibody dynamics classes. The instructor can make available pieces of code that the students may use during the developement of their own software. This will help the students not to get lost in coding the all software, but to concentrate on the overall structure of the program. Guidelines on the input/output of data and on the overall structure of the code should also be discussed by the instructor.
The students should also be encouraged to use professionally tailored linear algebra and numerical integration subroutines available in packages such as LAPACK, IMSL, NAG, ODEPACK, etc.
- **Use of commercial multibody dynamics software**
At university level, how much emphasis should be given to the teaching of commercial software? Also in this case there are different answers. Some teachers argue that the students should be educated for a correct use of commercial software. Since user's interface and capabilities of a software changes almost every year, other teachers, during instruction, put more emphasis on the theoretical part of the methodology rather than on the practical use of the software.
We believe that both aspects are relevant. The current trend is for the development of software with friendly user's interface. Thus an engineer can make himself familiar very quickly with the solid modeling and multibody dynamics software. Many companies offer specialized training on the specific software being used at their site. This training usually focus much more on the practical use of the software and less on its theoretical bases or limitations. The training offered at university level should involve a type of knowledge valid in the long term. Students are usually attracted by the use of simulation tools and less by the theory. Thus a teacher must search for a compromise between the instruction of the use of commercial software and the teaching of multibody dynamics theory.

5 The ingredients of a basic course in multibody dynamics

The research in analytical mechanics and computational dynamics provides many approaches for the systematic and computerized dynamic analysis of machine systems. None of the techniques available can be considered *a priori* the best one. Thus, the advantages and disadvantages of each technique should be evaluated by the instructor.

In the following sections a brief overview is offered.

5.1 The choice of coordinates

A multibody system is constituted of a number of parts, subject to interconnections and constraints of various kind[1]. There are several ways of representing a rigid body in space, however a system of coordinates must give at any time a unique representation of the configuration and displacements of the multibody system.

If $\delta q_1, \delta q_2, \ldots, \delta q_n$ are arbitrary infinitesimal increments of the coordinates in a dynamical system, these will define a possible displacement if the system is *holonomic*, while, for *non-holonomic* systems, a certain number, say m, of equations must be satisfied between them in order that they may correspond to a possible displacement.

The number

$$F = n - m \tag{1}$$

is the *number of degrees-of-freedom* of the system [31].

Hence, system of coordinates can be classified in

- *independent coordinates*, when $n = F$;
- *dependent coordinates*, when $n > F$.

Independent coordinates determine only the position of some parts. The positions of the remaining parts must be numerically computed solving a nonlinear system of equations with multiple solutions. Thus, independent coordinates are not suitable of unequivocally determine the position of the multibody system. Another problem involved with the use of independent coordinates is the variation of F during simulation. This is not a remote possibility and may happen, for instance, in linkages with particular dimensions or when modeling stiction in kinematic pairs.

The most common types of coordinates currently used to describe the motion of multibody systems are:

- *Relative coordinates.*
  The position of each element is defined with respect to the previous one (e.g. Fig. 3). These coordinates allow numerical efficiency due to their reduced number, but lead to small order and expensive to evaluate dense matrices. They are specially suited for open kinematic chain systems. The control of movement between adjacent parts is easy. The choice of variables requires a preprocessing, whereas a postprocessing is needed to determine the absolute motion of all the parts. The set of coordinates used by the well known Denavit-Hartenberg notation [35] belong to this cathegory. Noteworthy multibody dynamics formulations based on relative coordinates are reported in [17, 32].
  Often dual numbers algebra is used to express the relative motion between two adjacent links. Although unfamiliar to most engineers, dual algebra is a powerful tool for kinematic analysis. Dual numbers were proposed by Clifford (1873), but the first engineering publications were due to Denavit (1958) [35], Keler (1959) [36], Yang (1963) [37] and Yang and Freudenstein (1964) [38].

[1] For the sake of simplicity, only rigid parts and planar systems will be herein considered.

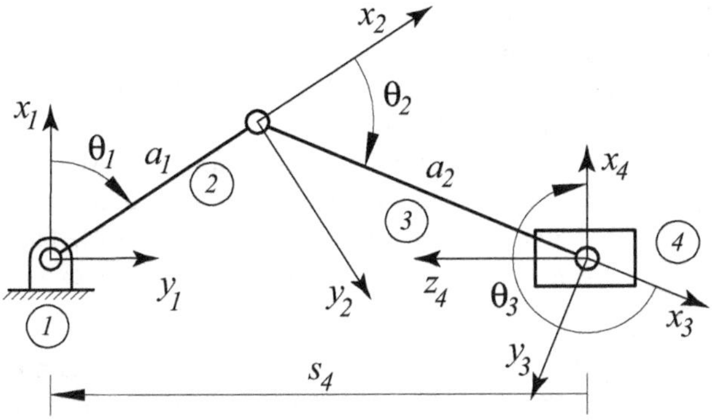

Fig. 3. Example of relative coordinates

At the textbook level, the works of Dimentberg [41], Beyer [40] and Fischer [34] are excellent introductory texts to the topic. Recent reviews of dual algebra kinematics are due to Wittenburg [42] and Angeles [43].

- *Cartesian generalized coordinates.*
  The absolute position of each body is independently located by a set of Cartesian generalized coordinates (3 for planar motion, 6 for spatial motion). Kinematic constraints between bodies are then introduced as algebraic equations among coordinates.

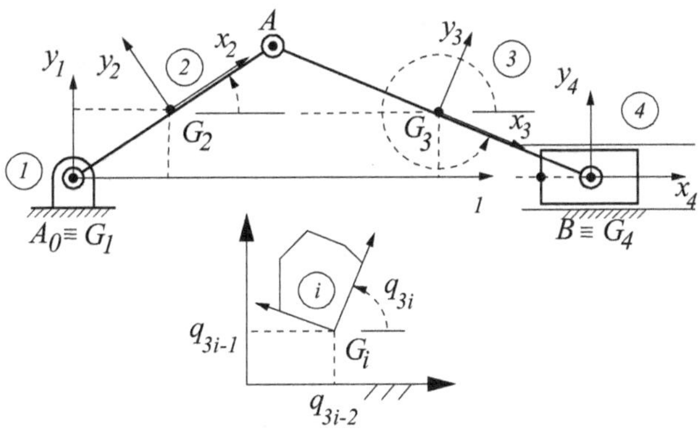

Fig. 4. Example of Cartesian generalized coordinates

Constraint expressions are numerous, but involve only the absolute coordinates of adjacent parts. (e.g. Fig. 4). With the purpose of avoiding singular configurations, some authors (*e.g.* [12, 14, 17, 24, 28]) prefer the definition of the spatial attitude of a body in terms of Euler parameters (4 coordinates) instead of Euler angles (3 coordinates).

Computer codes using these coordinates require only a minimal amount of pre and post processing. A substantial number of nonlinear constraint equations is involved. Coefficient matrices are large but sparse. It is advisable to take advantage of this condition in order to increase the numerical efficiency of the code. There are difficulties in prescribing the relative motion between adjacent bodies. The computer implementation is modular and library of components and standard joints can be defined and used in assembling a model. These coordinates are being used by general purpose dynamic simulation codes such as ADAMS [33] and DADS [14].

- *Natural coordinates*. Originally proposed by de Jalón and his coworkers, these coordinates are made by the Cartesian coordinates of a series of *basic points* of the mechanism. The points are chosen on the basis of the following criteria [18]:
  1. Each link must have at least two basic points.
  2. A basic point must be located at the center of revolute kinematic pairs. The point is shared by the two kinematic elements.
  3. In a prismatic pair one point is positioned on the line of action of the relative motion among the two kinematic elements. The other point is attached on the second kinematic element.

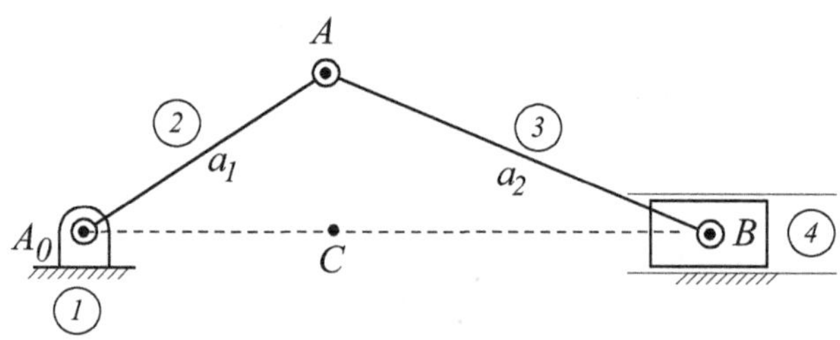

Fig. 5. Example of natural coordinates

  4. Other points whose motion needs to be monitored can be chosen as basic points.

  Fig. 5 shows the slider-crank mechanism modeled with natural coordinates.

Due to the elimination of angular coordinates, the number of natural coordinates required for the analysis is usually less than the number of Cartesian generalized coordinates.

de Jalón and Bayo [18] reported a thoughful discussion between these types of coordinates and their influence in the simulation process.

5.2 The modeling of kinematic constraints

The convenience of using sets of dependent coordinates has been already stated. These are related by the equations of constraints of the form

$$\Psi_i(q_1,\, q_2, \dots, q_n) = 0,\ (i = 1, 2, \dots, n) \tag{2}$$
$$\Psi_j(q_1,\, q_2, \dots, q_n,\, t) = 0,\ (j = 1, 2, \dots, n_t) \tag{3}$$

Equations (2) are the *spatial* or *scleronomic* constraints because only the space variables q appear as arguments. Equations (3) are said *driving* or *rheonomic* constraints because also the temporal variable t does appear explicitly.

Each multibody technique has a distinctive feature in the automatic generation and assembly of kinematic constraint equations. The algebraic structure of the constraint equations depends on the type of coordinates implemented.

The conditions used to generate the equations of constraints depend also on the type of coordinates used. For some set of coordinates previously discussed the constraint equations for the kinematic modeling of a slider-crank are reported in the following.

When relative coordinates $\{q\} = \left\{\, \theta_1\ \theta_3\ s_4 \,\right\}^T$ are used, *loop closure conditions* are often imposed. For example, with reference to the nomenclature of Fig. 3, the following equations can be written[2]

$$\Psi_1 \equiv a_1 \sin\theta_1 - a_2 \sin\theta_3 - s_4 = 0 \tag{4}$$
$$\Psi_2 \equiv a_1 \cos\theta_1 + a_2 \cos\theta_3 = 0 \tag{5}$$

Let a_i, α_i, θ_i, s_i be the Denavit-Hartenberg parameters, and

$$\widehat{\alpha}_i = \alpha_i + \varepsilon a_i \tag{6}$$
$$\widehat{\theta}_i = \theta_i + \varepsilon s_i \tag{7}$$

their dual counterparts ($\varepsilon^2 = 0$).

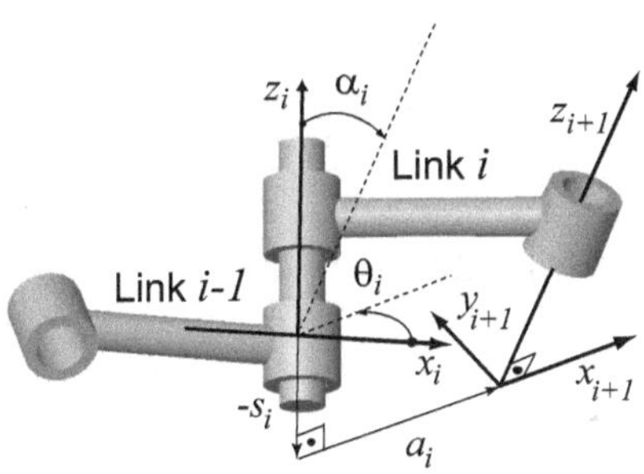

Fig. 6. Denavit-Hartenberg notation

With reference to Fig. 6, the links coordinate-transformation matrix takes the form [34]

---
[2] With this approach, the coordinate θ_2 is not involved.

$$\left[\widehat{T}\right]_{i+1}^{i} = \begin{bmatrix} \cos\widehat{\theta}_i & -\cos\widehat{\alpha}_i \sin\widehat{\theta}_i & \sin\widehat{\alpha}_i \sin\widehat{\theta}_i \\ \sin\widehat{\theta}_i & \cos\widehat{\alpha}_i \cos\widehat{\theta}_i & -\sin\widehat{\alpha}_i \cos\widehat{\theta}_i \\ 0 & \sin\widehat{\alpha}_i & \cos\widehat{\alpha}_i \end{bmatrix} \tag{8}$$

The closure condition of the slider-crank chain is expressed by the matrix product

$$\left[\widehat{T}\right]_2^1 \left[\widehat{T}\right]_3^2 \left[\widehat{T}\right]_4^3 \left[\widehat{T}\right]_1^4 = [I] \tag{9}$$

where $[I]$ is the unit matrix.

The constraint equations (4) and (5) follow by equating appropriate elements of the final matrix products.

Using Cartesian generalized coordinates, with reference to the nomenclature of Figure 4, the scleronomic constraints are expressed by the following equations

$$\Psi_1 \equiv X_{A_0}^{(1)} - X_{A_0}^{(4)} = 0 \tag{10}$$
$$\Psi_2 \equiv Y_{A_0}^{(1)} - Y_{A_0}^{(4)} = 0 \tag{11}$$
$$\Psi_3 \equiv X_A^{(2)} - X_A^{(3)} = 0 \tag{12}$$
$$\Psi_4 \equiv Y_A^{(2)} - Y_A^{(3)} = 0 \tag{13}$$
$$\Psi_5 \equiv X_B^{(3)} - X_B^{(4)} = 0 \tag{14}$$
$$\Psi_6 \equiv Y_B^{(3)} - Y_B^{(4)} = 0 \tag{15}$$
$$\Psi_7 \equiv Y_B^{(4)} = 0 \tag{16}$$

where $X_P^{(i)}, Y_P^{(i)}$ are the absolute coordinates of point P on the ith body. Such coordinates are related to the generalized Cartesian coordinates by the transform

$$\begin{Bmatrix} X_P^{(i)} \\ Y_P^{(i)} \end{Bmatrix} = \begin{bmatrix} \cos q_{3i} & -\sin q_{3i} \\ \sin q_{3i} & \cos q_{3i} \end{bmatrix} \begin{Bmatrix} x_P^{(i)} \\ y_P^{(i)} \end{Bmatrix} + \begin{Bmatrix} q_{3i-2} \\ q_{3i-1} \end{Bmatrix} \tag{17}$$

Using the natural coordinates (see Fig. 5, A_0, A, B and C are basic points. The coordinates of A_0 and C are fixed and known. Thus, $\{q\} = \{\, X_A\ Y_A\ X_B\ Y_B \,\}^T$ is the vector of variable coordinates. The constraints exquations are espressed by the following equations

$$\Psi_1 \equiv (X_A - X_{A_0})^2 + (Y_A - Y_{A_0})^2 - a_1^2 = 0 \tag{18}$$
$$\Psi_2 \equiv (X_A - X_B)^2 + (Y_A - Y_B)^2 - a_2^2 = 0 \tag{19}$$
$$\Psi_3 \equiv \det \begin{vmatrix} X_{A_0} & Y_{A_0} & 1 \\ X_B & Y_B & 1 \\ X_C & Y_C & 1 \end{vmatrix} = 0 \tag{20}$$

The constraints equations (2) e (3) are differentiated for velocity analysis[3]

[3] Dots denote differentiation w.r.t. time.

$$[\Psi_q]\{\dot{q}\} = -\{\Psi_t\} \tag{21}$$

and acceleration analysis,

$$[\Psi_q]\{\ddot{q}\} = \{\gamma\} \tag{22}$$

where $[\Psi_q]$ is the Jacobian of the constraint system and

$$\{\gamma\} = -\left([\Psi_q]\{\dot{q}\}\right)_q \{\dot{q}\} - 2[\Psi_{qt}]\{\dot{q}\} - \{\Psi_{tt}\} \tag{23}$$

5.3 Differential equation formulations

5.3.1 Newton-Euler equations

The Newton-Euler treatment is based on the consideration of a free rigid body, in the sense that, if constrained, the forces of constraint are included.

For each ith $(i = 1, 2, \ldots, nb)$ body in the system, this treatment leads to:

- three translational equations of motion of the center of mass

$$m_i\{\ddot{r}_i\} = \{F_i\} \tag{24}$$

  where
  - m_i is the mass of the body;
  - $\{r_i\}$ is the vector which locate the absolute position of center of mass G_i of the body;
  - $\{F_i\}$ is the vector of the resultant of forces acting on the body;
- three equations which determine the rotational motion of the body

$$[J_i]\{\dot{\omega}_i\} + [\widetilde{\omega}_i][J_i]\{\omega_i\} = \{\tau_{G_i}\} \tag{25}$$

  where
  $[J_i]$ is the inertia matrix;
  $\{\omega_i\}$ is the angular velocity vector;
  $\{\tau_{G_i}\}$ is the vector of the resultant of torques computed w.r.t. center of mass G_i.

5.3.2 Principle of virtual work

The combination of the principle of virtual work[4] and d'Alembert principle is expressed by the equation

$$\{\delta r\}^T\left([M]\{\ddot{r}\} - \{F\}\right) + \{\delta\pi\}^T\left([J]\{\dot{\omega}\} + [\widetilde{\omega}][J]\{\omega\} - \{\tau_G\}\right) = 0 \tag{26}$$

where

[4] Lagrange in his *Mécanique Analitique* used the term *Principle of virtual velocities*.

$$\{\delta r\} = \left\{ \{\delta r_1\}^T, \{\delta r_2\}^T, \cdots, \{\delta r_{nb}\}^T \right\}^T \tag{27}$$

$$\{\delta \pi\} = \left\{ \{\delta \pi_1\}^T, \{\delta \pi_2\}^T, \cdots, \{\delta \pi_{nb}\}^T \right\}^T \tag{28}$$

$$[M] = \text{diag}\left[m_1 [I_{3\times3}], m_2 [I_{3\times3}], \cdots, m_{nb} [I_{3\times3}] \right] \tag{29}$$

$$[J] = \text{diag}\left[[J_1], [J_2], \cdots, [J_3] \right] \tag{30}$$

$$\{F\} = \left\{ \{F_1\}^T, \{F_2\}^T, \cdots, \{F_{nb}\}^T \right\}^T \tag{31}$$

$$\{\tau_G\} = \left\{ \{\delta \tau_{G_1}\}^T, \{\delta \tau_{G_2}\}^T, \cdots, \{\delta \tau_{G_{nb}}\}^T \right\}^T \tag{32}$$

$$\{\omega\} = \left\{ \{\omega_1\}^T, \{\omega_2\}^T, \cdots, \{\omega_{nb}\}^T \right\}^T \tag{33}$$

The expression (26) is also known as the variational Newton-Euler equation [14].

5.3.3 Lagrange equations

The Lagrange's equations of motion are expressed by

$$\frac{\mathrm{d}}{\mathrm{d}t} \left\{ \frac{\partial T}{\partial \{\dot{q}\}} \right\}^T - \left\{ \frac{\partial T}{\partial \{q\}} \right\}^T = \{Q\} \tag{34}$$

where

$$T = \frac{1}{2} \sum \left(m_i \{\dot{r}_i\}^T \{\dot{r}_i\} + \{\omega_i\}^T [J_i] \{\omega_i\} \right) \tag{35}$$

is the kinetic energy of the system, and $\{Q\}$ the vector of generalized forces.

The effect of constraints can be included by using the Lagrage's multiplier technique. In this case the equation are applied to the extended form of kinetic energy

$$T^\star = T - \{\lambda\}^T \{\Psi\} \tag{36}$$

where $\{\lambda\}$ is the vector of Lagrange's multipliers.

5.3.4 Gauss principle

Gauss's principle asserts that among all the accelerations $\{a\}$ that a system of particles of masses m_1, m_2,. . ., m_{nb} can have at time t which are compatible with the constraints, the actual ones $\{\ddot{r}\}$ are those that minimize the quantity

$$G(\ddot{r}) = \{\ddot{r} - a\}^T [M] \{\ddot{r} - a\} \tag{37}$$

where

$$\{a_i\} = \frac{\{F_i\}}{m_i} \quad (i = 1, 2, \ldots, nb) \tag{38}$$

are the accelerations of particles without constraints.

A modern treatment of Gauss principle is reported in [21].

5.3.5 Gibbs-Appell equations

The Jourdain principle, for a system of particles, can be written in the form

$$\sum_{j=1}^{n}\left[\sum_{k=1}^{nb}(F_k^e - m_k\ddot{r}_k)\frac{\partial \ddot{r}_k}{\partial \ddot{q}_j}\right]\delta\dot{q}_j = 0 \tag{39}$$

where F_k^e is the kth external force.

If we let

$$S = \frac{1}{2}\sum_{k=1}^{nb} m_k\ddot{r}_k \cdot \ddot{r}_k \tag{40}$$

then, introduced the *quasi coordinates* u,

$$\frac{\partial S}{\partial \ddot{u}_i} = \sum_{k=1}^{nb} m_k\ddot{r}_k \cdot \frac{\partial \ddot{r}_k}{\partial \ddot{u}_i} \tag{41}$$

and (39) can be rewritten as follows

$$\sum_{j=1}^{n}\left[\frac{\partial S}{\partial \ddot{u}_i} - \left(\sum_{k=1}^{nb} F_k^e \cdot \frac{\partial \ddot{r}_k}{\partial \ddot{u}_i}\right)\right]\delta\dot{u}_i = 0 \tag{42}$$

More concisely, introduced the generalized forces

$$Q_i = \sum_{k=1}^{nb} F_k^e \cdot \frac{\partial \ddot{r}_k}{\partial \ddot{u}_i} = \sum_{k=1}^{N} F_k^e \cdot \frac{\partial \dot{r}_k}{\partial \dot{u}_i} \tag{43}$$

from (42) one obtains the Gibbs-Appell equations

$$\frac{\partial S}{\partial \ddot{u}_i} - Q_i = 0 \ (i = 1, 2, \ldots, n) \tag{44}$$

The extension of the Gibbs-Appell equations to systems of rigid bodies can be found in textbooks (e.g. [28]).

5.4 Computation of generalized forces

Libraries for the computation of generalized force elements due to external forces and torques are developed for a ready use when assembling the equations of motion. The most common element is the spring-damper-actuator.

5.5 Methodologies of numerical integration

Due to the use of redundant set of coordinates, a differential-algebraic equations (DAE) system of differential index 3 is composed of

$$\{\Psi(q,t)\} = 0 \tag{45}$$

together with Eqs. (22), (23) and

$$[M]\{\ddot{q}\} + [\Psi_q]^T\{\lambda\} = \{Q\} \tag{46}$$

often appear during the modeling process of multibody systems. The presence of actively controlled components may also require DAE for mathematical modeling.

There exists a large amount of literature on computational algorithms on DAE solving (e.g. [48, 49, 23]) and an exhaustive outline is not herein attempted.

The most straightforward approach requires the reduction of the original DAE to differential index 1. Thus only the simultaneous integration of Eqs (46) and (23) is herein considered. Since after numerical integration Eqs. (45) and (22) fail to be satisfied, the right side of the acceleration constraint is altered as follows

$$\{\bar{\gamma}\} = \{\gamma\} - 2\alpha\{\Psi\} - \beta\left\{\dot{\Psi}\right\} \tag{47}$$

where α and β have to be properly chosen. The DAE system to be integrated is thus transformed to

$$\begin{bmatrix} M & \Psi_q^T \\ \Psi_q & 0 \end{bmatrix}\begin{Bmatrix} \ddot{q} \\ \lambda \end{Bmatrix} = \begin{Bmatrix} Q \\ \bar{\gamma} \end{Bmatrix} \tag{48}$$

The coordinate partitioning method is an historically important and efficient computational scheme due to Wehage and Haug [47].

The set of coordinates $q \in \Re^n$ is partitioned into two sets $v \in \Re^F$ and $u \in \Re^m$ of independent and dependent coordinates, respectively. Thus, by definition, the sub-Jacobian $[\Psi_u]$ is non singular. Based on this partitioning and the DAE equations can be rewritten in the form[5]

$$[M^{vv}]\{\ddot{v}\} + [M^{vu}]\{\ddot{u}\} + [\Psi_v]^T\{\lambda\} = \{Q^v\} \tag{49}$$

$$[M^{uv}]\{\ddot{v}\} + [M^{uu}]\{\ddot{u}\} + [\Psi_u]^T\{\lambda\} = \{Q^u\} \tag{50}$$

$$\{\Psi(u,v)\} = 0 \tag{51}$$

$$[\Psi_u]\{\dot{u}\} + [\Psi_v]\{\dot{v}\} = 0 \tag{52}$$

$$[\Psi_u]\{\ddot{u}\} + [\Psi_v]\{\ddot{v}\} = \{\gamma\} \tag{53}$$

The non singularity of $[\Psi_u]$ and the implicit function theorem guarantee that $\{u\}$ con be locally computed as a function of $\{v\}$, i.e.

$$\{u\} = \{h(v)\} \tag{54}$$

With this the solution of the DAE system is reduced to a set of ODE system through the sequence of steps listed below

[5] It is assumed that the Jacobian $[\Psi_q]$ has full row rank.

1. Partition the vector $\{q\}$ of coordinates;
2. Determine $\{\dot{u}\}$ and $\{u\}$ at time t by means of Eqs. (52) and (54), respectively;
3. Solve Eqs. (49), (50, (53) w.r.t. $\{\ddot{u}\}$, $\{\ddot{v}\}$ and $\{\lambda\}$;
4. Integrate and compute $\{u\}$, $\{v\}$, $\{\dot{u}\}$, $\{\dot{v}\}$ at time $t + \Delta t$

A critical review of different dynamic formulations is offered in [50].

In order to reduce the DAE system to an ordinary differential equations (ODE) the elimination of the Jacobian matrix Ψ_q of the constraint equations from (46) is necessary. This approach offers the following advantages:

- The elimination of Lagrange's multipliers when solving equations:
- The possibility to partition the entire set of generalized coordinates into independent variables and dependent ones;
- The transform of the DAE system into a ODE gives the opportunity of a wider choice of numerical integration subroutines;
- Mechanical systems with a redundant number of constraints or with changing d.o.f. can be analysed.

For this purpose it is required to introduce a minimum set v of F independent coordinates. Let us append to the constraint vector $\{\Psi\}$ the equations $\{\Phi\}$ that can be established between v and q. Thus, we obtain

$$\{\Gamma\} = \left\{ \begin{array}{c} \Psi(q) \\ \Phi(v, q) \end{array} \right\} = 0 \tag{55}$$

The time derivative of (55) leads to

$$[\Gamma_v] \{\dot{p}\} + [\Gamma_q] \{\dot{q}\} = 0 \tag{56}$$

If we let

$$[V] = - [\Gamma_q]^{-1} [\Gamma_v] \tag{57}$$

one obtains

$$\{\dot{q}\} = [V] \{\dot{v}\} \tag{58}$$

When there is not any explicit dependence on time of constraints equations, the following *orthogonality condition* is deduced

$$[\Psi_q] [V] = 0 \tag{59}$$

and the accelerations $\ddot{q}$ can be expressed in the form

$$\{\ddot{q}\} = [V] \{\ddot{v}\} + \left[\dot{V}\right] \{\dot{v}\} \tag{60}$$

Premultiplying both sides of the dynamic equation of system (46) and taking into account (59) and (60), the vector of Lagrange's multipliers is eliminated from the differential equations of equilibrium and the following ODE is obtained

$$[V]^T [M] [V] \{\ddot{v}\} = [V]^T \{Q\} + [V]^T [M] \left[\dot{V}\right] \{\dot{v}\} \tag{61}$$

The matrix $[V]$ is not unique. The singular value decomposition algorithm [51] and the QR decomposition [52] are often used. In particular, when using the singular value decomposition, the Jacobian matrix is decomposed in the form

$$\begin{aligned}[\Psi_q]^T &= \left[[W_d] \; [W_i] \right] \begin{bmatrix} [\Lambda_1] \\ [0] \end{bmatrix} [U]^T \\ &= [W_d] \, [\Lambda_1] \, [U]^T \;, \end{aligned} \tag{62}$$

and

$$[V] = [W_i] \; . \tag{63}$$

When using the QR decomposition, the Jacobian matrix is decomposed in the form

$$[\Psi_q]^T = \left[[Q_1] \; [Q_2] \right] \begin{bmatrix} [R_1] \\ [0] \end{bmatrix} = [Q_1] \, [R_1] \;, \tag{64}$$

and

$$[V] = [Q_2] \; . \tag{65}$$

5.6 Interdisciplinary effects

The teaching of multibody dynamics involves disciplines like numerical analysis and computer graphics. Moreover, as previously mentioned, the knowledge of the particular field where multibody dynamics is being applied is required for a correct interpretation of results. The applications of multibody dynamics in human movement analysis and biomechanics are growing. This demonstrate the usefulness of multibody dynamics instruction in engineering curricula such as biomedical, control and computer science.

5.7 Computer programming and code organization

A moderate exposure of the student in computer programming of multibody dynamics algorithms seems appropriate for an effective learning. In order to reduce the burden of computer programming the instructor can make available to the students software modules for different main functions required to a multibody dynamics software. Thus the task of the student reduces to the correct assembly of parts and execution of the entire program. In some cases the instructor may encourage the introduction of improvements such as constraints stabilization or the testing of different integration algorithms settings.

The understanding of the overall structure of a multibody dynamics software (see Figure 7) and of basic functioning of its modules surely strength the knowledge of multibody dynamics theory and of the limits of the models developed.

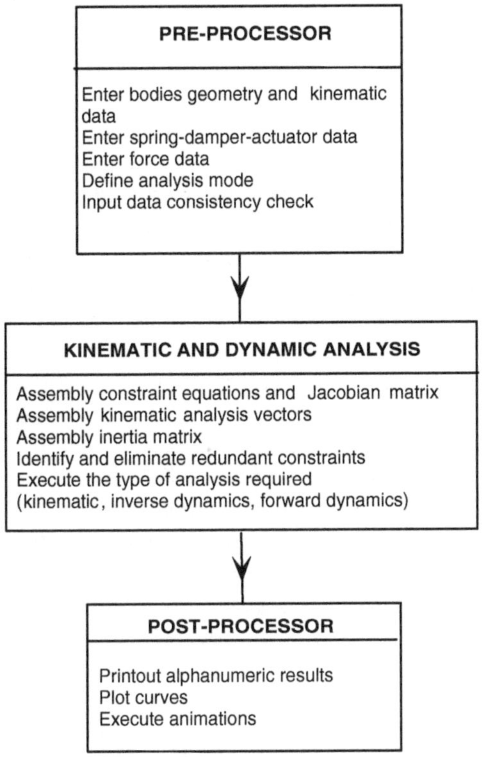

Fig. 7. Computational flow of a multibody dynamics simulation software

5.8 Home assignements and projects

In multibody dynamics instruction, the theoretical instruction and the practice can be variously combined. Practice in the form of homeworks and medium term projects are strongly recommended. At the conclusion of each theory topic, the instructor should require the students to work out autonomously by hand, or with the use of appropriate software tools, applications of the explained theory.

6 The author's experience

The first author (E.P.) recently introduced a course on multibody dynamics in the engineering curricula at Università di Roma Tor Vergata. The curricula were compliant with the Bologna agreement. The course is currently mandatory for fourth year mechanical and automatic control engineering students. This section describes some of the choices and may offer guidelines for similar initiatives.

6.1 Course syllabus

The method of constraints for planar kinematic analysis. Revolute, prismatic, gear and cam pairs are considered together with other 2 degrees-of-freedom types of constraints. The automatic assembly of the systems of equations for position, velocity and acceleration analysis. Iterative solution of systems of non linear equations. Geometry of masses. The principle of virtual work and Lagrange's equations. Dynamics of planar systems. Systematic computation and assembly of mass matrix. Computation of planar generalized forces for external forces and for actuator-spring-damper element. Simple applications of inverse and forward dynamic analysis. Numerical integration of first-order initial-value problems. The method of Baumgarte for the solution of mixed differential-algebraic equations of motion. The use of coordinates partitioning, QR and SVD decomposition for the orthogonalization of constraints. Kinematics of rigid bodies in space. Reference frames for the location of a body in space. Euler angles and Euler parameters. The formula of Rodrigues. Screw motion in space. Velocity, acceleration and angular velocity. Relationship between the angular velocity vector and the time derivatives of Euler parameters. Kinematic analysis of spatial systems. Basic kinematic constraints. Joint definition frames. The constraints required for the description in space of common kinematic pairs (revolute, prismatic, cylindrical, spherical). Equations of motion of constrained spatial systems. Computation of spatial generalized forces for external forces and for actuator-spring-damper element. Computation of reaction forces from Lagrange's multipliers.

Considered the introductory level of the course and the lack of funds for the renting of licenses, hands-on practice with multibody dynamics commercial software was not included in the course.

6.2 Homework and computer assignements

The homeworks requested are the development of computer codes for:

- the kinematic analysis of planar mechanisms with lower pairs
- the forward dynamic analysis of planar mechanisms with lower pairs, spring-damper-actuator elements, external forces;
- dynamic analysis of a 3D mechanical system composed of two rigid bodies with revolute, prismatic or spherical pairs.

The students were also requested to document their code to the best of their capabilities. The time allowed for each project was about three weeks. For this purpose, different pieces of software modules were made available to the students.

6.3 Course grading

The overall grading is based for 50% on the quality of the computer assignements handed during the course, 25% from written test and a 25% on a brief oral exam on different parts of the theory.

6.4 Response and comments from the students

The students seems enjoying the study of multibody dynamics. What they like most is the sistematicity of the approach. The computer oriented modeling of mechanical systems makes them more confident about the formal correctness of the governing equations deduced. Most of the complains arise from the limited time allowed for computer homework. In fact, they claim that the theory is easy to learn and understand, but debugging of the software takes most of their time. They find also useful the experience made with specialized software for linear algebra and numerical integration of differential equations. The software tools usually chosen for computer programming were Fortran90, C++, Matlab, Maple and Mathematica.

7 Needs

1. Increase the number of credits allocated for multibody dynamics courses in engineering curricula.
2. Multibody dynamics computer codes with open architecture.
3. Standardised input-output of data between multibody dynamics codes.
4. High-level programming languages geared toward multibody dynamics programming (mixed capabilities: numerical and symbolical).
5. Multibody dynamics software with a pre and post processing capabilities using web browsers only.
6. Centralized web resources for the exchange of informations between teachers, reseracher, students.

8 Conclusions

Different issues involved in multibody dynamics training of engineering students were discussed. On the basis of their experience, the authors recommend that multibody dynamics courses should be preferentially offered during or after the third year of an engineering curriculum. This will ensure a minimum of background. The question on the *most* effective syllabus is still open. It would be interesting to compare the proficiency in modeling mechanical systems of the students exposed to different multibody dynamics methodologies.

The inclusion of more advanced multibody dynamics topics in engineering curricula seems is not widespread in European universities. For instance flexible multibody dynamics is still perceived like a research topic rather than an established discipline. The scarsity of textbooks, usually very expensive, and of ready-to-use didactic material does not help the diffusion of courses. Due to the lack of funds, the renting and maintenance of commercial software licenses is also a problem. However, from the didactic point of view, the development of open source multibody dynamics software, made freely available to teachers and students or under a nominal fee, would greatly help the spread of the culture of multibody dynamic in the engineering curricula. Centralized web resources, where students and educators may find links to reports, tutorials, software on the different branches of multibody dynamics, are also useful for the above purposes.

The efforts in the developement of didactic tools and teaching methodologies in the field of multibody dynamics are worthwhile. Beside the already mentioned advantages of informed software users, the research in multibody dynamics will surely benefit of a large base of graduate students familiar with the basic techniques.

References

1. Medea Scientific Committee, Executive summary, September 2002
2. Shigley JE, Uicker JJ (1995) Theory of Machines and Mechanisms. McGraw-Hill Inc., New York, 2nd edition
3. Norton RL (1992) Design of Machinery. McGraw-Hill Inc., New York
4. Valasek M (2003) Multibody dynamics without analytical mechanics. In: Proceedings of the ECCOMAS Thematic Conference on Advances in Computational Multibody Dynamics, Paper MB2003-014, Lisbon
5. Freeman JS (1995) Multibody dynamics and Vehicle Simulation: Panacea or Problem. Fraunhofer USA Best of German/American Automotive Engine conference.
6. Shiakolas PS, Chandra V, Kebrle J, Wilhite D (2002) Design, Analysis, and Simulation of Engineering Systems Over the World Wide Web for Educational Use. Proc. of ASME 2002 Design Engineering Technical Conferences, Paper DETC02/CIE-34410, Montreal
7. Fisette P, Samin JC (2003) A Student Project in Multibody Dynamics for Training in Fundamental Mechanics. Procceedings of the ECCOMAS Thematic Conference on Advances in Computational Multibody Dynamics, Paper MB2003-042, Lisbon
8. Fanghella P, Galletti C, Giannotti E (2003) Teaching Multibody System Simulation to Postgraduate Students in Mechanical Engineering. Proceedings of the ECCOMAS Thematic Conference on Advances in Computational Multibody Dynamics, Paper MB2003-020, Lisbon
9. Masten MK (1995) A Strategy for Industry's Continuing Education Needs. Control Engineering Practice 3:717–721
10. Wittenburg J (1977) Dynamics of Systems of Rigid Bodies. B.G. Teubner, Stuttgart
11. Kane TR, Levinson DA (1985) Dynamics: Theory and Applications. McGraw-Hill Book Co., New York
12. Nikravesh PE (1988) Computer Aided Analysis of Mechanical Systems. Prentice-Hall Inc., Englewood Cliffs, NJ
13. Roberson RE, Schwertassek R (1988) Dynamics of Multibody Systems. Springer-Verlag. Berlin Heidelberg New York
14. Haug EJ (1989) Computer-Aided Kinematics and Dynamics of Mechanical Systems-Basic Methods. Allyn and Bacon, Boston
15. Huston RL (1990) Multibody Dynamics. Butterworth-Heinemann, Boston London Singapore Sydney Toronto Wellington
16. Schielen W *ed.* (1990) Multibody Systems Handbook. Springer-Verlag, Berlin Heidelberg New York
17. Haug EJ (1992) Intermediate Dynamics. Prentice-Hall Inc., Englewood Cliffs NJ

18. de Jalón J.C., Bayo E., Kinematic and Dynamic Simulation of Multibody Systems. Springer Verlag, Berlin Heidelberg New York
19. Shabana AA (1994) Computational Dynamics. John Wiley & Sons, New York
20. Pfeiffer F, Glocker C (1996) Multibody Dynamics with Unilateral Constaints. John Wiley & Sons, New York
21. Udwadia FE, Kalaba RE (1996) Analytical Dynamics - A New Approach. Cambridge University Press, Cambridge UK
22. Stejskal V Valasek M (1996) Kinematics and dynamics of machinery. Marcel Dekker, New York
23. Eich-Soellner E, Führer C (1998) Numerical Methods in Multibody Dynamics. B.G. Teubner, Stuttgart
24. Shabana AA (1998) Dynamics of Multibody Systems. Cambridge University Press, Cambridge UK
25. von Schwerin R (1999) Multibody System Simulation - Numerical Methods, Algorithms and Software. Springer Verlag, Berlin Heidelberg New York
26. Geradin M, Cardona A (2001) Flexible Multibody Dynamics: A Finite Element Approach. John Wiley and Sons, New York
27. Coutinho M (2001) Dynamic Simulation of Multibody Systems. Birkäuser, Boston
28. Pennestrì E (2002) Dinamica Tecnica e Computazionale. Casa Editrice Ambrosiana, Milano *in italian*
29. Hahn H (2002) Rigid Body Dynamics of Mechanisms - Theoretical basis. Springer Verlag, Berlin Heidelberg New York
30. Hahn H (2003) Rigid Body Dynamics of Mechanisms - Applications. Springer Verlag, Berlin Heidelberg New York
31. Whittaker ET (1937) A Treatise on the Analytical Dynamics of Particles and Rigid Bodies. Cambridge University Press, Cambridge
32. Bae DS (1986) A Recursive Formulation for Constrained Mechanical System Dynamics. Doctoral Diss., The University of Iowa, Iowa City
33. Orlandea N (1973) Development and Application of Node-Analogous Sparsity-Oriented Methods for Simulation of Mechanical Dynamic Systems, Doctoral Diss., University of Michigan, Ann Arbor MI
34. Fischer IS (1999) Dual-Number Methods in Kinematics, Statics and Dynamics. CRC Press, Boca Raton London New York Washington D.C.
35. Denavit J (1958) Displacement Analysis of Mechanisms on 2×2 Matrices of Dual Numbers VDI-Berichte 29:81–88
36. Keler M (1958) Analyse und Synthese der Raumkurbelgetriebe mittels Raumliniengeometrie und dualer Größen. Doctoral Diss., München University, München
37. Yang AT (1963) Application of Quaternion Algebra and Dual Numbers to the Analysis of Spatial Mechanisms. Doctoral Diss., Columbia University, New York
38. Yang AT, Freudenstein F (1964) Application of Dual-Number Quaternion Algebra to the Analysis of Spatial Mechanisms. ASME Journal of Applied Mechanics 86:300–308
39. Denavit J, Hartenberg RS (1955) A Kinematic Notation for Lower-Pair Mechanisms Based on Matrices. ASME Journal of Applied Mechanics 22:215–221
40. Beyer R (1963) Technische Raumkinematik. Springer, Berlin, Göttingen, Heidelberg
41. Dimentberg FM (1965) The Screw Calculus and Its Applications in Mechanics. Izdat. Nauka. Moscow, English Translation: AD680993, Clearinghaous for Federal and Scientific Technical Information
42. Wittenburg J (1984) Dual Quaternions in the Kinematics of Spatial Mechanisms. In: Haug EJ. (ed) Computer Aided Analysis and Optimization of Mechanical System Dynamics. Springer Verlag, Berlin, Heidelberg, New York

43. Angeles J (1998) The Application of Dual Algebra to Kinematic Analysis. In: Angeles J, Zakhariev E (eds) Computational Methods in Mechanical Systems. Springer Verlag, Berlin, Heidelberg, New York
44. Buffinton KW, Chang VW (1996) An Integrated Approach to Teaching Mechanism Analysis using Working Model. Proc. of ASME Design Engineering Technical Conferences, Paper 96-DETC/CIE-1431, Irvine, CA
45. Woods RO (2002) Drawing on Experience. Mechanical Engineering 11:57–58
46. Schielen W (1997) Multibody System Dynamics: Roots and Prospectives. Multibody System Dynamics 1:149–188
47. Wehage RA, Haug EJ (1982) Generalized Coordinate Partitioning for Dimension Reduction in Analysis of Constrained Dynamic Systems. ASME Journal of Mechanical Design 104:247–255
48. Brenan KE, Campbell SL, Petzold LR (1989) Numerical Solution of Initial Value Problems in Differential-Algebraic Equations. North-Holland, New York; reprinted in 1996 as Classics in Applied Mathematics 14, SIAM, Philadelphia
49. Hairer E, Wanner G (1996) Solving Ordinary Differential Equations II: Stiff and Differential-Algebraic Problems. Springer-Verlag, Berlin Heidelberg New York
50. Unda J, de Jalón J, Losantos F, Enparantza R (1987) A Comparative Study of Some Different Formulations of the Dynamic Equations of Constrained Mechanical Systems. ASME Journal of Mechanisms, Transmissions and Automation in Design 109:466-474
51. Mani NK, Haug EJ, Atkinson, KE (1985) Singular Value Decomposition for Analysis of Mechanical System Dynamics. ASME Journal of Mechanisms, Transmissions, and Automation in Design 107:82-87
52. Kim SS, Vanderploeg MJ (1986) QR decomposition for State Space Representation of Constrained Mechanical Dynamic Systems. ASME Journal of Mechanisms, Transmissions, and Automation in Design 108:176-182

FSC
www.fsc.org
MIX
Papier | Fördert gute Waldnutzung
FSC® C083411

Kolibri 360 GmbH
Ferdinand-Jühlke-Straße 7
99095 Erfurt, Deutschland
produktsicherheit@kolibri360.de